Wireless Network Performance Enhancement via Directional Antennas

Models, Protocols, and Systems

OTHER COMMUNICATIONS BOOKS FROM AUERBACH

Analytical Evaluation of Nonlinear Distortion Effects on Multicarrier Signals
Theresa Araújo
ISBN 978-1-4822-1594-6

Architecting Software Intensive Systems: A Practitioners Guide
Anthony J. Lattanze
ISBN 978-1-4200-4569-7

Cognitive Radio Networks: Efficient Resource Allocation in Cooperative Sensing, Cellular Communications, High-Speed Vehicles, and Smart Grid
Tao Jiang, Zhiqiang Wang, and Yang Cao
ISBN 978-1-4987-2113-4

Complex Networks: An Algorithmic Perspective
Kayhan Erciyes
ISBN 978-1-4665-7166-2

Data Privacy for the Smart Grid
Rebecca Herold and Christine Hertzog
ISBN 978-1-4665-7337-6

Generic and Energy-Efficient Context-Aware Mobile Sensing
Ozgur Yurur and Chi Harold Liu
ISBN 978-1-4987-0010-8

Machine-to-Machine Communications: Architectures, Technology, Standards, and Applications
Vojislav B. Misic and Jelena Misic
ISBN 978-1-4665-6123-6

Managing the PSTN Transformation: A Blueprint for a Successful Migration to IP-Based Networks
Sandra Dornheim
ISBN 978-1-4987-0103-7

MIMO Processing for 4G and Beyond: Fundamentals and Evolution
Edited by Mário Marques da Silva and Francisco A. Monteiro
ISBN 978-1-4665-9807-2

Mobile Evolution: Insights on Connectivity and Service
Sebastian Thalanany
ISBN 978-1-4822-2480-1

Network Innovation through OpenFlow and SDN: Principles and Design
Edited by Fei Hu
ISBN 978-1-4665-7209-6

Neural Networks for Applied Sciences and Engineering: From Fundamentals to Complex Pattern Recognition
Sandhya Samarasinghe
ISBN 978-0-8493-3375-0

Rare Earth Materials: Properties and Applications
A.R. Jha
ISBN 978-1-4665-6402-2

Requirements Engineering for Software and Systems, Second Edition
Phillip A. Laplante
ISBN 978-1-4665-6081-9

Security for Multihop Wireless Networks
Edited by Shafiullah Khan and Jaime Lloret Mauri
ISBN 9781466578036

Software Testing: A Craftsman's Approach, Fourth Edition
Paul C. Jorgensen
ISBN 978-1-46656068-0

The Future of Wireless Networks: Architectures, Protocols, and Services
Edited by Mohesen Guizani, Hsiao-Hwa Chen, and Chonggang Wang
ISBN 978-1-4822-2094-0

The Internet of Things in the Cloud: A Middleware Perspective
Honbo Zhou
ISBN 978-1-4398-9299-2

The State of the Art in Intrusion Prevention and Detection
Al-Sakib Khan Pathan
ISBN 978-1-4822-0351-6

ZigBee® Network Protocols and Applications
Edited by Chonggang Wang, Tao Jiang, and Qian Zhang
ISBN 978-1-4398-1601-1

Wireless Network Performance Enhancement via Directional Antennas

Models, Protocols, and Systems

Edited by

John D. Matyjas • Fei Hu • Sunil Kumar

CRC Press is an imprint of the
Taylor & Francis Group, an **informa** business

CRC Press
Taylor & Francis Group
6000 Broken Sound Parkway NW, Suite 300
Boca Raton, FL 33487-2742

Printed on acid-free paper
Version Date: 20150612

International Standard Book Number-13: 978-1-4987-0753-4 (Hardback)

Library of Congress Cataloging-in-Publication Data

Wireless network performance enhancement via directional antennas : models, protocols, and systems / editors, John D. Matyjas, Fei Hu, and Sunil Kumar.
pages cm
Includes bibliographical references and index.
ISBN 978-1-4987-0753-4 (alk. paper)
1. Antennas (Electronics)--Design and construction. 2. Wireless communication sytems--Equipment and supplies. 3. Computer network protocols. 4. Network performance (Telecommunication) 5. Beamforming. 6. Antenna radiation patterns. I. Matyjas, John D. II. Hu, Fei, 1972-

TK7871.6.W57 2016
621.3841'35--dc23 2015020752

Visit the Taylor & Francis Web site at
http://www.taylorandfrancis.com

and the CRC Press Web site at
http://www.crcpress.com

Contents

Preface

Antenna technologies have made significant advancements in the last decade; not only in direct support of communication systems, but also to include support of multiple simultaneous radio frequency (RF) functions such as radar. With the rapid development of electronic materials, electric circuits, and electromagnetic theories, many powerful antenna systems have been invented and employed across multiple spectrum bands. For example, switched or steered antennas can be used in WiFi, millimeter wave, centimeter wave, Ku-band, and other frequencies. Modes of operation for antennas can be a single beam (each time the antenna faces a specific direction) and/or multibeam (multiple links can deliver data simultaneously). And, multiple-input, multiple-output (MIMO) antenna configurations provide the opportunity to greatly cancel noise/interference and increase communication capacity by exploiting spatial diversity.

These antenna system advances further open the door to many exciting new design opportunities for wireless networks to enhance quality of service (QoS), performance, and network capacity. For example, directional antennas enable transmissions for a much larger distance/range over omni-directional antennas for the same total power budget, while also achieving higher data rates and less delay/latency. MIMO antenna configurations can significantly improve the packet arrival success rate and enable large-scale network node interconnections. Multibeam antennas can concurrently receive or transmit data in multiple radio links and greatly improve network throughput and efficiency. And "smart" (or adaptive) antennas can automatically adjust their parameters based on network context or feedback.

However, effective deployment of directional networks also faces many system- and operational-level design challenges. As an example, directional antennas often cause deafness issues, where a node may not be able to receive the intended (productive) packets for itself if its beam is facing another direction and/or listening to unintended (unproductive) packets (i.e., packets intended for a different destination). In the case of MIMO antennas, they need timely feedback from the receiver side to accurately adjust their beamforming matrix and enable reliable receipt. Furthermore, the conventional medium access control (MAC) and routing protocols for omni-directional antennas have to be modified or redesigned for directional networks since the control messages may only be received in certain directions. One mitigation strategy is to augment directional networks with an omni-based discovery/control channel; yet, this comes at the sacrifice of spectrum usage and network capacity (increased overhead) along with a reliance on a dedicated, synchronized control channel with security concerns therein.

This compilation is the first technical book canvasing research and development on wireless networks with directional antennas. Contributions go deep into lower-layer aspects (with models, algorithms, and protocols) of wireless and mobile networks with different types of antennas (MIMO, single beam, multibeam, etc.), while operating over different frequencies (2.4 GHz,

30 GHz, 60 GHz, contiguous and noncontiguous dynamic spectrum, etc.). This compilation also covers some important applications of directional networks.

Targeted audiences for this book include: (1) Researchers: The book identifies some interesting research problems in this important field with an opportunity to learn about some solid solutions to those issues. (2) Engineers: Industry designers/practitioners may find some practical hardware designs for deployment of next-generation antennas, as well as efficient network protocols for exploitation of directional communications. (3) Marketplace: Investors/venture capitalists will be able to baseline the state of the art/state of the practice in directional networking and track developing trends in this field.

We have invited subject matter experts from all over the world to provide chapter contributions. They have spent over one year on the writing and editing of their chapters. The book editors have carefully arranged the chapters into the following six sections:

- Section I. Directional Antennas: This section consists of three chapters on the hardware design of different types of antennas. We explain the differences between switched and steered antennas, wideband antenna arrays, and directional radio circuits.
- Section II. Directional MAC: This section focuses on the basic principles of designing MAC protocols for directional networks. We explain how the deafness and capture issues could be overcome through efficient medium access coordination strategies.
- Section III. Millimeter Wave: Millimeter wave (mmW) apertures operate in a high-frequency band (such as 60 GHz). They have good signal directionality (i.e., signals do not diffuse to wide angles). This means that we can use directional antennas to easily limit the signal direction within a narrow angle or beamwidth. Node coordination is more important than interference avoidance since mmW signals are not as susceptible to line-of-sight (LOS) blocking. If no LOS path is found, the nodes need to coordinate well with each other to find a non-LOS path. We include four chapters on different design aspects of mmW systems.
- Section IV. MIMO: This section explains how a MIMO system can be established. We also explain how a MIMO system operates in a cognitive radio network.
- Section V. Advanced Topics: This section includes some additional topics on directional networking such as beamforming in cognitive radio networks, multicast algorithm development, network topology management for connectivity, and sensor network lifetime issues.
- Section VI. Applications: In this section, we illustrate some important applications, such as military networks and airborne networking that desire the spectral benefits afforded by directional networking designs.

In summary, this book provides detailed technical content for both academia and industry on the research and development of network protocols with directional antennas. It is also a great reference for beginners who want to understand the basic principles, engineering design, and challenges to fully exploit directional networking in light of increased market demand for multifunction RF capabilities, spectrum efficiency, and/or network capacity and performance.

Editors

John D. Matyjas earned his PhD in electrical engineering from State University of New York at Buffalo in 2004. Currently, he is serving as the Connectivity & Dissemination Core Technical Competency Lead at the Air Force Research Laboratory (AFRL) in Rome, New York. His research interests include dynamic multiple-access communications and networking, spectrum mutability, statistical signal processing and optimization, and neural networks. He serves on the IEEE Transactions on Wireless Communications Editorial Advisory Board. Dr. Matyjas is the recipient of the 2012 IEEE R1 Technology Innovation Award, 2012 AFRL Harry Davis Award for "Excellence in Basic Research," and the 2010 IEEE International Communications Conference. Best Paper Award. He is an IEEE senior member, chair of the IEEE Mohawk Valley Signal Processing Society, and member of Tau Beta Pi and Eta Kappa Nu.

Fei Hu is currently an associate professor in the Department of Electrical and Computer Engineering at the University of Alabama, Tuscaloosa. He earned his PhD at Tongji University (Shanghai, China) in the field of signal processing (1999), and at Clarkson University (New York) in electrical and computer engineering (2002). He has published over 200 journal/conference papers and books. Dr. Hu's research has been supported by U.S. National Science Foundation, Cisco, Sprint, and other sources. His research expertise can be summarized as 3S: Security, Signals, Sensors: (1) Security: How to overcome different cyber attacks in a complex wireless or wired network. Dr. Hu's recent research focuses on cyber-physical system security and medical security issues. (2) Signals: This mainly refers to intelligent signal processing, that is, using machine learning algorithms to process sensing signals in a smart way in order to extract patterns (i.e., pattern recognition). (3) Sensors: This includes microsensor design and wireless sensor networking issues.

Sunil Kumar is currently a professor and Thomas G. Pine Faculty Fellow in the Electrical and Computer Engineering Department at San Diego State University (SDSU), San Diego, California. He earned his PhD in electrical and electronics engineering from the Birla Institute of Technology and Science, Pilani (India) in 1997. From 1997 to 2002, Dr. Kumar was a postdoctoral researcher and adjunct faculty member at the University of Southern California, Los Angeles. Prior to joining SDSU, Dr. Kumar was an assistant professor at Clarkson University, Potsdam, New York (2002–2006). He was a visiting professor (fall 2014) and ASEE summer faculty fellow (summer of 2007 and 2008) at the Air Force Research Lab in Rome, New York, where he conducted research in airborne wireless networks. Dr. Kumar is a senior member of IEEE and has published 130 research articles in international journals and conferences, seven books/book chapters, and three U.S. invention disclosures. His research has been supported by grants/awards from the National Science Foundation, U.S. Air Force Research Lab, Department of Energy, California Energy Commission, and Industry. His research areas include directional wireless networks, cross-layer and QoS-aware wireless protocols, and error-resilient video compression.

Contributors

David Abbott
CSIRO
New South Wales, Australia

Juan M. Alonso
Facultad de Ciencias Físico Matemáticas y Naturales
Universidad Nacional de San Luis
San Luis, Argentina

and

Facultad de Ciencias Exactas y Naturales
Universidad Nacional de Cuyo
Mendoza, Argentina

İlker Bekmezci
Computer Engineering Department
Turkish Air Force Academy
Istanbul, Turkey

Ritu Chadha
Knowledge-Based Systems Research
Applied Communication Sciences
Basking Ridge, New Jersey

Kishor Chandra
Department of Mathematics and Computer Science
Delft University of Technology
Delft, The Netherlands

Abdelaali Chaoub
Department of Telecommunication
National Institute of Posts and Telecommunications
Rabat, Morocco

Chi-Kin Chau
Department of Electrical Engineering and Computer Science
Madsar Institute of Science and Technology
Abu Dhabi, United Arab Emirates

Qian Chen
Internet of Things Connectivity Department
Institute for Infocomm Research
Agency for Science, Technology and Research
Singapore

Wen-Tsuen Chen
Institute of Information Science
Academia Sinica
Taipei, Taiwan

Xiang Chen
School of Information Science and Technology
Sun Yat-sen University
Guangdong Province, China

Dave Chester
Aerospace Electronics
Harris Corporation
Government Communications Systems Division
Palm Bay, Florida

Francois Chin
Internet of Things Connectivity Department
Institute for Infocomm Research
Agency for Science, Technology and Research
Singapore

Pai H. Chou
Department of Electrical Engineering and Computer Science (EECS)
Center for Embedded Cyber-Physical Systems (CECS)
Irvine, California

Stephen M. Dudley
L3 Communications
Communication Systems-West
Salt Lake City, Utah

Wei Feng
Department of Electronic Engineering
Tsinghua University
Beijing, China

Richard J. Gibbens
Computer Laboratory
University of Cambridge
Cambridge, United Kingdom

Y. Thomas Hou
Bradley Department of Electrical and Computer Engineering
Virginia Tech
Blacksburg, Virginia

Elhassane Ibn-Elhaj
Department of Telecommunication
National Institute of Posts and Telecommunications
Rabat, Morocco

Depeng Jin
Department of Electronic Engineering
Tsinghua University
Beijing, China

Raja Jurdak
CSIRO
Pullenvale, Australia

Latha Kant
Knowledge-Based Systems Research
Applied Communication Sciences
Basking Ridge, New Jersey

Pavlos I. Lazaridis
Department of Engineering and Technology
University of Huddersfield
West Yorkshire, United Kingdom

John Lee
Knowledge-Based Systems Research
Applied Communication Sciences
Red Bank, New Jersey

Yong Li
Department of Electronic Engineering
Tsinghua University
Beijing, China

Wenjing Lou
Department of Computer Science
Virginia Tech
Falls Church, Virginia

Shiwen Mao
Electrical and Computer Engineering
Auburn University
Auburn, Alabama

George Mastorakis
Department of Business Administration
Technological Educational Institute of Crete
Crete, Greece

Constandinos X. Mavromoustakis
Department of Computer Science
University of Nicosia
Nicosia, Cyprus

John McEachen
Department of Electrical and Computer Engineering
Naval Postgraduate School
Monterey, California

Todd Mcintyre
L3 Communications
Communication Systems-West
Salt Lake City, Utah

Tan Ngo
Department of Electrical and Computer Engineering
Naval Postgraduate School
Monterey, California

Amanda Nordhamn
Department of Information Technology
Uppsala University
Uppsala, Sweden

Keith Olds
Aerospace Electronics
Harris Corporation
Government Communications Systems Division
Palm Bay, Florida

Simon Olofsson
Department of Information Technology
Uppsala University
Uppsala, Sweden

Evangelos Pallis
Department of Informatics Engineering
Technological Educational Institute of Crete
Crete, Greece

Xiaoming Peng
Institute for Infocomm Research
(Satellite Department)
Agency for Science, Technology and
Research
Singapore

R. Venkatesha Prasad
Department of Mathematics and Computer
Science
Delft University of Technology
Delft, The Netherlands

Muhammad Irfan Rafique
Faculty of Electrical Engineering and
Information Technology
Technische Universität Chemnitz
Chemnitz, Germany

Theodore (Ted) S. Rappaport
Electrical and Computer Engineering
New York University Polytechnic School of
Engineering
and
Computer Science
Courant Institute of Mathematical Sciences
New York University
and
Radiology
School of Medicine
New York University
New York, New York

Marc J. Russon
L3 Communications
Communication Systems-West
Salt Lake City, Utah

Hanif D. Sherali
Grado Department of Industrial and Systems
Engineering
Virginia Tech
Blacksburg, Virginia

Matthew Sherman
BAE Systems-Electronic Systems (ES)
Wayne, New Jersey

Yi Shi
Bradley Department of Electrical and
Computer Engineering
Virginia Tech
Blacksburg, Virginia

Wen-Chan Shih
Institute of Information Science
Academia Sinica
Taipei, Taiwan

Christos Skeberis
Department of Electrical and Computer
Engineering
Aristotle University of Thessaloniki
Thessaloniki, Greece

Jerry Sonnenberg
Aerospace Electronics
Harris Corporation
Government Communications Systems Division
Palm Bay, Florida

Dimitrios I. Stratakis
Department of Informatics Engineering
Technological Educational Institute of Crete
Crete, Greece

Li Su
Department of Electronic Engineering
Tsinghua University
Beijing, China

Emil Svatik
Aerospace Electronics
Harris Corporation
Government Communications Systems Division
Palm Bay, Florida

Şamil Temel
Aeronautics and Space Technologies Institute
Turkish Air Force Academy
Istanbul, Turkey

Don Towsley
Department of Computer Science
University of Massachusetts
Amherst, Massachusetts

Murali Tummala
Department of Electrical and Computer Engineering
Naval Postgraduate School
Monterey, California

Thiemo Voigt
Department of Information Technology
Uppsala University
Uppsala, Sweden

and

SICS Swedish ICT
Stockholm, Sweden

T. Owens Walker III
Department of Electrical and Computer Engineering
United States Naval Academy
Annapolis, Maryland

Jing Wang
Research Institute of Information Technology, TNList
Tsinghua University
Beijing, China

Victor Wells
L3 Communications
Communication Systems-West
Salt Lake City, Utah

David Tung Chong Wong
Institute for Infocomm Research (Satellite Department)
Agency for Science, Technology and Research
Singapore

Thomas D. Xenos
Department of Electrical and Computer Engineering
Aristotle University of Thessaloniki
Thessaloniki, Greece

Seokhyun Yoon
Department of Electronics and Electrical Engineering
Dankook University
Gyeonggi-do, Republic of Korea

Zaharias D. Zaharis
Department of Electrical and Computer Engineering
Aristotle University of Thessaloniki
Thessaloniki, Greece

Xiujun Zhang
Research Institute of Information Technology
Tsinghua University
Beijing, China

Ming Zhao
Research Institute of Information Technology
Tsinghua University
Beijing, China

Shidong Zhou
Research Institute of Information Technology
Tsinghua University
Beijing, China

DIRECTIONAL ANTENNAS

Chapter 1

Introduction: Switched/ Steered Directional Antennas for Networking

Tan Ngo, Murali Tummala, and John McEachen

Contents

With the ever-increasing demand for capacity over wireless networks, the demand on communication systems are fast outgrowing their bandwidth allocations. Nowhere is this problem more evident than in mobile/cellular networks. Here, the increasing number of users coupled with their insatiable demands for data are quickly outgrowing the limits of their bandwidth allocations. To further exacerbate the problem, most networks operate in urban environments where multipath fading and delay spread heavily effect throughput. One solution to these problems is to employ directional antennas, such as phase arrays, ubiquitously across wireless networks. This type of antenna can efficiently direct its radiative power in a desired direction, enable spatial reuse, and help to maximize a network's throughput. Also, with the use of adaptive algorithms these antennas can compensate for many channel-related issues such as multipath fading and delay spread.

In this chapter, we begin with an overview of microstrip/patch antennas as they pertain to beamforming for networking. We discuss the fundamentals of switched/steered directional antennas and the errors common to their application in ad hoc networks. Finally, we present a case study covering the design of a 16-element millimeter-wave microstrip phase array antenna.

1.1 Microstrip/Patch Antenna Elements for Directional Antenna Systems

A microstrip antenna, sometimes called a patch antenna, is composed of a small resonant element often placed above a ground plane. The two elements are separated by a dielectric substrate whose thickness is much less than the operating wavelength, λ [1]. Commonly found in low power microwave applications, they are often found in directional network applications because of their small size and ease of fabrication [2]. Microstrip antennas other favorable qualities are light weight, conformable, low cost, compatibility with solid-state devices, and can be linearly or circularly polarized [1,3,4]. Of course, no antenna is without flaws; due to their relatively small size, microstrip antennas suffer from a number of issues including poor impedance bandwidth, polarization purity, efficiency, and surface wave excitation [3,4].

1.2 Basic Components and Theory of a Microstrip Antenna and Array

The three basic components of a side-fed microwave rectangular patch antenna are shown in Figure 1.1. The three basic components are the feed structure, dielectric substrate, and the radiator. It is the design/selection of each component that dictates the overall antenna performance. For example, the shape and material of the radiator will affect its polarization, gain, and bandwidth. The thickness, dielectric constant ε_r, and loss tangent tan δ of the substrate will affect its radiation performance. For more details regarding the design/selection of each component (see References 1, 3, and 5).

Microstrip antennas can be implemented with a number of different feed structures. The feed structure of a microstrip antenna is of key design consideration because it dictates the impedance matching, modes, surface wave excitation, and overall array geometry. They can be fed from below using a coaxial probe through the ground plan or between the radiator and ground plan with a proximity-coupled structure, from the side using a coplanar microstrip, and below the ground plan using an aperture-coupled feed structure [3]. The following theory will only consider a basic rectangular microstrip patch antenna.

For a simple side-fed rectangular patch antenna, the lowest order resonant frequency is given in Reference 1 as

$$f_r = \frac{c}{2d\sqrt{\varepsilon_r}} \tag{1.1}$$

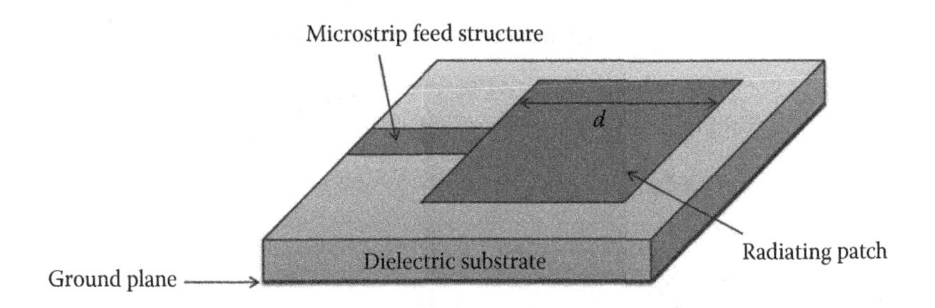

Figure 1.1 Rectangular patch antenna with coplanar microstrip feed structure.

where c is the speed of light, d is the distance from the feed point to the opposite end of the patch, and ε_r is the dielectric constant. Note for the proper resonance d must be on the order of $\lambda_e/2$ where λ_e is the equivalent operating wavelength.

A more complicated feed structure is the probe-fed rectangular patch antenna [6]. This configuration feeds the radiating element from below allowing for more flexibility in array geometry. Illustrated in Figure 1.2, this microstrip patch antenna is fed from below through the ground plane. The lowest resonant frequency for the probe-fed rectangular patch antenna is given in Reference 3 as

$$f_r = \frac{c}{2(L + \Delta L)\sqrt{\varepsilon_{\text{reff}}}} \tag{1.2}$$

where L is the length of the rectangular patch, ΔL is the equivalent length after taking into account the fringing fields, and $\varepsilon_{\text{reff}}$ is the effective relative dielectric constant.

1.2.1 Single-Element Radiation Pattern

One of the most common radiators for millimeter-wave phase arrays are microstrip patch antennas. In Figure 1.3, the radiation pattern for an ideal rectangular patch antenna of dimensions 76 × 50 mm is shown. From this illustration we can see that each element has a fairly wide single lobe beam. When four such elements are placed in a uniformly spaced array, as illustrated in Figure 1.4, we can see that the radiation pattern (also called beam pattern or array factor) is

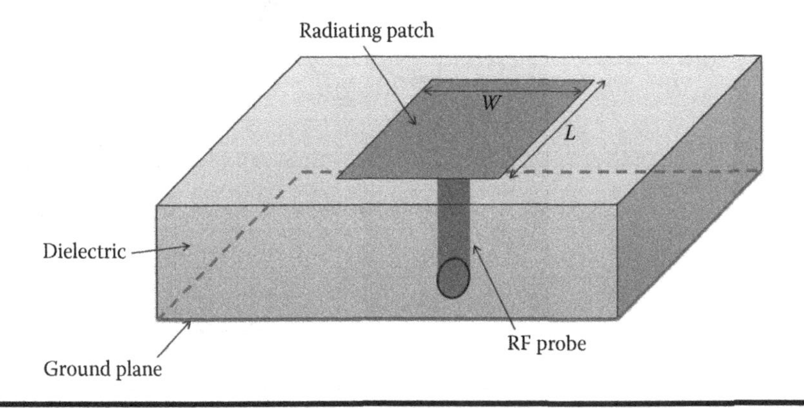

Figure 1.2 Radio frequency probe-fed rectangular microstrip antenna.

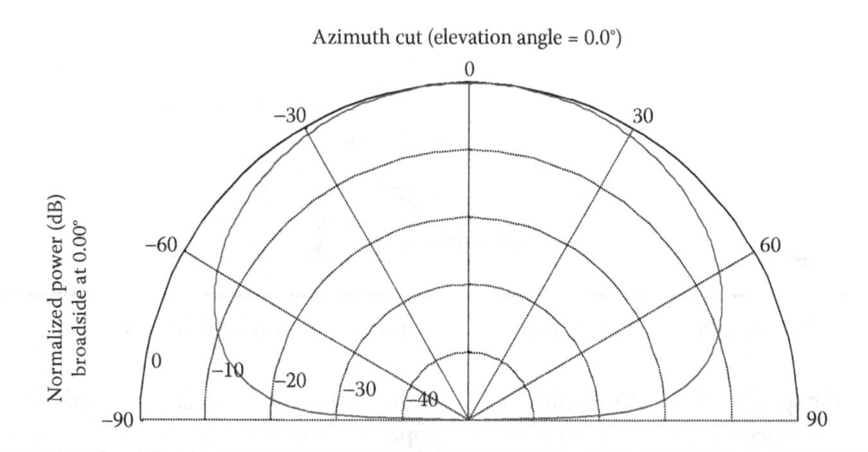

Figure 1.3 Single-element microstrip patch antenna radiation pattern.

drastically different. The beam pattern now has multiple lobes consisting of one main lobe and multiple side lobes. Note that side lobes are any lobes other than the main lobe. Also, note that the main lobe is now much narrower compared to the single antenna case. Later, we show that the main lobe beam width decreases as the number of elements increases.

It is important to note that Figures 1.3 and 1.4 are 2D only for illustrational purposes and that all array beam patterns are naturally 3D. To better illustrate this point, we show the full 3D radiation pattern for a 16-element rectangular uniform array in Figure 1.5.

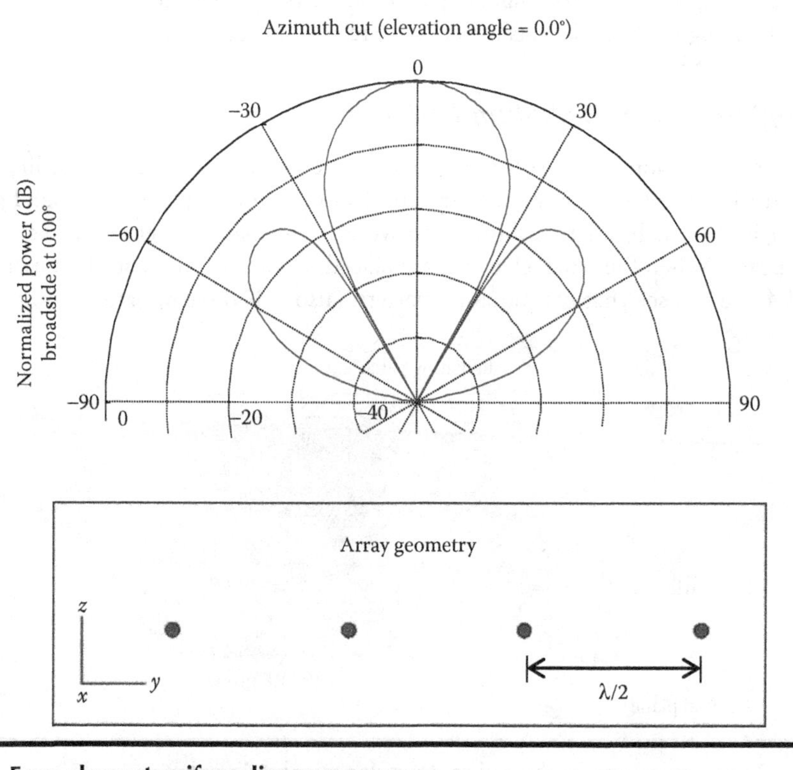

Figure 1.4 Four-element uniform linear array.

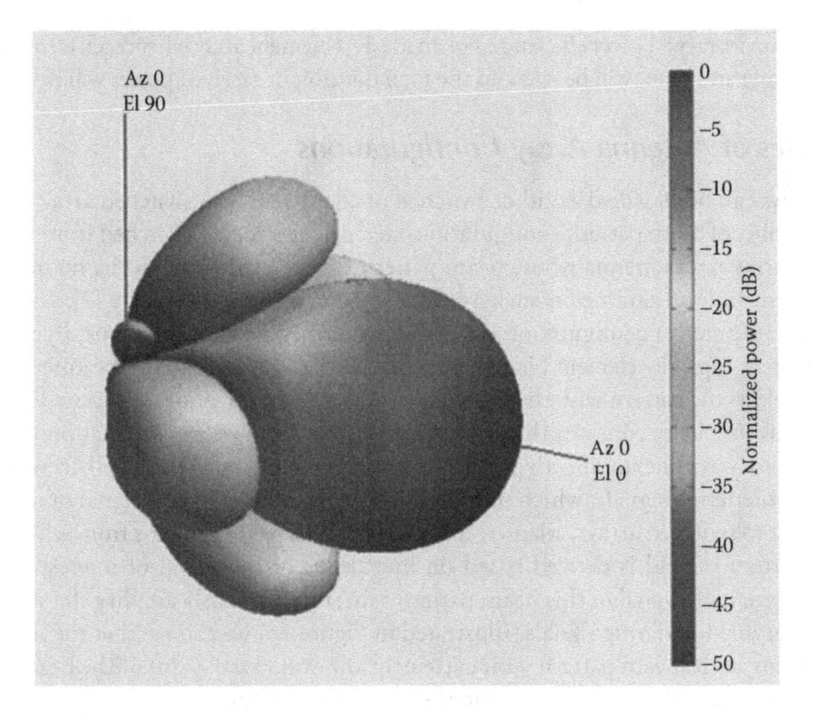

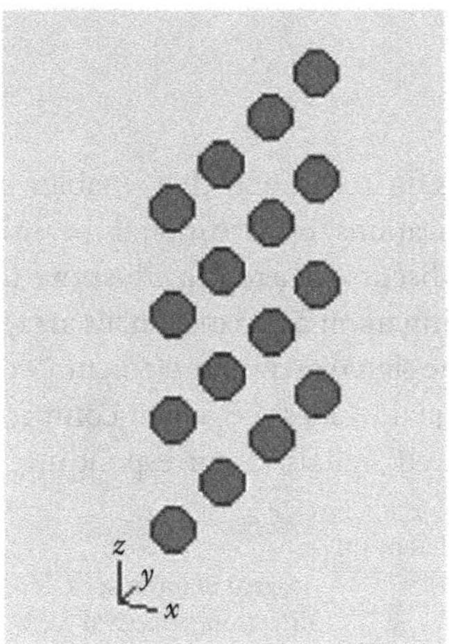

Figure 1.5 Four-by-four element uniform rectangular array with interelement spacing of λ/2.

1.3 Directional Antennas

Many modern wireless communication systems require an antenna with directive gain capabilities. Typically, increasing the gain of an antenna can be obtained by increasing the electrical size of the antenna or constructively combining a number of radiative elements to yield the required directional gain. The latter approach is termed an antenna array [7,8]. An antenna array consists of multiple radiating/receiving elements that are used collectively to create a single radiation pattern, that is, an array beam pattern. This unique capability allows the array to control its directivity, ultimately enabling the spatial diversity required in directional networks for space-division multiple access [9]. Another key benefit of antenna arrays is power efficiency. For an ideal M-element array where each element has a gain of ρ, the resulting array gain will be $M\rho$ and the total transmit or received power will be $(M\rho)^2$ [8,10].

1.3.1 Types of Antenna Array Configurations

Antenna arrays can be classified as either switched or adaptive [11]. A switched array is an antenna with a set number of beam pattern configurations that can be selected/switched from Reference 12. An adaptive array is an antenna whose beam pattern is generated through the manipulation of a number of element-level parameters such as phase, magnitude, and geometry [7].

The switched antenna configuration is the simpler of the two configurations. By preselecting a set of parameters, typically element phase, a set of beam patterns is created. The antenna controller then simply selects the pattern with the most gain in the desired direction. Illustrated in Figure 1.6, we see that this simplicity comes at the price of optimality. Here we can see that preselected beam patterns do not always line up directly with the signal of interest. Also, switched arrays do not take into account interfering signals, which may also be present along with the signal of interest.

In contrast to switched arrays, adaptive arrays have no set beam patterns from which to choose. The beam pattern instead is derived based on knowledge of the signal of interest and incident interference signals. The goal of this beam pattern is to simultaneously amplify the signal of interest while nullifying interfering signals. Illustrated in Figure 1.7, we can see that the main lobe and nulls (directions in the beam pattern with extremely low or no array gain) of the beam pattern are adapted to best deal with the signal and interference environment.

1.4 Beamforming

Adaptive beamforming is an efficient signal processing technique that enables space-division multiple access [9]. The benefits and various applications of traditional beamforming have been well documented in Reference 10 and are not discussed here. To illustrate the concept of beamforming we consider a plane wave signal impinging upon a three-element array as illustrated in Figure 1.8. From this illustration we can see that the signal arrives at element two with a phase lag equal to $d \sin \theta_s$ and at element three with a lag equal to $2d \sin \theta_s$ when compared to element one. In order to coherently combine the three received signals, a corresponding phase-shift is

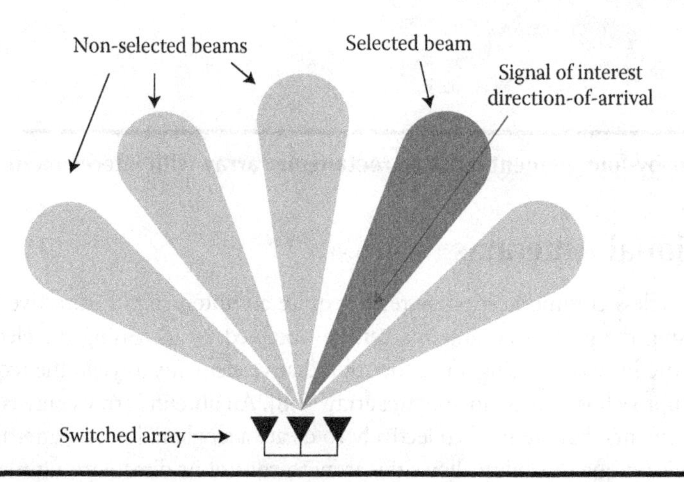

Figure 1.6 Switched array concept.

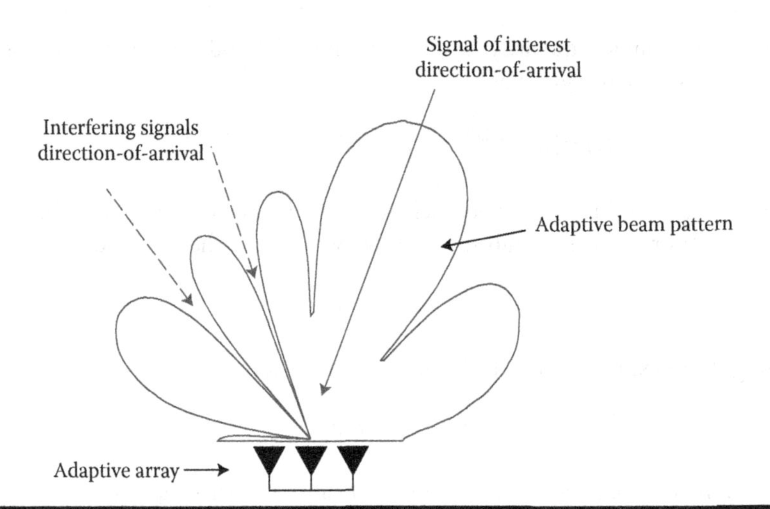

Figure 1.7 Adaptive array concept.

applied to elements two and three before all the signals are summed together. Since the applied phase-shift is equal to the phase delay at each element, this beamformer is considered a phase-shift beamformer. It is important to note that, in order to avoid performance loss due to grating lobes or mutual coupling, d is typically set to $\lambda/2$ [8].

1.4.1 Phase-Shift Beamforming

The phase-shift beamformer can be represented with two complex vectors, the steering vector and the signal vector. With the phase of the incoming signal at element $M = 1$ set to zero, the signal vector is given in Reference 8 as

$$\boldsymbol{u}(\theta_s) = [1, e^{jk\sin\theta_s}, \ldots, e^{j(M-2)k\sin\theta_s}, e^{j(M-1)k\sin\theta_s}]^T \tag{1.3}$$

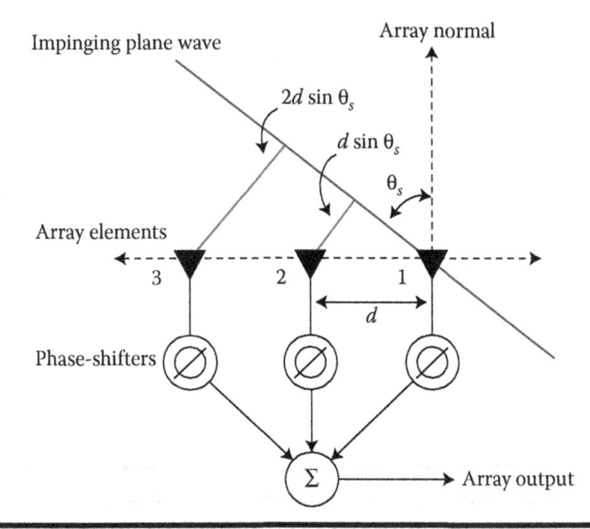

Figure 1.8 Plane wave impinging on a three-element uniform linear array.

where j indicates a complex quantity, $k = 2\pi d/\lambda$, and $[*]^T$ indicates the transpose operation. The steering vector is similarly given as

$$v(\theta_a) = [1, e^{-jk\sin\theta_a}, \ldots, e^{-j(M-2)k\sin\theta_a}, e^{-j(M-1)k\sin\theta_a}] \tag{1.4}$$

where θ_a is the array steering angle and each term in the vector represents the corresponding elements' phase-shift or complex weighting. The array output is then represented as

$$A_{\text{out}} = \boldsymbol{uv}. \tag{1.5}$$

By scanning θ_s, an expression for the beam pattern is expressed as

$$AF(\theta_s) = \boldsymbol{uv}(\theta_s) \tag{1.6}$$

The key parameters that dictate the resulting beam pattern of an array are the individual element radiation pattern, their configuration/geometry, the number of elements, and their complex weight and magnitude [8,10]. The effects of these parameters on a six-element uniform linear array are illustrated in Figure 1.9.

In Figure 1.9a, each element is spaced $\lambda/2$ distance apart with uniform phase and magnitude. This results in a beam pattern with the main beam pointed broad side. Note that increasing the number of elements above two will cause multiple sidelobes to appear in the beam pattern. In Figure 1.9b, the same array is implemented with phase-shift beamforming to steer the main beam 45° off broadside. In Figure 1.9c, the interelement spacing is increased to 2λ as opposed to $\lambda/2$ causing multiple grating lobes to appear in the beam pattern. Finally, in Figure 1.9d, the

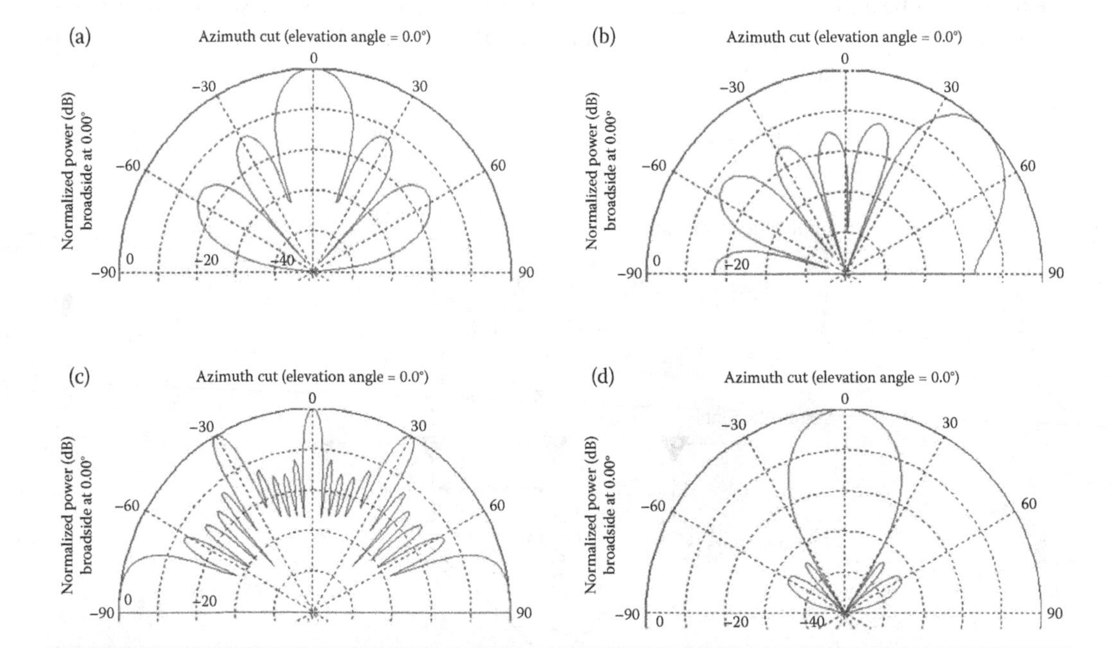

Figure 1.9 Different beam pattern configurations for a six-element uniform linear array: (a) Six element uniform linear array; (b) Array steered to 45° off boardside; (c) Array with inter-element spaced = 2; (d) Array with a 30 dB array taper.

magnitude of each element has been altered in order to reduce the sidelobe response; this is referred to as array tapering/windowing and is discussed later in this chapter [7].

1.4.2 Adaptive Beamforming

Adaptive beamforming is the process of adapting an antenna's array factor in a manner that both simultaneously increases the gain of a target signal while decreasing the gain of all interfering signals. This is accomplished by selecting the proper phase-shift, that is, complex phase weight, for each antenna element that will result in a desired beam pattern. These weights modify the phase and allow for the coherent amplification (reduction) of a target signal (undesired signals). In this section, we illustrate the adaptive minimum variance distortionless response (MVDR) beamformer and compare it to the classical phase-shift beamformer [13,14].

The MVDR beamformer combats the effects of N_u number of interfering signals by placing nulls in their direction. This is achieved by selecting the array weights that minimize the array's output noise variance. The array output is expressed in Reference 8 as

$$y(t) = \boldsymbol{w}^H s + \boldsymbol{w}^H \boldsymbol{u} \tag{1.7}$$

where $\boldsymbol{w}$ is an $M \times 1$ vector of the complex weights to be determined, $\boldsymbol{u}$ is an $M \times 1$ vector of the sum of all interfering signal vectors $\eta_i = [1, e^{i\beta d \sin\theta_i}, \ldots, e^{i\beta(M-1)d\sin\theta_i}]^T$ represented as

$$\boldsymbol{u} = \sum_{i=1}^{N_u} u_i(t)\eta_i \tag{1.8}$$

and $\boldsymbol{s}$ is an $M \times 1$ the signal vector represented by

$$s = s(t)\boldsymbol{v} \tag{1.9}$$

with $s(t)$ being the desired signal. By minimizing the output noise, that is, contributions of the interfering signals in the output, we can determine the optimal weights $\boldsymbol{w}_o$. The output variance can be shown as

$$\sigma_y^2 = w^H R_s w + w^H R_u w \tag{1.10}$$

Assuming that $\boldsymbol{R}_s = E\{\boldsymbol{ss}^H\}$ and $\boldsymbol{R}_u = E\{\boldsymbol{uu}^H\}$ are known, we observe that the minimization of the output variance is reduced to the minimization of $\boldsymbol{w}^H\boldsymbol{R}_u\boldsymbol{w}$. This results in the expression for the optimal array weights

$$w_o = \frac{1}{v^H R_u^{-1} v} R_u^{-1} v \tag{1.11}$$

To illustrate and compare both the phase-shift and MVDR beamforming concepts, we consider a scenario where a target signal with two interfering signals are received by a five-element uniform linear array similar to the one shown in Figure 1.4. The target signal-of-interest is shown in Figure 1.10, with a direction-of-arrival (DOA) of 35° and signal amplitude of one. The two random interfering signals are arriving at a DOA of 0 and 60°, each with maximum amplitude of 10.

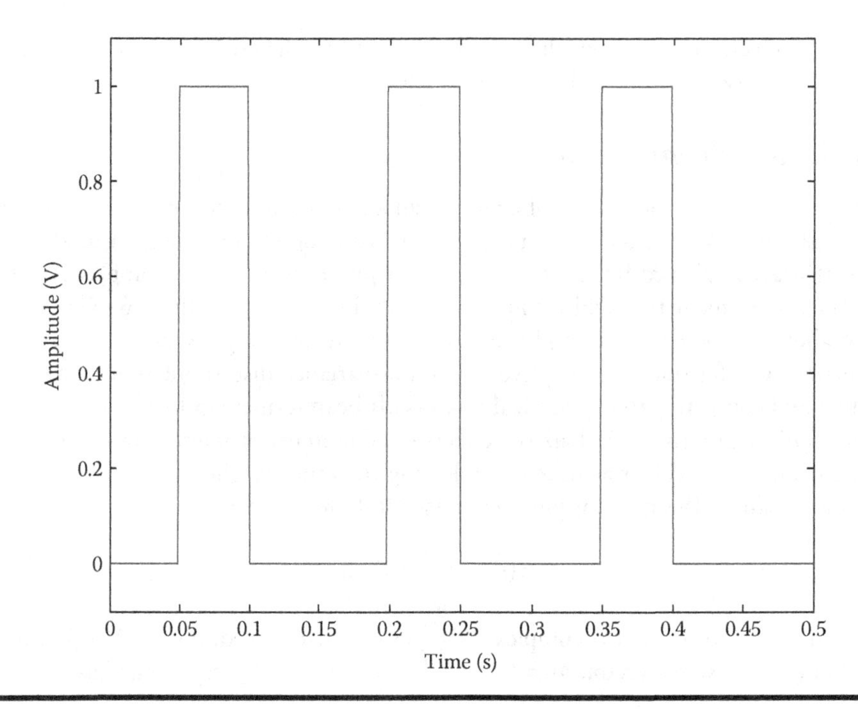

Figure 1.10 Transmitted signal-of-interest.

The phase-shift beamformer's beam pattern and its response with and without interference are shown in Figure 1.11. Since a phase-shift beamformer does not take into account interfering signals we can see in Figure 1.11a that the array still has a strong response at 0° and 60° (the direction of the two interfering signals). As a result, we can see in Figure 1.11c that the two interfering signals have fully corrupted the target signal. In contrast to the phase-shift beamformer, the adaptive MVDR beamformer's response and associated beam pattern are shown in Figure 1.12. Since the MVDR beamformer has taken into account the interference by placing a null in their direction, we see in Figure 1.12a that the signal of interest has been successfully recovered.

1.5 Error Effects in Beamforming

Two key errors to consider in beamforming are pointing errors and unintended sidelobe response. Pointing errors are defined as the difference between the array's actual steering direction and that of the maximum response. The standard deviation of the pointing error is given by Reference 4 as

$$\sigma_{PE} \approx \frac{\sigma_{\phi}(B_w)}{\sqrt{M}} \tag{1.12}$$

where σ_{ϕ} is the standard deviation of the phase errors. With the beamwidth for a uniform linear array as defined as $B_w \approx 0.88\lambda/(dM)$, the pointing errors can be expressed as

$$\sigma_{PE,ULA} \approx \frac{0.88\lambda\sigma_{\phi}}{dM\sqrt{M}} \tag{1.13}$$

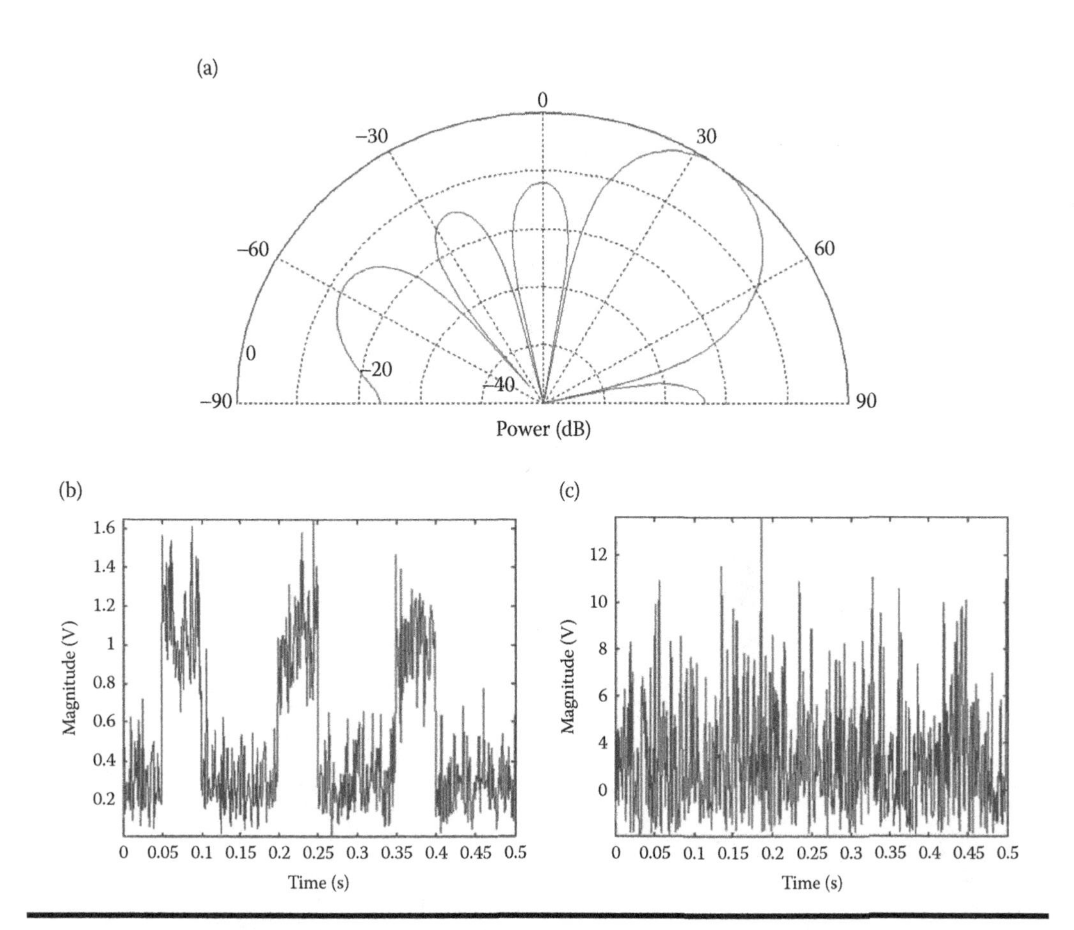

Figure 1.11 Phase-shift beamformer beam pattern and response with and without interfering signals present. (a) Phase-shift beamformer beam pattern; (b) Array response without interference; (c) Array response with interference.

The expression for the expected sidelobe level is given by Reference 8 as

$$\Delta_{\mathrm{sl}} = \frac{a(1-A^2)+\sigma_{\Delta a}^2}{Ma^2A^2} \tag{1.14}$$

where a is the error free main beam gain, $\sigma_{\Delta a}^2$ is the variance of amplitude error, and $A^2 = e^{-\sigma_{\Delta\phi}^2}$ for zero mean Gaussian phase error.

1.6 Sidelobe Control via Array Tapering

For a uniform linear array, the sidelobe response can be controlled by adjusting the gain/magnitude of each element's response [15]. This process is termed tapering and is akin to windowing in digital filtering [16]. Tapering an array provides a trade-off between the beamwidth and the sidelobe levels. With this in mind, we briefly examine the characteristics of four tapering methods: Uniform (no taper), Dolph–Chebyshev [17], Taylor–Kaiser [18], and binomial [19].

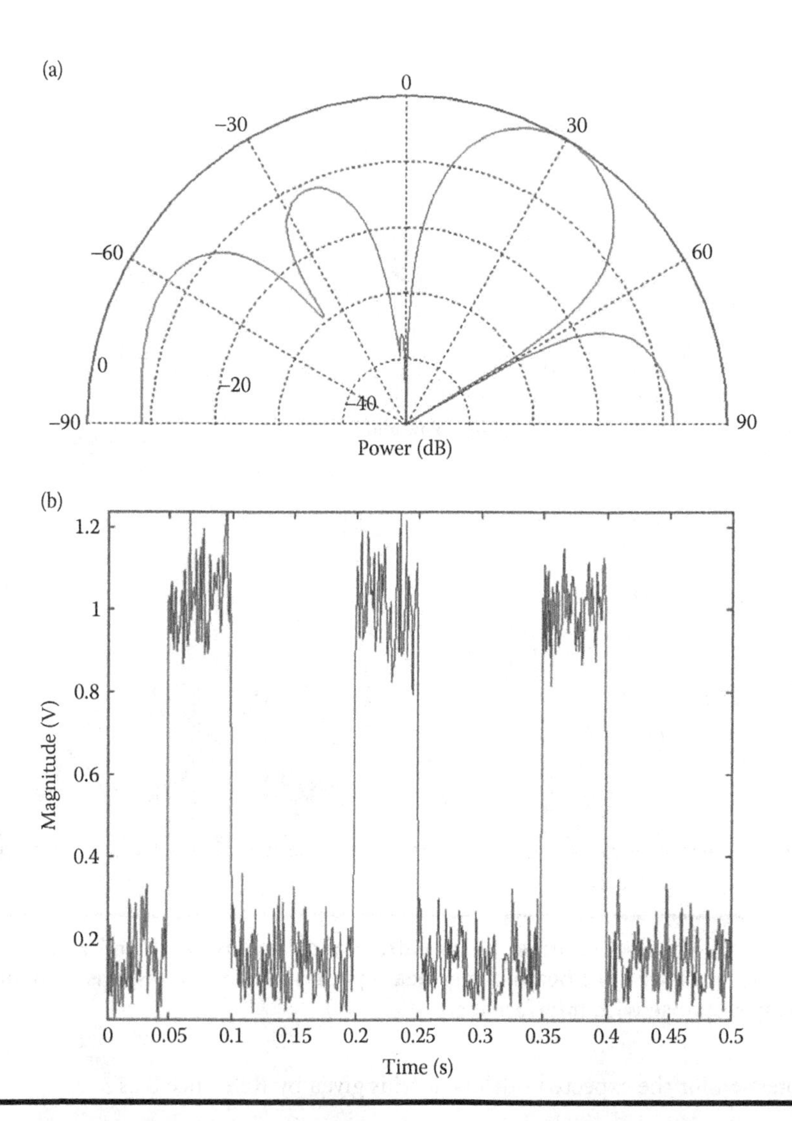

Figure 1.12 Beam pattern and array response of the minimum variance distortionless response beamformer in the presence of a target signal at 35° and two interfering signals at 0° and 60°: (a) MVDR beamformer beam pattern; (b) MVDR beamformer array response.

1.6.1 Tapering Methods for Uniform Linear Arrays

The uniform tapered array is actually a nontapered array as all nodes are uniformly weighted equally, which results in the standard phase-shift beamformer with an array factor of

$$A_{FL}(\theta) = \sum_{m=1}^{M} e^{j(m-1)\beta d \sin\theta} \quad (1.15)$$

This tapering method makes no consideration to sidelobe response and has the highest sidelobe level when compared to other methods.

The binomial tapered array is unique because it has zero sidelobes, but this is at the cost of a wider beamwidth. The weights of an M-node binomial uniform linear array are calculated from the binomial coefficients as follows in Reference 20 as

$$w_{BN}(m) = \frac{(M-1)!}{m!(M-1-m)!}, \quad m = 0,1,\ldots,M-1 \tag{1.16}$$

The corresponding array factor is given as

$$A_{FB}(\theta) = \left(2\cos\left(\frac{\theta}{2}\right)\right)^{M-1} \tag{1.17}$$

From this expression, we can see that the array factor decreases to zero as θ approaches $\pm\pi$.

The Dolph–Chebyshev [17] taper method is best applied when the sidelobe response at all angles must be kept below a specific value R_{DC}. The taper weights for the Dolph–Chebyshev uniform linear array are calculated using the Dolph–Chebyshev window transform [21]

$$W_{\text{Cheb}}(\omega_k) = \frac{\cos(M\cos^{-1}[R_0\cos(\pi i/M)])}{\cosh(M\cosh^{-1}(R_0))}, \quad i = 0,1,\ldots,M-1 \tag{1.18}$$

where the scaling factor R_0 is defined as

$$R_0 = \cosh\left(\frac{\cosh^{-1}(10^{R_{DC}/20})}{M}\right) \tag{1.19}$$

The array weights $w_{DC}(m)$ are then computed as the inverse discrete Fourier transform of $W_{\text{Cheb}}(\omega_k)$. The corresponding array factor for an array with an even number of nodes is expressed in Reference 20 as

$$A_{FD}(\theta) = \cos\left(2M\cos^{-1}\left(R_0\cos\left(\frac{\theta}{2}\right)\right)\right) \tag{1.20}$$

The Taylor–Kaiser tapering method is similar to the Dolph–Chebyshev method but has an exponentially decreasing sidelobe response. The array weights can be computed as Reference 21

$$w_{TK}(m) = I_0\left(\pi\alpha_{TK}\sqrt{\frac{1-m^2}{M^2}}\right), \quad m = 1,\ldots,M \tag{1.21}$$

where I_0 is the zeroth-order modified Bessel function of the first kind and α_{TK} is the parameter used to control the sidelobe level response. The Taylor–Kaiser array factor can be expressed as

$$A_{FT}(\theta) = \frac{M}{I_0(\pi\alpha_{TK})}\frac{\sinh\left(\sqrt{\alpha_{TK}^2\pi^2-(M\theta/2)^2}\right)}{\sqrt{\alpha_{TK}^2\pi^2-(M\theta/2)^2}} \tag{1.22}$$

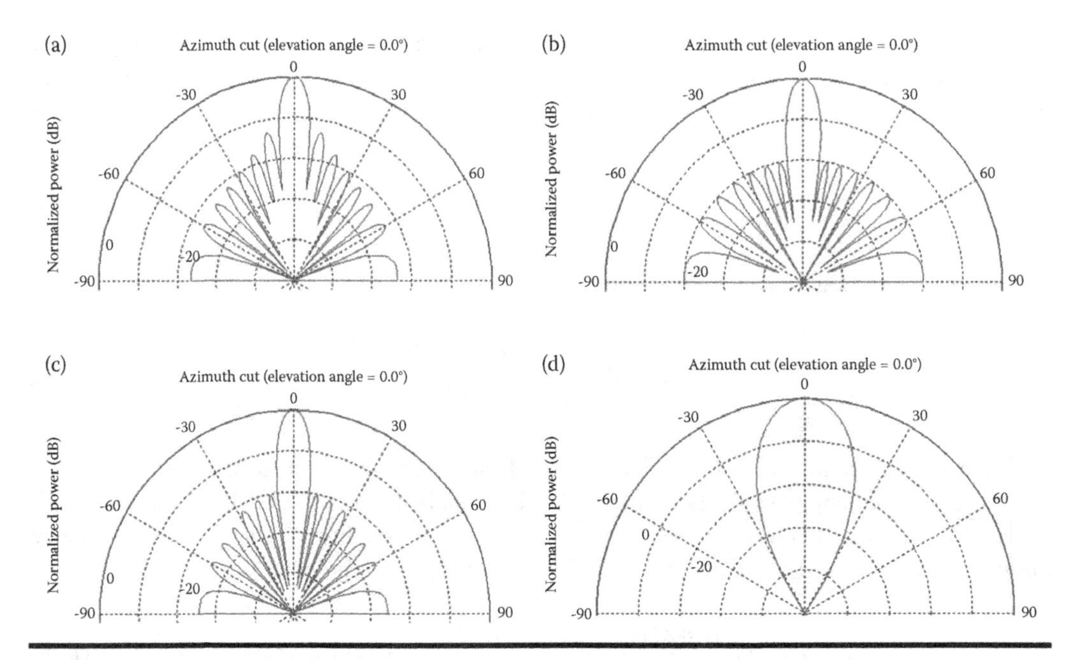

Figure 1.13 Sidelobe level trade-off for four different tapering methods using a 15-element uniform linear array. (a) Uniform (non-tapered); (b) Dolph–Chebyshev taper; (c) Taylor–Kaiser taper; (d) Binomial taper.

1.6.2 Comparison of the Taper Methods

A comparison of the four taper methods is illustrated in Figure 1.13. Here, we see that the beamwidth is inversely proportional to the sidelobe response. We can see that the uniform tapering achieves the narrowest beamwidth with the highest sidelobe response, whereas the binomial tapering has the widest beamwidth and zero sidelobe response. The Dolph–Chebyshev has the unique ability to maintain a given maximum sidelobe response, while the Taylor–Kaiser taper shows a better sidelobe response at the cost of a slightly wider beamwidth.

1.7 Directional Antenna Case Study: A Wafer Scale Fully Integrated 16-Element Phase Array Transmitter

In this section the design of a fully integrated wafer scale 16-element phase array transmitter is introduced. Designed for use with the IEEE 802.15.3c protocol, this array is a prime example of a directional antenna built for directional networks. Implemented in SiGe BiCMOS, this mmWave array is capable of steering a multiple-Gigabits per second (Gbps) directional link while only occupying an area of 44 mm^2 with a maximum power consumption of 6.2 W [22].

The extremely high carrier frequency of the IEEE 802.15.3c protocol pushes radiator elements down to the millimeter-wave scale, and at these frequencies path loss becomes a significant issue. This results in an increased reliance on phased array antennas that can efficiently deliver a high effective isotropic radiated power (EIRP) [3] to deal with path losses and provide directivity to combat a challenging non-line-of-sight environment. With the added bonus of being fully characterized on a wafer, this phase array design is a perfect example of directional antennas built to complement/enhance directional networks.

To fully explore this design, we start with a set of design requirements. For this design, it is the band limited average and maximum power density, which is set at 9 and 18 μW/cm^2 by the U.S. Federal Communication Commission [23]. This translates to an average and max EIRP of 40 and 43 dBm [22]. To determine the required output power of each element in the array we first examine the array's EIRP. The EIRP of a phased array is given in Reference 10 as

$$T_{\mathrm{EIRP}} = P_{NE}(MG_a) \tag{1.23}$$

where P_{NE} is the total power in watts, G_a is the per-element antenna gain in decibel isotropic, and M is the number of elements. Equation 1.23 is then rewritten to account for the required single-element output power as

$$T_{\mathrm{EIRP}} = P_{1E} + G_a + 20\log(M) \tag{1.24}$$

where P_{1E} is the required per-element output power in decibel-milliwatts and total EIRP is now in units of decibel-milliwatts. Assuming a per-element antenna gain of 4 dBi [24], the trade-off between the number of elements and per-element output power is shown in Figure 1.14. In Reference 22, the design selected 16 elements which resulted in a per-element output of 10–12 dBm.

1.7.1 Phased-Array System Architecture

Following the architecture laid out in Reference 25, this design uses a double-conversion architecture with a frequency tripler. The radio frequency (RF) signal path begins with a sliding

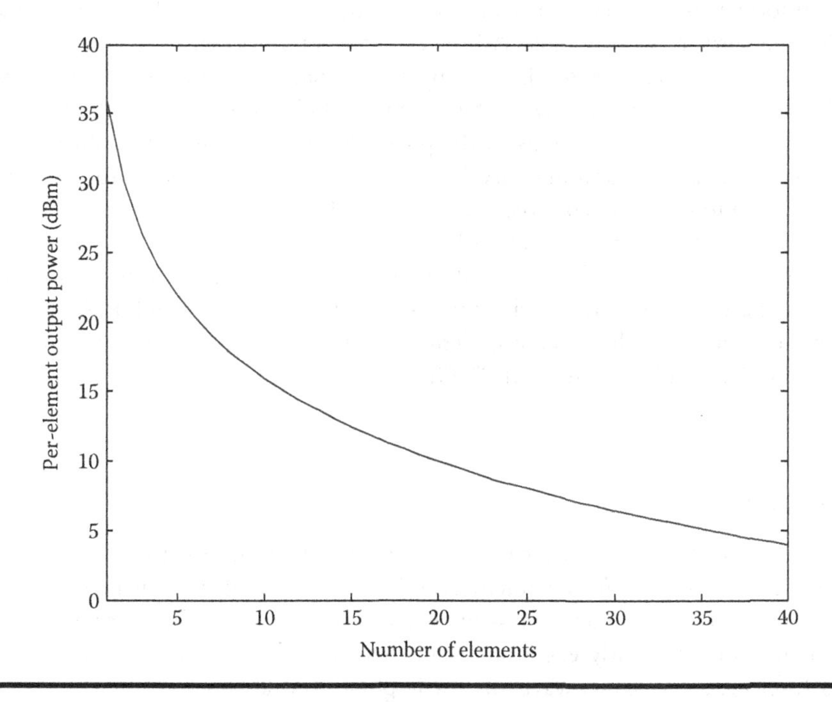

Figure 1.14 Trade-off between number of elements per array versus per-element output power.

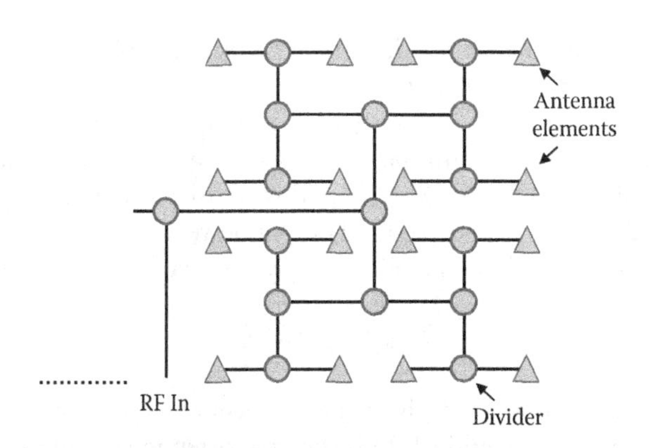

Figure 1.15 Right half of a 32-element corporate feed structure.

intermediate frequency of 8.331–9.257 GHz which is then tripled to a RF center frequency of 58.32–64.8 GHz. This is followed by a 1–16 power distribution tree which leads to the phase-shifter and power amplifier and ends at the radiating element.

Power distribution for phase arrays implemented in the millimeter-wave scale is a critical design consideration, as power loss in transmission lines is proportional to the operating frequency. Besides consideration in power loss, the distribution design must also account for linearity, wafer area, and bandwidth [3,10,22]. To accommodate all these considerations, this design chose to implement a corporate feed architecture. Shown in Figure 1.15, this architecture uses a hybrid passive/active approach to minimize the power loss while maintaining strong isolation, linearity, and bandwidth.

The beamformer for each element consists of a phase shifter, variable three-stage amplifier, and a digital memory array beam table. All the transmission lines consist of copper side-shielded microstrip lines with a ground plane also in copper. The phase shifters in this design have five bits of resolution and consist of the passive voltage-controlled reflection-type phase shifters found in Reference 26. The main considerations in the phase shifter design are phase shift range, power consumption, low signal loss, and linearity. This implementation was shown to provide >200° of phase-shift with an insertion loss between 5.5 and 9.5 dB. The power amplifiers are designed in a cascade topology to ultimately deliver up to 15 dBm of power.

The last component of the RF signal path is the radiating element. In this design, the radiators are the balanced-fed aperture-coupled microstrip patch antennas found in Reference 24. The measured performance of such an antenna elements was shown to have a peak gain of 8 dBi with 10 dB return loss bandwidth greater than 10 GHz.

1.8 Summary

This chapter examined various aspects of directional antennas and their capabilities. We began with an overview of microstrip/patch antennas and how they can be implemented in a directional antenna. We discussed the fundamentals of switched and adaptive phased array antennas and some common errors frequently encountered in their implementation. Finally, we covered the design of a 16-element millimeter-wave microstrip phase array antenna.

References

1. Howell, J., 1975. Microstrip antennas, *IEEE Trans. Antennas Propag.*, 23(1), 90–93.
2. Bellofiore, S., J. Foutz, C. A. Balanis, and A. S. Spanias, 2002. Smart-antenna systems for mobile communication networks Part 2: Beamforming and network throughput, *IEEE Antennas Propagat. Mag.*, 44(4), 106–114.
3. Chen, Z. N. and M. Y. Chia, 2006. *Broadband Planar Antennas: Design and Applications*, New York: John Wiley & Sons.
4. Lo, Y. T., D. Solomon, and W. Richards, 1979. Theory and experiment on microstrip antennas, *IEEE Trans. Antennas Propag.*, 27(2), 137–145.
5. Gupta, K. C., P. S. Hall, and C. Wood, 1988. *Microstrip Antenna Design*, Boston, MA: Artech House.
6. Bahl, I., P. Bhartia, and S. Stuchly, 1981. Design of microstrip antennas covered with a dielectric layer, *IEEE Trans. Antennas Propag.*, 29(3), 314–318.
7. Trees, H. L. V., 2002. *Optimum Array Processing*, New York: John Wiley & Sons Inc.
8. Litva, J. and T. K. Lo, 1996. *Digital Beamforming in Wireless Communications*, Boston, MA: Artech House.
9. Veen, B. D. V. and K. M. Buckley, 1986. Beamforming: A versatile approach to spatial filtering, *IEEE ASSP Mag*, 5(2), 4–24.
10. Mailloux, R. J., 1994. *Phased Array Antenna Handbook*, Boston, MA: Artech House.
11. Balanis, C. A. and P. I. Ioannides, 2007. *Introduction to Smart Antennas*, Arizona: Morgan & Claypool.
12. Ramanathan, R., J. Redi, C. Santivanez, D. Wiggins, and S. Polit, 2005. Ad hoc networking with directional antennas: A complete system solution, *IEEE J. Sel. Areas Commun.*, 23(3), 496–506.
13. Pan, C., J. Chen, and J. Benesty, 2014. Performance study of the MVDR beamformer as a function of the source incidence angle, *IEEE Trans. Speech Audio Process.*, 22(1), 67–79.
14. Souden, M., J. Benesty, and S. Affes, 2010. A study of the LCMV and MVDR noise reduction filters, *IEEE Trans. Signal Process.*, 58(9), 4925–4935.
15. Holzman, E. L., 2003. A different perspective on taper efficiency for array antennas, *IEEE Trans. Antennas Propag.*, 51(10), 2963–2967.
16. Oppenheim, A. V., R. W. Schafer, and J. R. Buck, 1999. *Discrete-Time Signal Processing*, Upper Saddle River, NJ: Prentice-Hall.
17. Nuttal, A., 1974. Generation of Dolph–Chebyshev weights via a fast fourier transform, *IEEE Proc.*, 62(10), 1396.
18. Kaiser, J. F. and R. W. Schafer, 1980. On the use of the I0-sinh window for spectrum analysis, *IEEE Trans. Acoust., Speech, Signal Process.*, 28(1), 105–107.
19. McCormack, C. and R. Haupt, 1991. Antenna pattern synthesis using partially tapered arrays, *IEEE Trans. Magn.*, 27(5), 3902–3904.
20. Orfanidis, S. J., 1996. *Introduction to Signal Processing*, New York: Prentice-Hall.
21. Harris, F. J., 1978. On the use of windows for harmonic analysis with the discrete Fourier transform, *IEEE Proc.*, 66(1), 51–83.
22. Valdes-Garcia, A., S. T. Nicolson, J. W. Lai, A. Natarajan, P. Y. Chen, S. K. Reynolds, J. H. C. Zhan, D. G. Kam, D. Liu, and B. Floyd, 2010. A fully integrated 16-element phased-array transmitter in SiGe BiCMOS for 60-GHz communications, *IEEE J. Solid-State Circuits*, 45(12), 2757–2773.
23. FCC, 2004. *Code of Federal Regulation, Title 47 Telecommunication*, FCC, Washington, DC, http://www.fcc.gov
24. Liu, D., J. Akkermans, H. Chen, and B. Floyd, 2011. Packages with integrated 60-GHz aperture-coupled patch antennas, *IEEE Trans. Antennas Propag.*, 59(10), 3607–3616.
25. Reynolds, S., A. Valdes-Garcia, B. Floyd, B. Gaucher, D. Liu, and N. Hivik, 2007. Second generation transceiver chipset supporting multiple modulations at Gb/s data rates. In *Proc. IEEE Bipolar/BiCMOS Circuits and Technology Meeting*, October, Boston, MA, USA.
26. Hardin, R. N., E. J. Downey, and J. Munushian, 1960. Electronically variable phase shifter utilizing variable capacitance diodes, *Proc. IRE*, 48, 944–945.

Chapter 2

Design and Optimization of Wideband Log-Periodic Dipole Arrays under Requirements for High Gain, High Front-to-Back Ratio, Optimal Gain Flatness, and Low Side Lobe Level: The Application of Invasive Weed Optimization

Zaharias D. Zaharis, Pavlos I. Lazaridis, Christos Skeberis, George Mastorakis, Constandinos X. Mavromoustakis, Evangelos Pallis, Dimitrios I. Stratakis, and Thomas D. Xenos

Contents

Abstract

The design and optimization of wideband log-periodic dipole arrays (LPDAs) are presented in this chapter. The LPDAs are expected to simultaneously satisfy several requirements inside a wide operating frequency range. In particular, the optimized LPDA has to provide standing wave ratio (SWR) below a predefined value, gain values as high as possible, gain flatness (GF) below a desired value, both side lobe level (SLL) and front-to-back ratio (FBR) below a desired value, and all these requirements must be satisfied at the same time inside the entire frequency range of operation. Since the design problem is nonlinear and inherently multiobjective, the simultaneous realization of all the above requirements can only be achieved by applying global optimization algorithms. These algorithms are usually based on evolutionary optimization methods and have been proved to be capable of solving complex nonlinear problems with great success. Such an evolutionary method with high potential in antenna design is presented in this chapter. The method is called invasive weed optimization (IWO) and is applied in conjunction with the method of moments (MoM) to optimize LPDAs under the above-mentioned requirements. The MoM is a well-known full-wave analysis method and is utilized here to extract the radiation characteristics of the LPDA required by the IWO algorithm. Several design cases are studied concerning the LPDA geometry and the operating bandwidth. The derived LPDA geometries exhibit a behavior close to the desired one and therefore are able to enhance the performance of a wireless network in practical applications.

2.1 Introduction

The performance of a wireless network is seriously affected by the radiation characteristics of the antenna utilized by the base stations of this network. An optimized antenna, which provides all the desired characteristics demanded by the network, is mostly required in practice. Therefore, a carefully designed antenna is a practically interesting issue. LPDAs belong to a special antenna array category that exhibit wideband behavior [1]. In particular, they are composed of dipoles of gradually increasing length and two booms that connect the dipoles in such a way that the dipole feeding is inverted when passing from one dipole to the next one. Therefore, not all the dipoles radiate at a certain frequency but only those being in resonance condition. These are dipoles of proper length with significant current distribution along their axis compared to the rest dipoles of the array. At the certain frequency, these dipoles are considered as active, while the rest ones are considered as parasitic. In particular, the larger parasitic dipoles act as reflectors, while the shorter ones act as directors, which mean that the radiated power is directed along the LPDA axis toward the shorter dipoles. To ensure that the radiation is directed toward shorter dipoles at every frequency, phase delay must be applied to the dipole excitation when moving from shorter to larger dipoles. This is achieved by applying a feeding source in the middle point of the shortest dipole of the array. As the frequency moves to higher values, the active part of the array moves to shorter dipoles. This behavior makes an LPDA operate in a frequency range, which is much greater than that of a single dipole of specified length. The increased operating bandwidth makes LPDAs become important structures in wireless networks. In literature, the operating bandwidth is usually defined as the frequency range where the SWR at the antenna input is equal to or less than 2. Therefore, the operating bandwidth is defined from the upper limit of SWR. Of course, based on the researcher's preferences and the application requirements, a limit of SWR greater or less than 2 can be used to define the bandwidth.

As mentioned before—when the frequency moves to higher values—the active part of the array moves to shorter dipoles, which means that the current distribution gradually becomes significant on shorter dipoles (they start to act as active dipoles), and—at the same time—the current diminishes on dipoles that were previously active. However, at certain frequencies, a significant current distribution may arise on large dipoles, which are normally considered as parasitic ones, and make them radiate together with the active elements of the array. This effect is due to high-order resonance developed on these dipoles and usually results in a degradation of electromagnetic characteristics of the LPDA in these frequencies, such as the antenna gain. The problem is solved by using a short-circuited stub of proper length behind the largest dipole of the array. The stub has the ability to choke the current developed on large dipoles due to high-order resonances induced on them. This is a simple way to keep such resonances suppressed and help the LPDA to preserve its behavior throughout the entire operating bandwidth.

The performance of a wireless network depends not only on the antenna operating bandwidth itself, but also on far-field radiation characteristics inside this bandwidth, such as the gain (G), the FBR, and the SLL of the antenna, as well as the antenna GF defined as the difference between the maximum (G_{max}) and the minimum (G_{min}) gain values found inside the entire bandwidth (i.e., $GF = G_{max} - G_{min}$). A directional antenna—that is, an antenna with high gain values inside the whole operating bandwidth—is mostly desired in practice since it may significantly increase the performance of a wireless network. However, the performance is also affected by the values of three additional parameters. These are the values of GF, SLL, and FBR discussed earlier. The value of GF determines the gain degradation over the operating bandwidth. A good GF means that the gain maintains its values over the entire frequency range. Also, both low SLL and high FBR mitigate the performance degradation, which may occur due to multipath fading inside the network service area. Therefore, an antenna, which achieves high gain, FBR above a predefined value, and SWR, GF, and SLL below respective specified values over the operating bandwidth, seems to be the best solution for base stations of wireless networks.

The most complete method of LPDA design has been proposed by Carrel [1,2]. Later, this method was corrected in Reference 3 since Carrel did not describe the current distribution on the LPDA elements in the right way. According to this method, all the dipoles are assumed to be located inside the same angular sector, as shown in Figure 2.1. Also, the two booms used to feed the dipoles are modeled as a transmission line of two conductive wires, which are inverted when passing from one dipole to the next one. The dipole lengths L_m ($m = 1, \dots, M$), the dipole radii r_m ($m = 1, \dots, M$), and the distances S_m ($m = 1, \dots, M{-}1$) between adjacent dipoles, the length S_m of the short-circuited stub optionally located behind the largest (Mth) dipole, and finally the characteristic impedance Z_0 of the line that models the booms constitute the geometry of an M-element LPDA (Figure 2.1). It has to be noted that the line considered for boom modeling must not be confused with the main transmission line, which is used to feed the whole antenna and is assumed to have a characteristic impedance of 50 Ω as shown in Figure 2.1. According to Carrel's method, the values of L_m ($m = 1, \dots, M$), r_m ($m = 1, \dots, M$) and S_m ($m = 1, \dots, M{-}1$) can easily be

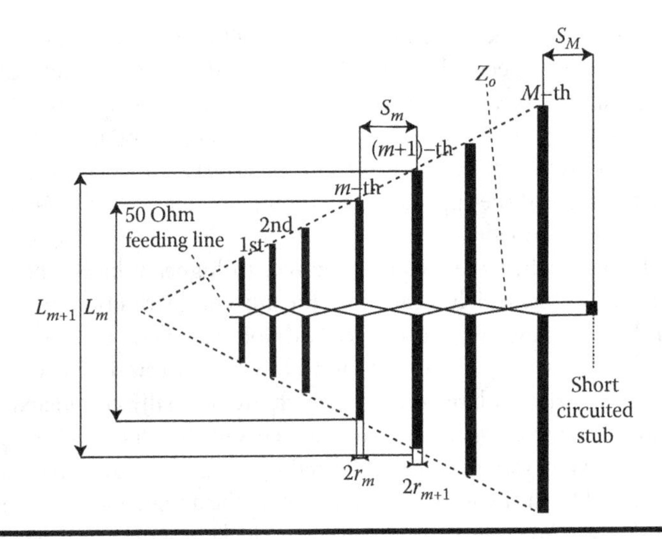

Figure 2.1 LPDA geometry based on Carrel's method.

found by using two special parameters known as *scale factor* τ and *relative spacing* σ, respectively defined by the following expressions:

$$\tau = \frac{L_m}{L_{m+1}} = \frac{r_m}{r_{m+1}}, \quad m = 1, \ldots, M-1 \tag{2.1}$$

$$\sigma = \frac{S_m}{2L_{m+1}}, \quad m = 1, \ldots, M-1 \tag{2.2}$$

The appropriate values of τ and σ for various predefined average values of the antenna directivity were estimated by Carrel and are graphically presented in a constant directivity contour curve graph [1–3].

To apply Carrel's method, the upper and the lower operating frequency as well as the desired average directivity of the LPDA must initially be defined. From the desired average directivity, we can find the appropriate values of τ and σ, by using the constant directivity contour curve graph mentioned earlier. Then, the length L_M and the radius r_M of the largest dipole are estimated. The value of L_M is set equal to half-wavelength at the lower operating frequency. In this way, it is considered that the largest dipole becomes active (i.e., in resonance condition) at the lower operating frequency. To be active in practice, the above value of L_M must be reduced (e.g., by 4%), since the dipole thickness affects the resonance condition. Also, values easily found in practice are used for r_M (e.g., 5 or 6 mm). Finally, the values of L_m, r_m, and S_m are estimated by applying the following expressions, which are extracted from Equations 2.1 and 2.2:

$$L_m = \tau^{M-m} L_M, \quad m = 1, \ldots, M \tag{2.3}$$

$$r_m = \tau^{M-m} r_M, \quad m = 1, \ldots, M \tag{2.4}$$

$$S_m = 2\sigma\tau^{M-m-1} L_M, \quad m = 1, \ldots, M-1 \tag{2.5}$$

If the LPDA extracted in the way explained earlier does not produce the expected behavior, for example, an increase in SWR or rapid gain drops are observed in some frequencies, then a short-circuited stub with proper length S_M may be used to mitigate such effects. These effects are mostly caused due to high-order resonances developed on some dipoles and the use of a stub helps to suppress these resonances.

However, due to simplifications made by Carrel, this method is rather an approximate than a full-wave analysis technique, and therefore it cannot accurately estimate the behavior of the antenna array over the entire operating bandwidth. Also, the method is almost capable of controlling the SWR as well as the values of G and FBR inside a predefined bandwidth; however, it totally fails to control the values of GF and SLL, which are essential for increased network performance. A global optimization method that makes use of a full-wave analysis technique would be more appropriate to accurately control all the above electromagnetic characteristics including the GF and SLL. As the literature (given in Section 2.2) shows, the global optimization methods are capable of solving various complex nonlinear problems under multiple requirements. Therefore, such a method is suitable for the problem described here. On the other hand, the use of a full-wave analysis technique is the best way to approximate the exact behavior of the antenna at any given frequency.

In this chapter, such a global optimization method, called invasive weed optimization (IWO), is applied as a design tool of LPDAs. The IWO is an evolutionary method that simulates the invasive nature of weeds and so far it has successfully been used to solve several problems of telecommunications and electromagnetics [4–11]. The full-wave analysis engine employed by the IWO to extract the electromagnetic characteristics of any LPDA geometry is the numerical electromagnetics code (NEC) [12], which implements the well-known MoM [13]. In order to increase the degrees of freedom in comparison to Carrel's method, the dipoles of the LPDA are not considered inside the same angular sector, which means that in the proposed method no τ and σ parameters are used (Figure 2.2). Therefore, the parameters L_m ($m = 1, \ldots, M$), r_m ($m = 1, \ldots, M$), S_m ($m = 1, \ldots, M$), and Z_0 are independently optimized by the IWO algorithm.

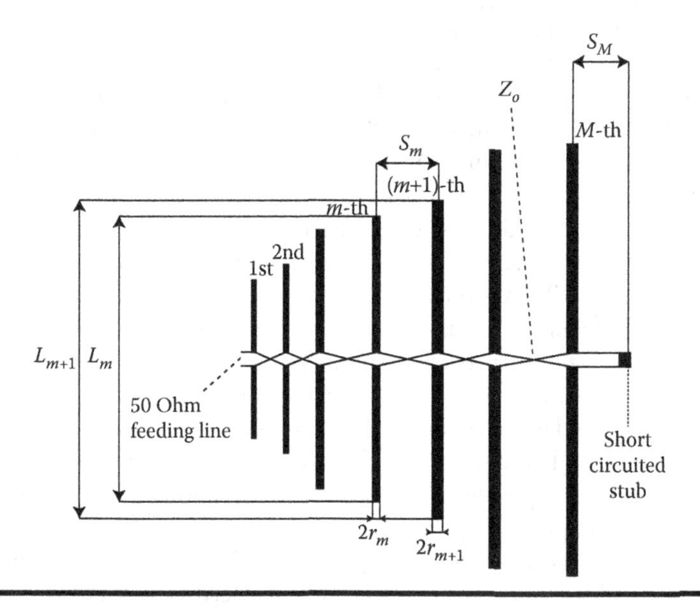

Figure 2.2 Proposed LPDA geometry.

The rest of the chapter is organized as follows: Section 2.2 is a description of the prior art related to log-periodic antenna optimization under various requirements. The IWO method is described in Section 2.3, while the design requirements and the fitness function to be minimized by the IWO method are thoroughly analyzed in Section 2.4. Several cases of LPDA optimization and design are presented and discussed in Section 2.5, while the conclusions are presented in Section 2.6. We conclude this chapter by citing relevant references and giving definitions of key terms used in the chapter.

2.2 Prior Art Related to Log-Periodic Antenna Optimization

The antenna optimization is a complex nonlinear problem, especially when multiple requirements have to be satisfied. Usually, deterministic methods—such as gradient-search methods—are not capable of providing good results, mostly when the number of requirements increases. Therefore, evolutionary optimization algorithms are a good choice for such problems. The literature contains a variety of evolutionary methods, which are properly utilized to optimize various antenna configurations. Such methods are the genetic algorithms (GAs), the differential evolution (DE), the particle swarm optimization (PSO), Taguchi optimization, the IWO, and many other [14–21]. Most of them are stochastic methods; however, there are also nonstochastic ones, like Taguchi optimization, with similar performance and efficiency. We now present a literature related to log-periodic antenna optimization under requirements.

A log-periodic monopole array (LPMA) is optimized in Reference 22 by applying a Pareto GA. This algorithm is a multiobjective version of GAs that has the ability to combine multiple competing goals. In this study, four goals are required to be achieved over the entire operating bandwidth: (a) the half-power beam-width (HPBW) of the main lobe on the E-plane must be as close as possible to its desired value, (b) the FBR must be above a desired value, (c) the SWR must be below a desired value, and (d) the percentage of the input power, which is radiated by the antenna, must be above a desired value. The first goal that concerns specific value of HPBW is directly related to a specific value of antenna gain, since the gain increases by decreasing the value of HPBW. Therefore, the HPBW requirement can be considered as a requirement for specific gain value. In order to exhibit the efficiency of the proposed method, an LPMA composed of eight monopoles is optimized for operation in the range 25–88 MHz, by applying the Pareto GA, a simple GA, and the simplex method. The antenna is required to provide HPBW close to 30°, $FBR > 10$ dB, $SWR < 1.8$, and percentage of the input power radiated by the antenna greater than 50%. The MoM is employed by all the optimizers to extract the radiation characteristics of the antenna. The comparative results show that the Pareto GA outperforms the other two methods, since it is capable of achieving better LPMA geometries.

The study given in Reference 23 is an effort to optimize the lengths, the radii, and the distances of the dipoles of an LPDA in order to achieve four specific goals over the entire operating bandwidth: (a) maximization of the mean gain value, (b) minimization of the mean value of SWR, (c) minimization of the standard deviation of the gain values over the whole frequency band, and (d) minimization of the standard deviation of the SWR values over the whole frequency band. In fact, the third requirement is equivalent to a requirement for optimal GF. However, in the fourth requirement, the authors require an optimal SWR flatness as well. Therefore, the second and the fourth requirements try to keep the SWR at low values. A similar optimization would be performed, if the second and the fourth requirements were replaced by only one concerning the maximum value of SWR over the entire bandwidth. To achieve these goals, a GA, the Nelder–Mead downhill simplex method, and a hybrid combination of GA and the Nelder–Mead method are utilized. The Nelder–Mead algorithm is a local optimizer, whereas GA is a global optimizer. It is expected that the hybrid combination between the two methods will give better convergence than that of each method alone. Two antenna configurations are considered. The first one is a 7-dipole LPDA operating over the range 800–1600 MHz, and the second is a 20-dipole LPDA operating over the range 200–1300 MHz. 6000 iterations are used for each method. The results show that the hybrid GA/Nelder–Mead method gives better results than those derived from the GA or the Nelder–Mead method alone.

A GA in conjunction with the NEC software is applied in Reference 24 in order to reduce the size of an LPDA antenna while preserving the average values of gain and SWR. The idea here is to design an LPDA with specific average values of gain and SWR by using the traditional design technique (i.e., Carrel's method) and then apply the GA/NEC method to design an LPDA that has nearly the same average values of gain and SWR but with less number of elements and shorter length. Two design examples are provided in this regard. In the first one, a five-element LPDA is designed for operation in the range 300–400 MHz by applying the traditional design method. The LPDA has a length of 59.37 cm, and provides an average gain of 7.46 dBi and an average SWR of 1.9. By applying the GA/NEC method, a three-element LPDA is optimized for operation in the same frequency range. The derived LPDA has a length of 20 cm, while the average values of gain and SWR are found to be respectively decreased by around 1.26 dB and 0.6 , which means that the optimized LPDA has a definitely shorter length without seriously affecting the average values of gain and SWR. In the second design example, a 10-element LPDA is designed for operation in the range 400–1800 MHz by applying the traditional design method. This LPDA is more wideband than the previous one, has a length of 39 cm, and provides average values of gain and SWR respectively equal to 7.24 dBi and 1.9 . By applying the GA/NEC method, an effort is made to reduce the total antenna length and the number of elements from 10 to 5. The optimized LPDA consists of five dipoles, has a length of 21 cm, and provides average values of gain and SWR respectively equal to 6.4 dBi and 1.8, which means that the antenna length is reduced by about 46% without seriously affecting the average gain and the average SWR.

In Reference 25, an effort is made to maximize the gain and simultaneously minimize the SWR and the length of LPDAs, which are considered to be operating in the range 3–30 MHz. The optimization is performed by applying a multiobjective version of GA called nondominated sorting genetic algorithm II (NSGA-II). The electromagnetic characteristics of the LPDA needed by the optimization procedure are extracted by applying the graphic numerical electromagnetics code (GNEC). A special interface called iSIGHT is used to establish automatic communication between the NSGA-II and GNEC. Despite the fact that the LPDA is considered to be operating in the range 3–30 MHz, the antenna is actually analyzed and optimized only in three frequencies, that is, 3, 15, and 27 MHz, regarded as samples of the above range. In these frequencies, sufficient results are acquired concerning the SWR and the gain. However, it is explicitly shown that the optimized LPDA geometry does not provide low SLL or high FBR. This fact is predictable since the optimization requirements do not include desired values of SLL or FBR.

A combination of the PSO and the NEC software is utilized in Reference 26 in order to design optimal LPDAs in terms of several radiation characteristics, that is, the bandwidth (considered as the width of the frequency range where $SWR < 2$ with reference to a characteristic impedance of 75 Ω) and the mean values of SWR, G, and FBR over this bandwidth. This study adopts Carrel's consideration where all the dipoles are located inside the same angle and therefore τ and σ are treated as optimization parameters. A parametric study in terms of τ and σ is performed by the authors, where τ ranges from 0.82 to 0.96 and σ ranges from 0.10 to 0.22. From this study, contour graphs are extracted in terms of τ and σ showing the variation of the above radiation characteristics (bandwidth and mean values of SWR, G, and FBR) including also the mean values of HPBW of the radiation pattern, respectively, on the E- and H-planes ($HPBW_E$ and $HPBW_H$). The advantage of these graphs is that they approximate the real LPDA behavior better than the constant directivity contour curve graph [1–3] derived from Carrel's method, since they are derived from full-wave analysis of the LPDA by applying the NEC software. Therefore, the gain contour graph can be used instead of Carrel's directivity contour graph. Finally, a more specific design example is presented, where the optimized LPDA is assumed to be composed of 10 dipoles and is required to operate in the range 450–1350 MHz with mean $G \geq 8.2$ dBi, mean $SWR \leq 1.5$, and $FBR \geq 20$ dB. The LPDA extracted from the optimization procedure satisfies all the above requirements.

A circular switched parasitic array composed of four LPDAs is optimized in Reference 27. At any given time, only one LPDA is connected to the signal source, while the other LPDAs act as reflectors. By changing the feeding from one LPDA to another, beam-steering is achieved and four identical radiation patterns rotated by a proper angle are produced. This array of the four LPDAs is optimized in terms of G and SWR by applying a GA. The radiation characteristics of the array are extracted by performing full-wave analysis with the SuperNEC software, which is a hybrid combination of MoM and the unified theory of diffraction (UTD). The desired value of SWR with reference to a characteristic impedance of 50 Ω is 1, which means that the SWR is required to be as low as possible (since the desired value is actually the theoretically lower limit of SWR for any antenna). On the other hand, the desired value of G is 8 dBi. The optimization is performed in the range 3.1–10.6 GHz at steps of 100 MHz. Due to the wide frequency steps, information about the variation of G and SWR may be missing and thus may not be evaluated by the optimization procedure (due to high-order resonances generated at some frequencies, rapid changes in the values of G or SWR may arise). The optimized array does not exhibit satisfying performance in the lower region of the predefined frequency range, since sharp increases in the values of SWR and sharp drops in the values of G are observed at 3.6, 4.9, and 6.1 GHz. Since these rapid changes are located in the lower region of the operating bandwidth, they are probably due to high-order resonances developed on some dipoles of the LPDAs. In the rest of the operating band, the optimized structure exhibits sufficient behavior in terms of SWR and G.

Another hybrid combination of GA and the NEC-2 software is proposed in Reference 28. The geometry optimized here is an inverted-V LPDA, which is composed of thin metal wires mounted over lossy ground. The optimum antenna geometry is considered to be operating in the range 6–30 MHz and is required to achieve values of $SWR < 1.5$, gain close to 8 dBi, and $SLL < -6$ dB under restrictions for the total size and height of the antenna. The upper limit of the SLL is not so demanding and as stated by the authors it has been set just to prevent the main lobe from splitting in the elevation plane. The geometry derived from the optimization procedure achieves gain values from 3 to 10 dBi and provides values of SWR below 2 inside the entire bandwidth. Therefore, the antenna does not satisfy at all the requirement for $SWR < 1.5$, but it provides impedance matching over the whole bandwidth since $SWR < 2$. Also, due to the restriction in the size and height of the antenna, the upper SLL limit (1.5) defined above is difficult to be reached especially at higher frequencies.

A new method is described in Reference 29 to predict the input reflection coefficient of a large log-periodic array by taking into account only a small part of the array, which is called a partial array. The method calculates the S-parameters between the elements of the partial array and then utilizes these parameters to predict the reflection coefficient of the full array. The method is applied in a special antenna structure called Eleven antenna. The basic geometry of an Eleven antenna is two parallel folded half-wave dipoles located above a ground plane at a distance between each other equal to half of their resonant length. This basic geometry is extended with a log-periodic scaling factor to create a number of dipole pairs and these pairs are cascaded one after another to form two LPDAs located opposite to each other. These two arrays form a linearly polarized Eleven antenna. An Eleven antenna composed of 14 dipole pairs is optimized for operation in the range 2–13 GHz. The only requirement that has to be satisfied by the optimized antenna over the above range is to minimize the input reflection coefficient. Actually, this requirement is equivalent to low SWR requirement, since a low reflection coefficient corresponds to a low SWR value. The optimization procedure is performed by applying a GA. The reflection coefficient evaluations needed by the GA are obtained by employing the partial array approach. The optimized geometry undergoes full-wave analysis by using the CST Microwave Studio in order to calculate the input reflection coefficient over the entire operating bandwidth and check in this way the validity of the partial array approach. This coefficient is found to be less than −9.4 dB over the whole frequency band. Finally, the optimized geometry of the linearly polarized Eleven antenna is utilized to construct a dual polarized Eleven antenna, which is actually an arrangement of two linearly polarized Eleven antennas with the symmetry plane of the first antenna rotated by 90° with reference to the symmetry plane of the second one. The reflection coefficient of the experimental model of the antenna is found to be below −10 dB in the range 2.2–10 GHz and below −8 dB in the range 10–13 GHz.

A PSO algorithm is employed in Reference 30 to optimize planar LPDAs in terms of SWR and gain. The fitness function is estimated by summing quantities of the form $|\mathrm{Real}(Z_{in}) - 50| + |\mathrm{Imag}(Z_{in})|$ (Z_{in} is the antenna input impedance and 50 Ω is considered to be the characteristic impedance of the line that feeds the antenna) in all the frequency samples inside the operating bandwidth, and if this sum becomes less than a predefined value (which means that the SWR has acceptable values over the entire bandwidth) then the gain values extracted in all the frequency samples are additionally included in the fitness function. In this way, the optimizer gives priority to the antenna matching, and if the SWR has acceptable values over the entire bandwidth, for example, $SWR < 2$, then the optimizer performs gain maximization. The variables to be optimized are the lengths and the widths of the LPDA elements as well as the distances between them. To obtain a design with reduced size, limitations are applied to the upper boundaries of the optimization variables. The radiation characteristics, which are necessary for fitness calculations, are extracted by applying full-wave analysis of the antenna with the FEKO software, which employs the MoM. The optimized planar LPDA is suitable for operation in the *S*-band. In particular, it provides $SWR < 2$ from 2 to 5.75 GHz, while in the range 2–4 GHz it achieves gain between 8.5 and 10 dBi and cross-polarization ratio (which actually was not optimized) less than −20 dB. At frequencies greater than 4 GHz, the gain is rapidly reduced and the cross polarization increases.

An LPDA is optimized in Reference 31 for operation in global system for mobile (GSM, I and II), WiMAX, Bluetooth, Wi-Fi, and third-generation (3G) mobile communication bands. The optimization is performed by applying a PSO algorithm. The optimized LPDA is required to achieve lower SWR and greater gain than those of an LPDA derived from the conventional LPDA design method. The LPDA designed by the conventional method consists of 13 dipoles with length-to-diameter ratio equal to 125, $\tau = 0.90$, $\sigma = 0.16$, and length of the larger dipole equal to half-wavelength in the lower operating frequency of 400 MHz, while the characteristic impedance Z_0 of the line that models the booms is equal to 100 Ω. The LPDA derived from the optimization procedure is found to have smaller size compared to the conventional LPDA design, although the small antenna size is not included in the optimization requirements. Also, the gain of the optimized LPDA has been improved up to 0.6, 0.7, 0.6, and 0.8 dB, respectively in WiMAX, GSM, Bluetooth, and 3G mobile communication bands, compared to the conventional LPDA. However, the SWR of the optimized geometry does not seem to be noticeably improved in comparison to the conventional design. Certainly, the values of SWR of both geometries are less than 2 in all the above operating bands (i.e., both geometries satisfy the impedance matching condition).

A similar optimization procedure is performed in Reference 32 by using a GA. The LPDA is optimized for operation in WiMAX, GSM-I, GSM-II, and Wi-Fi bands. The variables to be optimized are the length and the diameter of the LPDA elements as well as the distances between these elements. The fitness function is actually the average gain found over the four frequency bands mentioned earlier. Therefore, the maximization of the average gain is the only requirement that has to be satisfied. Initially, an LPDA is designed by applying the conventional method in order to be used for comparison with the optimized antenna. The conventional LPDA geometry consists of 10 dipoles and uses the same parameter values as the conventional LPDA in Reference 31 (i.e., length-to-diameter ratio equal to 125, $\tau = 0.90$, $\sigma = 0.16$, $Z_0 = 100\ \Omega$, and length of the larger dipole equal to half-wavelength in the reference frequency of 400 MHz). The average gain of the conventional antenna is derived equal to 9.1, 9.5, 9.2, and 8.5 dBi, respectively in WiMAX, GSM-I, GSM-II, and Wi-Fi bands. The average gain of the optimized LPDA is derived equal to 10.7, 11.2, 9.9, and 9.1 dBi, respectively in each one of the above four bands. The comparative gain graphs, provided for each one of these bands, exhibit the superiority of the optimized LPDA in comparison to the conventional LPDA and the LPDA optimized in Reference 31. The physical size of the antenna optimized in Reference 32 is found to be reduced by 12% in comparison to the conventional LPDA size, although the size reduction is not included in the optimization requirements.

Finally, a promising optimization algorithm called bacteria foraging algorithm (BFA) is employed in Reference 33 to optimize LPDAs for operation in the UHF-TV band (470–870 MHz). The array elements are considered inside the same angle, which means that Carrel's assumption is adopted by the authors. Therefore, τ and σ are treated as optimization parameters. The BFA aims at minimizing a fitness function of the form $a \times |Z_{\mathrm{inAv}} - 50| + b \times |G_{\mathrm{Av}} - 9| + c \times |FBR_{\mathrm{Av}} - 40| + d \times |SLL_{\mathrm{Av}} - 40| + e \times |SWR_{\mathrm{Av}} - 1.1|$, where Z_{inAv}, G_{Av}, $\mathrm{FBR}_{\mathrm{Av}}$, $\mathrm{SLL}_{\mathrm{Av}}$, and $\mathrm{SWR}_{\mathrm{Av}}$ are the average values of the input impedance in Ohms, the gain in decibels isotropic, the FBR in decibels, the absolute value of SLL in decibels, and the SWR of the LPDA, respectively, over the UHF-TV band. Also, a, b, c, and d are positive coefficients used to balance the minimization of the five terms that compose the fitness function. The values used for the optimization procedure are $a = 0.003$, $b = 0.02$, $c = 0.004$, $d = 0.004$, and $e = 0.4$. From the above expression of the fitness function, it seems that five requirements have to be satisfied by the optimized LPDA: (1) the average input impedance must be as close as possible to the characteristic impedance of 50 Ω, (2) the average gain must be maximized and reach as close as possible 9 dBi, (3) the average FBR must be maximized and reach the desired value of 40 dB, (4) the average SLL must be minimized and reach as close as possible −40 dB, and (5) the average SWR must be minimized to the value of 1.1. It is easy to realize that both the first and fifth requirements refer to the impedance matching condition but they aim at different values of SWR. The first requirement aims at decreasing $\mathrm{SWR}_{\mathrm{Av}}$ until it becomes equal to 1. The fifth requirement does the same until $\mathrm{SWR}_{\mathrm{Av}}$ becomes 1.1. Therefore, when $\mathrm{SWR}_{\mathrm{Av}}$ equals 1.1, the fifth requirement is satisfied but not the first one. Therefore, the optimization is continued so as to decrease $\mathrm{SWR}_{\mathrm{Av}}$ below 1.1 in order to satisfy the first requirement; however, the fifth requirement again becomes not satisfied. This means that when $SWR_{Av} < 1.1$, then the first and the fifth requirements become competitive to each other and none of them is satisfied. The impedance matching condition is well defined by the fifth requirement and therefore the first one should be removed from the fitness formula. Three LPDA configurations—respectively composed of 6, 9, and 12 dipoles—are considered. The optimized LPDAs are verified by applying the MININEC software and seem to be capable of satisfying most of the requirements. However, the desired SLL value of −40 dB is not satisfied by any of the above antenna configurations at any frequency. This value is a very hard goal for an antenna to accomplish and the difficulty increases due to the small number of optimization variables (τ and σ).

A comparative look at the above literature reveals that in almost all the research the antennas are optimized in terms of SWR and gain. Exceptions to this observation are the study given in Reference 29, where the antenna is optimized only in terms of SWR, and the study given in Reference 32, where requirement is defined only for the gain values. Also, in some studies, additional requirements are set concerning FBR, GF, or the antenna size. As far as FBR is concerned it has to be mentioned that retaining FBR at high values is not enough to enhance the performance of the antenna, since unnecessary spatial spread of radiated power or degradation in the signal reception due to multipath fading may happen by significant side lobes. Therefore, the antenna performance can only be increased if the SLL is kept below a desired value together with keeping FBR above a respective limit. Minimization of SLL is performed only in Reference 28 and 33. As mentioned in the analysis of Reference 28, the desired SLL value of −6 dB is not so demanding and has been set just to prevent the main lobe from splitting in the elevation plane. On the other hand, the desired SLL value of −40 dB is too demanding to be reached by any of the three LPDAs optimized in Reference 33, although the operating bandwidth is not wide enough (the upper-to-lower frequency ratio of UHF-TV band is less than 2:1). An SLL value of −20 dB would be sufficient enough to enhance the antenna performance and this value has been reached by the two LPDAs of Reference 33 respectively composed of 9 and 12 elements. Finally, the necessity to reduce the antenna size depends on the specific application where the antenna is going to be used. In cases of using antennas in base stations of wireless networks, no such necessity for size reduction arises. However—as discussed earlier—a base station antenna has to achieve high gain; FBR above a predefined value; and SWR, GF, and SLL below the respective specified values over the entire operating band. We design and optimize such antennas in the following sections.

2.3 The Invasive Weed Optimization Method

Most of the evolutionary optimization methods are based on mathematical models of mechanisms that determine the behavior of biological or ecological systems in nature. A common feature of these methods is their iterative logic, which means that a main procedure is repeated until a certain number of iterations is reached or a specific criterion is satisfied. In fact, evolution of any system in nature can be achieved only through iterations of a basic procedure or a combination of procedures. Such iterative methods are the GAs, the DE method, the PSO, and the IWO. The GAs and DE are based on biological processes, which are responsible for the reproduction and evolution of chromosomes. The PSO is inspired by movements made by the members of a swarm (e.g., bees, birds, or fishes) in nature during their effort to find the place with the most food than any other place. The IWO, which we discuss shortly, is a recent method based on processes that are responsible for the reproduction and spreading of weeds in nature. It was introduced in Reference 34. Since then, many complex nonlinear problems have been solved by applying this method and several studies found in the literature exhibit the effectiveness of the IWO in comparison to other methods [4–11]. The IWO has already been used in antenna design problems [4–8,11]. However, it has not been applied so far in design and optimization of LPDAs.

The weeds are species that can easy adapt to any environment, reproduce quickly, disperse widely, and in this way they resist eradication in any ecosystem. Due to this behavior, they find the most fertile region (or point) of the space where they spread out. This *invasive* behavior is analyzed in three basic phases called *reproduction*, *spatial dispersion*, and *competitive exclusion*. Reproduction is the phase according to which a weed produces seeds. The number of seeds generated by a weed depends on the *fertility* of the weed. The more fertile a weed is, the more seeds this weed generates. Therefore, fertility is a very crucial attribute, since it specifies the way of weed spreading in space. Also, fertility is not a hereditary attribute but it depends on the weed position, which means that the position or the region where a weed lies determines how fertile the weed is. Spatial dispersion is the phase according to which a weed disperses its seeds generated in the previous phase. The seeds are dispersed around this weed following a normal distribution with mean value equal to the weed position. These seeds grow up and become the new weeds of the colony. Finally, the competitive exclusion is the phase that mainly justifies the invasive behavior of weeds. Due to this behavior, not all the weeds (old and new ones) are able to survive. Weed fertility is the criterion which determines which ones are going to survive and keep reproducing themselves and which ones are going to be eliminated. Therefore, a certain number of the more fertile weeds is going to survive while the rest are terminated. By mathematically modeling these three phases and by repeating them in the above-mentioned order up to a maximum number of iterations I, we result in the main structure of the IWO algorithm. We now discuss this algorithm in detail.

Before applying an optimization method, we have to define the actual problem. To do this we have to

- Set the requirements that have to be satisfied.
- Construct an algebraic term for each requirement in such a way that the requirement is satisfied when the respective term is minimized.
- Compose a *fitness function* f as a linear combination of the terms constructed in the above step.
- Determine the *optimization variables* x_n ($n = 1, \ldots, N$), that is, the variables that have to be set to proper values to satisfy all the above-defined requirements.
- Determine the lower and upper boundaries, respectively denoted as $x_{\min,n}$ ($n = 1, \ldots, N$) and $x_{\max,n}$ ($n = 1, \ldots, N$), of the optimization variables.

It is obvious that the problem definition refers to a properly defined fitness function f. The global minimization of f provides the best solution to the problem. If f reaches its global minimum point, then all the terms that compose this function reach their respective minimum values, which means that all the requirements are satisfied. It is easy to realize that the fitness function depends on the optimization variables x_n ($n = 1, \ldots, N$). These variables are changed by the optimization algorithm in a particular way (which depends on the fundamental principles of the algorithm) in order to find their optimum values $x_{opt,n}$ ($n = 1, \ldots, N$) that correspond to the global minimum fitness value. This value is indicated as $F_{\min,I}$, since it is achieved when the maximum number of iterations I is reached. Therefore, we can generally write the expression

$$F = f(x_1, x_2, \ldots, x_N) \tag{2.6}$$

where F is the value of f for given values of x_n ($n = 1, \ldots, N$), while for the global minimum fitness value, we can also write

$$F_{\min,I} = f(x_{\text{opt},1}, x_{\text{opt},2}, \ldots, x_{\text{opt},N}) \tag{2.7}$$

The boundaries $x_{\min,n}$ and $x_{\max,n}$ determine the area of permissible values of the variable x_n. This area is the interval $[x_{\min,n}, x_{\max,n}]$ and is called *search space* of the nth optimization variable. By taking into account all the optimization variables x_n ($n = 1, \ldots, N$), the *total search space* is an

N-dimensional hypercube of the form $[x_{\min,1}, x_{\max,1}] \times [x_{\min,2}, x_{\max,2}] \times \ldots \times [x_{\min,N}, x_{\max,N}]$. Finally, it has to be noted that the problem is considered to be solved by minimizing the fitness function, since the optimization algorithm aims at finding the global minimum of a function. However, the algorithm may have such a structure that performs function maximization (i.e., the algorithm aims at finding the global maximum of a function). In this case, each term that composes the fitness function must be constructed in such a way that the respective requirement is satisfied when the term is maximized.

2.3.1 Initialization

Before the three above-mentioned phases of the IWO algorithm start to be repeatedly applied, the algorithm has to be initialized. This is done by distributing a population of W weeds inside the N-dimensional search space, which is specified by the boundaries $x_{\min,n}$ $(n = 1, \ldots, N)$ and $x_{\max,n}$ $(n = 1, \ldots, N)$. Therefore, each wth weed $(w = 1, \ldots, W)$ has a position defined by the vector

$$X(w) = [x_1(w) \quad x_2(w) \quad \ldots \quad x_N(w)] \tag{2.8}$$

where $x_n(w)$ $(n = 1, \ldots, N)$ are the values of variables x_n that correspond to the wth weed. When the algorithm is initialized, each value $x_n(w)$ is produced by a uniform random number generator inside the interval $[x_{\min,n}, x_{\max,n}]$. It is easy to realize that each vector $X(w)$ is a potential solution to the optimization problem. The optimum values $x_{\text{opt},n}$ $(n = 1, \ldots, N)$ correspond to the best weed position (i.e., the solution of the problem), which is defined by the vector

$$X_{\text{opt}} = [x_{\text{opt},1} \quad x_{\text{opt},2} \quad \ldots \quad x_{\text{opt},N}] \tag{2.9}$$

The next step of initialization is the estimation of the fitness value $F(w)$ for every wth weed. This value depends on the coordinates $x_n(w)$ $(n = 1, \ldots, N)$ of the weed position, as shown hereunder:

$$F(w) = f(x_1(w), x_2(w), \ldots, x_N(w)), \quad w = 1, \ldots, W \tag{2.10}$$

It seems that there will be positions that correspond to high or low fitness values. In this manner, every weed can be evaluated in terms of its fitness and is considered to be better than other weeds if a lower fitness value has been estimated for this weed. In a manner of speaking, the fitness function is the mathematical equivalent of fertility, which is a basic attribute of weeds as mentioned above. Therefore, a weed with lower fitness value is a weed with increased fertility compared to other weeds. On the contrary, higher fitness values are equivalent to decreased fertility. The equivalence between fitness and fertility is confirmed by Equation 2.10, which is a mathematical expression of the fact that the weed fitness depends on the weed position, just as fertility does.

The initialization phase ends by sorting the weeds in ascending order of their fitness. Then, the fitness value of the first weed $(w = 1)$ will be the lowest one in the whole weed colony, that is, $F_0(1) = F_{\min,0}$, while the fitness of the Wth weed will be the highest one, that is, $F_0(W) = F_{\max,0}$ (the subscript 0 denotes that the fitness values have been calculated during the initialization phase).

2.3.2 Reproduction

In order to spread out in space, every weed produces seeds. As mentioned earlier, the more fertile a weed is, the more seeds it generates. Due to the equivalence between fitness and fertility, a weed with low fitness value must be able to produce more seeds compared to a weed that corresponds to high fitness. Therefore, the number of seeds $ns_i(w)$ produced by any wth weed at the ith iteration must be a function of the weed fitness $F_{i-1}(w)$ estimated at the previous $(i-1)$ iteration. This function follows a linear law, as shown hereunder:

$$ns_i(w) = \text{int}\left[\frac{F_{\max,i-1} - F_{i-1}(w)}{F_{\max,i-1} - F_{\min,i-1}}(ns_{\max} - ns_{\min}) + ns_{\min}\right], \quad w = 1, \ldots, W \quad \text{and} \quad i = 1, \ldots, I \qquad (2.11)$$

where $F_{\min,i-1}$ and $F_{\max,i-1}$ are respectively the lower and the upper weed fitness found at the previous iteration, $ns_{\min}$ and $ns_{\max}$ are respectively the minimum and the maximum number of seeds that any weed is allowed to produce, and int[•] rounds a real number to its nearest integer (ns_i is an integer parameter). Therefore, the ability to explore the search space for better positions is mainly due to fertile weeds, since their offspring are going to continue exploring at the next iterations.

2.3.3 Spatial Dispersion

In this phase, the seeds generated by a weed are dispersed around the position of this weed. In nature, the dispersion is performed in such a way that most of the seeds are thrown close to the parent weed, while the seed distribution decreases rapidly as the distance from the weed increases. This type of dispersion can mathematically be modeled by using a normal distribution with mean value equal to the position of the parent weed. In order to make the algorithm converge after a predefined number I of iterations, the standard deviation (sd) of the seed distribution must not remain constant at every iteration but it should be reduced with increasing the number of iterations. Such a reduction can be achieved by applying the expression given hereunder:

$$sd_i = \left(\frac{I-i}{I-1}\right)^{mi}(sd_{\max} - sd_{\min}) + sd_{\min}, \quad i = 1, \ldots, I \qquad (2.12)$$

where sd_i is the standard deviation at the ith iteration, $sd_{\min}$ and $sd_{\max}$ are the standard deviation boundaries, while mi is a positive real number called *nonlinear modulation index* and is used to control the decreasing rate of sd_i. If $mi = 1$, sd_i is decreased according to the linear law, while if $mi > 1$, then the decreasing rate of sd_i becomes greater than that of the linear decrease. It seems from Equation 2.12 that in the first iterations the algorithm has increased exploration ability because the weeds are able to disperse their seeds far away from their position. On the contrary, at the end of the optimization procedure, the algorithm reaches a convergence state, since the seeds are thrown close to the parent weed and therefore every weed can only fine-tune the position that has already been found.

However, the above expression considers the same value of sd_i for every optimization variable x_n and does not take into account the different width $x_{\max,n} - x_{\min,n}$ of the search space of different variables. Therefore, the direct application of Equation 2.12 to every variable would result in convergence failure of the IWO algorithm, since any value of sd_i gives narrow seed dispersion in cases where the search space of a variable is wide and wide dispersion in cases where the search space of some other variable is narrow. In fact, sd_i should be considered only as *normalized standard deviation*, which means that it should concern dispersions inside a unitary-width search space, and thus the values of $sd_{\min}$ and $sd_{\max}$ must be within the interval [0,1]. Consequently, the standard deviation should have a different value for each variable and must be calculated by taking into account the search space width of the variable. This is done by applying the expression given hereunder:

$$sd_{i,n} = (x_{\max,n} - x_{\min,n})sd_i, \quad n = 1, \ldots, N \quad \text{and} \quad i = 1, \ldots, I \tag{2.13}$$

After the positions have been calculated for all the dispersed seeds, the fitness value is estimated for every seed, except for those which have variables with values being outside their search space. Since these seeds do not represent acceptable solutions for the problem under consideration, we have to find a way to exclude them from the rest of the optimization procedure without having to estimate their fitness and thus spend unnecessary computational time. This is achieved by assigning a very large fitness value to these seeds. This value can be considered as a penalty value assigned to these seeds, since their position is outside the N-dimensional search space. Actually, the exclusion of these seeds is performed in the next phase (i.e., the competitive exclusion phase), which is explained in the following paragraph. After a fitness value (either calculated or given as penalty) has been assigned to a seed, this seed becomes a new weed. All the weeds (old and new ones) are going to be evaluated in the next phase of the IWO algorithm.

2.3.4 Competitive Exclusion

Due to their invasive behavior, not all the weeds are going to survive. In nature, the criterion that determines which weeds are going to survive and keep reproducing themselves and which ones are going to be terminated is the weed fertility. Due to the equivalence between fitness and fertility, a weed with low fitness value is a fertile weed and therefore it survives, while a weed with high fitness (i.e., a weed with low fertility) is likely to be eliminated. In order to simulate what happens in nature, we have to define a particular parameter $W_{\max}$, which expresses how many weeds are going to survive. These are $W_{\max}$ weeds with the lower fitness values. Therefore, all the weeds (old and new ones) must be sorted in ascending order of their fitness. Then, the first $W_{\max}$ ones with the lower fitness values (i.e., from $F_i(1)$ to $F_i(W_{\max})$) are considered to survive and keep reproducing themselves in the next iteration, while the rest are eliminated. The extreme fitness values, $F_i(1)$ and $F_i(W_{\max})$, of the surviving weeds are used in Equation 2.11 during the reproduction phase of the next iteration. Also, the weeds, that were assigned in the previous phase a very high fitness value since they lie outside the N-dimensional search space, are not included in the $W_{\max}$ weeds with the lowest fitness values and hence they are deleted. In this way, nonacceptable solutions are rejected in the same iteration they are created, without any waste of computational time. Finally, it has to be noted that, in general, W and $W_{\max}$ are two parameters different from each other. The parameter W is used only in the initialization phase, whereas $W_{\max}$ is used in all the iterations of the algorithm. However, it can optionally be set $W = W_{\max}$.

All the above description of the IWO algorithm is graphically shown in the flowchart of Figure 2.3.

2.3.5 User-Defined Parameters

To run the IWO algorithm, some parameters must be defined by the user. These parameters have already been mentioned in the various phases of the algorithm, and are summarized hereunder:

- I: total number of iterations of the IWO algorithm
- W: initial number of weeds distributed in the search space during the initialization phase

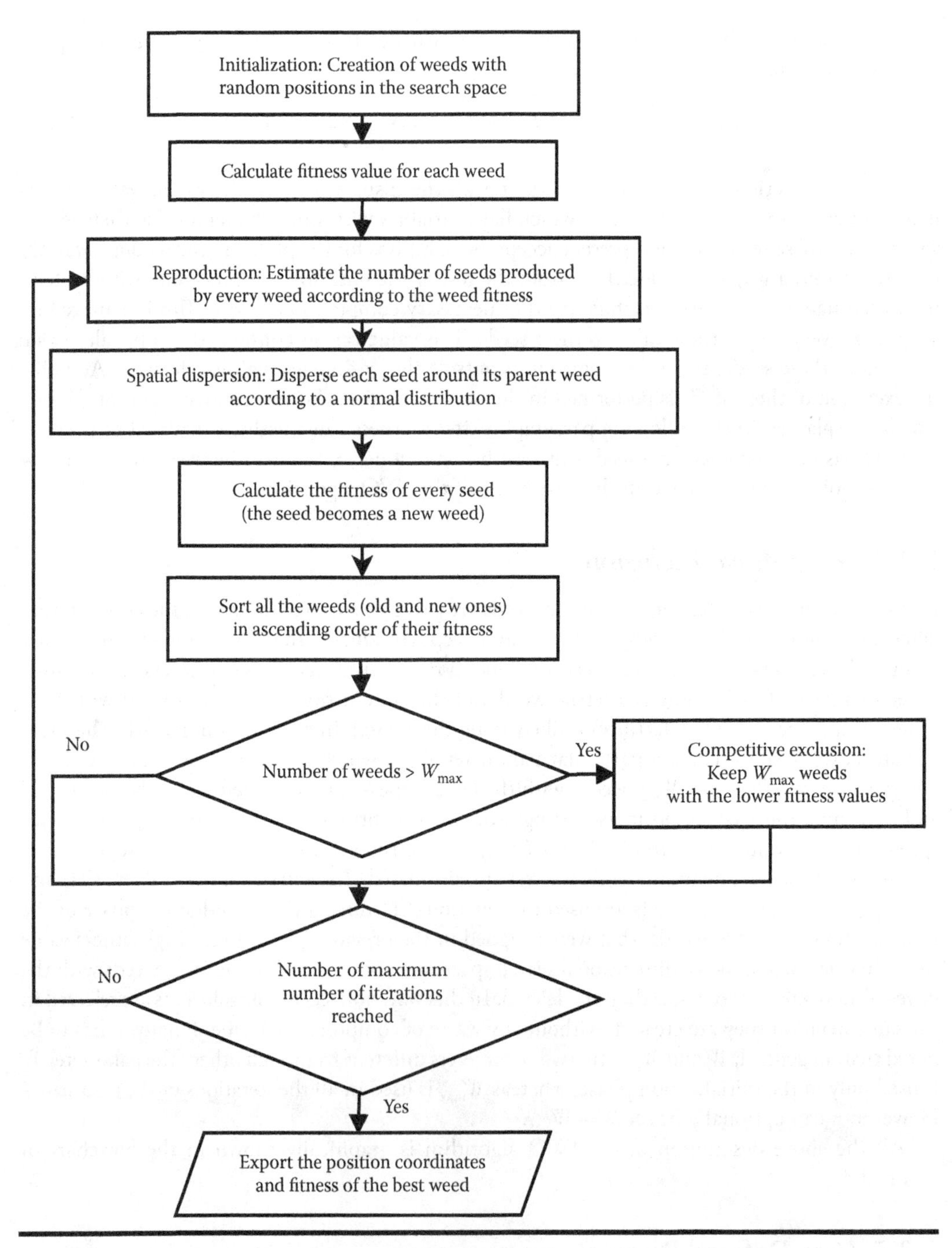

Figure 2.3 Flowchart of invasive weed optimization.

- W_{max}: maximum number of weeds allowed to continue reproducing themselves in any iteration
- ns_{min} and ns_{max}: respectively the minimum and the maximum number of seeds that any weed is allowed to produce in any iteration
- sd_{min} and sd_{max}: respectively the minimum and the maximum value of the normalized standard deviation (these values must be within the interval [0,1])
- mi: nonlinear modulation index used to control the decreasing rate of the normalized standard deviation

The values of these parameters are the same in every design case studied hereunder. These values are: $I = 2000$, $W = W_{max} = 20$, $ns_{min} = 0$, $ns_{max} = 5$, $sd_{min} = 0$, $sd_{max} = 0.10$, and $mi = 2.5$.

What has been left before running the IWO algorithm is to set the design requirements and compose a suitable fitness function according to these requirements. These issues are discussed in Section 2.4.

2.4 Design Requirements and the Fitness Function Definition

The design cases studied hereunder concern different LPDA geometries and different operating bandwidths. However, the requirements demanded in all the cases are the same and have to be satisfied by the optimized LPDAs in every frequency inside the operating bandwidth. These requirements are listed hereunder.

1. $SWR \leq 1.8$
2. G as high as possible
3. $GF \leq 2$ dB
4. $SLL \leq -20$ dB
5. $FBR \geq 20$ dB

Although the operating bandwidth is usually defined as the frequency range where $SWR \leq 2$ (as mentioned in Section 2.1 of this chapter), a more strict upper limit is used for the SWR to define the bandwidth in all the cases. Therefore, if the simulated LPDA achieves values of $SWR \leq 1.8$ inside the operating bandwidth, then we can be quite sure that the constructed LPDA will have $SWR \leq 2$ inside the same bandwidth. In other words, a harder upper limit of SWR is used for the simulated antenna in order to be sure that the real (constructed) antenna will obtain in practice values of $SWR \leq 2$.

The above electromagnetic characteristics can be calculated with sufficient accuracy if we apply a full-wave analysis software to the antenna geometry. Such software is the NEC [12], which we use here. Since it is difficult to analyze the antenna at any frequency inside the operating bandwidth, we can make the analysis at quite close frequency samples in order to be sure that no important information about the antenna behavior will be missed. Therefore, the frequencies (i.e., the frequency samples) are chosen at steps of 10 MHz and for each frequency the LPDA is analyzed with the NEC software, extracting thus the SWR and the radiation pattern of the antenna. From this pattern, it is easy to calculate the values of G, SLL, and FBR. Also, by scanning the gain values for all the frequency samples, we can find the maximum and the minimum gain values, respectively G_{max} and G_{min}, inside the entire bandwidth, and then calculate the value of GF ($GF = G_{max} - G_{min}$). By making a similar scan in terms of SWR and SLL, we can find their maximum values SWR_{max} and SLL_{max}, respectively. It is easy to realize that if $SWR_{max} \leq 1.8$, then $SWR \leq 1.8$ for every frequency and so the first requirement is satisfied. Thus, it is easier to formulate this requirement by using only the maximum SWR instead of using all the SWR values at all the frequency samples. In the same manner, G_{min} can be used instead of G in the second requirement. If G_{min} has achieved the highest possible value, then G is considered to have achieved its highest possible value at every frequency. In the fourth requirement, SLL_{max} can similarly be used instead of SLL.

Both the last two requirements concern the secondary lobe level compared to the main lobe peak. In particular, the SLL compares the gain in the direction of the maximum side lobe with the antenna gain (which corresponds to the main lobe peak), while the FBR compares the gain in the direction of the back lobe with the antenna gain. In practice, we are interested in obtaining simultaneously low SLL and high FBR, since either a high SLL or a low FBR results in unreasonable waste of radiated power and also in degradation of signal reception due to multipath fading effect. Therefore, it is practically more efficient to use only one parameter, which takes into account all the secondary lobes. To do that, we change the definition and also the way of calculating the SLL. From now on, the SLL is defined as the gain in the direction of the maximum secondary lobe (either side or back lobe) in relation to the antenna gain. Therefore, if $SLL_{max} \leq -20$ dB, all the secondary lobes will be at least 20 dB below the main lobe peak at every frequency, and then the last two requirements are simultaneously satisfied. Therefore, the last two requirements can be merged into a single one, that is, $SLL_{max} \leq -20$ dB, provided that the calculation of SLL takes into account all the secondary lobes and not only the side lobes. In view of all the above considerations, the requirements demanded by the optimized LPDA can be converted into the following

1. $SWR_{max} \leq 1.8$
2. G_{min} as high as possible
3. $GF \leq 2$ dB
4. $SLL_{max} \leq -20$ dB

Due to multiple requirements that have to be satisfied, the LPDA design is an inherently multiobjective optimization problem. In order to be solved by applying a global optimization method (like IWO), it has to be converted into a single-objective problem by combining all the requirements into a single mathematical expression, which is the fitness function mentioned in Section 2.3. The above combination into a single expression is due to the fact that global optimization algorithms are usually structured as maximizers or minimizers of a single function, that is, the fitness function (similar names with the same meaning are cost function and objective function). Therefore, the fitness function must be formed as a linear combination of terms, where each term takes into account a respective requirement. When the fitness function achieves its global optimum (i.e., minimum in our case), all the terms reach their respective minimum values, which means that all the requirements have been satisfied.

By taking into account the above four requirements, the fitness function can be formulated as follows:

$$f(x_1, x_2, \ldots, x_N) = b_1[\max(SWR_{max}, 1.8) - 1.8] - b_2 G_{min} + b_3[\max(GF, 2) - 2] + b_4[\max(SLL_{max}, -20) + 20] \quad (2.14)$$

The first term of Equation 2.14 is based on the first requirement. As SWR_{max} decreases, this term decreases as well, until it becomes equal to zero when $SWR_{max} = 1.8$. For $SWR_{max} < 1.8$, the term does not decrease any more but remains equal to zero. The reason for creating a term with such a behavior is that our intention is to decrease SWR_{max} until it becomes equal to 1.8. Any value of SWR_{max} less than 1.8 is considered to satisfy the first requirement and therefore it must not affect any more the value of the term. The last two terms are based respectively on the third and the fourth requirements and they have been created in a similar way like the first term. Therefore, as GF and SLL_{max} decrease (i.e., improved), these two terms decrease as well, until they become equal to zero when $GF = 2$ dB and $SLL_{max} = -20$ dB, respectively. If $GF < 2$ dB or $SLL_{max} < -20$ dB, these terms do not decrease any more but remain equal to zero. The second term has different behavior than the other three ones. This is due to the fact that the second requirement does not aim at a desired gain value, and so the second term cannot converge to any final value. The negative sign in front of G_{min} makes this term continuously decrease as G_{min} increases. In this way, even when the other three terms have become equal to zero (and so the respective three requirements have been satisfied), the second term may still decrease. In order to balance the minimization of the above four terms, four respective coefficients b_i ($i = 1, 2, 3, 4$) are used. During the minimization of f, some terms may not be minimized as faster than the rest ones. To accelerate the minimization of such a term, we may increase the value of the respective coefficient b_i. After several trials on cases of LPDA design, it was found that SLL_{max} is the most difficult parameter to be minimized, a little easier is SWR_{max}, next to this is GF and finally G_{min} is considered as the easiest parameter to be minimized since it is not required to converge to any final value. According to the above order of difficulty, b_4 must be greater than all the other coefficients, b_1 must be less than b_4, next to b_1 is b_3, and finally b_2 must be the coefficient with the smallest value. The trials made on LPDA design showed that convenient values that balance the minimization of the four terms of Equation 2.14 are the following: $b_1 = 12$, $b_2 = 6$, $b_3 = 8$, and $b_4 = 20$.

The procedure to estimate the fitness function, each time this is required by the IWO algorithm, is listed hereunder.

1. For given values of optimization variables x_n ($n = 1, \ldots, N$) set by the algorithm, the NEC software is applied at steps of 10 MHz inside the whole operating bandwidth, resulting in the SWR value and the antenna radiation pattern for every frequency.
2. From the radiation pattern at every frequency, the values of G and SLL are extracted (in the estimation of SLL all the secondary lobes are taken into account).
3. By scanning the values of SWR, G, and SLL for all the frequency samples inside the bandwidth, we find the values of SWR_{max}, G_{max}, G_{min}, and SLL_{max}.
4. From G_{max} and G_{min}, we calculate the GF ($GF = G_{max} - G_{min}$).
5. From SWR_{max}, G_{min}, GF, and SLL_{max}, we estimate the fitness value by using Equation 2.14.

It has to be noted that the most computational time is spent by the optimization algorithm to make fitness function calculations, since these calculations are directly related to the antenna full-wave analysis, which obviously is a time-consuming procedure. The only solution to reduce the computational time is to increase the distance between the frequency samples, for example, apply the NEC at steps of 30 MHz instead of 10 MHz. If the NEC is applied for less frequency samples, the optimization procedure is faster, but useful information concerning the antenna behavior may be lost. Therefore, the frequency step must be defined by taking into account not only the total computational time of the optimization process but also the risk of losing information since the antenna is analyzed in less frequencies.

2.5 Case Study of LPDA Optimization and Design

Six cases are studied in this section. In the first two cases, the antenna consists of 10 dipoles and is optimized under the above-mentioned requirements for operation in the range 450–900 MHz. In the first case, a short-circuited stub is used to suppress high-order resonances and thus help the LPDA achieve the desired radiation characteristics. In the second case, no stub is used in order to examine the ability of the IWO method to find an optimal geometry that satisfies the above requirements without the help of a stub. In the next two cases, the same 10-dipole LPDA is optimized, respectively with and without a short-circuited stub, for operation in the same frequency range. In these cases, however, the optimization variables have been restricted, making thus the design problem more difficult. In fact, the radii of the dipoles are not considered as optimization variables, but they have fixed and practically convenient values. Therefore, the advantage of these two cases is that the optimized antennas can easily be constructed in practice. All the above antennas are suitable for digital television (TV) broadcasting. The fifth case concerns the optimization under the same requirements of a 12-dipole LPDA that uses a short-circuited stub and operates in the range 700–3300 MHz. This frequency range covers most of the wireless network services recently used. Finally, in the sixth case, the same antenna geometry is optimized for operation in the same frequency range, but this time no stub is used. This case is studied to examine the ability of the IWO method to find an optimal antenna geometry that provides the desired radiation characteristics inside a wide operating bandwidth without using any stub. These cases are analyzed hereunder.

Case 1

In the first case, the LPDA is considered to be composed of 10 wire dipoles ($M = 10$) and is optimized for operation in the range 450–900 MHz. This antenna can be used for digital TV broadcasting. The LPDA geometry is defined by the dipole lengths L_m ($m = 1, \ldots, M$), the dipole radii r_m ($m = 1, \ldots, M$), the distances between adjacent dipoles S_m ($m = 1, \ldots, M - 1$), the length S_M of the short-circuited stub optionally located behind the largest (Mth) dipole, and finally the characteristic impedance Z_0 of the transmission line that models the booms of the LPDA (Figure 2.2). All the above geometry parameters are considered as optimization variables x_n ($n = 1, \ldots, N$ where $N = 3M + 1 = 31$), which means that the IWO algorithm has to find the appropriate values of all these variables that minimize the fitness function given in Equation 2.14 and thus satisfy the above-mentioned requirements. Therefore, the weed positions used by the algorithm are defined by vectors of 31 coordinates, as given hereunder:

$$X = [x_1 x_2 \ldots x_{31}] = [L_1 L_2 \ldots L_{10} r_1 r_2 \ldots r_{10} S_1 S_2 \ldots S_{10} Z_0] \tag{2.15}$$

After the optimization variables have been determined, we have to specify their lower and upper boundaries ($x_{\min,n}$ and $x_{\max,n}$, $n = 1, \ldots, N$). The values of Z_0 are considered to be between $Z_{\min 0} = 50\ \Omega$ and $Z_{\max 0} = 200\ \Omega$. Also, the same boundaries, $r_{\min} = 1$ mm and $r_{\max} = 5$ mm, are used for all the dipole radii. However, if we use the same boundaries for every dipole length L_m and the same boundaries for every dipole distance S_m, then the optimization procedure undergoes unnecessary search, since it is well known from Carrel's method that, as we move from the front to the back side of the LPDA (i.e., from the left to the right of Figures 2.1 and 2.2), the dipoles become larger with larger distances from their adjacent dipoles. Therefore, it is better to use different boundaries for the lengths and the distances. In this way, the total search space is restricted and thus the optimization procedure converges faster. In fact, these boundaries should become larger as we move from the front to the back side of the LPDA. To estimate the boundaries, first we have to find typical length and distance values as those extracted from Carrel's method described in Section 2.1 of the chapter. The idea to estimate the boundaries in such a way is based on the assumption that the optimized lengths and distances will slightly deviate from typical length and distance values. By considering average antenna directivity with a typical value of 8 dBi and by using the constant directivity contour curve graph given by Carrel, we find the optimum values of τ and σ parameters, which are $\tau = 0.89$ and $\sigma = 0.165$. Next, we find the length L_{10} of the largest dipole (10th dipole), which is set equal to half-wavelength at the lower operating frequency, that is, at 450 MHz, and is then reduced by 4% since the thickness of the dipole affects its resonance condition. Therefore, its length is calculated as follows:

$$L_{10} = (1 - 0.04)\frac{\lambda_{450\text{MHz}}}{2} = 0.32\,\text{m} \tag{2.16}$$

The values of L_m ($m = 1, \ldots, M - 1$) and S_m ($m = 1, \ldots, M - 1$) are then estimated by respectively using Equations 2.3 and 2.5. These typical lengths and distances are respectively shown in the second column of Tables 2.1 and 2.2.

From the typical lengths of two adjacent dipoles, mth and ($m + 1$)-th, an average length value $(L_m + L_{m+1})/2$ can be estimated. This value is considered as the upper length boundary $L_{\max,m}$ of the mth dipole and, at the same time, as the lower length boundary $L_{\min,m+1}$ of the ($m + 1$)-th dipole. In this way, all the length boundaries are calculated, except for $L_{\min,1}$ and $L_{\max,M}$. Therefore, $L_{\min,1}$ can be found by considering L_1 as the average of $L_{\min,1}$ and $L_{\max,1}$. In the same manner, $L_{\max,M}$ can be estimated by considering L_M as the average of $L_{\min,M}$ and $L_{\max,M}$. Therefore, L_1 and L_M can respectively be expressed as

$$L_1 = \frac{L_{\min,1} + L_{\max,1}}{2} \tag{2.17}$$

$$L_M = \frac{L_{\min,M} + L_{\max,M}}{2} \tag{2.18}$$

Table 2.1 Typical Length Values Extracted from Carrel's Method and Length Boundaries for the Frequency Range 450–900 MHz

m	*Typical Length L_m (m)*	*Minimum Length $L_{min,m}$ (m)*	*Maximum Length $L_{max,m}$ (m)*
1	0.1121	0.1052	0.1190
2	0.1260	0.1190	0.1338
3	0.1415	0.1338	0.1503
4	0.1590	0.1503	0.1689
5	0.1787	0.1689	0.1897
6	0.2008	0.1897	0.2132
7	0.2256	0.2132	0.2395
8	0.2535	0.2395	0.2691
9	0.2848	0.2691	0.3024
10	0.3200	0.3024	0.3376

Table 2.2 Typical Distance Values Extracted from Carrel's Method and Distance Boundaries for the Frequency Range 450–900 MHz

m	*Typical Distance S_m (m)*	*Minimum Distance $S_{min,m}$ (m)*	*Maximum Distance $S_{max,m}$ (m)*
1	0.0416	0.0291	0.0540
2	0.0467	0.0327	0.0607
3	0.0525	0.0367	0.0682
4	0.0590	0.0413	0.0767
5	0.0663	0.0464	0.0861
6	0.0744	0.0521	0.0968
7	0.0836	0.0586	0.1087
8	0.0940	0.0658	0.1222
9	0.1056	0.0739	0.1373
10	–	0.0120	0.1667

The above two expressions are used to find the values of $L_{min,1}$ and $L_{max,M}$. These are

$$L_{min,1} = 2L_1 - L_{max,1} \tag{2.19}$$

$$L_{max,M} = 2L_M - L_{min,M} \tag{2.20}$$

For the Case 1 studied here, where $M = 10$, the above expressions are used to calculate $L_{min,1}$ and $L_{max,10}$. Finally, all the length boundaries calculated from the procedure described earlier are shown in the third and fourth columns of Table 2.1.

From the typical distances S_m ($m = 1, \ldots, M - 1$) given in the second column of Table 2.2, we can find the boundaries of the dipole distances as follows:

$$S_{min,m} = 0.70 S_m, \quad m = 1, \ldots, M-1 \tag{2.21}$$

$$S_{max,m} = 1.30 S_m, \quad m = 1, \ldots, M-1 \tag{2.22}$$

In fact, we consider that the lower boundary is derived from the typical distance decreased by 30%, while the upper boundary is derived from the typical distance increased by 30%.

Since a short-circuited stub behind the largest (Mth) dipole is used in this case, we have to define the lower and the upper limit of the length of the stub. In the NEC model of the antenna, the stub is modeled by a short wire segment (e.g., 0.01 m in length) with radius r_M, that is, equal to the radius of the largest dipole. In order to avoid this segment and the largest dipole touching each other for any radius value, the lower length of the stub must be

$$S_{min,M} = 2r_{max} + 0.002[\text{meters}] \tag{2.23}$$

where a minimum distance of 2 mm is considered between the surface of the largest dipole and the surface of the wire segment used to model the stub. Since $r_{max} = 5$ mm, the minimum distance is derived equal to 12 mm. The upper length of the stub is defined by considering that the maximum variation of impedance, voltage, or current takes place along a quarter of the maximum wavelength (i.e., the wavelength at the lower frequency). Therefore, the upper length is defined as

$$S_{\max,M} = \frac{\lambda_{\max}}{4} \tag{2.24}$$

Since the lowest frequency in this case is 450 MHz, we get

$$S_{\max,M} = \frac{\lambda_{450\,\text{MHz}}}{4} = 0.1667\,\text{m} \tag{2.25}$$

All the distance boundaries defined by the above procedure, including the length limits of the stub, are shown in the third and fourth columns of Table 2.2.

The LPDA geometry derived from the optimization procedure is described in Table 2.3. The values of the geometry parameters shown in Table 2.3 are given as input to the NEC software to evaluate the variations of SWR, G, and SLL with increasing frequency inside the operating bandwidth. These variations are displayed in Figures 2.4 through 2.6. Finally, from Figures 2.4 through 2.6 we extract the extreme values of SWR, G, and SLL as well as the value of GF, which are given in Table 2.4.

It seems that the optimized LPDA geometry satisfies all the requirements defined earlier, since SWR remains below 1.8 and SLL is below −20 dB over the entire operating bandwidth, while GF is better than the required value of 2 dB. In addition, the gain values are high enough compared to the typical average antenna directivity of 8 dBi considered in the beginning of the design.

Case 2

In this case, the LPDA is considered to be composed of 10 wire dipoles ($M = 10$) and is optimized for operation in the same frequency range, that is, 450–900 MHz. However, no stub is used in the

Table 2.3 Optimized LPDA Geometry of Case 1

m	$L_{opt,m}$ (m)	$r_{opt,m}$ (m)	$S_{opt,m}$ (m)
1	0.1119	0.0022	0.0313
2	0.1196	0.0037	0.0448
3	0.1448	0.0021	0.0604
4	0.1613	0.0038	0.0650
5	0.1801	0.0041	0.0847
6	0.1924	0.0047	0.0900
7	0.2242	0.0021	0.0760
8	0.2405	0.0042	0.1186
9	0.2695	0.0041	0.0959
10	0.3206	0.0033	0.0121

$Z_{opt,0} = 95.3\ \Omega$.

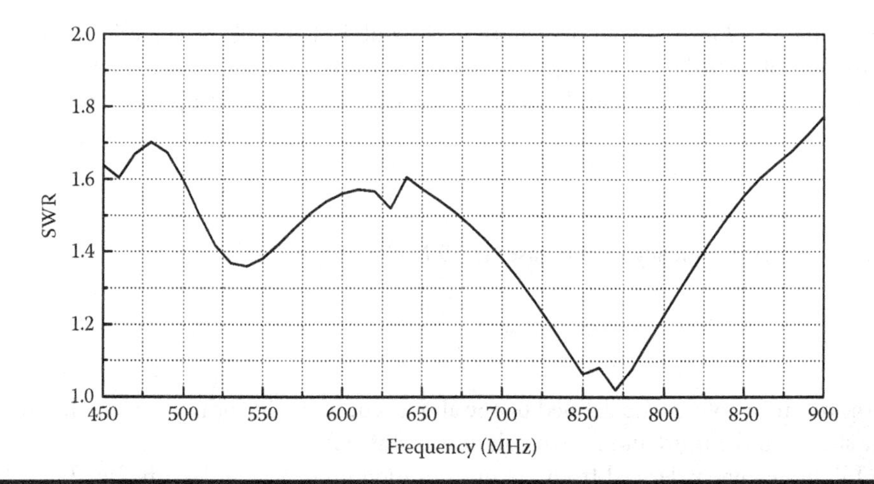

Figure 2.4 SWR graph of Case 1.

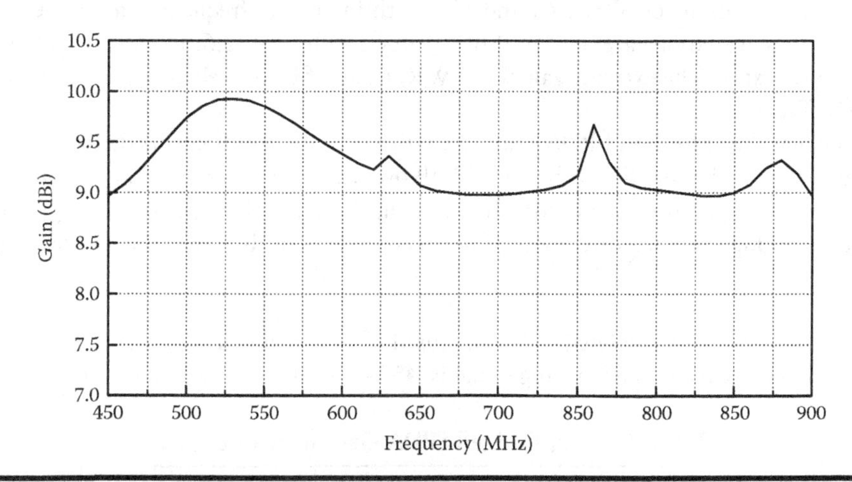

Figure 2.5 Gain graph of Case 1.

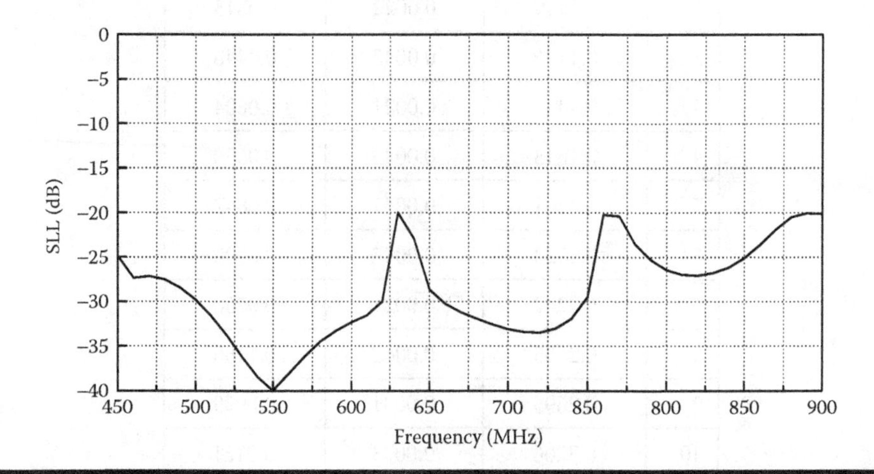

Figure 2.6 SLL graph of Case 1.

Table 2.4 Antenna Characteristics of Case 1

SWR_{min}	1.02
SWR_{max}	1.77
G_{min} (dBi)	8.97
G_{max} (dBi)	9.93
GF (dB)	0.96
SLL_{min} (dB)	–39.89
SLL_{max} (dB)	–20.00

antenna geometry. This case is studied in order to examine the ability of the IWO algorithm to find a geometry that satisfies all the above requirements without the help of a stub. Therefore, the LPDA geometry is defined by the dipole lengths L_m ($m = 1, \ldots, 10$), the dipole radii r_m ($m = 1, \ldots, 10$), the distances between adjacent dipoles S_m ($m = 1, \ldots, 9$), and the characteristic impedance Z_0 of the transmission line that models the booms of the antenna. All these parameters are considered as optimization variables, which are 30 in total ($N = 3M = 30$). Therefore, the weed positions are defined by vectors of 30 coordinates, as given hereunder:

$$X = [x_1 x_2 \ldots x_{30}] = [L_1 L_2 \ldots L_{10} r_1 r_2 \ldots r_{10} S_1 S_2 \ldots S_9 Z_0] \tag{2.26}$$

It is obvious that the boundaries of optimization variables are the same as those of Case 1, since the LPDA consists of 10 dipoles and is considered to operate in the same frequency range, that is, just like in Case 1. Therefore, $Z_{min0} = 50\ \Omega$, $Z_{max0} = 200\ \Omega$, $r_{min} = 1$ mm, and $r_{max} = 5$ mm. Moreover, by considering average antenna directivity with a typical value of 8 dBi, the length boundaries are derived as those given in the third and fourth columns of Table 2.1, while the distance boundaries are derived as those given in the third and fourth columns of Table 2.2, except for the limits given in the last row since they refer to the stub, which does not exist in this case.

The optimized geometry is shown in Table 2.5. This geometry is given as an input to the NEC software in order to find the variations of SWR, G, and SLL inside the operating bandwidth. These variations are depicted in Figures 2.7 through 2.9 and are then used to extract the extreme values of SWR, G, and SLL as well as the value of GF, which are given in Table 2.6.

Although no stub is used in this case, the optimized LPDA satisfies all the requirements defined earlier. In particular, SWR remains explicitly below 1.8 and SLL is kept under –20 dB over the entire bandwidth, while GF is better than the required value of 2 dB. In addition, all the gain values are high enough compared to the typical average antenna directivity of 8 dBi defined in the beginning of the design case.

Case 3

A 10-dipole LPDA is also considered in this case and is optimized for operation in the same frequency range, that is, 450–900 MHz. A short-circuited stub is placed behind the larger dipole (just like in Case 1). Therefore, the LPDA geometry is defined by the lengths L_m ($m = 1, \ldots, 10$), the radii r_m ($m = 1, \ldots, 10$), the distances S_m ($m = 1, \ldots, 10$) including the length S_{10} of the stub, and the characteristic impedance Z_0. However, not all the above parameters are considered as optimization variables. Actually, the dipole radii are assumed to have fixed values, which are

Table 2.5 Optimized LPDA Geometry of Case 2

m	$L_{opt,m}$ (m)	$r_{opt,m}$ (m)	$S_{opt,m}$ (m)
1	0.1087	0.0027	0.0309
2	0.1206	0.0048	0.0447
3	0.1388	0.0034	0.0582
4	0.1677	0.0045	0.0735
5	0.1824	0.0046	0.0750
6	0.2047	0.0045	0.0930
7	0.2169	0.0050	0.1050
8	0.2490	0.0027	0.0816
9	0.2826	0.0049	0.1349
10	0.3045	0.0049	–

$Z_{opt,0} = 100\ \Omega$.

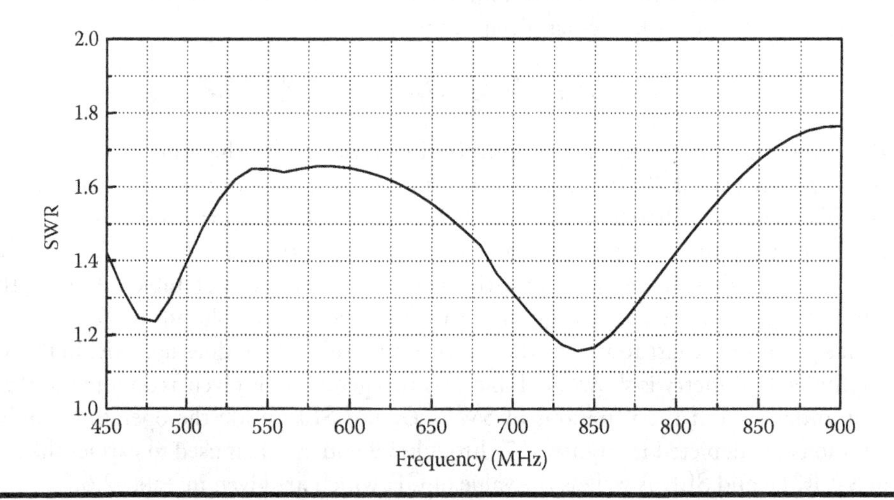

Figure 2.7 SWR graph of Case 2.

$r_m = 2$ mm ($m = 1, 2, 3$), $r_m = 3$ mm ($m = 4, 5, 6$), and $r_m = 4$ mm ($m = 7, 8, 9, 10$). The advantage of this LPDA is that it can easily be constructed in practice, since the radii values are practically convenient. Therefore, this case makes use of 21 optimization variables ($N = 2M + 1 = 21$). These are L_m ($m = 1, \ldots, 10$), S_m ($m = 1, \ldots, 10$), and Z_0. Therefore, the weed positions are defined by vectors of 21 coordinates, as given hereunder:

$$X = [x_1 x_2 \ldots x_{21}] = [L_1 L_2 \ldots L_{10} S_1 S_2 \ldots S_{10} Z_0] \tag{2.27}$$

It is obvious that the boundaries of optimization variables are the same as those of Case 1, except for the radii, which have fixed values and are no more variables. Therefore, by considering average antenna directivity with a typical value of 8 dBi, the length and distance boundaries

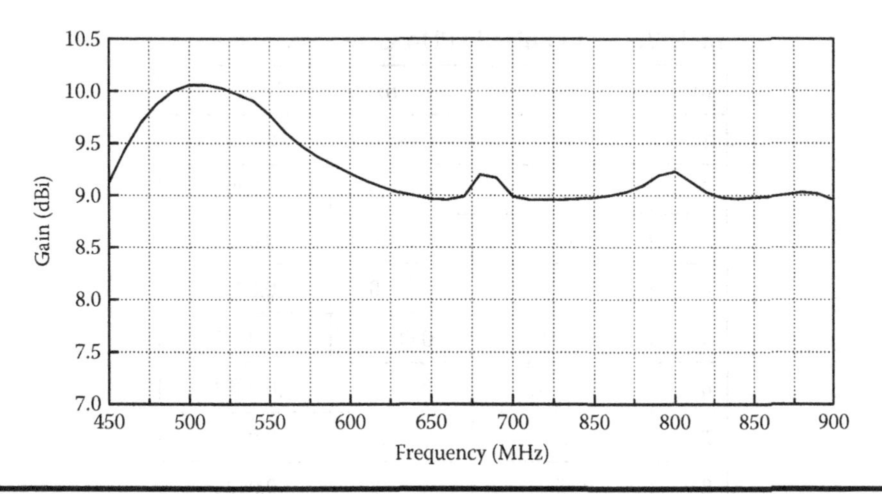

Figure 2.8 Gain graph of Case 2.

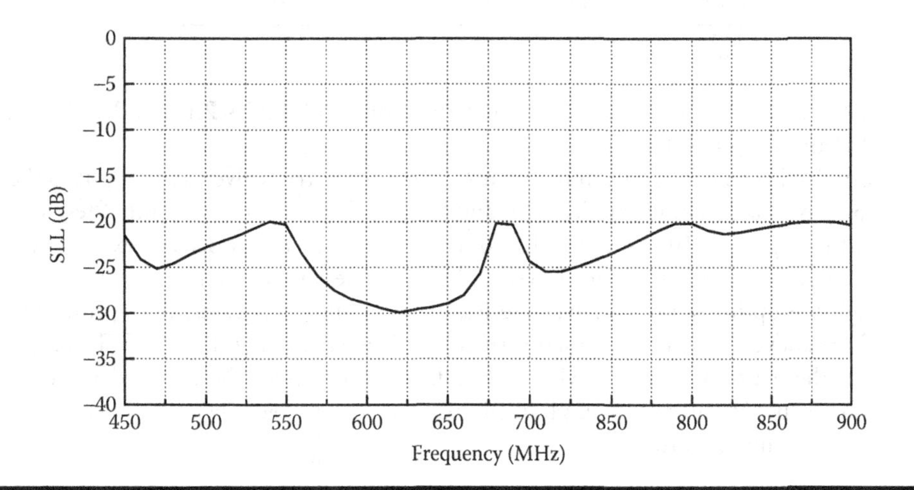

Figure 2.9 SLL graph of Case 2.

Table 2.6 Antenna Characteristics of Case 2

SWR_{min}	1.16
SWR_{max}	1.76
G_{min} (dBi)	8.96
G_{max} (dBi)	10.06
GF (dB)	1.10
SLL_{min} (dB)	−29.91
SLL_{max} (dB)	−20.02

Table 2.7 Optimized LPDA Geometry of Case 3

m	$L_{opt,m}$ (m)	r_m (m)	$S_{opt,m}$ (m)
1	0.1093	0.0020	0.0354
2	0.1264	0.0020	0.0426
3	0.1409	0.0020	0.0662
4	0.1591	0.0030	0.0755
5	0.1698	0.0030	0.0515
6	0.2018	0.0030	0.0853
7	0.2251	0.0040	0.0935
8	0.2559	0.0040	0.0675
9	0.2935	0.0040	0.0761
10	0.3240	0.0040	0.0101

$Z_{opt,0} = 76.8\ \Omega$.

are derived as those given in the third and fourth columns of Tables 2.1 and 2.2. In addition, $Z_{\min 0} = 50\ \Omega$ and $Z_{\max 0} = 200\ \Omega$.

The optimized geometry is shown in Table 2.7. This geometry is given as an input to the NEC software to derive the variations of SWR, G, and SLL inside the operating band. These variations are depicted in Figures 2.10 through 2.12, and are then used to extract the extreme values of SWR, G, and SLL as well as the value of GF. All these values are shown in Table 2.8.

Although the optimization variables are less than the previous cases, the optimized LPDA satisfies all the above-defined requirements, since SWR remains below 1.8 and SLL is kept under −20 dB over the entire bandwidth, while GF is much better than the required value of 2 dB. Also, the gain values are sufficiently high in comparison to the typical average directivity of 8 dBi defined in the beginning of the case.

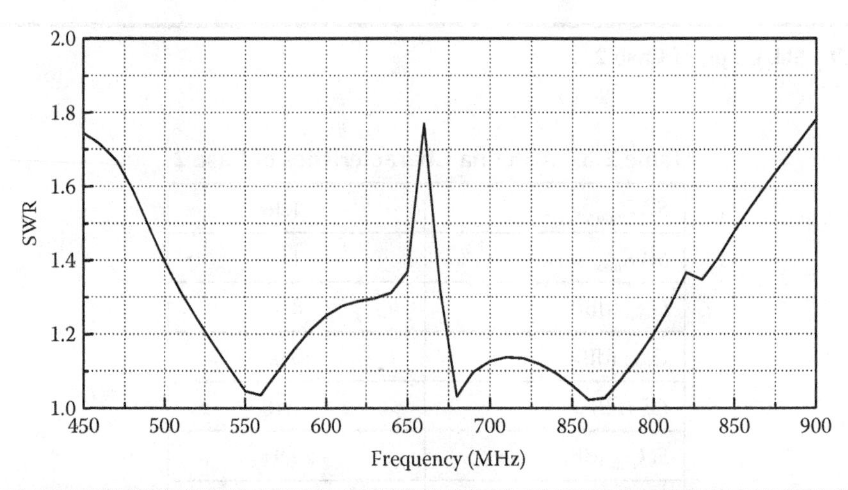

Figure 2.10 SWR graph of Case 3.

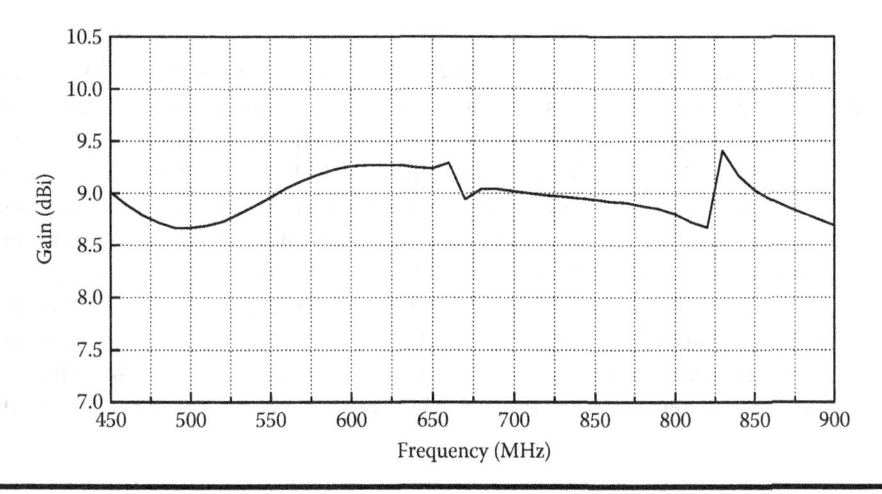

Figure 2.11 Gain graph of Case 3.

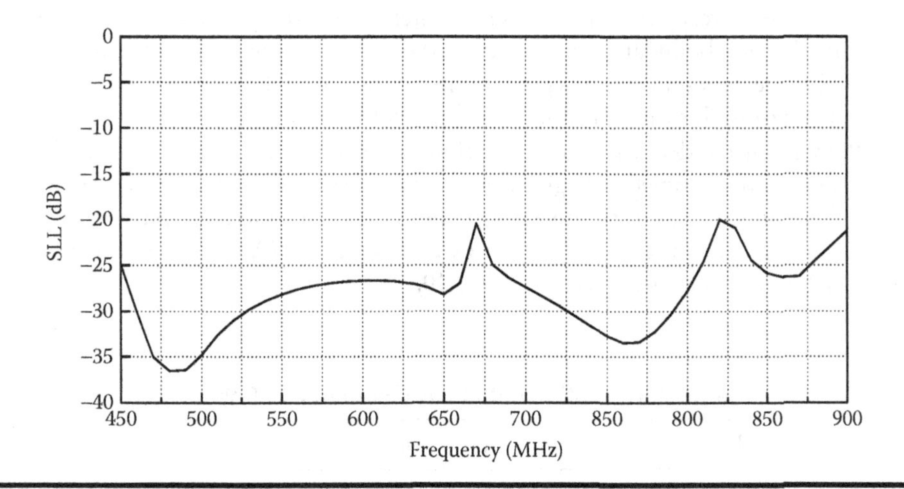

Figure 2.12 SLL graph of Case 3.

Table 2.8 Antenna Characteristics of Case 3

SWR_{min}	1.02
SWR_{max}	1.78
G_{min} (dBi)	8.67
G_{max} (dBi)	9.41
GF (dB)	0.74
SLL_{min} (dB)	−36.55
SLL_{max} (dB)	−20.01

Case 4

This case is more difficult than the previous three cases. The LPDA consists of 10 dipoles and is optimized for operation in the range 450–900 MHz. However, no stub is used here to mitigate high-order resonances that may appear on some dipoles of the array. Also, the dipole radii are assumed to have fixed values, which are: $r_m = 2$ mm ($m = 1, 2, 3$), $r_m = 3$ mm ($m = 4, 5, 6$), and $r_m = 4$ mm ($m = 7, 8, 9, 10$). This LPDA has easier construction than the previous ones, since the radii values are practically convenient and also no stub needs to be added. Although the LPDA geometry is defined by the lengths L_m ($m = 1, \ldots, 10$), the radii r_m ($m = 1, \ldots, 10$), the distances S_m ($m = 1, \ldots, 9$) between adjacent dipoles, and the characteristic impedance Z_0, the optimization variables are only L_m ($m = 1, \ldots, 10$), S_m ($m = 1, \ldots, 9$), and Z_0. Therefore, this case makes use of 20 optimization variables ($N = 2M = 20$), which are less than those used in all the previous cases. Therefore, the weed positions are defined by vectors of 20 coordinates, as follows:

$$X = [x_1 x_2 \ldots x_{20}] = [L_1 L_2 \ldots L_{10} S_1 S_2 \ldots S_9 Z_0] \tag{2.28}$$

It is obvious that the boundaries of L_m ($m = 1, \ldots, 10$), S_m ($m = 1, \ldots, 9$), and Z_0 are the same as those of Case 1. Therefore, by considering average antenna directivity with a typical value of 8 dBi, the length and distance boundaries are derived as those given in the third and fourth columns of Tables 2.1 and 2.2, except for the limits given in the last row of Table 2, since they refer to the stub, which does not exist in this case. In addition, $Z_{min0} = 50\ \Omega$ and $Z_{max0} = 200\ \Omega$.

The LPDA geometry derived from the optimization procedure is shown in Table 2.9. The parameter values of this geometry are given as an input to the NEC software to extract the variations of SWR, G, and SLL inside the operating band. These variations are depicted in Figures 2.13 through 2.15, and are then used to extract the extreme values of SWR, G, and SLL as well as the value of GF. All these values are given in Table 2.10.

Table 2.9 Optimized LPDA Geometry of Case 4

m	$L_{opt,m}$ (m)	r_m (m)	$S_{opt,m}$ (m)
1	0.1054	0.0020	0.0293
2	0.1263	0.0020	0.0435
3	0.1437	0.0020	0.0547
4	0.1612	0.0030	0.0750
5	0.1809	0.0030	0.0773
6	0.2004	0.0030	0.0813
7	0.2253	0.0040	0.0940
8	0.2584	0.0040	0.0816
9	0.3004	0.0040	0.1366
10	0.3054	0.0040	–

$Z_{opt,0} = 112.9\ \Omega$.

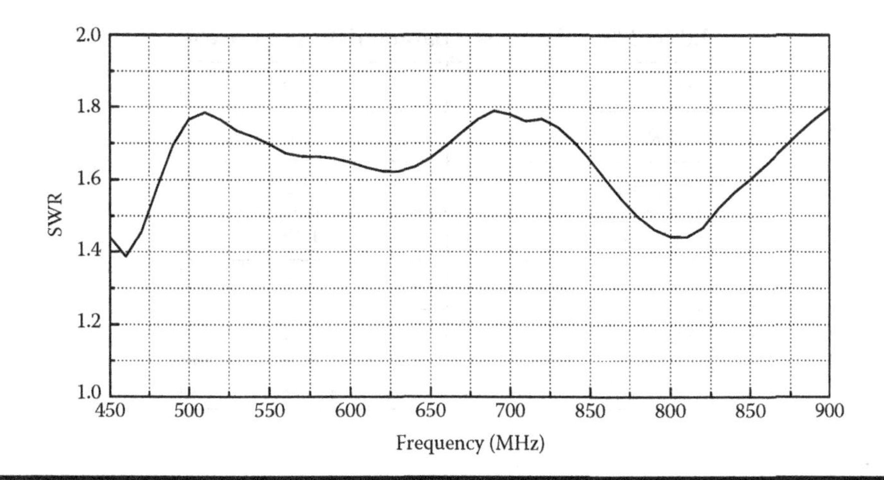

Figure 2.13 SWR graph of Case 4.

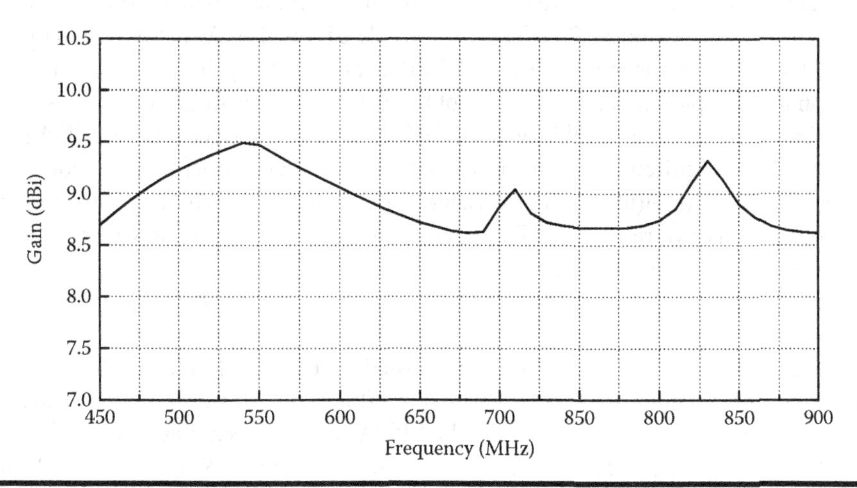

Figure 2.14 Gain graph of Case 4.

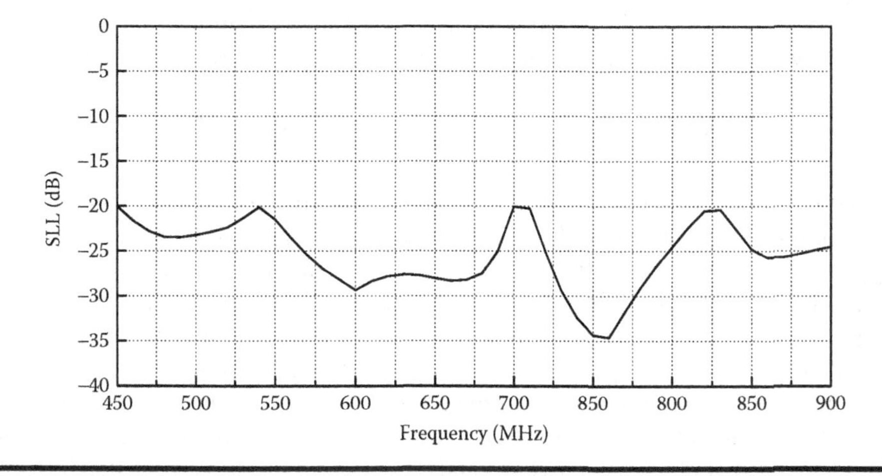

Figure 2.15 SLL graph of Case 4.

Table 2.10 Antenna Characteristics of Case 4

SWR_{min}	1.39
SWR_{max}	1.80
G_{min} (dBi)	8.62
G_{max} (dBi)	9.49
GF (dB)	0.87
SLL_{min} (dB)	–34.65
SLL_{max} (dB)	–20.03

Although this case makes use of less optimization variables than all the previous cases, the optimized LPDA satisfies again all the above-defined requirements. As shown in Table 2.10, SWR remains below 1.8 and SLL is kept under –20 dB over the whole bandwidth. Also, GF is equal to 0.87 dB, which is much better than the required value of 2 dB. Moreover, the gain values are sufficiently high in comparison to the typical average directivity of 8 dBi defined in the beginning of this case.

All the above cases exhibit the ability of the IWO method to extract various LPDA geometries that meet predefined requirements in the range 450–900 MHz by either using or not using a stub and by using dipoles with either fixed or variable radii. In the next two cases, we optimize under the same requirements two respective LPDAs, which are going to operate in wider bands than the band used in the previous four cases.

Case 5

In this case, the LPDA is considered to be composed of 12 wire dipoles ($M = 12$) and is optimized for operation in the range 700–3300 MHz. This range has been chosen since it covers most of the recent wireless network services. The LPDA geometry is defined by the dipole lengths L_m ($m = 1, \ldots, 12$), the dipole radii r_m ($m = 1, \ldots, 12$), the distances S_m ($m = 1, \ldots, 12$) including the length S_{12} of the short-circuited stub located behind the largest dipole, and finally the characteristic impedance Z_0 of the transmission line that models the booms of the LPDA. All the above parameters are considered as optimization variables, which are 37 in total ($N = 3M + 1 = 37$). Therefore, the weed positions are defined by vectors of 37 coordinates, as given hereunder:

$$X = [x_1 x_2 \ldots x_{37}] = [L_1 L_2 \ldots L_{12} r_1 r_2 \ldots r_{12} S_1 S_2 \ldots S_{12} Z_0] \quad (2.29)$$

The lower and upper boundaries of the optimization variables are determined in the same way as in the previous cases. First, the values of Z_0 are considered to be between $Z_{min0} = 50\ \Omega$ and $Z_{max0} = 200\ \Omega$, while the same boundaries, $r_{min} = 1$ mm and $r_{max} = 5$ mm, are used for all the dipole radii. Nevertheless, different boundaries are used for the lengths and the distances. By assuming average antenna directivity with a typical value of 7.5 dBi and by using the constant directivity contour curve graph given by Carrel, we find the optimum values of τ and σ parameters, which are $\tau = 0.86$ and $\sigma = 0.16$. It has to be noted that the average directivity was chosen to be less than 8 dBi (this value was chosen in the previous cases) because the operating bandwidth is much wider than that of the previous cases and the directivity is thus expected to be reduced as the frequency increases. Afterwards, we find the typical length of the largest dipole (12th dipole), which is set equal to half-wavelength of the lowest operating frequency (700 MHz), and is then reduced by 4% due to the dipole thickness. Therefore, this length is calculated as follows:

$$L_{12} = (1 - 0.04)\frac{\lambda_{700\text{MHz}}}{2} = 0.2057\,\text{m} \tag{2.30}$$

Then, the values of the typical lengths and distances are estimated by using Equations 2.3 and 2.5 and are respectively shown in the second column of Tables 2.11 and 2.12.

From the average values, that is,

$$\frac{L_m + L_{m+1}}{2}$$

of the typical lengths of every two adjacent dipoles, we estimate the lower length boundaries $L_{\text{min},m}$, $m = 2, \ldots, 12$, and the upper length boundaries $L_{\text{max},m}$, $m = 1, \ldots, 11$, while $L_{\text{min},1}$ and $L_{\text{max},12}$ are respectively calculated from Equations 2.19 and 2.20. All the length boundaries are shown in the third and fourth columns of Table 2.11. From the typical distances given in the second column of Table 2.12, we can find the boundaries of the dipole distances by using Equations 2.21 and 2.22. Also, the length limits of the stub are estimated by Equations 2.23 and 2.24. All these distance boundaries (including the length limits of the stub) are shown in the third and fourth columns of Table 2.12.

The LPDA geometry extracted from the optimization procedure is described in Table 2.13. This geometry is given as an input to the NEC software to evaluate the variations of SWR, G, and SLL with increasing frequency inside the operating band. These variations are shown in Figures 2.16 through 2.18, which are then used to extract the extreme values of SWR, G, and SLL as well as the value of GF. These values are given in Table 2.14.

Table 2.11 Typical Length Values Extracted from Carrel's Method and Length Boundaries for the Frequency Range 700–3300 MHz

m	Typical Length, L_m (m)	Minimum Length, $L_{min,m}$ (m)	Maximum Length, $L_{max,m}$ (m)
1	0.0392	0.0360	0.0423
2	0.0455	0.0423	0.0492
3	0.0529	0.0492	0.0572
4	0.0616	0.0572	0.0666
5	0.0716	0.0666	0.0774
6	0.0832	0.0774	0.0900
7	0.0968	0.0900	0.1047
8	0.1125	0.1047	0.1217
9	0.1308	0.1217	0.1415
10	0.1521	0.1415	0.1645
11	0.1769	0.1645	0.1913
12	0.2057	0.1913	0.2201

Table 2.12 Typical Distance Values Extracted from Carrel's Method and Distance Boundaries for the Frequency Range 700–3300 MHz

m	Typical Distance, S_m (m)	Minimum Distance, $S_{min,m}$ (m)	Maximum Distance, $S_{max,m}$ (m)
1	0.0146	0.0120	0.0189
2	0.0169	0.0120	0.0220
3	0.0197	0.0138	0.0256
4	0.0229	0.0160	0.0298
5	0.0266	0.0186	0.0346
6	0.0310	0.0217	0.0403
7	0.0360	0.0252	0.0468
8	0.0419	0.0293	0.0544
9	0.0487	0.0341	0.0633
10	0.0566	0.0396	0.0736
11	0.0658	0.0461	0.0856
12	–	0.0120	0.1071

Table 2.13 Optimized LPDA Geometry of Case 5

m	$L_{opt,m}$ (m)	$r_{opt,m}$ (m)	$S_{opt,m}$ (m)
1	0.0362	0.0010	0.0127
2	0.0424	0.0013	0.0167
3	0.0499	0.0020	0.0195
4	0.0648	0.0026	0.0238
5	0.0725	0.0045	0.0331
6	0.0860	0.0047	0.0400
7	0.1000	0.0048	0.0403
8	0.1111	0.0042	0.0362
9	0.1316	0.0039	0.0547
10	0.1431	0.0037	0.0409
11	0.1854	0.0038	0.0677
12	0.1990	0.0047	0.0483

$Z_{opt,0} = 92.8\ \Omega$.

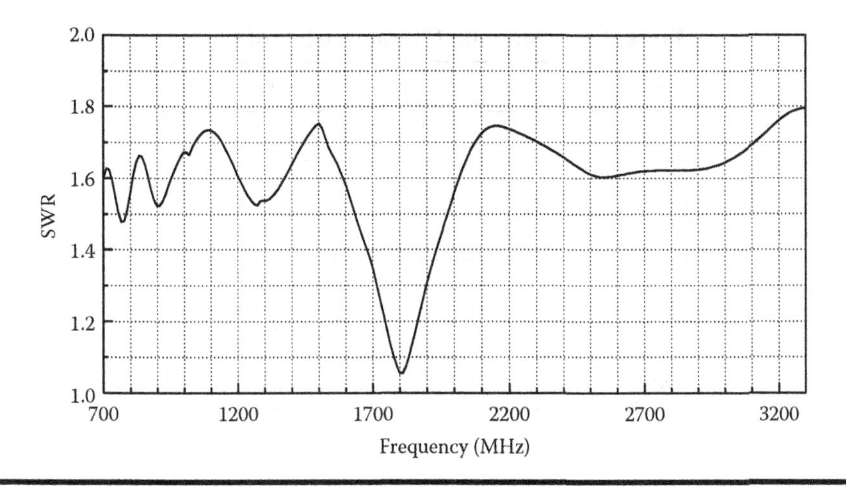

Figure 2.16 SWR graph of Case 5.

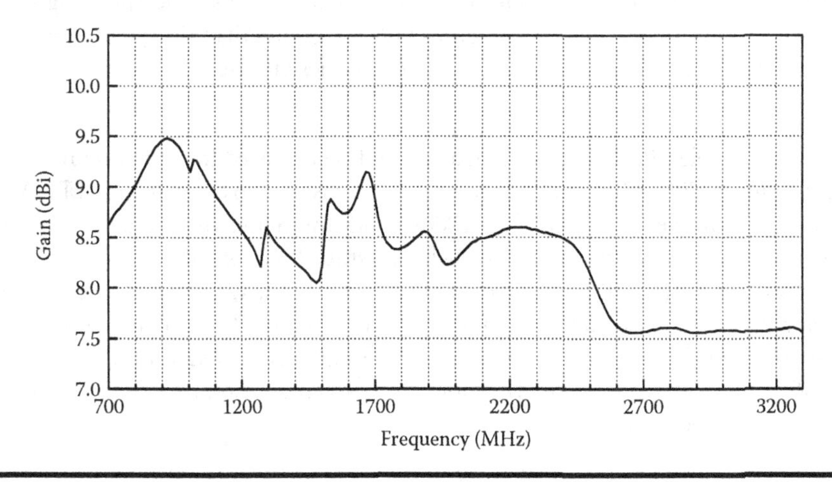

Figure 2.17 Gain graph of Case 5.

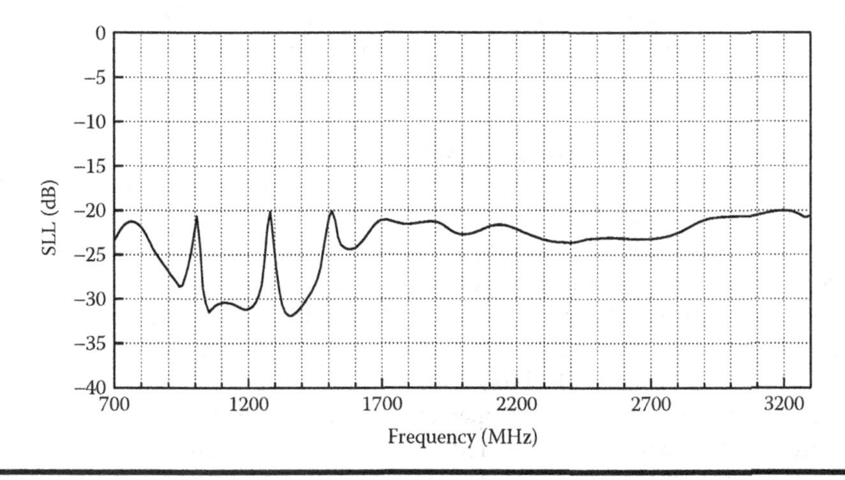

Figure 2.18 SLL graph of Case 5.

Table 2.14 Antenna Characteristics of Case 5

SWR_{min}	1.05
SWR_{max}	1.80
G_{min} (dBi)	7.56
G_{max} (dBi)	9.48
GF (dB)	1.92
SLL_{min} (dB)	−31.88
SLL_{max} (dB)	−20.03

The optimized LPDA geometry satisfies all the requirements defined earlier. In particular, SWR and SLL remain respectively below 1.8 and −20 dB over this wide operating band, while GF is better than 2 dB. In addition, all the gain values are greater than the typical average antenna directivity of 7.5 dBi considered in the beginning of this design case.

Case 6

In this case, a 12-dipole LPDA is considered again and is optimized for operation in the same frequency range, that is, 700–3300 MHz. However, no stub is used to construct the LPDA. This case is studied here to examine the ability of the IWO algorithm to find an optimal geometry that satisfies all the above requirements without the help of a stub. Therefore, the LPDA is defined by the dipole lengths L_m ($m = 1, \ldots, 12$), the dipole radii r_m ($m = 1, \ldots, 12$), the distances S_m ($m = 1, \ldots, 11$) between adjacent dipoles, and the characteristic impedance Z_0 of the transmission line that models the booms of the antenna. All these parameters are considered as optimization variables, which are 36 in total ($N = 3M = 36$). Therefore, every weed position is defined by a vector of 36 coordinates, as given hereunder:

$$X = [x_1 x_2 \ldots x_{36}] = [L_1 L_2 \ldots L_{12} r_1 r_2 \ldots r_{12} S_1 S_2 \ldots S_{11} Z_0] \tag{2.31}$$

It is obvious that the boundaries of the optimization variables are the same as those of Case 5, except for the boundaries that concern the length of the stub, since the stub does not exist in this case. Therefore, $Z_{min0} = 50\ \Omega$, $Z_{max0} = 200\ \Omega$, $r_{min} = 1$ mm, and $r_{max} = 5$ mm. Moreover, by considering average antenna directivity with a typical value of 7.5 dBi, the length boundaries are derived as those given in the third and fourth columns of Table 2.11, while the distance boundaries are derived as those given in the third and fourth columns of Table 2.12, except for the limits given in the last row of Table 2.12 since they refer to the stub.

The optimized geometry is displayed in Table 2.15. The geometry parameters shown in Table 2.15 are given as an input to the NEC software and thereby the variations of SWR, G, and SLL inside the operating band are extracted. These variations are displayed in Figures 2.19 through 2.21 and are then used to extract the extreme values of SWR, G, and SLL as well as the value of GF. These values are given in Table 2.16.

Although the LPDA is required to operate inside a wide band without using any stub, the optimized LPDA satisfies all the requirements defined earlier. In particular, SWR remains explicitly below 1.8 and SLL is kept under −20 dB over this wide frequency range,

Table 2.15 Optimized LPDA Geometry of Case 6

m	$L_{opt,m}$ (m)	$r_{opt,m}$ (m)	$S_{opt,m}$ (m)
1	0.0360	0.0010	0.0122
2	0.0424	0.0014	0.0177
3	0.0502	0.0022	0.0199
4	0.0634	0.0029	0.0256
5	0.0739	0.0042	0.0319
6	0.0850	0.0044	0.0367
7	0.0943	0.0050	0.0379
8	0.1089	0.0043	0.0330
9	0.1355	0.0035	0.0462
10	0.1448	0.0048	0.0430
11	0.1811	0.0050	0.0568
12	0.1986	0.0027	–

$Z_{opt,0} = 99.8\ \Omega$.

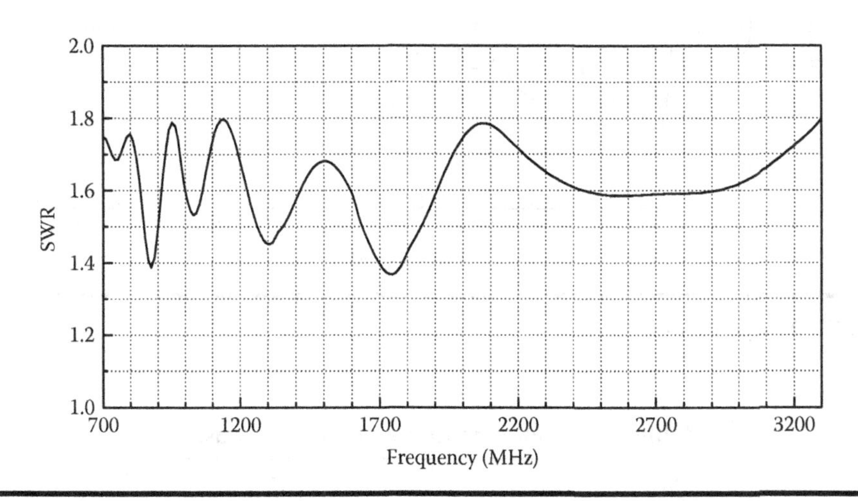

Figure 2.19 SWR graph of Case 6.

while GF is close to its required value. As shown in Figure 2.20, there is a small frequency region where the gain is less than the typical average antenna directivity of 7.5 dBi defined in the beginning of this case. However, in the major part of the bandwidth, all the gain values are much greater than this typical directivity value, resulting in an average gain greater than 7.5 dBi.

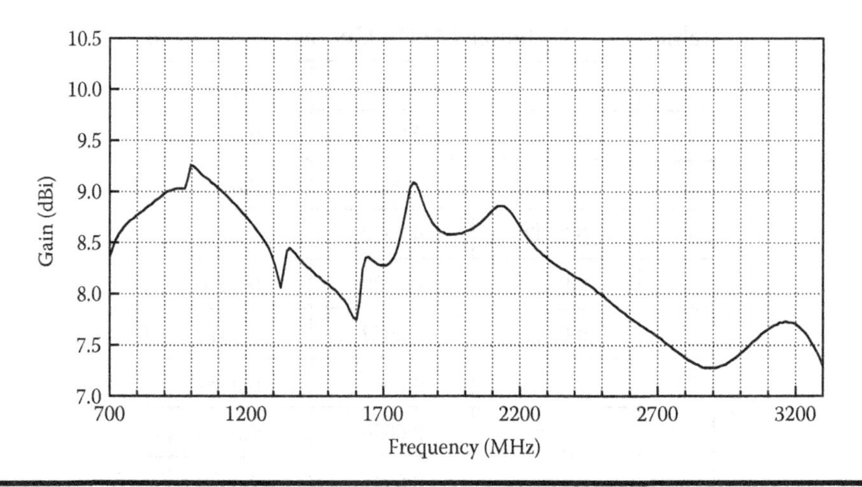

Figure 2.20 Gain graph of Case 6.

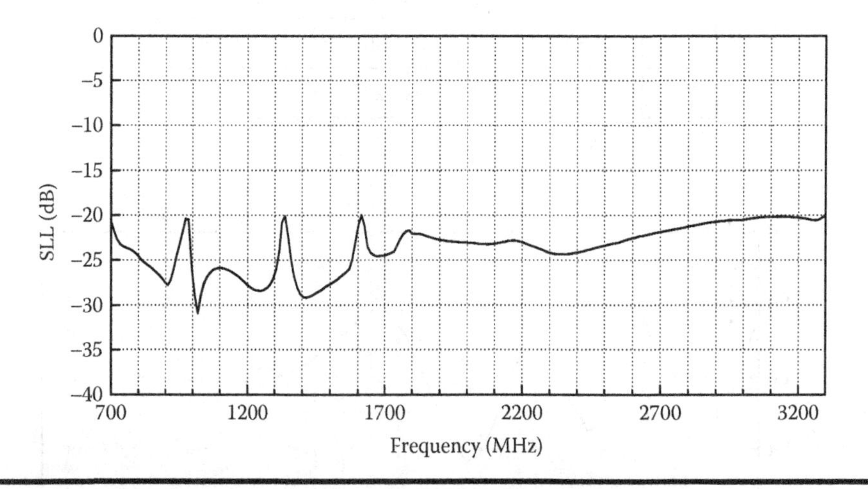

Figure 2.21 SLL graph of Case 6.

Table 2.16 Antenna Characteristics of Case 6

SWR_{min}	1.37
SWR_{max}	1.80
G_{min} (dBi)	7.28
G_{max} (dBi)	9.27
GF (dB)	1.99
SLL_{min} (dB)	−30.85
SLL_{max} (dB)	−20.01

2.6 Conclusions

The IWO method has been applied in this chapter to design optimal LPDAs, which satisfy multiple predefined requirements. This method aims at providing effective solutions to complex nonlinear problems, like the LPDA design concerned here, especially when many requirements have to be met at the same time. The LPDA design methods applied so far have shown limited ability to control some of the radiation characteristics of the LPDA, such as SWR, *G*, and FBR. The IWO method has been proved to be capable of controlling and optimizing the values of SWR, *G*, FBR, GF, and SLL at the same time over a wide frequency range. In this way, the optimized LPDA is practically suitable for many wireless network applications.

The use of the IWO method in antenna optimization problems demands to thoroughly define the boundaries of the optimization variables and to carefully formulate the fitness function. The cases studied in this chapter concern LPDAs operating in different frequency ranges with different bandwidths, LPDAs that either use or not use a short-circuited stub, and LPDAs that use either dipoles with radii under optimization or dipoles with fixed radius values. In all the cases, the IWO method satisfies all the predefined requirements, providing thus directional antennas with gain greater than a predefined average directivity value and with optimal values of SWR, GF, FBR, and SLL over wide operating bands. Most of these antennas can be suitably applied in base stations of several wireless network services.

Key Terms

Antenna Directivity or Antenna Maximum Directivity: Ratio of the maximum radiation intensity from the antenna (which corresponds to the main lobe peak) to the radiation intensity averaged over all directions.

Antenna Gain or Antenna Maximum Gain: Ratio of the maximum radiation intensity from the antenna (which corresponds to the main lobe peak) to the radiation intensity that would be obtained if the power at the input of the antenna were radiated isotropically.

Characteristic Impedance: Ratio of the voltage to the current developed along a transmission line due to a single wave propagating in one direction of the line (i.e., without reflections in the other direction).

E-Plane: Plane that contains the electric field vector and the direction of maximum radiation.

Front-to-Back Ratio: Ratio of the antenna maximum gain to the gain in the direction of the back lobe.

Gain Flatness: Difference between the maximum and the minimum value of the antenna gain found over the entire frequency range of operation.

H-Plane: Plane that contains the magnetic field vector and the direction of maximum radiation.

Half-Power Beam-Width: Angle between two directions where the radiation intensity from the antenna is half the maximum radiation intensity (which corresponds to the main lobe peak).

Radiation Pattern: Variation of the power radiated by an antenna as a function of the direction away from the antenna.

Reflection Coefficient: Complex ratio of the electric field strength of the reflected wave to that of the incident wave.

S-Parameters or Scattering Parameters: Frequency-dependent parameters that describe the input–output relationship between ports in a high-frequency system. In a communication system with two ports, S_{11} parameter is the reflection coefficient at port 1 with port 2 terminated by a matched load, S_{22} parameter is the reflection coefficient at port 2 with port 1 terminated by a matched load, S_{21} parameter is the forward transmission gain (from port 1 to 2) with port 2 terminated in a matched load, and S_{12} parameter is the reverse transmission gain (from port 2 to 1) with port 1 terminated in a matched load.

Side Lobe Level: Ratio of the gain in the direction of the maximum side lobe to the antenna maximum gain.

Standing Wave Ratio: Ratio of the maximum to the minimum voltage amplitude along a transmission line that feeds an antenna.

References

1. Balanis, C. A., 2005. *Antenna Theory, Analysis and Design*. 3rd ed., New Jersey: John Wiley & Sons.
2. Carrel, R. L., 1961. *Analysis and design of the log-periodic dipole antenna*. PhD thesis. University Microfilms, Inc., Ann Arbor, MI.
3. Butson, P. C. and G. T. Thompson, 1976. A note on the calculation of the gain of log-periodic dipole antennas, *IEEE Transactions on Antennas and Propagation*, 24(1), 105–106.
4. Karimkashi, S. and A. A. Kishk, 2010. Invasive weed optimization and its features in electromagnetics, *IEEE Transactions on Antennas and Propagation*, 58(4), 1269–1278.
5. Karimkashi, S., A. A. Kishk, and D. Kajfez, 2011. Antenna array optimization using dipole models for MIMO applications, *IEEE Transactions on Antennas and Propagation*, 59(8), 3112–3116.
6. Mallahzadeh, A. R., S. Es'haghi, and A. Alipour, 2009. Design of an e-shaped MIMO antenna using IWO algorithm for wireless application at 5.8 GHz, *Progress in Electromagnetics Research*, 90, 187–203.
7. Monavar, F. M. and N. Komjani, 2011. Bandwidth enhancement of microstrip patch antenna using Jerusalem cross-shaped frequency selective surfaces by invasive weed optimization approach, *Progress in Electromagnetics Research*, 121, 103–120.
8. Pal, S., A. Basak, S. Das, and A. Abraham, 2009. Linear antenna array synthesis with invasive weed optimization algorithm. In *2009 International Conference of Soft Computing and Pattern Recognition*, December 4–7, Malacca, pp. 161–166.
9. Zaharis, Z. D., C. Skeberis, and T. D. Xenos, 2012. Improved antenna array adaptive beamforming with low side lobe level using a novel adaptive invasive weed optimization method, *Progress in Electromagnetics Research*, 124, 137–150.
10. Zaharis, Z. D., C. Skeberis, T. D. Xenos, P. I. Lazaridis, and J. Cosmas, 2013. Design of a novel antenna array beamformer using neural networks trained by modified adaptive dispersion invasive weed optimization based data, *IEEE Transactions on Broadcasting*, 59(3), 455–460.
11. Zaharis, Z. D., P. I. Lazaridis, J. Cosmas, C. Skeberis, and T. D. Xenos, 2014. Synthesis of a near-optimal high-gain antenna array with main lobe tilting and null filling using Taguchi initialized invasive weed optimization, *IEEE Transactions on Broadcasting*, 60(1), 120–127.
12. Burke, G. J. and A. J. Poggio, 1981. *Numerical Electromagnetics Code (NEC)—Method of Moments*. San Diego, CA: Naval Ocean Systems Center, Tech. Doc. 116.
13. Hansen, R. C., 1990. *Moment Methods in Antennas and Scattering*, Norwood, MA: Artech House.
14. Goudos, S. K., Z. D. Zaharis, and T. V. Yioultsis, 2010. Application of a differential evolution algorithm with strategy adaptation to the design of multi-band microwave filters for wireless communications, *Progress in Electromagnetics Research*, 109, 123–137.
15. Kampitaki, D., A. Hatzigaidas, A. Papastergiou, P. Lazaridis, and Z. Zaharis, 2006. Dual-frequency splitter synthesis suitable for practical RF applications, *WSEAS Transactions on Communications*, 5(10), 1885–1891.
16. Kampitaki, D. G., A. T. Hatzigaidas, A. I. Papastergiou, and Z. D. Zaharis, 2007. On the design of a dual-band unequal power divider useful for mobile communications, *Electrical Engineering*, 89(6), 443–450.

17. Lazaridis, P. I., Z. D. Zaharis, C. Skeberis, T. Xenos, E. Tziris, and P. Gallion, 2014. Optimal design of UHF TV band log-periodic antenna using invasive weed optimization. In *4th International Conference on Wireless Communications, Vehicular Technology, Information Theory and Aerospace & Electronic Systems*, May 11–14, Aalborg, pp. 1–5.
18. Zaharis, Z. D., D. G. Kampitaki, P. I. Lazaridis, A. I. Papastergiou, A. T. Hatzigaidas, and P. B. Gallion, 2007. Improving the radiation characteristics of a base station antenna array using a particle swarm optimizer, *Microwave and Optical Technology Letters*, 49(7), 1690–1698.
19. Zaharis, Z. D., D. G. Kampitaki, P. I. Lazaridis, A. I. Papastergiou, and P. B. Gallion, 2007. On the design of multifrequency dividers suitable for GSM/DCS/PCS/UMTS applications by using a particle swarm optimization-based technique, *Microwave and Optical Technology Letters*, 49(9), 2138–2144.
20. Zaharis, Z. D., 2012. A modified Taguchi's optimization algorithm for beamforming applications, *Progress in Electromagnetics Research*, 127, 553–569.
21. Zaharis, Z. D., C. Skeberis, P. I. Lazaridis, D. I. Stratakis, and T. D. Xenos, 2014. IWO-based synthesis of log-periodic dipole array. In *2014 International Conference on Telecommunications and Multimedia*, July 28–30, Heraklion, pp. 150–154.
22. Fisher, S. E., D. S. Weile, E. Michielssen, and W. Woody, 1999. Pareto genetic algorithm based optimization of log-periodic monopole arrays mounted on realistic platforms, *Journal of Electromagnetic Waves and Applications*, 13(5), 571–598.
23. Chung, Y. C. and R. Haupt, 2001. Log-periodic dipole array optimization, *Journal of Electromagnetic Waves and Applications*, 15(9), 1269–1280.
24. Mangoud, M. A., M. A. Aboul-Dahab, A. I. Zaki, and S. E. El-Khamy, 2003. Genetic algorithm design of compressed log periodic dipole array. In *46th IEEE Midwest Symposium on Circuits and Systems*, December 27–30, Cairo, Vol. 3, pp. 1194–1197.
25. Pitzer, T. L., A. James, G. B. Lamont, and A. J. Terzuoli, 2006. Linear ensemble antennas resulting from the optimization of log periodic dipole arrays using genetic algorithms. In *2006 IEEE Congress on Evolutionary Computation*, July 16–21, Vancouver, pp. 3189–3196.
26. Fernandez Pantoja, M., A. R. Bretones, F. Garcia Ruiz, S. G. Garcia, and R. G. Martin, 2007. Particle-Swarm optimization in antenna design: Optimization of log-periodic dipole arrays, *IEEE Antennas and Propagation Magazine*, 49(4), 34–47.
27. Tsitouri, C. I., S. C. Panagiotou, T. D. Dimousios, and C. N. Capsalis, 2008. A circular switched parasitic array of log-periodic antennas with enhanced directivity and beam steering capability for ultra wideband communications applications. In *2008 Loughborough Antennas and Propagation Conference*, March 17–18, Loughborough, pp. 281–284.
28. Zhang, X.-L. and H.-T. Gao, 2011. An optimum design of miniaturized high frequency inverted-V log-periodic dipole antenna. In *2011 IEEE CIE International Conference on Radar*, October 24–27, Chengdu, Vol. 2, pp. 1185–1188.
29. Yang, J. and P.-S. Kildal, 2011. Optimization of reflection coefficient of large log-periodic array by computing only a small part of it, *IEEE Transactions on Antennas and Propagation*, 59(6), 1790–1797.
30. Hashemi, S. M., V. Nayyeri, M. Soleimani, and A. R. Mallahzadeh, 2011. Designing a compact-optimized planar dipole array antenna, *IEEE Antennas and Wireless Propagation Letters*, 10, 243–246.
31. Haq, M. A., M. T. Rana, U. Rafique, Q. D. Memon, M. A. Khan, and M. M. Ahmed, 2012. Log periodic dipole antenna design using particle swarm optimization, *International Journal of Electromagnetics and Applications*, 2(4), 65–68.
32. Touseef, M., M. Aziz-ul-Haq, U. Rafique, M. A. Khan, and M. M. Ahmed, 2012. Genetic algorithm optimization of log-periodic dipole array, *International Journal of Electromagnetics and Applications*, 2(6), 169–173.
33. Mangaraj, B. B., I. S. Misra, and S. K. Sanyal, 2013. Application of bacteria foraging algorithm in designing log periodic dipole array for entire UHF TV spectrum, *International Journal of RF and Microwave Computer-Aided Engineering*, 23(2), 157–171.
34. Mehrabian, A. R. and C. Lucas, 2006. A novel numerical optimization algorithm inspired from weed colonization, *Ecological Informatics*, 1(4), 355–366.

DIRECTIONAL MAC

Chapter 3

Discovery Strategies for a Directional Wake-Up Radio in Mobile Networks

Wen-Chan Shih, Raja Jurdak, David Abbott,
Pai H. Chou, and Wen-Tsuen Chen

Contents

Abstract

This chapter describes a long-range directional wake-up radio (LDWuR) for wireless mobile networks. In contrast to most of the wake-up radios (WuRs) available to date, which are of short range, the proposed LDWuR is applicable to long-range deployments. Existing studies achieved long communication distance using modulation and coding schemes or directional antennas, although the latter require exploring the direction of the transmitter. To address this issue, the proposed LDWuR adopts both static and dynamic antennas, where the static ones are directional, while the dynamic ones are omnidirectional for beamforming. Theoretical analysis shows that our design can explore the transmitter's direction with the same average power consumption, thereby enhancing the node performance.

3.1 Introduction

Long-range directional wake-up radios (LDWuR) can be an important technology is important for reducing the energy consumption of nodes and improving the communication range in wireless mobile networks. For nodes that must keep their radios on for accepting incoming commands or packet relaying, idle listening can dominate the total energy consumption of sensor nodes. Although low-power-listening media access control (MAC) protocols have been proposed to reduce the energy consumed during idle listening using duty cycling of the receiver, they incur long wake-up latency; therefore, the energy consumption is still lower bounded by the duty cycle. Wake-up radios (WuRs) provide a promising solution to completely eliminate the idle listening cost and significantly extend the lifetime of WuRs.

We illustrate the requirements of long-range communication through three examples. First, the Clemson's Intelligent River project [1] deploys sensor nodes along the 312-mile Savannah River from the headwaters in North Carolina to the port in Savannah in a large scale. The sensor network provides real-time data on water quality and quantity. The sensor data are important for improving water resource management regarding drinking water and industrial pollutants. The second example is IBM smart water project [2], which distributes a sensor network along the 315-mile Hudson River to collect and analyze data continuously and in real time. The final example is the flying fox tracking project [3], in which a resourceful base node communicates with resource-constrained collar nodes on demand at a maximum possible range. These applications exemplify long-range sensor network deployments.

To achieve long-range communication using WuRs, we adopt directional antennas to extend communication distance by increasing the antenna gain. As the existing low-power WuRs provide short-range communication, they use modulation and coding schemes to improve the communication distance by reducing the packet error rate (PER). However, they provide a limited improvement in communication range. A directional antenna can be used to achieve long communication distance. A directional antenna focuses the radiation beam in one direction to obtain much antenna gain. We make use of the advantage of direction antennas for long-range communication in practice.

The challenge in adopting directional antennas is searching for the appropriate direction in which the transmitter should be located. As the directional antenna focuses the radiation beam in one direction, the transmitter might not be located in the same direction. This incurs the overhead of a motor to rotate the antenna in all directions [4]. To address this issue, our proposed LDWuR adopts static and dynamic strategies. Static strategy means that LDWuR uses multiple-directional antennas in different directions. Dynamic strategy means that LDWuR uses multiple omnidirectional antennas that form beams in different directions. To evaluate these strategies, we design MAC protocol and system model to investigate salient characteristics in terms of total power consumption and latency.

First, we propose adopting directional antennas to enhance the communication range of an LDWuR for long-range applications. Second, we design strategies to search for the transmitter's location for extending directional antenna's cover range. Third, we develop a system model and a MAC protocol to evaluate the performance of the proposed strategies of LDWuR in terms of total power consumption and latency.

The remainder of this chapter is organized as follows. Section 3.2 surveys related literature. In Section 3.3, we introduce the proposed LDWuR and describe the strategies to explore the transmitter's direction. Then, we carry out the theoretical analysis of our system model and salient characteristics of MAC protocol.

3.2 Related Work

This section reviews communication technologies and design metrics used in the existing WuRs. WuRs have been proposed to improve PER and communication range for the Internet of Things (IoT) and wireless sensor network (WSN) applications. We review modulation, addressing, channel coding, and beamforming techniques.

3.2.1 Modulation and Addressing

Some studies have explored solutions to improve the PER of WuRs through different modulation techniques, such as pulse width modulation (PWM) [5] and pulse position modulation (PPM) [6], while other studies investigate address coding schemes [7,8]. However, they achieved limited modulation gain and high power consumption by the longer wake-up sequence of 31 to 48 bits to maintain a probability of detection of 90%.

3.2.2 Channel Coding

Some previous studies propose a variety of channel coding and forward error correction (FEC) methods to improve BER and PER for wireless and optical communication. Reference 9 used the Hadamard codes to improve the probability of bit error instead of block orthogonal code on a Rayleigh fading channel with a bandwidth constraint. The Hadamard code achieves better performance than the block orthogonal code does at the cost of increase in decoding complexity. Reference 10 proposed two novel classes of optical orthogonal code (OOC) for synchronous and asynchronous incoherent optical code division multiple access CDMA (OCDMA). The proposed OOC scales well with the number of users. Reference 11 proposed the error control coding using orthogonal codes. From the bandwidth efficiency analysis, a specific length of orthogonal codes provides an error correction scheme without an increase in bandwidth by partitioning data into blocks at the expense of complexity. Reference 12 proposed the orthogonal on–off keying (OOOK) for free-space laser communication, which is explained by the same author later. A block of data are mapped into a block of biorthogonal codes, which provides error correction capability through a correlation process. However, these designs do not consider the sensitivity, power consumption, and latency.

Reference 13 explored the improved interference rejection capability for channels with multiple access or multipath interference. It is clear from the results that the 64-ary orthogonal modulation shows better performance than binary phase shift keying (BPSK) does for the turbo product code of rate of 0.793 when compared to PER. However, the use of orthogonal modulation and turbo product code increases the implementation complexity. All of the abovementioned studies are about wireless communication in general, but are not specific to WuRs. The generality increases implementation complexity and may be impractical for WuRs.

3.2.3 Beamforming

Several studies have been proposed to optimize beamforming vectors and power configurations [14,15,16,17] for an improvement in information and energy transfer. However, these studies did not take WuRs into account. The most relevant work is Reference 18. The authors of this study had taken into account the advantage of multiple ultra-wideband impulse radios (UWB-IRs) statically mounted on the wall that form a circle and cover all semi-active sensor nodes inside. Multiple UWB-IRs can be used for beamforming to trigger the wake-up detector of semi-active sensor node. However, they do not consider a scenario for mobile UWB-IRs and sensor nodes. To our knowledge, till date no research exists that addresses WuRs and beamforming for wireless mobile networks.

3.3 Theoretical Analysis

We first present a system model analysis to determine PER, throughput, power consumption, and latency of the block orthogonal code. Second, we analyze two strategies to detect the transmitter's direction for mobile networks in the case study.

3.3.1 System Model

The system is modeled such that the transmitter transmits the input signal X of the block orthogonal codes through an additive white Gaussian noise (AWGN) channel and the receiver receives the signal Y. One purpose of this model is to find the symbol error probability of the block orthogonal codes. From symbol error probability, we can find the packet error probability of the block orthogonal codes (Section 3.3.2) and compare it with the existing schemes. The received signal Y through AWGN channel is expressed as follows:

$$Y(i) = \max\{\beta X(i) \otimes S_{\text{pattern}}^{\text{ideal}}(\tau - i)\} \tag{3.1}$$

where β is the attenuation from the AWGN channel with Gaussian distribution, $S_{\text{pattern}}^{\text{ideal}}$ is the ideal pattern of the block orthogonal codes' symbols for data bits zero and one, τ is the correlation time period, and $\otimes$ represents the correlation operation.

The block orthogonal codes use *chips per symbol* (N_{cps}) and *oversampling factor* (N_{os}) to suppress noise. The N_{os} reduces the noise variance (noise power) by a factor of $1/N_{\text{os}}$. The block orthogonal codes can be represented as the M-ary block orthogonal codes $O(N_{\text{cps}}, k)$ with diversity $L = N_{\text{cps}}/M$, where $M = 2^k$ and N_{cps} is the number of *chips per symbol.*

The *average signal-to-noise ratio per chip*, SNR_c, is given by

$$\text{SNR}_c = \frac{k}{L}\text{SNR}_b N_{\text{os}} = \frac{kMN_{\text{os}}\text{SNR}_b}{N_{\text{cps}}} \tag{3.2}$$

where SNR_b is the *signal-to-noise ratio per bit.* The *symbol* (code word) *error probability,* $p_{\text{es}}^{\text{orth}}$ for M-ary block orthogonal signaling is expressed as follows:

$$p_{\text{es}}^{\text{orth}} = 1 - \sum_{i=0}^{N_{\text{cc}}} \binom{N_{\text{cps}}}{i} p_{\text{ec}}^{\text{ook}^i} (1 - p_{\text{ec}}^{\text{ook}})^{(N_{\text{cps}}-i)} \tag{3.3}$$

where N_{cc} is the *number of chips that can be corrected* by M-ary block orthogonal signaling:

$$N_{cc} = \left\lfloor \frac{d_{\min} - 1}{2} \right\rfloor = \left\lfloor \frac{N_{cps}}{M} - \frac{1}{2} \right\rfloor \tag{3.4}$$

The *minimum Hamming distance*, $d_{\min}$, with diversity L, is expressed as follows:

$$d_{\min} = 2L = 2\frac{N_{cps}}{M} \tag{3.5}$$

The *chip error probability of coherent OOK modulation*, p_{ec}^{ook}, through a matched filter can be expressed as follows [19]:

$$p_{ec}^{ook} = Q(\sqrt{\text{SNR}_c}) \tag{3.6}$$

The value of SNR_c can be expressed as follows:

$$\text{SNR}_c = \frac{kMN_{os}\text{SNR}B_s}{N_{cps}R} \tag{3.7}$$

3.3.2 Symbol Error Rate (SER) and Packet Error Rate (PER)

To compare the block orthogonal codes with other schemes, we analyze the packet error probability from Equation 3.3. The *packet error probability of the block orthogonal codes*, p_{ep}^{orth}, can be expressed as follows:

$$p_{ep}^{orth} = 1 - (1 - p_{es}^{orth})^{L_{wurp}/k}, \tag{3.8}$$

where L_{wurp} denotes the *wake-up packet length* in bits. To explore the performance of varying modulation and encoding schemes, we analyze both SER and PER. For PWM, we use the BER derived in Reference 5. In terms of other encoding schemes, we apply the system model used in Reference 20 to analyze SER and PER. The relationship between the *symbol error probability*, p_{es}^{ook}, and *bit error probability*, p_{eb}^{ook}, of binary coherent OOK modulation through a matched filter can be expressed as follows:

$$p_{es}^{ook} = p_{eb}^{ook} = Q(\sqrt{\text{SNR}_b}) = Q\left(\sqrt{\frac{\text{SNR}B_s}{R}}\right) \tag{3.9}$$

The *packet error probability* of OOK modulation *with 8B10B coding*, p_{ep}^{8B10B}, can be calculated as follows:

$$p_{ep}^{8B10B} = 1 - (1 - p_{es}^{ook})^{1.25L_{wurp}} \tag{3.10}$$

The *packet error probability* of OOK modulation *with Hamming* (7, 4) *code*, p_{ep}^{h74}, that can detect and correct one error, can be represented as follows:

$$p_{ep}^{h74} = 1 - ((1 - p_{es}^{ook})^7 + 7 p_{es}^{ook} (1 - p_{es}^{ook})^6)^{L_{wurp}/4} \tag{3.11}$$

3.3.3 Throughput

For a performance comparison of the block orthogonal codes with other modulation schemes and encodings, we now analyze the respective throughput values. We assume the system bandwidth as the noise bandwidth B_s, the wake-up packet length is L_{wurp} bits, the transmitter sends L_{packet} wake-up packets, and the packet reception rate (PRR) of the WuR can be calculated as follows:

$$\text{PRR}_{wur} = 1 - p_{ep} \tag{3.12}$$

where PRR_{wur} is the WuR's PRR and p_{ep} is the WuR's packet error probability with varying modulation types and encodings. The throughput S_{wur} can be expressed as follows:

$$S_{wur} = \frac{\text{PRR}_{wur} B_s}{L_{packet} L_{wurp}} \tag{3.13}$$

3.3.4 Power Consumption

Our approach to evaluate power consumption, for the purpose of comparing WuRs, is to analyze how effectively they implement the given wake-up protocol [21] in terms of total power consumption and latency. The wake-up protocol as illustrated in Figure 3.1 shows that the PER is an important metric. We find the optimal preamble time T_{pre}^{opt} to achieve the minimum power consumption through a single-hop network with neighbors.

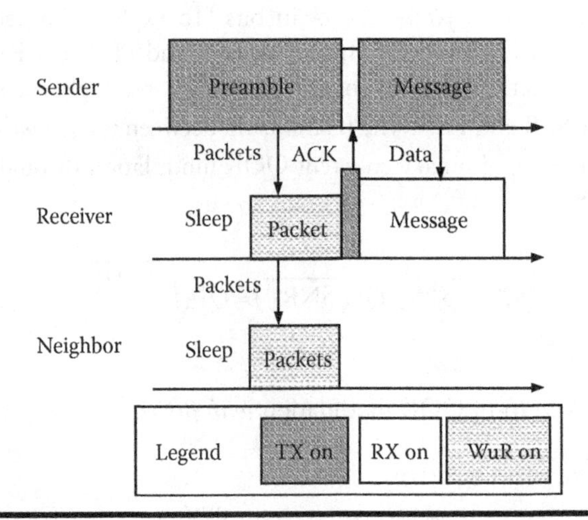

Figure 3.1 Wake-up protocol.

We assume the number of WuRs N_{wur} to be the *number of neighbor nodes* N_{n}. The optimal preamble time duration, $T_{\mathrm{pre}}^{\mathrm{opt}}$, is expressed as follows [21]:

$$T_{\mathrm{pre}}^{\mathrm{opt}} = \sqrt{\frac{P_{\mathrm{wur}} T_{\mathrm{wurp}} N_{\mathrm{wur}}}{\left((V_{\mathrm{ccd}} I_{\mathrm{td}}) + \frac{N_{\mathrm{wur}} P_{\mathrm{wur}}}{2}\right) \mathrm{PR} E[N_{\mathrm{tx}}]}} \tag{3.14}$$

where $T_{\mathrm{pre}}^{\mathrm{opt}}$ includes consecutive wake-up packets to the receiver's WuR, P_{wur} is the WuR's power consumption, $V_{\mathrm{ccd}}\ I_{\mathrm{td}}$ is the data radio's transmission power, PR is the packet rate, $E[N_{\mathrm{tx}}]$ is the expected transmission time, and T_{wurp} is the wake-up packet's time duration.

The total power consumption, P_{t}, can be calculated as follows [21]:

$$P_{\mathrm{t}} = (q_1 P_1 + q_2 P_2 + q_3 P_3) E[N_{\mathrm{tx}}] + P_{\mathrm{idle}} \tag{3.15}$$

where q_1, q_2, and q_3 represent the probabilities of three cases: WuR with packet error, data radio with packet error, and WuR and data radio both with all correct packets, respectively; P_1, P_2, and P_3 represent the respective power consumption; and P_{idle} is the *idle listening power consumption.*

Equation 3.15 shows that the high PER_{wur} will have more *expected transmission times* $E[N_{\mathrm{tx}}]$; then, P_t will increase. Higher-PER WuRs will increase power consumption significantly as they have more $E[N_{\mathrm{tx}}]$ and wake-up packets' *retransmission* $Re_{\mathrm{tx}}^{\mathrm{wur}}$. Moreover, they wake up the respective higher-power data radios.

3.3.5 Latency

To compare the latencies of the existing WuRs, we implement the given wake-up protocol to determine the total latency instead of component's propagation delay time in the WuR circuit. Therefore, the latency emphasizes the importance of the metric PER again. *Latency* is defined as the time duration between transmission and successful reception of the WuR packet. The optimal preamble time period $T_{\mathrm{pre}}^{\mathrm{opt}}$ for our wake-up protocol [21] is designed to achieve the lowest power dissipation for the respective existing WuRs. Moreover, $T_{\mathrm{pre}}^{\mathrm{opt}}$ should include at least eight symbols to represent the wake-up address of 16 bits for addressing 65,536 nodes.

We assume the latencies are the same when the false and successful wake-up signals occur. The WuR's *total latency*, T_{t}, can be expressed as follows [21]:

$$T_{\mathrm{t}} = E[N_{\mathrm{tx}}](T_{\mathrm{pre}}^{\mathrm{opt}} + T_{\mathrm{msg}}) \tag{3.16}$$

where T_{msg} is the *message time duration*. When the transmitter sends a message to the receiver's data radio, the receiver spends T_{msg} for receiving the message from the transmitter.

For low-power WuRs with the higher PER, they have larger $E[N_{\mathrm{tx}}]$ as they require more $Re_{\mathrm{tx}}^{\mathrm{wur}}$ of wake-up packets. Therefore, they increase the total latency significantly.

3.3.6 Case Study: Transmitter Localization by LDWuR for Mobile Network

We propose two strategies to detect the transmitter's direction for mobile networks. One strategy is static antennas. The other strategy is dynamic antennas.

3.3.6.1 Static Antennas

The receiver adopts the number of LDWuRs in different directions. The receiver polls each LDWuR for each direction to explore the transmitter's direction. The average total power consumption is the same as WuRs Equation 3.15 with larger cover range. The average total power consumption, P_t^{sta}, is expressed as follows:

$$P_{\text{t}}^{\text{sta}} = \frac{\sum_{i=1}^{N_{\text{dir}}} P_t T_{\text{scan}}}{N_{\text{dir}} T_{\text{scan}}} = P_{\text{t}} \quad (3.17)$$

where T_{scan} represents *scan time for each direction* and N_{dir} is the *number of different directions* that means the *number of LDWuRs*. For example, the receiver adopts four LDWuRs in different directions, such as east, south, west, and north. Equation 3.17 shows that LDWuRs consume the same power consumption to explore the transmitter's direction. Figure 3.2 shows that one of four LDWuRs scans the north direction with 90°.

The total latency is N_{dir} times the total latency of WuRs in Equation 3.16 as the receiver spends more time on scan of each direction. The total latency is expressed as follows:

$$T_{\text{t}}^{\text{sta}} = N_{\text{dir}} T_{\text{t}} \quad (3.18)$$

3.3.6.2 Dynamic Antennas

In this strategy, the receiver adopts multiple LDWuRs. Each LDWuR consists of a WuR with a built-in omnidirectional antenna. Multiple LDWuRs cooperate to steer a beam in different directions in turn. Therefore, the receiver can use a beam to detect the transmitter's direction.

The average total power consumption is the same as WuRs in Equation 3.15 with larger cover range. The average total power consumption, P_t^{dyn}, is expressed as follows:

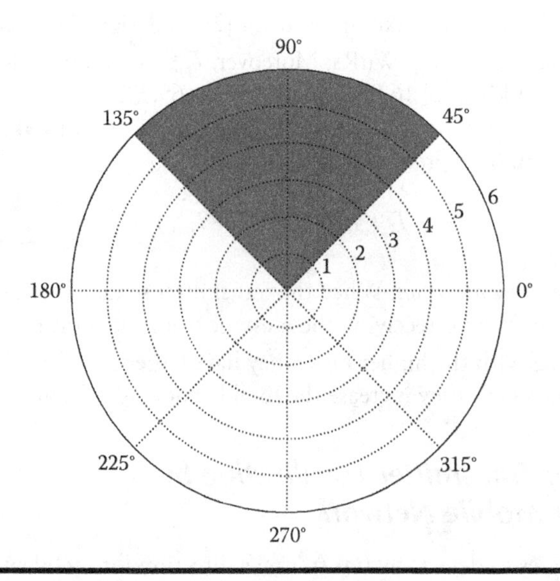

Figure 3.2 One of four LDWuRs scans the north direction.

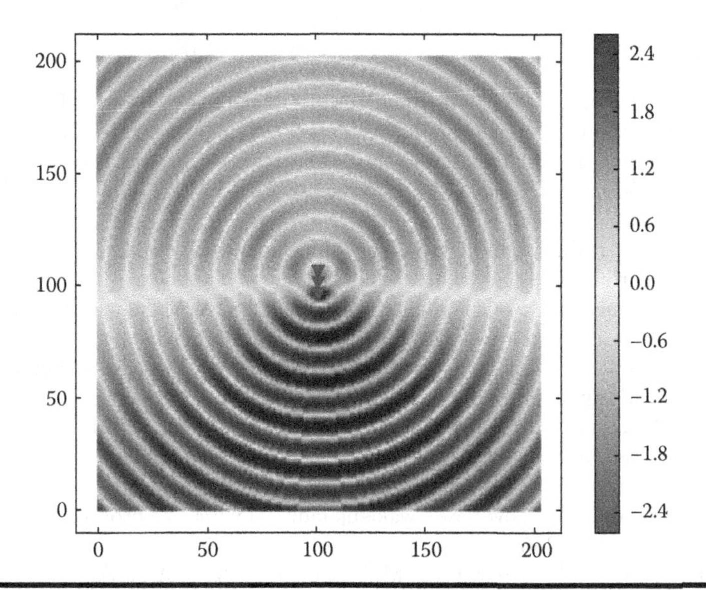

Figure 3.3 Three LDWuRs cooperate to steer a beam in the south direction.

$$P_t^{dyn} = \frac{\sum_{i=1}^{N_{dir}^{beam}} N_{ant}^{omn} (1/N_{ant}^{omn}) P_t T_{scan}^{beam}}{N_{dir}^{beam} T_{scan}^{beam}} = P_t \tag{3.19}$$

where T_{scan}^{beam} represents *scan time for each direction of beam*, N_{ant}^{omn} is the *number of omnidirectional antennas*, and N_{dir}^{beam} is the *number of different directions of beams*. For example, the receiver adopts three LDWuRs to cooperate to form beams in different directions with a certain degree of cover range. Equation 3.19 shows that the receiver adopt N_{ant}^{omn} LDWuRs with N_{ant}^{omn} antennas of $1/N_{ant}^{omn}$ times power consumption to explore the transmitter's direction. Figure 3.3 shows that three LDWuRs cooperate to steer a beam in the south direction with the spacing of 5 units, related phase delay of 7 units, and beam pattern size of 90° at a signal strength of 70% maximum signal strength.

The total latency is N_{dir}^{beam} times the total latency of WuRs in Equation 3.16 as the receiver spends more time on scan of each direction. The total latency is expressed as follows:

$$T_t^{dyn} = N_{dir}^{beam} T_t \tag{3.20}$$

3.4 Conclusion

We proposed two strategies to search for the transmitter's direction to concentrate the receiver's resources on the right location. This means that the proposed LDWuR can improve the receiver's cover range and power saving. We designed a system model to analyze the power consumption and latency for our LDWuR. From the theoretical analysis, we observed that using the same power consumption as in previous WuRs can detect the transmitter's position and enhance the nodes' communication range.

References

1. White, D. L., S. Esswein, J. O. Hallstrom, F. Ali, S. Parab, G. Eidson, J. Gemmill, and C. Post, 2010. The intelligent river: Implementation of sensor web enablement technologies across three tiers of system architecture: Fabric, middleware, and application. In *Proceedings of the 2010 International Symposium on Collaborative Technologies and Systems (CTS)*, Chicago, Illinois, USA, May 17–21, 2010, ACM, pp. 340–348.
2. IBM. *Smarter Water Management*. Available online: http://www.ibm.com/smarterplanet/au/en/water_management/visions (accessed on 23 July 2015).
3. Jurdak, R., P. Sommer, B. Kusy, N. Kottege, C. Crossman, A. Mckeown, and D. Westcott, 2013. Camazotz: Multimodal activity-based GPS sampling. In *Proceedings of the 12th International Conference on Information Processing in Sensor Networks, IPSN '13*, Philadelphia, PA, USA, April 08–11, 2013, ACM, pp. 67–78.
4. Bhatti, N. A., A. A. Syed, and M. H. Alizai, 2014. Sensors with lasers: Building a WSN power grid. In *Proceedings of the 13th International Symposium on Information Processing in Sensor Networks, IPSN '14*, Berlin Germany, April 15–17, 2014, pp. 261–272. IEEE Press.
5. Le-Huy, P. and S. Roy, 2010. Low-power wake-up radio for wireless sensor networks. *Mobile Networks and Applications*, 15, 226–236.
6. Drago, S., F. Sebastiano, L. J. Breems, D. M. W. Leenaerts, K. A. A. Makinwa, and B. Nauta, 2009. Impulse based scheme for crystal-less ULP radios. *IEEE Transactions on Circuits and Systems I: Regular papers*, 56(5), 1041–1052.
7. Ansari, J., D. Pankin, and P. Mähönen, 2008. Radio-triggered wake-ups with addressing capabilities for extremely low power sensor network applications. In *Proceedings of IEEE 19th International Symposium on Personal, Indoor and Mobile Radio Communications (PIMRC)*, Cannes, French Riviera, France, September 15–18, 2008, pp. 1–5.
8. Pletcher, N. M., S. Gambini, and J. M. Rabaey, 2009. A 52 μW wake-up receiver with −72 dBm sensitivity using an uncertain-IF architecture. *IEEE Journal of Solid-State Circuits*, 44(1), 269–280.
9. Proakis, J. G. and I. Rahman, 1979. Performance of concatenated dual-k codes on a rayleigh fading channel with a bandwidth constraint. *IEEE Transactions on Communications*, 27(5), 801–806.
10. Djordjevic, I. B. and B. Vasic, 2004. Combinatorial constructions of optical orthogonal codes for OCDMA systems. *IEEE Communications Letters*, 8(6), 391–393.
11. Faruque, S. 2003. Investigation of error control properties of orthogonal codes. In *Proceedings of IEEE Military Communications Conference (MILCOM)*, Boston, MA, USA, October 13–16, 2003, Vol. 2, pp. 791–795.
12. Faruque, S. 2011. Free space laser communications based on Orthogonal On-Off Keying (O3K). In *Proceedings of the 2011 IEEE International Conference on Electro/Information Technology (EIT)*, Mankato, MN, USA, May 15–17, 2011, pp. 1–4.
13. Pursley, M. B. and T. C. Royster, 2006. High-rate direct-sequence spread spectrum with error-control coding. *IEEE Transactions on Communications*, 54(9), 1693–1702.
14. Morsi, R., D. S. Michalopoulos, and R. Schober, 2014. Multi-user scheduling schemes for simultaneous wireless information and power transfer. In *Proceedings of IEEE International Conference on Communications, ICC 2014*, Sydney, Australia, June 10–14, 2014, pp. 4994–4999.
15. Ng, D. W. K., R. Schober, and H. Alnuweiri, 2014. Secure layered transmission in multicast systems with wireless information and power transfer. In *Proceedings of IEEE International Conference on Communications, ICC 2014*, Sydney, Australia, June 10–14, pp. 5389–5395.
16. Shih, W-C., P. H. Chou, and W-T. Chen, 2014. Empirical validation of Energy-Neutral operation on wearable devices by MISO beamforming of IEEE 802.11ac. In *Proceedings of the 2nd International Workshop on Energy Neutral Sensing Systems (ENSsys), ENSsys '14*, Memphis, TN, USA, November 6, 2014, pp. 49–54. ACM.
17. Timotheou, S., I. Krikidis, G. Zheng, and B. Ottersten, 2014. Beamforming for MISO interference channels with QoS and RF energy transfer. *IEEE Transactions on Wireless Communications*, 13(5), 2646–2658.

18. Trösch, F. and A. Wittneben, 2006. A simple ultra-wideband wake-up scheme for semi-active sensor nodes. In *Ultra-Wideband, The 2006 IEEE 2006 International Conference on*, Waltham, MA, September 24–27, 2006, pp. 663–668.
19. Leon W. C. 1992. *Digital and Analog Communication Systems*, Upper Saddle River, NJ: Prentice Hall PTR.
20. Zuniga, M. and B. Krishnamachari, 2004. Analyzing the transitional region in low power wireless links. In *Proceedings of the First Annual IEEE Communications Society Conference on Sensor and Ad Hoc Communications and Networks. (SECON)*, Santa Clara, California, October 4–7, 2004, pp. 517–526.
21. Shih, W-C., R. Jurdak, B-H. Lee, and D. Abbott. 2011. High sensitivity wake-up radio using spreading codes: Design, evaluation and applications. *EURASIP Journal on Wireless Communications and Networking (EURASIP JWCN)*, 2011:26.

Chapter 4

Medium Access Control for Wireless Networks with Directional Antennas

T. Owens Walker III, Murali Tummala, and John McEachen

Contents

Abstract

The use of directional antennas in wireless networks offers a wide variety of potential gains, including increased capacity and range, improved reliability and energy efficiency, and enhanced security. To realize these potential gains, the medium access layer must be designed to effectively utilize the unique capabilities of directional antennas. In this chapter, we review the fundamentals of medium access for wireless networks, highlight the opportunities presented by directional antennas as they apply to medium access, and discuss the medium access control (MAC) layer design issues relevant to effective utilization of directional antennas. We close with a sampling of medium access control protocols that have been proposed for directional antennas, discussing both their operation and their performance.

4.1 Introduction

The wireless medium is a shared, broadcast medium and requires a mechanism to mediate access. A wireless channel is time-varying and asymmetric due to multipath propagation and fading. Accordingly, the wireless medium can be thought of as a set of error-prone half-duplex links. A wireless network typically has no clear network boundaries and the topology is often dynamic due to node mobility, node state, and channel state. In this section, we discuss functions of the MAC layer, contention versus contention-free medium access, and power management at the MAC layer. We close this section by presenting the highlights of the ubiquitous IEEE 802.11 (Wi-Fi) standard upon which much of the directional antenna medium access work, to date, has been based.

4.1.1 Functions of the Medium Access Control Layer

The MAC layer is a sublayer of the data link layer defined by the Transmission Control Protocol/ Internet Protocol (TCP/IP) model and further refined by the IEEE 802 Standard [1] as in Figure 4.1. The goal of the MAC layer is to effectively and efficiently manage access to the transmission medium. Examples of IEEE 802 protocols that address the MAC layer include IEEE 802.3 (Ethernet), IEEE 802.11 (WiFi), and IEEE 802.15.4 (Bluetooth). Common performance metrics for the MAC layer include throughput, delay, and energy consumption.

4.1.2 Contention versus Contention-Free

Wireless medium access solutions generally fall into two categories: contention-based and scheduled (contention-free). It has been well established that the collision-free approach of scheduled schemes [2] provides high throughput in high demand scenarios at the expense of overhead and packet delay.

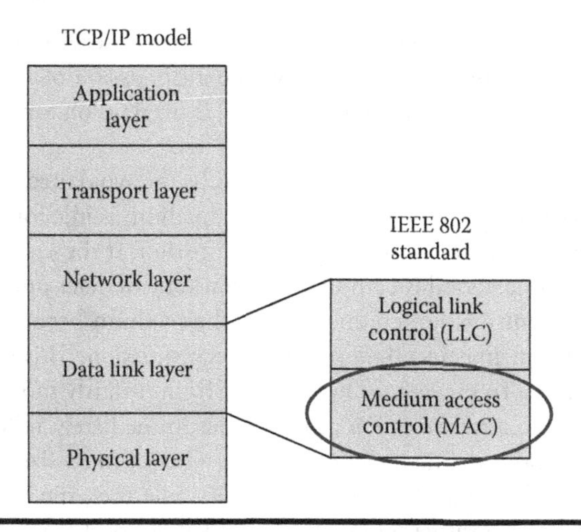

Figure 4.1 Medium access layer as defined by the IEEE 802 standard.

In comparison, contention-based approaches [3] provide low delay times at low-to-moderate network loads, but performance begins to degrade rapidly as the network becomes saturated. Hybrid solutions have also been proposed to combine the benefits of both approaches in response to changing network load. These hybrid solutions include the multiple operating modes of well-established wireless standards such as IEEE 802.11 (Wi-Fi) [1] and IEEE 802.15.4 (Bluetooth) [4].

4.1.3 Power Management

Power consumption is a significant challenge in untethered wireless networks due to the limited battery power available. Ye et al. [5] and Demirkol et al. [6] identify five major sources of energy waste in wireless communications. Packets must be discarded and retransmitted when they experience collisions. The *retransmissions* result in an increase in both power consumption and latency. *Overhearing* occurs when nodes receive and process packets for which they are not the intended receiver. *Control packet overhead* stems from the use of dedicated control packets to coordinate transmissions. *Idle listening* occurs when nodes listen for packets while the channel is idle. Finally, *overemitting* occurs when a message is transmitted to a destination that is not ready to receive it.

Power management approaches can be divided into two categories. *Power save* techniques attempt to minimize the power loss due to the communication issues described in the previous paragraph. These approaches typically involve powering down the transceiver in a "sleep" state. *Power control* techniques, in contrast, attempt to minimize power consumption by directly addressing the transmit power used in each transmission. Metrics commonly used for power consumption include total energy consumption, per packet energy consumption, node lifetime, and network lifetime.

4.1.4 IEEE 802.11 (Wi-Fi)

The medium access specification in the IEEE 802.11 standard [1] comprises three components: the distributed coordination function (DCF), the point coordination function (PCF), and the hybrid coordination function (HCF). DCF is a fully distributed, contention-based access scheme while PCF is a centralized, contention-free approach that is laid on top of DCF. HCF combines the functionality of both DCF and PCF to support quality-of-service or QoS-capable stations. In this subsection, we will focus on DCF, but include a discussion of PCF and HCF to understand how the contention-free mechanism operates and is overlaid on top of the contention-based foundation.

DCF utilizes carrier sense multiple access with collision avoidance (CSMA/CA). A station wishing to transmit a packet senses the medium. If the medium is idle for the duration of a specified minimum time period, the station will begin transmission. If the medium is busy or becomes busy before the specified time period is complete, the station will defer until the end of the current transmission. When the transmission completes and the medium becomes idle, the station will back off a random length of time and then reattempt transmission. This can be seen as a version of nonpersistent CSMA [3]. To prevent a single station from unfairly monopolizing the medium, a station also defers prior to a transmission attempt that immediately follows a successful transmission. The back-off period is defined by a random number of fixed-size slots and the associated counter is decremented only while the medium remains idle (i.e., the countdown is suspended while the medium is busy). A positive acknowledgment (ACK) packet is included to confirm successful packet delivery. In an optional extension, Request-To-Send (RTS)/Clear-to-Send (CTS) control packets [1] may be used to acquire the channel and reduce the probability of data packet collisions. This RTS/CTS mechanism is discussed in more detail in Section 4.3.1.

The length of the random back-off time in slots is determined by the selection of a random integer from the uniformly distributed interval [0,CW] where CW is known as the contention window and its size is bounded by the parameters CW_{min} and CW_{max}. Initially, CW is set to CW_{min}. With each subsequent unsuccessful transmission attempt, the size of the contention window exponentially increases (i.e., doubles), up to the maximum value of CW_{max}. At the completion of a successful transmission, the size of the contention window is reset to CW_{min}.

Prioritized access to the medium is provided by a mechanism called the interframe space (IFS). The IFS is the period that the medium must be idle before the station can either transmit or begin/resume the back-off process. Thus, a station with a shorter IFS can seize the medium before a station with a longer IFS, and will have priority over that station. An overview of the DCF access method can be seen in Figure 4.2. When utilizing the DCF contention mode, nodes contend for

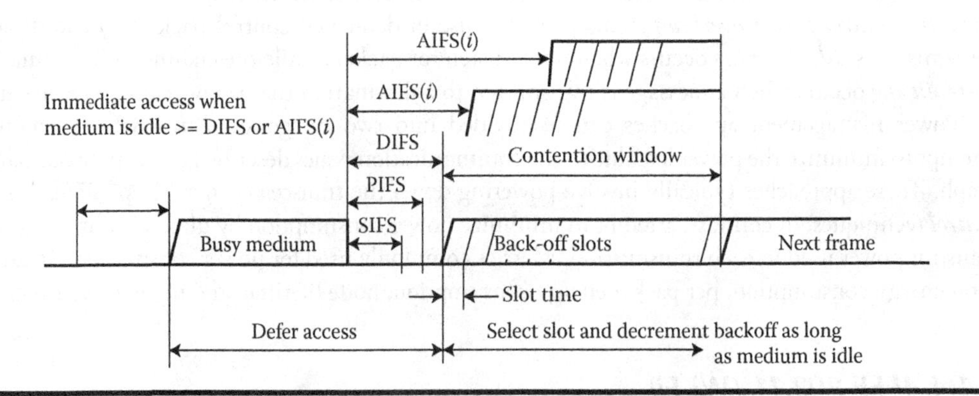

Figure 4.2 IEEE 802.11 DCF access method. (Adapted from IEEE Standard for Information Technology-Telecommunications and Information Exchange between Systems-Local and Metropolitan Area Networks-Specific Requirements—Part 11: Wireless LAN Medium Access Control (MAC) and Physical Layer (PHY) Specifications, IEEE Standard 802.11-2012 (Revision of IEEE Standard 802.11-1999), 2012.)

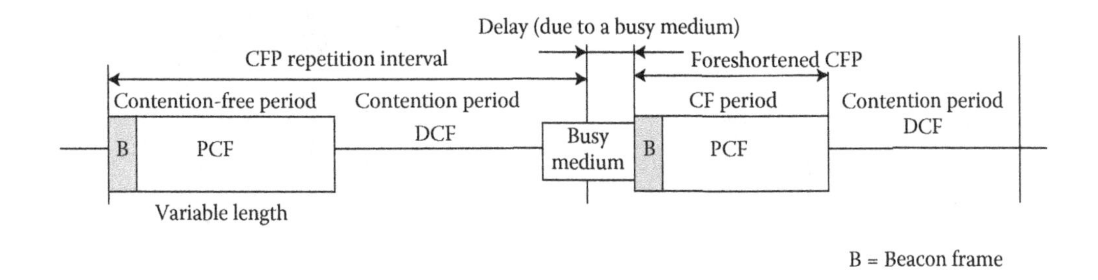

Figure 4.3 IEEE 802.11 Superframe. (Adapted from IEEE Standard for Information Technology-Telecommunications and Information Exchange between Systems-Local and Metropolitan Area Networks-Specific Requirements—Part 11: Wireless LAN Medium Access Control (MAC) and Physical Layer (PHY) Specifications, IEEE Standard 802.11-2012 (Revision of IEEE Standard 802.11-1999), 2012.)

the medium using a Distributed IFS (DIFS). Once a station has captured the medium, it uses Short IFS (SIFS) delays to maintain the medium through the complete data exchange. Enhanced QoS was provided in the IEEE 802.11 standard by the introduction of the enhanced distributed channel access (EDCA) and its use of additional (varied length) IFS parameters (AIFS[i] in Figure 4.2).

PCF provides centralized, contention-free access to the medium through the use of a polling mechanism. The polling master is known as the point coordinator and resides at the access point. PCF is overlaid on top of DCF and priority access to the medium is provided to the point coordinator by the PCF IFS (PIFS) mechanism. Note from Figure 4.2 that PIFS is shorter than DIFS and thus provides PCF priority over DCF. In HCF, a contention-free period (CFP) is combined with a contention period (CP) in a superframe, as shown in Figure 4.3. HCCA (HCF controlled channel access) is an enhanced version of HCF in which the individual transmission time of the polled stations is capped to improve fairness in the PCF scheme.

IEEE 802.11 includes a power management mechanism to power down a node into a reduced power sleep state called the power save (PS) mode. Specifically, the transceiver is powered down in the PS mode and a node can neither transmit nor receive. This PS mode is implemented as follows. In the infrastructure mode, the access point (AP) is always on and coordinates traffic for the mobile host in the PS mode. The mobile host informs the AP when it is powering down and the AP then buffers packets for the host in PS mode.

4.2 Directional Antennas: Opportunities at the Medium Access Layer

Directional antennas have the potential to provide a number of well-understood benefits at the medium access layer. The most commonly cited are the potential for increased capacity through spatial reuse and increased transmission range. The first is a product of the reduced interference "footprint" when utilizing directional antennas, and the second is the result of the increased signal-to-noise ratio available due to the gain provided by the directional antennas. In this section, we discuss these benefits as well as a number of related opportunities provided by directional antennas.

First, though, a word on the assumptions and conventions we use in this chapter. Unless otherwise specified, we assume a spherical transmission model and do not address the presence of side and rear lobes in the directional model. The notable exception to the latter assumption is when we

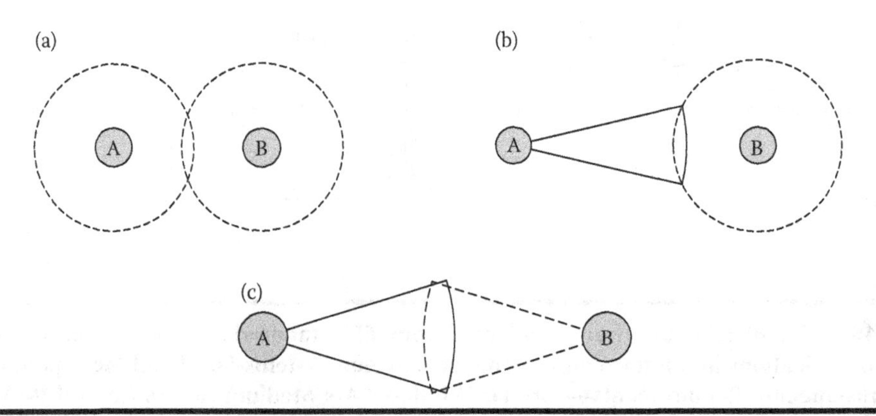

Figure 4.4 Connectivity convention in this chapter. (a) Omni–omni (OO) connectivity. (b) Directional–omni (DO) connectivity. (c) Directional–directional (DD) connectivity.

explore the minor lobe problem in Section 4.3.5. We use the following convention to reflect node connectivity (or, similarly, node interference) in our accompanying figures: two nodes are considered to be connected (i.e., within transmission range of each other), if their transmission diagrams overlap. In Figure 4.4a, nodes A and B are both in omnidirectional mode and are connected. In Figure 4.4b, node A is in directional mode and node B is in omnidirectional mode. Again, they are both connected. Finally, in Figure 4.4c, nodes A and B are both in directional mode and are connected. This is slightly different from the convention typically used in the literature and has the advantage that it can clearly represent the three connectivity cases seen in wireless networks using directional antennas.

4.2.1 Increased Capacity through Spatial Reuse

The potential increase in capacity is due to the reduced "footprint" of directional antennas. For example, consider the topology of Figure 4.5 in which three data transfers are pending. In Figure 4.5a, the data transfer from node A to node B blocks those of C to D and E to F because node A's transmission is omnidirectional and will be detected by nodes D and F, which are listening

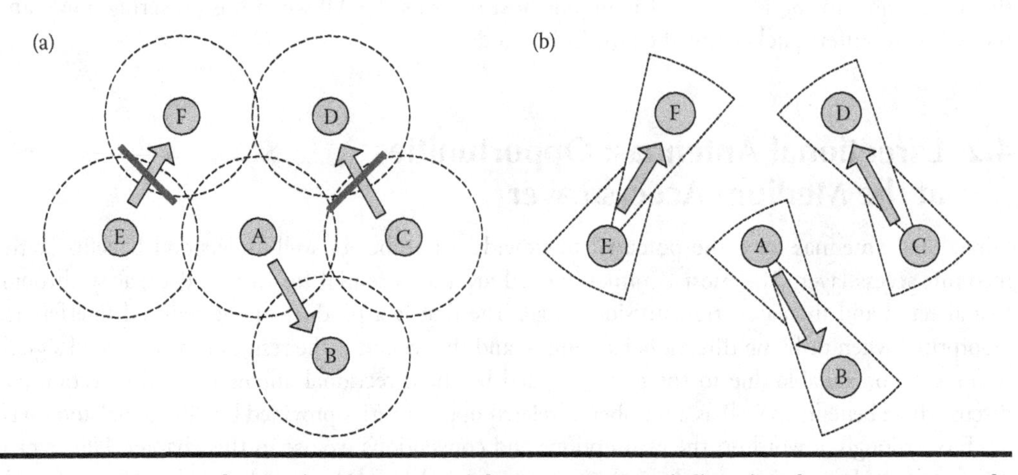

Figure 4.5 Increased capacity due to spatial reuse. (a) Omni-directional antennas case. (b) Directional antennas case (capacity is increased).

omnidirectionally. This would cause a collision at nodes D and F. If, instead, we use directional antennas, all three transmissions can occur simultaneously without interfering with each other, as in Figure 4.5b. While admittedly it is topology-dependent (which we shall see later), this simple example presents a potential three-fold increase in capacity by using directional antennas.

4.2.2 Transmission Range Extension

Transmission range extension can be developed from the combination of the Friis transmission equation and the log distance path loss model in which the propagation loss of a signal as a function of distance, $L(d)$, is given by

$$L(d) \equiv \frac{P_T}{P_R} = \left(\frac{1}{G_T G_R}\right)\left(\frac{4\pi f d_0}{c}\right)^2 \left(\frac{d}{d_0}\right)^n \tag{4.1}$$

where

P_T is the transmitted power
P_R is the received power
G_T is the gain of the transmitting antenna
G_R is the gain of the receiving antenna
d is the transmission range
d_0 is the reference distance
f is the frequency
c is the speed of light
n is the path loss exponent ($n = 2$ for free space, typically 4–6 in buildings with obstructions).

Antenna gain is one for an isotropic (omnidirectional) antenna and greater than one for a directional antenna that concentrates the transmitted power in a given direction. Thus, antenna gain is increased by increasing the directivity of the antenna. Rearranging Equation 4.1, we can clearly see that for fixed power, transmission range is a function of antenna gain as in

$$d = \left\{ d_0 \left[\frac{P_T}{P_R} \left(\frac{c}{4\pi f d_0} \right)^2 \right]^{1/n} \right\} (G_T G_R)^{1/n}. \tag{4.2}$$

Thus, an increase in the gain (through directivity) of either the transmitter, the receiver, or both, will result in an increase in transmission range.

4.2.3 Improved Energy Efficiency

In the section above, we discussed the potential transmission range increase associated with directional antennas. If, instead, we fix the transmission range (as well as the received power required) and rearrange Equation 4.1 as in

$$P_T = \left[P_R \left(\frac{4\pi f d_0}{c} \right)^2 \left(\frac{d}{d_0} \right)^n \right] \left(\frac{1}{G_T G_R} \right) \tag{4.3}$$

we can see that the required transmission power can be reduced with an increase in the gain (through directivity) of either the transmitter, the receiver, or both. Thus, with directional antennas, we can use less transmission power to achieve the same transmission range (when compared to the omnidirectional case).

4.2.4 Enhanced Reliability

The reliability of a channel can be defined by the connectivity of the channel, which is a function of the signal-to-noise ratio (SNR) at the receiver. As the SNR falls off, the bit error rate (BER) increases, and, thus, the channel quality is compromised. The SNR is given by

$$SNR \text{ at the receiver} = \frac{\text{Signal power at the receiver}}{\text{Noise power at the receiver}} = \frac{P_R}{P_N}. \tag{4.4}$$

We can again rearrange Equation 4.1 and substitute it into Equation 4.4, to arrive at

$$SNR = \left[\frac{P_T}{P_N}\left(\frac{c}{4\pi f d_0}\right)^2 \left(\frac{d_0}{d}\right)^n\right](G_T G_R). \tag{4.5}$$

From this, we see that SNR (and thus reliability) can be increased by increasing the gain (through directivity) of either the transmitter, the receiver, or both.

4.2.5 Increased Capacity Due to Higher Data Rates

The increase in SNR available with directional antennas can also lead to an improvement in achievable data rates. This can be seen as a direct result of Shannon's Capacity Theorem [7] in which the maximum achievable data rate (C in bps) is given by

$$C = B\log_2(SNR + 1) \tag{4.6}$$

where B is the available bandwidth. Substituting Equation 4.5 into Equation 4.6, we can see that capacity is a function antenna gain (directivity) and an increase in gain will result in an increase in capacity as in

$$C = B\log_2\left(\left[\frac{P_T}{P_N}\left(\frac{c}{4\pi f d_0}\right)^2 \left(\frac{d_0}{d}\right)^n\right](G_T G_R) + 1\right). \tag{4.7}$$

4.2.6 Enhanced Security

The use of directional antennas has been shown to potentially improve the security of a wireless network in the presence of both jamming [8] and eavesdropping [9]. Jamming appears as interference at the intended receiver and manifests itself in the same manner as the noise power term of Equation 4.5. Thus, the presence of jamming represents a decrease in SNR which, from Equation 4.5 can potentially be offset by an increase in the gain of the transmitting and/or receiving antennas. The receiver can also steer a null towards the jammer (if its direction is known). This has the effect of specifically reducing the gain in the direction of the jammer. In both cases, the effectiveness of directed antennas in the presence of jamming depends on the relative positions of the transmitter and the jammer. If they are both in the same direction from the receiver, the gain will affect both the jamming signal and desired signal similarly.

The probability of eavesdropping depends on both the density of eavesdropping nodes and the exposure region of the transmitting node [9], where the exposure region can be thought of as the transmission "footprint" in Figure 4.5. Clearly, a directional antenna provides the opportunity to steer a transmission exposure region away from an area of high eavesdropping node density, but the actual size of the exposure region may not be reduced due to the increased range provided by the directional antenna [9]. The combination of power control (and hence reduced transmission range) combined with directional antennas is a promising approach to reducing the probability of eavesdropping in a wireless networks.

4.3 Design Issues of MAC Using Directional Antennas

In this section, we examine the design issues associated with a wireless medium access protocol that utilizes directional antennas. We begin with the two most commonly addressed challenges in wireless medium access with directional antennas: (1) the hidden node problem and (2) deafness. We then examine a series of additional design issues and conclude with a discussion of relevant power management issues. In all cases, we highlight the fundamental issues, provide illustrative examples, and explore some of the proposed solutions in the literature.

4.3.1 Hidden Node Problem (Hidden Terminals)

The hidden node problem is well understood in omnidirectional wireless networks [10]. It occurs when a node cannot "hear" an ongoing conversation and therefore transmits, causing a collision. In Figure 4.6, node A is transmitting data to node B. Node C has a packet transmission for node B as well and senses the medium prior to transmission. Node C, which is out of the range of node A, finds the medium free and transmits causing a packet collision at node C.

The fundamental issue here is that the medium sensing occurs around the transmitting node rather than around the receiving node (which is where the collision occurs). To rectify this, a Request-to-Send (RTS) and Clear-to-Send (CTS) handshake [11] has been proposed and widely adopted (including in the IEEE 802.11 standard). In this scheme, an intended transmitter sends

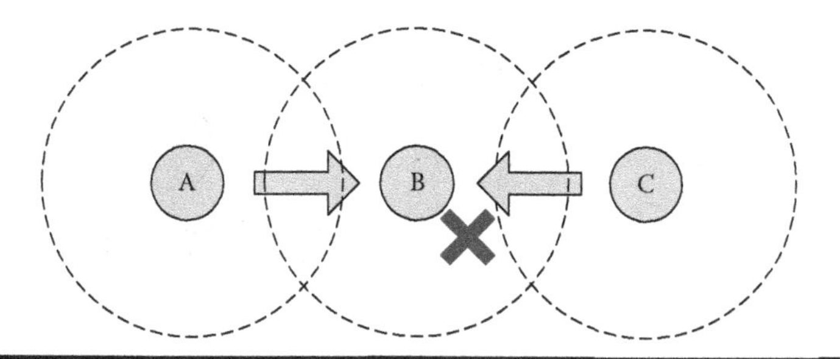

Figure 4.6 Hidden node problem in omnidirectional antennas.

a small RTS control packet to its intended destination. The destination replies with a small CTS control packet. Each control packet includes a field containing the anticipated length of the subsequent transmission. This field is known as the Network Allocation Vector (NAV) and is used by nodes that overhear the RTS/CTS exchange to defer until the transmission (RTS/CTS/Data/ACK) is complete. The NAV is typically stored in a countdown timer that maintains the state of the medium. The NAV is updated based on message duration fields contained in overheard RTS and CTS packets and counts down at a uniform rate. This technique is often referred to as "virtual carrier sensing" since it is performed in conjunction with physical carrier sensing. The medium is considered idle when no transmission is detected by the physical carrier sensing mechanism and the NAV is zero. The RTS transmission has the effect of clearing the medium around the transmitter, while the CTS transmission has the effect of clearing the medium around the receiver. In the example of Figure 4.6, node C would not be able to hear the RTS transmission from node A, but would hear the corresponding CTS transmission from B, and, thus, defer.

This virtual carrier sensing mechanism was modified in Basic Directional MAC (Basic DMAC) [12] to support directional antennas (and, similarly, in Reference 13). Idle nodes monitor the medium in omnidirectional mode, but RTS, CTS, Data, and ACK transmissions are all directional. A Directional NAV (DNAV) table maintains the virtual carrier sensing state of the medium for every available direction of arrival. Nodes overhearing RTS or CTS transmissions in a given directional of arrival update the corresponding entry in their DNAV table. Prior to an RTS transmission, nodes perform both physical and virtual carrier sensing in the desired direction of transmission. When a node is backing off (either before its initial RTS transmission or before subsequent RTS retransmissions, as in IEEE 802.11), the node continues to sense the medium in directional mode (in the desired direction of transmission). In favorable topologies, Basic DMAC has been shown to provide more than twice the throughput of IEEE 802.11. Basic DMAC assumes that "transceiver profiles" are available that provide communication parameters (e.g., direction of transmission) for all of a node's neighbors. This requirement to perform neighbor discovery is nontrivial and is investigated later in this chapter.

The use of directional antennas introduces two additional hidden node (or "hidden terminal") cases. The first is caused by the asymmetry in gain between the omnidirectional mode and the directional mode [14]. Consider the example of Figure 4.8, where node A and node C are not within range of one another unless they are both pointed toward each other in directional mode. In the example of Figure 4.7, Node B has a packet transmission for node A and sends a directional RTS to node A (numeral 1 inside circle in Figure 4.7), which responds with a directional CTS. Node C cannot hear this exchange because it is idle and listening in its omnidirectional mode. Once the directional RTS/CTS handshake is complete, node A begins receiving node B's

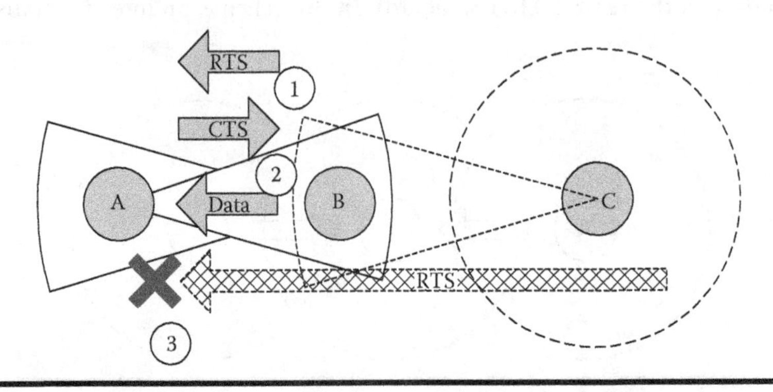

Figure 4.7 Hidden node problem caused by the asymmetry in gain.

directional data transfer ②. Subsequently, node C has a packet for node B. It will directionally sense in the direction of node B but will not hear the ongoing transmission to node A (because it is pointed away from node C). Perceiving the medium to be free, node C will send a directional RTS to node B. Since node A is in the same direction as node B (relative to node C), nodes A and C are now pointing to each other in directional mode and are now within range of each other. Hence, the RTS from node C causes a collision at node A ③.

The second case occurs when a neighboring node misses the CTS transmission because it is currently "pointed away" from the intended receiver [14]. Upon completing its directed data transmission, the neighboring node no longer has the current state of the medium (as defined by the NAV) and it is susceptible to transmission if it finds the medium around it free. In the example of Figure 4.7, node B sets up a data transmission with node B using the directional RTS/CTS mechanism. Node C does not "hear" the directional CTS from node A because node C is pointing towards node D (numeral 1 inside circle in Figure 4.7). Subsequently, node C has a packet for node A and sends a directional RTS packet to node A (because it is unaware of the ongoing transmission between node A and node B). This causes a collision at node A ②.

The hidden node problem that arises due to the asymmetry in gain between the directional RTS/CTS/Data/ACK exchange and the omnidirectional idle listening has been addressed through the use of directional idle listening, which allows DD (directional–directional) communication range during the RTS/CTS reservation exchange [15–17]. In these solutions, idle nodes sweep through all potential directions listening for a directional RTS or CTS transmission. Nodes subsequently update their DNAV tables based on this DD information to prevent future collisions. The asymmetry gain problem has also been addressed by sending a directional RTS "upstream" to the potential interferer (e.g., from node B to node C in Figure 4.8) [18]. This has the disadvantage that a transmission might be blocked if the upstream node is engaged in an ongoing transmission pointed away from the sending node and cannot hear the RTS. To accommodate this, [18] only sends this upstream RTS to a node that has proven itself to be a persistent interferer.

Multi-channel, busy-tone solutions have also been proposed [19] to address the hidden node problem. Here, control signals are separated from data transmissions and nodes sense the control (busy-tone) channel to determine if the medium is busy (rather than sensing the data channel). The busy tones, which are unmodulated carrier waves that do not need to be decoded, can be transmitted throughout the duration of the accompanying RTS/CTS/Data/ACK exchange. To prevent collisions with ongoing transmissions, a node (i.e., node C in Figure 4.6) senses the control channel for the presence of a tone (in Figure 4.6, node C will hear the tone from the receiver, node B). In Dual Busy Tone Multiple Access (DBTMA) [19], two controls channels are provided, one for the transmitter busy-tone and one for the receiver busy-tone, which are used in conjunction with an RTS packet to initiate the transmission. With the introduction of directed antennas, these schemes have been modified to incorporate directional busy tones, which are specifically designed to protect the directed data transmissions [20]. A single channel tone-based solution has also been proposed to address hidden terminals due to unheard CTS transmissions [21]. Transmissions are broken up into smaller fragments and a directional tone is transmitted between the fragments. The tone is transmitted in the opposite direction of the data fragments (sent to upstream nodes).

4.3.2 Deafness

Deafness is unique to directional antennas and occurs when the receiver is "pointed away" from the transmitter [22]. In Figure 4.8, nodes A and B are engaged in a directional data transfer (① in Figure 4.8). If node C subsequently has a packet transmission for node A, its RTS transmission

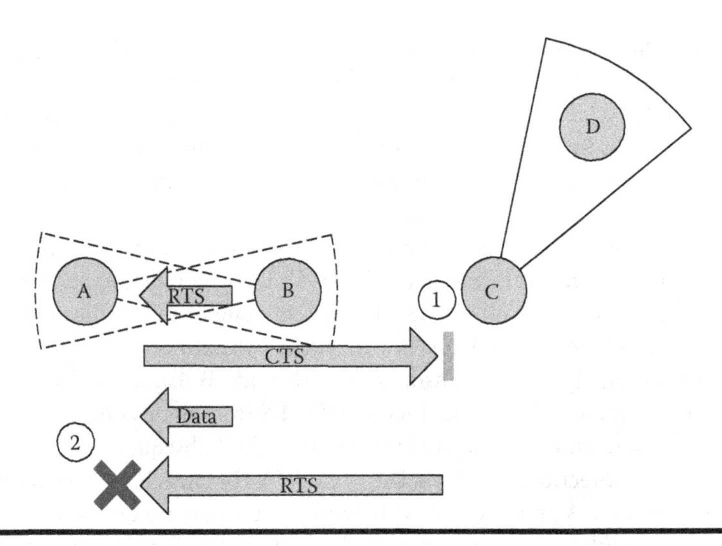

Figure 4.8 Hidden node problem caused by unheard CTS transmission.

will not be heard by node A since node A is pointed towards node B ②. This problem is compounded by the binary exponential back-off approach in typical CSMA-based schemes. Node C may attempt multiple failed transmissions to node A with the contention window doubling each time (along with the mean back-off duration). In worst case scenarios, the deafness can be "chained" and lead to deadlock, as shown in Figure 4.9.

The fundamental issue at the heart of the deafness problem is that a node (i.e., node C in Figure 4.8) is unaware that its target receiver (i.e., node A in Figure 4.8) is currently engaged in a directional exchange pointed in another direction. Accordingly, solutions have been proposed to inform nodes outside the directional footprint of the upcoming transmission and its duration. A common example is the use of sweeping (or circular) RTS and CTS transmissions [23,24]. These are transmitted after the initial directional RTS and CTS packets, which are used to protect the directional footprint of the data exchange. An outlying node (not involved in the data exchange) will record the presence (and its associated duration) of a directional RTS or CTS in its DNAV table, while it will record the presence of a circular RTS or CTS (and its associated duration) in its deafness table. Prior to sending an RTS to initiate a data exchange, a node will check its DNAV

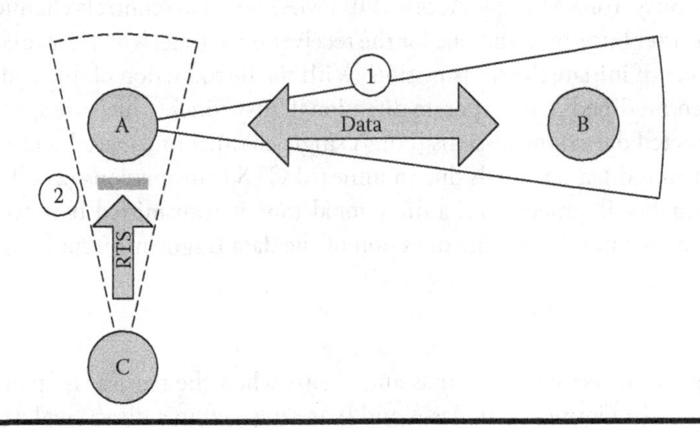

Figure 4.9 Deafness problem with directional antennas.

table to see if it is in the footprint of an ongoing transmission and its deafness table to see if its intended target is participating in an ongoing transmission. The DNAV table is referenced by direction while the deafness table is referenced by intended receiver.

Circular RTS/CTS transmissions incur significant additional control transmission overhead. An alternate approach was proposed in Takata et al. [25] in which a node maintains a table (called a polling table) of neighbor nodes with packets that are addressed to it. This table is populated by the neighbor nodes transmission as follows. When a node is preparing to transmit a packet to a destination, it checks its own transmission queue to see if there are any other packets destined for the same node. If there are, it appends that packet's size to the header of the current packet to be transmitted. At the receiver, the node identification (ID) and the packet size are added to the receiver's polling table along with a time-stamp indicating when it was received. Upon completion of a data exchange, a node checks its polling table to determine if there are any nodes potentially suffering from deafness. If there are multiple candidates, the node will prioritize them by waiting time and then send a Ready-to-Receive (RTR) packet to that node to initiate a data exchange. Tone-based solutions have also been used to solve the deafness problem in which the tones are modified to provide node ID information, which can then be used to prevent deafness [22,26].

Several proposals exist to mitigate the impact of the deafness problem by modifying the binary exponential back-off (BEB) scheme in the IEEE 802.11 standard. In [22], a node resets its contention window to the minimum value when it learns that the node it is trying to reach has completed an ongoing transmission, and Ramanathan et al. [27] proposes different back-off mechanisms for different events (i.e., busy channel, missed CTS, missed ACK). The missed CTS is indicative of the deafness problem and results in a linear increase of the contention window size (vice exponential). A common solution to the deafness deadlock problem is to return nodes to omnidirectional listening when they are backing off [22,25]. This not only prevents deadlock to deafness, it also has the added benefit of allowing a node to receive an RTS and participate in a data exchange while its own queued data transmission is waiting for the direction or receiver to become available.

4.3.3 Communication Range Utilization

Maximum transmission range is realized when two nodes are exchanging data in the DD mode (as seen in Figure 4.4c). The challenge lies in coordinating the directional connection between the two nodes and reserving the medium to prevent collisions. For example, idle nodes monitor the medium in omnidirectional mode in Basic DMAC [14]. This has the advantage of allowing nodes to accept directional RTS invitations from any direction. However, it limits the subsequent data transmission range to the directional-omni (DO) case of Figure 4.4b since the receiving node (node B in the figure) will not hear the initial RTS from node A if it is outside this range.

Basic DMAC [14] proposed a multihop RTS mechanism to overcome this shortcoming and set up the DD data transfer. As in Basic DMAC, nodes still monitor the medium in omni mode, but utilize a *DO-neighbor route* to transmit a multihop RTS to an intended destination that is outside the DO transmission range (but within the DD transmission range). DNAV entries are maintained by all nodes in range of initial directional RTS transmission from the source node which reserves the medium for transmission around the sender. Subsequent *forwarding-RTS* transmissions, which are used along the remainder of the *DO-neighbor route,* do not trigger a DNAV entry. Upon receiving the *forwarding-RTS* packet, the intended destination will point its directional antenna towards the source node and transmit a directional CTS packet (the source node is already pointed towards the destination awaiting this CTS). Nodes overhearing this directional CTS transmission will update their DNAV entries accordingly, which reserves the medium around the destination. This solution assumes the presence of a neighbor discovery mechanism and a link layer routing mechanism. It also assumes the existence of a stable *DO-neighbor route* between the source and destination nodes.

A number of authors have proposed solutions to achieve DD data transfers in which idle nodes monitor the medium in directional mode instead of omnidirectional mode such as those identified in the hidden node discussion above [15–17]. In these solutions, idle nodes sweep through all potential directions listening for a directional RTS transmission that is intended to initiate a directional data exchange. Once received, the target nodes respond with a directional CTS to confirm the exchange and clear the medium in their direction. This can be viewed as a preamble sampling problem [28] in which the sending node must transmit its RTS long enough to be detected by the receiving node during its sweep (which defines the sampling period). Jakllari et al. [29] use a contention-free polling approach in which DD neighbors are discovered and transmissions are reserved in synchronous slots. Again, idle nodes monitor the medium in directional mode during the neighbor search (discovery) phase.

4.3.4 Exposed Node Problem

The exposed node problem is also well understood in omnidirectional wireless networks [12]. It occurs when a node defers transmission when its transmission would not cause a collision. Consider the example in Figure 4.10 in which node B is transmitting to node A. If node C subsequently has traffic for node D (outside the range of either node A or node B), it will refrain from transmitting, because it will find the medium busy, due to node B's transmission. However, in this case, since node C is out of range of node A, and node D is out of the range of node B, the concurrent transmissions of B-to-A and C-to-D can occur without collision. Instead, though, node C will defer until node B's transmission is complete and overall capacity will be reduced.

A similar exposed node problem exists in wireless networks utilizing directional antennas [30]. In Figure 4.11, node A used a directional RTS/CTS exchange to transmit directional data to node B. Subsequently, node C has traffic for node D but will refrain from transmission (since it heard node A's RTS) even though the C-to-D transmission will not cause a collision at either node B or node D.

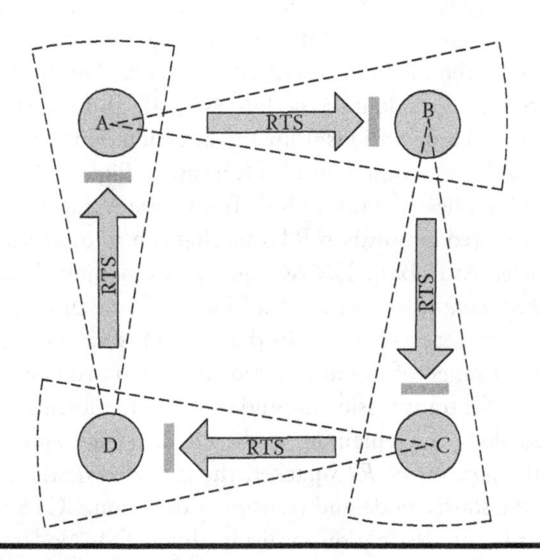

Figure 4.10 Deadlocked due to deafness.

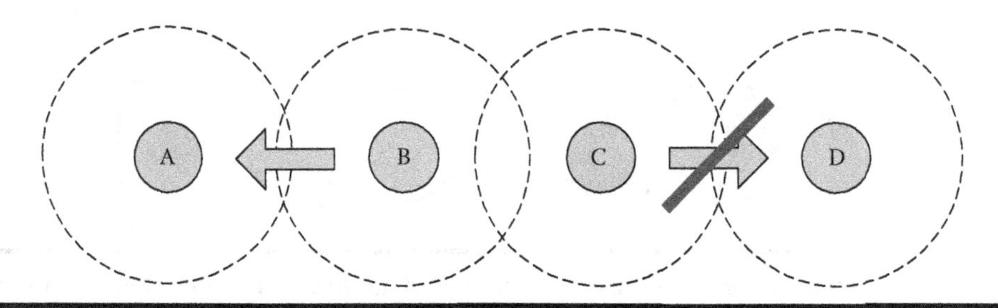

Figure 4.11 Exposed node problem in omnidirectional wireless networks.

In both cases, the simultaneous transmission of B-to-A and C-to-D can occur if the transmissions are synchronized. If this is not the case, the transmission from node C may collide with either the CTS or the ACK from node A to node B. If they are synchronized, then the RTS/CTS mechanism can be used to allow the concurrent transmissions as follows. If node C hears the RTS from node B but not the CTS from node A (after a sufficient time-out period), it may assume that its transmission will not interfere with node A's reception, and, therefore, it is free to transmit to note D. Furthermore, node C can use the duration information in the RTS from node A to tailor its transmission to node D to match up with node B's transmission to node A. This approach applies to both the omnidirectional case of Figure 4.10 and the directional case of Figure 4.11.

4.3.5 Minor Lobe Problem

The minor problem stems from the fact that directional antennas are not ideal. In additional to the main lobes, both side and rear lobes are often present and non-negligible, as shown in Figure 4.12. Medium access layer models often either ignore these minor lobes using the cone model seen in Figure 4.4b and Figure 4.4c or simplify it as a cone plus sphere [31], as shown in Figure 4.12.

These side and rear lobes can cause both unintentional interference with and experience interference from other ongoing transmissions that are within its side or rear lobes. This is particularly troublesome when the interferer's main lobe is pointed at the side or rear lobe of the unintended

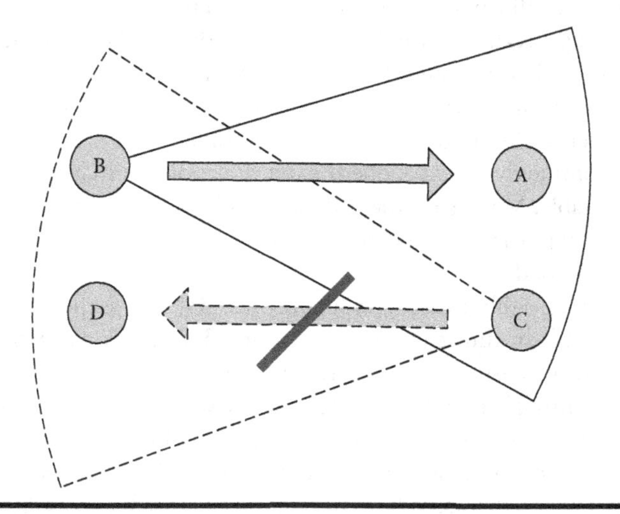

Figure 4.12 Exposed node problem using directional antennas.

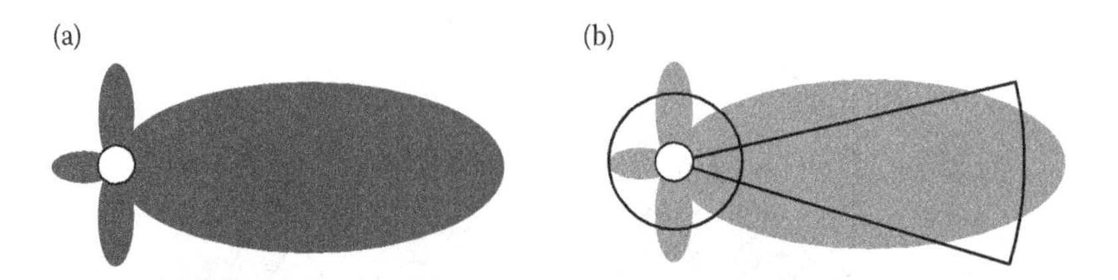

Figure 4.13 Directed antenna beam pattern and model. (a) Example of directed antenna beam pattern. (b) Cone plus sphere model.

receiver. As an example, consider Figure 4.13a in which a typical medium access model will assume that node B is deaf towards node A's transmission to node C. In reality, though, the side lobes present at node B will cause node A's transmission to be received as potentially non-negligible interference at node B, as in Figure 4.13b. Recent results [32] demonstrate that the use of directional antennas in conjunction with virtual carrier sensing may actually increase interference over the omnidirectional case if the beamwidths are not sufficiently small. Solutions to mitigate the effect of minor lobes have been proposed that include minor lobe information in the DNAV table [31].

4.3.6 Miss–Hit Problem and Neighbor Discovery

The miss–hit problem occurs when a node uses an incorrect (or outdated) transmission profile to transmit a directional RTS to an intended receiver. The effect is similar to what is seen with the deafness problem (including the additional detrimental effect of the binary exponential back-off algorithm), but the root of the problem is that the sending node no longer knows where the intended receiver is located (directionally). This brings us to an exploration of neighbor discovery techniques. Although not necessarily MAC layer function (and often performed by higher layers), it is nonetheless critical to the effective performance of directional antennas at the MAC layer and is, therefore, briefly discussed here.

The goal of a neighbor discovery algorithm is to identify all nodes in the one-hop neighborhood and it is applicable to all wireless networks, omnidirectional and directional. The directional case is more challenging because of the limited beamwidth available in the directional mode. Accordingly, one can see that mapping the one-hop DD neighborhood can potentially be more difficult than mapping the one-hop OO (omnidirectional-omnidirectional) neighborhood or even the one-hop DO (directional-omnidirectional) neighborhood. The information recorded for the neighborhood maps commonly include node ID and direction but may also include location and required power, if available. Metrics for these algorithms include the time to discover all (or some percentage of) the one-hop neighbors and percentage of the total number of neighbors discovered in the one-hop neighborhood.

Synchronous, slotted neighbor discovery algorithms [33] and asynchronous algorithms [34] have both been proposed. In direct discovery versions [35], nodes discover their neighbors by receiving successful direct transmissions from them. In gossip-based approaches [35], nodes learn about their neighbors through indirect communication with other nodes. Algorithms can utilize one-way or two-way (or more) communication schemes [36]. In the former, nodes announce their presence and the discovery algorithm terminates when at least one successful transmission is received from all one-hop neighbors. While this is simple and reduces overhead, it has the disadvantage that there is no confirmation of discovery provided. To overcome this, two-way exchanges have been proposed, particularly in synchronous algorithms, where it is easier to coordinate the handshake. The full DD one-hop neighborhood can be probed using directional transmissions, but the performance of the discovery algorithm is tied to the effectiveness of the scanning technique utilized. Acknowledging that deterministic scan techniques tend to perform better in low-node-density environments, and random scanning tends to perform better as the node density (and subsequent collisions) increases, Liu et al. [36] have proposed a scanning-based algorithm that combines the merits of both by introducing a random idle period into the deterministic scanning schedule.

4.3.7 Head-of-Line Blocking

Head-of-line blocking occurs when packets to be transmitted in a direction that is currently in use block the transmission of packets that will utilize a direction currently not in use (as in a First-In-First-Out or FIFO queue arrangement) [37]. This can be viewed as a multi-channel blocking problem in which each potential direction of transmission is a separate channel. A packet at the head of the queue that uses a currently busy channel will block the potential transmission of a packet behind it in the queue that could utilize a free channel. This problem is not seen in omnidirectional antenna systems because the transmissions treat the entire medium as a single channel (in the direction sense). Solutions for the head-of-line blocking problem are queue-management-based, such as the one proposed by Kolar et al. [37] in which the DNAV entries are used to choose the packet in the queue with the least waiting time.

4.3.8 Power Management

While power control research in wireless networks with directed antennas is limited, several proposals exist in the literature, that utilize transmission power settings to trade off transmission range for power consumption [17]. Additionally, a number of protocols rely on a reduction of transmission power to ensure that the transmission range in the directional mode is the same as that in the omnidirectional mode [31,38].

The overhearing problem identified in Section 4.1.3 is further compounded in directional antennas and is referred to as the MAC-layer Capture problem [39]. This is because a node that overhears a transmission for which it is not the intended receiver will not only expend energy receiving and decoding it, it will also come out of omnidirectional idle listening mode and point towards the transmitter to receive it. This will result in the deafness problem discussed earlier, but can be avoided by providing additional control over the antenna in the idle state. An example is proposed by Choudhury and Vaidya [39] in which a node's physical layer will turn off the antenna for any direction in which an ongoing transmission is underway (as indicated by the reception of an RTS or CTS packet and the subsequent DNAV entry).

4.4 Selected MAC Protocols for Directional Antennas

In this section, we offer several example protocols that have been proposed in the literature. This short list is designed to give the reader a sampling of some of the current art in the field. Having said that, we begin with an often cited single-channel protocol that has served as the foundation for much of the work in wireless medium access with directional antennas. We then present a single-channel protocol that modifies the binary exponential back-off (BEB) mechanism found in

IEEE 802.11 as well as both multi-channel tone- and non-tone-based solutions. We conclude this section with a single-channel, non-contention-based protocol that utilizes time division multiple access (TDMA).

4.4.1 Single-Channel (Foundational): Basic DMAC [12]

Basic Directional MAC (Basic DMAC) [12] is a foundational protocol in the field of wireless medium access control using directional antennas. Proposed in 2002, it makes use of directional RTS and CTS control signaling, omnidirectional idle listening, directional data transmission, and a directional Network Allocation Vector (DNAV) table.

When a node has a packet to transmit, it performs physical carrier sensing by directionally listening in the direction of the intended receiver and virtual carrier sensing by checking its DNAV table for the intended direction of transmission. If the medium is free, the node will back off as in IEEE 802.11 and then, when complete, it sends a directional RTS to the intended receiver. Note that during the back-off phase, a node listens in the directional mode towards the intended receiver.

In the idle state, all nodes listen in omnidirectional mode. Upon receiving an RTS in which it is the intended receiver, a node performs physical carrier and virtual carrier sensing using its own DNAV table in the direction towards the sender for the length of an SIFS (short interframe space, as in IEEE 802.11). If the medium remains clear, the intended receiver will transmit a directional CTS back to the sender. Once the sender receives the CTS and awaits an IFS, it will directionally transmit the data and then wait for the acknowledgment. The acknowledgment is sent by the receiver upon successful reception of the data and, again, after awaiting an SIFS. If a sender fails to receive a CTS or ACK, it will terminate the transmission and schedule a retransmission using the back-off algorithm in IEEE 802.11.

If nodes other than the receiver overhear a directional RTS or CTS, they update their DNAV tables accordingly, with both the direction of the intended transmission (determined by the angle of arrival of the transmission) and the duration (contained in the RTS or CTS, as in IEEE 802.11). A node defers if its intended transmission direction is within some prespecified threshold of an entry in its DNAV table.

While the performance of Basic DMAC is dependent on the physical topology of the network, results indicate that Basic DMAC is capable of providing more than double the throughput of IEEE 802.11. As explained above, Basic DMAC suffers from the hidden node problem, deafness, and is capable of only DO data exchanges. It assumes that a transceiver profile accompanies a packet when it is delivered by an upper layer and does not consider either side lobes or node mobility.

4.4.2 Single-Channel, Contention-Based: EDMAC [40]

Enhanced Directional MAC (EDMAC) [40] is designed to mitigate the unfairness and low channel utilization caused by deafness that is exacerbated by the binary exponential back-off (BEB) mechanism found in IEEE 802.11. Rather than having individual nodes independently back off exponentially, EDMAC modifies the typical directional MAC algorithm to ensure that nodes sending to a common receiver all use a common (and constant) contention window size. Each node maintains a neighbor table that includes the direction to the neighbor, the contention window size that should be used for transmissions to that neighbor, and an indication of whether or not the node has recently heard from that neighbor. To populate this table, the contention

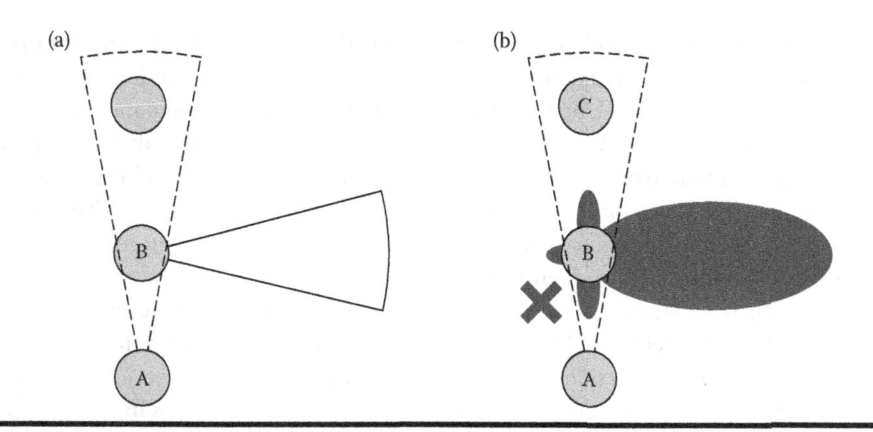

Figure 4.14 Minor lobe problem. (a) Transmission model. (b) Actual transmission.

window size for a given receiver is placed in the receiver's ACK packet whenever it completes a data exchange with a neighbor.

Analysis is provided to arrive at an optimal contention window size, W, that is a function of the duration of an RTS transmission, T_{RTS}, the duration of a back-off slot, T_{slot}, and the number of transmitters n as in

$$W = \frac{4T_{RTS}}{T_{slot}} n - 1. \tag{4.8}$$

A receiver calculates this value by counting the number of active neighbors found in its neighbor table. Results provided demonstrate that EDMAC is capable of providing at least a 25% improvement in throughput over Basic DMAC [12].

4.4.3 Multi-Channel, Tone-Based: DSDMAC [41]

Dual Sensing Directional MAC Protocol (DSDMAC) [41] is a two-channel, tone-based protocol that is designed to address the hidden node, deafness, and exposed node problems. One channel is reserved for the busy tone while the other is used for the directional RTS/CTS/Data/ACK exchange. Two different directional tones are defined, BT_1 and BT_2, as in Figure 4.14, and they are capable of being rotational-transmitted across all desired directions (Figure 4.15).

When a node receives a packet for transmission, it performs directional physical carrier sensing on both the data and busy-tone channels as well as virtual carrier sensing, using its DNAV table. If the directional data channel is clear, no BT_1 is present, and the direction is not blocked in the

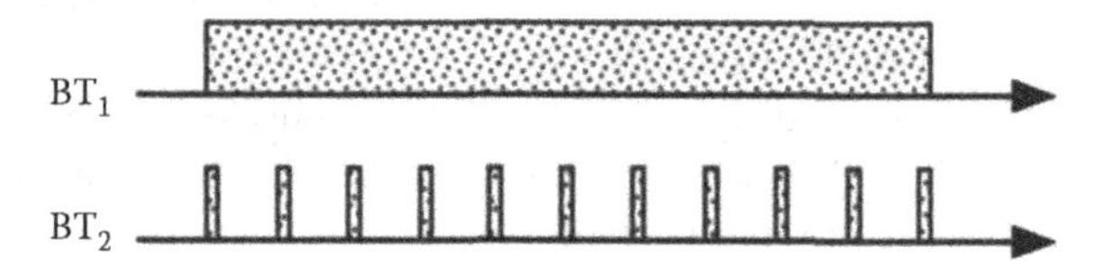

Figure 4.15 Busy-tone signals defined in DSDMAC. (Adapted from Abdullah, A. et al., 2012. "DSDMAC: Dual Sensing Directional MAC protocol for ad hoc networks with directional antennas antennas." In *IEEE Trans. Veh. Technol.*, 61(3), 1266–1275.)

DNAV table, the node immediately sends a directional RTS on the data channel and transmits a BT_1 in all other directions on the busy tone channel. Otherwise, the node will go through a back off similar to IEEE 802.11. Upon completion of the RTS transmission and after waiting an SIFS, the sending node switches from BT_1 to BT_2. Upon hearing the RTS and waiting an SIFS, the intended receiver responds with a directional CTS back to the sender and transmits a BT_2 in all other directions. The directional data transmission then follows and the receiver turns off its BT_2 after transmitting the directional ACK. The sender secures its BT_2 upon receiving the ACK.

BT_1 is used to overcome the hidden terminal problem (it is sent to the upstream nodes) and BT_2 is used to mitigate deafness. A node that sends an RTS to an intended receiver can check for a BT_2 signal if it fails to receive a CTS. If a BT_2 signal is present, it assumes the node is engaged in a transmission in another direction and schedules a retransmission without doubling its contention window size. If no BT_2 is present, it assumes a collision and backs off, as in IEEE 802.11. The DNAV table is maintained using overheard directional RTS and CTS transmissions, but a station that hears an RTS and does not hear a subsequent CTS will clear its DNAV table entry after a sufficient time-out period. This minimizes the exposed node problem, as discussed earlier.

Analysis is provided to validate DSDMAC and it is shown to significantly outperform both IEEE 802.11 and other directional antenna proposals (including ToneDMAC [22] and MMAC [14]) in throughput and delay by 15%–184%. It assumes no side lobes and does not address mobility.

4.4.4 Multi-Channel, Non-Tone-Based: CMDMAC [31]

Cooperative Multi-channel Directional Medium Access (CMDMAC) protocol [31] is a multi-channel solution that proposes to use cooperation among nodes to improve a node's ability to accurately determine the state of the medium. It comprises a control channel in which control packets are exchanged in omnidirectional mode and a set of data channels in which data packets are transmitted in directional mode. Transmission power control is used to ensure transmission range is the same in both channels, which has the effect of eliminating the hidden node problem that arises due to the asymmetry in gain. Each node maintains two tables: a DNAV table and a Neighbor Information Table (NIT). The latter maps the node's two-hop neighborhood by indicating the sector for each two-hop neighbor relative to the node's one-hop neighbors. It also includes a flag to indicate when a neighbor is close enough to cause interference through the minor (side or rear) lobes.

Upon receiving a packet for transmission, a sending node performs both physical carrier sensing in the direction of the intended transmission and virtual carrier sensing using its DNAV table. If the medium is clear in the intended direction, the sending node sends an RTS in the direction of the intended receiver which includes not only the duration of the exchange, but also the intended data channel and antenna sector to be used. Neighbors overhearing this omnidirectional transmission on the control channel, check their DNAV table and their NIT to determine if the intended transmission will interfere with any ongoing transmissions within the two-hop neighborhood that they are aware of. If a potential conflict exists, the neighbor will respond with a Deny Signal Type A (DYSA) packet that includes the node whose ongoing transmission triggered the conflict. To minimize collisions between multiple DYSA transmissions, the DYSA transmitters enter a Cooperation Back-off Period (CBP), which is a CSMA-based contention mechanism in which each DYSA transmitter selects a random back off.

Upon reception of the RTS, the intended receiver awaits the duration of the CBP and then checks the medium. If it is busy (due to the presence of a DYSA transmission), the exchange is terminated. Otherwise, the receiver sends an omnidirectional CTS packet which again contains the intended duration, data channel, and antenna sector. Upon hearing the CTS, the receiver's neighbors check their DNAV table and NIT for conflicts and report the presence of any using a Deny Signal Type B (DYSB) transmission in the same manner as the DYSA described above.

Upon reception of the CTS, the sending node awaits the duration of the CBP and then checks the medium. If it is busy (due to the presence of a DYSB transmission), the exchange is terminated. Otherwise, the sender transmits a Confirm Type A (CFA) packet which includes the duration of the transmission and allows the sending node's neighbors to update their DNAV tables. If the intended destination receives the CFA and did not receive any DYSB packets, it transmits a Confirm Type B (CFB) packet which includes the duration of the transmission and allows the receiving node's neighbors to update their DNAV tables. In the event that the sending node does not receive a CFB, it will send a Cancel Signal (CLS) to clear the DNAV entries of its neighbors.

Deafness is minimized because the DYSA/DYSB packet contains the node ID of the node engaged in the ongoing data exchange and the omnidirectional RTS and CTS transmissions address the hidden node problem. CMDMAC considers the minor lobes using the cone and sphere model and reflect their effect in the NIT. The throughput of CMDMAC is shown to outperform both IEEE 802.11 and a non-cooperative version of the protocol as well as Multi-hop MAC [14] and DMAC [13].

4.4.5 Single-Channel, Contention-Free, TDMA-Based: SDVCS [42]

Slotted Directional Virtual Carrier Sensing (SDVCS) [42] is a TDMA-based protocol for medium access control in wireless networks with directional antennas. It is a slotted approach that builds on the directional transmit and receive algorithm (DTRA) [33]. SDVCS improves on the performance of DTRA by considering interference with other transmissions and introducing a slotted directional network allocation vector (SDNAV).

As in DTRA [33], SDVCS comprises three phases: (1) Neighbor Discovery, (2) Reservation, and (3) Data. Each phase is divided into a set of slots, which are further divided into minislots. During the Neighbor Discovery phase, nodes discover the presence and direction of their one-hop neighbors. Each slot in this phase is subdivided into three minislots. In the first minislot, a node will directionally transmit a "hello" type message with a 50% probability. This initial message contains the node's ID and any node overhearing it will reply in the second minislot, indicating whether or not it wishes to establish a new communication with the sender, along with a list of available slots in its Reservation phase. If a new communication session is desired by either node, the original sender will respond in the third minislot with the slot to be used by the pair in the Reservation phase. The original sender rotates its transmitter and there are sufficient slots in the Neighbor Discovery phase to allow the sender to transmit in all possible directions. The algorithm is capable of "discovering" only one node (the closest) in a given direction.

The Reservation phase is utilized to reserve slots for data transmission in the Data phase and to share interference information used to populate the SDNAV. Data slots may be randomly added, randomly released, or selectively released in the Reservation phase. Reservation slots are subdivided into two minislots. The first slot is used by one node in the pair to perform slot maintenance, share a list of available transmission and reception data slots, and to share a list of slots where interference may occur either from another transmitting node or to another node in receiving mode. The second node then uses its slot allocation table and SDNAV to choose slots for transmission and reception with the first node as needed. It sends these choices in the second minislot along with its own interference list. In the next Reservation phase, the nodes will swap roles. Node pairs will continue to use the same Reservation slot in each Reservation phase until they experience multiple communication failures in that slot (in which case they will reschedule during the next Neighbor Discovery phase). Slots in the Data phase are optionally subdivided into two minislots to support an acknowledgement following the data transmission, if desired. Node pairs occupy a given data slot in subsequent Data phases until it is either released during the Reservation phase or multiple communication failures occur.

SDVCS makes use of two SDNAV tables, one for transmission and one for reception. They record potential interference caused by another node's previously scheduled reception or transmission, respectively. These tables are updated based on reservation exchanges and contain both the ID and direction of potential interferers. Accordingly, these SDNAV tables address both the hidden node and deafness problems. Results indicate that SDVCS throughput outperforms DTRA by as much as 40% in heavily loaded scenarios (DTRA was previously shown to significantly outperform IEEE 802.11 in both throughput and delay [33]). As a TDMA scheme, SDVCS requires global time synchronization and it does not address the presence of minor lobes.

4.5 Conclusion

In this chapter, we began by examining the fundamentals of medium access for wireless networks, including a discussion of the IEEE 802.11 (WiFi) standard. We next highlighted the opportunities presented by directional antennas. These included, but were not limited to, potential increases in both capacity and range. We discovered that with these opportunities come a number of design challenges. Much work has been done to address the hidden node and deafness problems, but other design issues include full utilization of the communication range available, overcoming the exposed node, head-of-the-line blocking, and miss–hit problems as well as accommodating the minor lobes that exist in directional antenna beam patterns. We looked at these design issues in detail and explored many of the solutions that have been proposed in the literature. We also briefly examined some of the power management issues present at the medium access control layer when utilizing directional antennas. We closed with a sampling of five medium access control protocols that have been proposed for directional antennas, including one of the foundational protocols and four recent proposals encompassing both single- and multi-channel solutions.

References

1. IEEE Standard for Information Technology-Telecommunications and Information Exchange between Systems-Local and Metropolitan Area Networks-Specific Requirements—Part 11: Wireless LAN Medium Access Control (MAC) and Physical Layer (PHY) Specifications, IEEE Standard 802.11-2012 (Revision of IEEE Standard 802.11-1999), 2012.
2. Rajendran, V. et al., 2006. "Energy-efficient, collision-free medium access control for wireless sensor networks." In *Wireless Networks*, 12(1), 163–78.
3. Kleinrock, L. and F. Tobagi, 1975. "Packet switching in radio channels: Part I—Carrier sense multiple-access modes and their throughput-delay characteristics." In *IEEE Trans. Commun.*, 23(12), 1400–1416.
4. IEEE Standard For Information Technology—Telecommunications and Information Exchange between Systems- Local and Metropolitan Area Networks- Specific Requirements Part 15.4: Wireless Medium Access Control (MAC) and Physical Layer (PHY) Specifications for Low-Rate Wireless Personal Area Networks (WPANs), IEEE Standard 802.15.4-2011 (Revision of IEEE Standard 802.15.4-2003), 2011.
5. Ye, W. et al., 2004. "Medium access control with coordinated adaptive sleeping for wireless sensor networks." In *IEEE/ACM Trans. Net.*, 12(3), 493–506.
6. Demirkol, I. et al., 2006. "MAC protocols for wireless sensor networks: A survey." In *IEEE Commun. Mag.*, 44(4), 115–121.

7. Shannon, C. E. 1948. "A mathematical theory of communication." In *Bell Syst. Tech. J.*, 27, 379–423, 623–656.
8. Noubir, G. 2004. "On connectivity in ad hoc network under jamming using directional antennas and mobility." In *Second Int. Conf. Wired/Wireless Internet Communications*, (WWIC), Frankfurt/Oder, Germany, February 4–6, 2004, pp. 521–532.
9. Qiu Wang et al., 2013. "Eavesdropping security in wireless ad hoc networks with directional antennas." In *22nd Wireless and Optical Communication Conference (WOCC)*, Chongqing, China, May 16–18, 2013, pp.687–692.
10. Tobagi, F. and L. Kleinrock, 1975. "Packet switching in radio channels: Part II—The hidden terminal problem in carrier sense multiple-access and the busy-tone solution." In *IEEE Trans. Commun.*, 23(12), 1417–1433.
11. Karn, P. 1990. "MACA—A new channel access method for packet radio." In *Proc. 9th ARRL Computer Networking Conf.*, London, Ontario, Canada, September 22, 1990, pp. 134–140.
12. Choudhury, R. et al., 2002. "Using directional antennas for medium access control in ad hoc networks." In *ACM Int. Conf. Mobile Computing and Networking (Mobicom)*, September, Atlanta, GA, pp. 59–70.
13. Takai, M. et al., 2002. "Directional virtual carrier sensing for directional antennas in mobile ad hoc networks." In *Third ACM Int. Symp. Mobile Ad Hoc Networking and Computing (MobiHoc)*, Lausanne, Switzerland, June 9–11, 2002, pp. 183–193.
14. Choudhury, R. R. et al., 2006. "On designing MAC protocols for wireless networks using directional antennas." In *IEEE Trans. Mobile Computing*, 5(5), 477–491.
15. Shihab, E. et al., 2009. "A distributed asynchronous directional-to-directional MAC protocol for wireless Ad Hoc networks." In *IEEE Trans. Veh. Technol.*, 58(9), 5124–5134.
16. Kulkarni, S. S. and C. Rosenberg, 2005. "DBSMA: A MAC protocol for multi-hop ad-hoc networks with directional antennas." In *IEEE Int. Symp. Personal, Indoor, and Mobile Radio Communications (PIMRC)*, September, Berlin, Germany, Vol. 12, pp. 1371–1377.
17. Takatsuka, Y. et al., 2008. "A MAC protocol for directional hidden terminal and minor lobe problems." In *IEEE Wireless Telecommunications Symposium (WTS)*, Pomona, CA, pp. 210–219.
18. Chin, K. W. 2007. "SpotMAC: A pencil-beam MAC for wireless mesh networks." In *IEEE Int. Conf. Computer Communications and Networks (ICCCN)*, August, Honolulu, Hawaii, pp. 81–88.
19. Haas, Z. J. and J. Deng, 2002. "Dual Busy Tone Multiple Access (DBTMA)-A multiple access control scheme for ad hoc networks." In *IEEE Trans. Commun.*, 50(6), 975–985.
20. Huang, Z. et al., 2002. "A busy-tone based directional MAC protocol for ad hoc networks." In *IEEE Military Communications Conference (Milcom)*, October, Anaheim, CA, Vol. 2, pp. 1233–1238.
21. RamMohan, V. A. et al., 2007. "A new protocol to mitigate the unheard RTS/CTS Problem in networks with switched beam antennas." In *IEEE Int. Symp. Wireless Pervasive Computing (ISWPC)*, February, San Juan, Puerto Rico, pp. 129–134.
22. Choudhury, R. and N. Vaidya, 2004. "Deafness: A MAC problem in ad hoc networks when using directional antennas." In *IEEE Int. Conf. Network Protocols (ICNP)*, October, Berlin, Germany, pp. 283–292.
23. Gossain, H. et al., 2005. "MDA: An efficient directional MAC scheme for wireless ad hoc networks." In *IEEE Global Telecommunications Conf. (GLOBECOM)*, November, St. Louis, MO, Vol. 6, pp. 3633–3637.
24. Li, Y. and A. M. Safwat, 2006. "DMAC-DACA: Enabling efficient medium access for wireless ad hoc networks with directional antennas." In *IEEE Int. Symp. Wireless Pervasive Computing (ISWPC)*, January, Phuket, Thailand, pp. 1–5.
25. Takata, M. et al., 2006. "A receiver-initiated directional MAC protocol for handling deafness in ad hoc networks." In *IEEE Int. Conf. Communications (ICC)*, Vol. 9, Istanbul, Turkey, June 11–15, 2006, 4089–4095.
26. Dai, H. N. et al., 2007. "A busy-tone based MAC Scheme for wireless ad hoc networks using directional antennas." In *IEEE Global Telecommunications Conf. (GLOBECOM)*, November, Washington, USA, pp. 4969–4973.
27. Ramanathan, R. et al., 2005. "Ad hoc networking with directional antennas: A complete system solution." In *IEEE J. Sel. Areas Commun.*, 23(3), 496–506.

28. Walker, T. O. et al., 2013. "Optimal preamble sampling for low power wireless networks." In *46th Hawaii Int. Conf. on System Sciences (HICSS)*, January 7–10, Maui, HI, pp. 5123–5131.
29. Jakllari, G. et al., 2007. "An integrated neighbor discovery and MAC protocol for ad hoc networks using directional antennas." In *IEEE Trans. Wireless Commun.*, 6(3), 11–21.
30. Wang, J. et al., 2009. Directional medium access control for ad hoc networks, *Springer Wireless Networks*, 15(8), 1059–1073.
31. Wang, Y. et al., 2014. "Multi-channel directional medium access control for ad hoc networks: A cooperative approach." In *IEEE Int. Conf. Communications (ICC)*, June 10–14, pp. 53–58.
32. Alabdulmohsin, I. 2014. "Interference in wireless ad hoc networks with smart antennas." In *10th Int. Wireless Communications and Mobile Computing Conference (IWCMC)*, Nicosia, Cyprus, August 4–8, 2014, pp. 666–671.
33. Zhang, Z. 2005. "DTRA: Directional transmission and reception algorithms in WLANs with directional antennas for QoS support." In *IEEE Network*, 19(3), 27–32.
34. Tian, F. et al., 2013. "Pure asynchronous neighbor discovery algorithms in ad hoc networks using directional antennas." In *IEEE Global Communications Conference (GLOBECOM)*, Atlanta, GA, December 9–13, 2013, pp. 498–503.
35. Vasudevan et al., 2005. "On neighbor discovery in wireless networks with directional antennas." In *Proc. IEEE 24th Annual Joint Conf. of the IEEE Computer and Communications Societies (INFOCOM)*, Vol. 4, Miami, FL, March 13–17, 2005. pp. 2502–2512.
36. Liu, Bo. et al., 2013. "Neighbor discovery algorithms in directional antenna based synchronous and asynchronous wireless ad hoc networks." In, *IEEE Wireless Commun.*, 20(6), 106–112.
37. Kolar, V. et al., 2004. "Avoiding head of line blocking in directional antenna." In *29th Annual IEEE Int. Conf. Local Computer Networks (LCN)*, Tampa, FL, November 16–18, 2004, pp. 385–392.
38. de Melo Guimaraes, L. and J. L. Bordim, 2013. "Directional pulse/tone based channel reservation." In *IEEE 27th Int. Conf. on Advanced Information Networking and Applications (AINA)*, Barcelona, Spain, March 25–28, 2013, pp. 276–283.
39. Choudhury, R. R. and N. Vaidya, 2007. "MAC-layer capture: A problem in wireless mesh networks using beamforming antennas." In *44th Annual IEEE Communications Society Conf. Sensor, Mesh and Ad Hoc Communications and Networks (SECON)*, San Diego, CA, June 18–21, 2007, pp. 401–410.
40. Chen, Z. et al., 2013. "EDMAC: An enhanced directional medium access control protocol for 60 GHz networks." In *IEEE 24th Int. Symp. Personal, Indoor, and Mobile Radio Communications (PIMRC)*, London, England, September 8–11, 2013, pp. 1726–1730.
41. Abdullah, A. et al. 2012. "DSDMAC: Dual Sensing Directional MAC protocol for ad hoc networks with directional antennas antennas." In *IEEE Trans. Veh. Technol.*, 61(3), 1266–1275.
42. Tu, Y. et al., 2013. "A novel MAC protocol for wireless ad hoc networks with directional antennas." In *15th IEEE Int. Conf. Communication Technology (ICCT)*, Guilin, China, November 17–19, 2013, pp. 494–499.

Chapter 5

IEEE 802.11ad Wireless Local Area Network and Its MAC Performance

David Tung Chong Wong, Francois Chin,
Xiaoming Peng, and Qian Chen

Contents

5.1 Introduction

High data rate is crucial for ensuring better quality-of-service (QoS) in wireless local area network (WLAN). The most popular WLAN deployed today is the IEEE 802.11 WLAN. IEEE 802.11b can support a data rate of up to 11 Mbps, while both IEEE 802.11a and IEEE 802.11g can support a data rate of up to 54 Mbps. IEEE 802.11b and IEEE 802.11g operate in the 2.4 GHz band, while IEEE 802.11a operates in the 5 GHz band. IEEE 802.11b uses direct-sequence spread spectrum (DSSS), while IEEE 802.11a and IEEE 802.11g use orthogonal frequency division multiplexing (OFDM) for the physical layer (PHY). IEEE 802.11n can support a data rate of up to 600 Mbps using OFDM PHY and multiple-input, multiple-output (MIMO) technology to enhance diversity. IEEE 802.11n can operate on both the 2.4 and 5 GHz bands. IEEE 802.11ac is a very high-throughput WLAN. IEEE 802.11ac WLAN can support a data rate of up to 6933.3 Mbps. IEEE 802.11ac operates below the 6 GHz band except for the 2.4 GHz band. The very high throughput PHY of IEEE 802.1ac uses OFDM PHY.

Another standardized distributed medium access control (MAC) that can deliver a data rate of up to 480 Mbps is the WiMedia MAC or ECMA-368 MAC [1]. There are also three other standards that are already standardized. One of them is the IEEE 802.15.3c [2], while the other one of them is the ECMA-387 [3]. Both of them can deliver even higher data rates than those in WiMedia. IEEE 802.15.3c can deliver a data rate of up to 6 or 7.35 Gbps at the PHY, while ECMA-387 can deliver a data rate of up to {6.35, 12.701, 19.051, 25.402} Gbps for one-channel, two-channel, three-channel, and four-channel bonding, respectively. These standards operate in the unlicensed band at 60 GHz. However, there is no vendor support for these standards from the authors' knowledge. The third emerged WLAN standard is the IEEE 802.11ad standard [4] which also operates in the 60 GHz band. It has a maximum data rate of up to 6756.75 Gbps. Qualcomm has acquired Wilocity so that it can combine traditional Wi-Fi and IEEE 802.11ad on a single chunk of silicon for mobile gadgets [5]. Wi-Fi is another name for WLAN. Our focus in this chapter is on the IEEE 802.11ad WLAN and its MAC performance.

In the MAC architecture of IEEE 802.11ad, the fast session transfer is used for multi-band operation. The new beacon interval (BI) in IEEE 802.11ad consists of beacon transmission interval, association beamforming training time (A-BFT), announcement transmission interval, contention-based access periods (CBAPs), and service periods (SPs). New directional MAC protocols include beamforming (BF), access point/personal basic service set central point (AP/PCP) clustering, spatial sharing, link switching relaying, and link cooperating relaying. BF allows devices, which are not the AP/PCP, to beamform with the AP/PCP. This is started by receiving a beacon from the AP/PCP. AP/PCP clustering allows multiple APs/PCPs in the IEEE 802.11ad wireless network. Spatial sharing allows multiple SPs to coexist concurrently by overlapping or some overlapping in the same band. Finally, link switching relaying and link cooperating relaying allow relaying in IEEE 802.11ad for the first time.

This chapter has four main sections—Sections 5.2 through 5.5. Section 5.2 covers the applications of IEEE 802.11ad WLAN. Section 5.3 covers the PHY of IEEE 802.11ad WLAN, while Section 5.4 covers the MAC protocols of IEEE 802.11ad WLAN. Section 5.5 covers the analysis and numerical results for IEEE 802.11ad MAC. The IEEE 802.11ad WLAN technology is clearly explained with illustrations to the necessary knowledge to better understand the rationale for the design of IEEE 802.11ad WLAN presented in detail. Section 5.2 lists a few of the applications for IEEE 802.11ad WLAN. The important enabling technologies for IEEE 802.11ad WLAN include control, single carrier and OFDM PHYs, and directional MAC protocols. A two-dimensional discrete-time Markov chain (MC) is used for analytical modeling of the CBAP. Numerical examples with typical parameters are used to show the performance of the saturated throughput of CBAP transmission devices in the presence of *directivity* and *multi-rate*. From the results, at small number of devices, the saturated throughput using omnidirectional antenna is better than that for those with directional antennas having a number of sectors. However, at large number of devices, the saturated throughputs of the antennas with two numbers of sectors can be better than that using omnidirectional antenna. A 4 Gbps saturated throughput is achievable for large data payload. Some conclusions and other new technologies to watch out for as well as the next extra-very high throughput WLAN are highlighted in Section 5.6. These new technologies are the next frontiers of IEEE 802.11 WLAN. These upcoming standards of IEEE 802.11 WLAN will stretch the limits that we know of WLANs today.

5.2 IEEE 802.11ad Applications

There are many applications for IEEE 802.11ad WLAN as follows:

- Wireless personal computer (PC) display for uncompressed video
- TV for uncompressed video
- Projector for uncompressed video
- Sync-and-go between handheld devices or downloading movies or pictures from a camera
- Internet access
- File transfer
- Data backup

Uncompressed video requires a high data rate to stream. IEEE 802.11ad WLAN has a high data rate of up to 6756.75 Gbps to enable uncompressed video for wireless transfer to PC display, TV, and projector. More information on the PHY data rates is presented in Section 5.3. Due to high data rate in IEEE 802.11ad WLAN, the transfer times to download movies from the Internet to a device and to transfer pictures from a camera to a PC are largely cut down and the user's experience for file transfer between devices is also better. Data backup to a thumb drive wirelessly using IEEE 802.11ad WLAN will also be much faster.

5.3 IEEE 802.11ad Physical Layer (PHY) Overview

There are three PHY types for different usages in IEEE 802.11ad WLAN [4]. They are the control PHY, the single carrier (SC) PHY, and the OFDM PHY [6]. A brief overview of these PHYs is presented in this section.

5.3.1 Control PHY

The control PHY is designed for low signal-to-noise ratio (SNR) operation prior to BF [6]. It uses a mandatory SC mode with data rate of about 27.5 Mbps (modulation and coding set (MCS 0)). A 32-sample Golay spreading sequence modulated by $\pi/2$-differential binary phase shift keying (BPSK) modulation mitigates against longer spread channels [6]. This is more robust in the presence of phase noise, allowing for shorter training field [6]. Rate 1/2 coding is also used, shortened from the common 3/4 low density parity check (LDPC) code [6]. Short LDPC code is used because it is more efficient for short packets and bits are evenly divided between code words to allow equal protection [6]. The maximum packet length is 1024 bytes.

5.3.2 Single Carrier (SC) PHY

The single carrier (SC) PHY enables low power/low complexity transceivers and low power SC PHY provides further support for additional reduction in implementation power with simpler coding and shorter symbol structure [6]. The SC PHY, modulations, and data rates are shown in Table 5.1 [6]. The data rates for SC PHY are 385, 770, 962.5, 1155, 1251.25, 1540, 1925, 2310, 2502.5, 3080, 3850, and 4620 Mbps for MCS indexes ranging from 1 to 12, respectively. N_{CBPS} is the number of coded bits per symbol in Table 5.1. π/2-BPSK, π/2-QPSK, and π/2-16QAM modulations are used in the SC PHY. π/2 rotation is applied to all modulations [6]. MCS indexes 1–4 are mandatory (Table 5.1). The low power SC PHY, modulations, and data rates are shown in Table 5.2 [6]. The data rates for low power SC PHY are 385, 770, 962.5, 1155, 1251.25, 1540, and 1925 Mbps for MCS indexes ranging from 25 to 31, respectively. A simple forward error correction (FEC) with Reed–Solomon (224, 208) and inner Hamming-like block-code (16, 8) is used in two cases of the low power SC PHY [6]. A simple equalizer is used for short multipath [6]. These satisfy the need for a low complexity low power mode [6]. N_{CPB} is the number of coded bits per block in Table 5.2. π/2-BPSK and π/2-QPSK modulations are used in the low-power SC PHY.

5.3.3 Orthogonal Frequency Division Multiplexing (OFDM) PHY

The OFDM PHY is used for high performance in frequency selective channels and can use maximum data rates up to 6756.75 Gbps using 64-quadrature amplitude modulation

Table 5.1 Single Carrier PHY

MCS Index	*Modulation*	N_{CBPS}	*Repetition*	*Code Rate*	*Data Rate (Mbps)*
1	π/2-BPSK	1	2	1/2	385
2	π/2-BPSK	1	1	1/2	770
3	π/2-BPSK	1	1	5/8	962.5
4	π/2-BPSK	1	1	3/4	1155
5	π/2-BPSK	1	1	13/16	1251.25
6	π/2-QPSK	2	1	1/2	1540
7	π/2-QPSK	2	1	5/8	1925
8	π/2-QPSK	2	1	3/4	2310
9	π/2-QPSK	2	1	13/16	2502.5
10	π/2-16QAM	4	1	1/2	3080
11	π/2-16QAM	4	1	5/8	3850
12	π/2-16QAM	4	1	3/4	4620

Table 5.2 Low-Power Single Carrier PHY

MCS Index	*Modulation*	*Effective Code Rate*	*Repetition*	N_{CPB}	*Data Rate (Mbps)*
25	π/2-BPSK	13/28	RS(224,208) + Block-Code(16,8)	392	385
26	π/2-BPSK	13/21	RS(224,208) + Block-Code(12,8)	392	770
27	π/2-BPSK	52/63	RS(224,208) + SPC(9,8)	392	962.5
28	π/2-QPSK	13/28	RS(224,208) + Block-Code(16,8)	392	1155
29	π/2-QPSK	13/21	RS(224,208) + Block-Code(12,8)	392	1251.25
30	π/2-QPSK	52/63	RS(224,208) + SPC(9,8)	392	1540
31	π/2-QPSK	13/14	RS(224,208) + Block-Code(8,8)	392	1925

Note: RS represents Reed–Solomon; SPC represents single parity check.

(64-QAM) [6]. The OFDM PHY, modulations, and data rates are shown in Table 5.3 [6]. The data rates for OFDM PHY are 693, 866.25, 1386, 1732.5, 2079, 2772, 3465, 4158, 4504.5, 5197.5, 6237, and 6756.75 Mbps for MCS indexes ranging from 13 to 24, respectively. N_{BPSC} is the number of coded bits per SC, N_{CBPS} is the number of coded bits per symbol, and N_{DBPS} is the number of data bits per symbol in Table 5.3, while SQPSK represents spread QPSK modulation in Table 5.3. SQPSK, QPSK, 16-QAM, and 64-QAM modulations are used in the OFDM PHY. LDPC is also used for low complexity/low latency encoding and high throughput/low power decoding [6].

Table 5.3 Orthogonal Frequency Division Multiplexing (OFDM) PHY

MCS Index	*Modulation*	*Code Rate*	N_{BPSC}	N_{CBPS}	N_{DBPS}	*Data Rate (Mbps)*
13	SQPSK	1/2	1	336	168	693.00
14	SQPSK	5/8	1	336	210	866.25
15	QPSK	1/2	2	672	336	1386.00
16	QPSK	5/8	2	672	420	1732.50
17	QPSK	3/4	2	672	504	2079.00
18	16-QAM	1/2	4	1344	672	2772.00
19	16-QAM	5/8	4	1344	840	3465.00
20	16-QAM	3/4	4	1344	1008	4158.00
21	16-QAM	13/16	4	1344	1092	4504.50
22	64-QAM	5/8	6	2016	1260	5197.50
23	64-QAM	3/4	6	2016	1512	6237.00
24	64-QAM	13/16	6	2016	1638	6756.75

5.4 IEEE 802.11ad Medium Access Control (MAC) Protocols

This section describes new MAC protocols for IEEE 802.11ad WLAN. Firstly, we look at the MAC architecture for IEEE 802.11ad WLAN. Secondly, we describe the new BI and new MAC protocols for IEEE 802.11ad WLAN. The new directional MAC protocols include BF, AP/PCP clustering, spatial sharing, link switching relaying, and link cooperating relaying.

5.4.1 MAC Architecture

The MAC architecture for legacy WLAN (IEEE 802.11a/b/g/n) is shown on the left side of Figure 5.1, while that for IEEE 802.11ad WLAN is shown on the right side of Figure 5.1.

The legacy WLANs operate in the 2.4/5 GHz band, while IEEE 802.11ad WLAN operates in the 60 GHz band. Distributed coordination function (DCF) uses carrier sense multiple access with collision avoidance (CSMA/CA) MAC protocol, while point coordination function (PCF) uses a reservation-like period for packet transmissions. Enhanced distributed channel access (EDCA) and hybrid coordination function controlled channel access (HCCA) are similar to DCF and PCF, respectively, except that EDCA have four access categories and eight priority classes and HCCA is an enhanced version of PCF. The four access categories are background, best effort, video, and voice. DCF, PCF, EDCA, and HCCA are for legacy WLANs. PCF is seldom implemented by vendors, if at all.

Under IEEE 802.11ad MAC, there are millimeter wave (mmWave) channel accesses like A-BFT access, announcement transmission interval (ATI) access, contention-based access, SP access, and polled access. These are shown on the right side of Figure 5.1. There is also a fast session

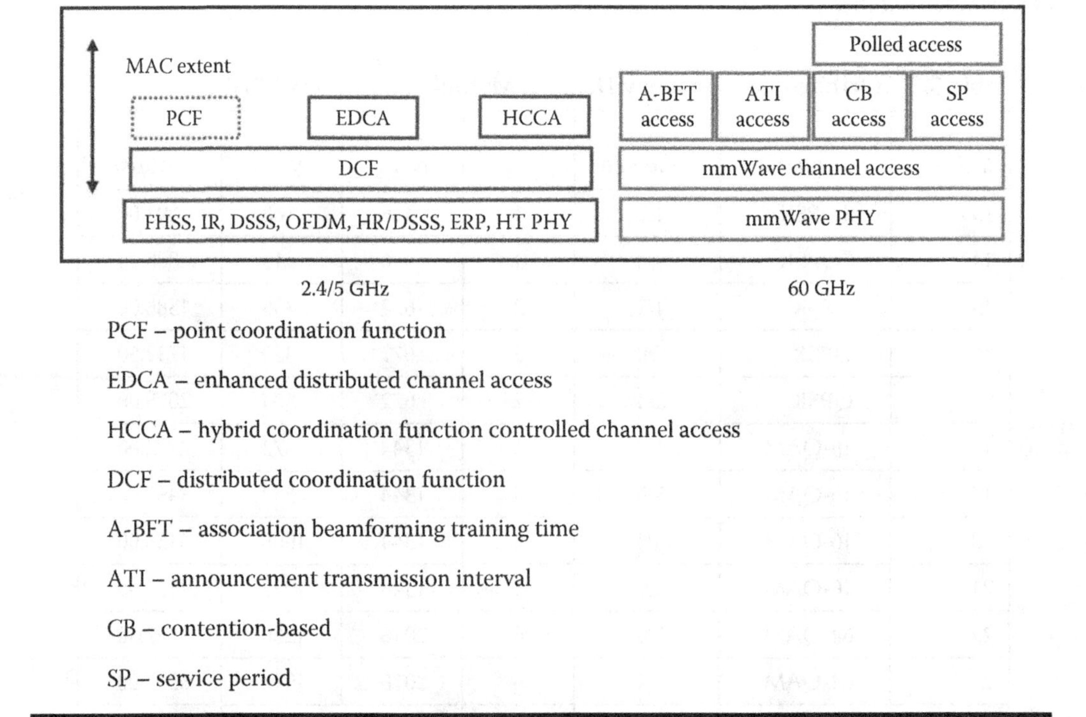

Figure 5.1 IEEE 802.11ad MAC architecture.

transfer MAC protocol to switch between legacy WLANs, including IEEE 802.11ac WLAN, at 2.4/5 GHz bands and IEEE 802.11ad WLAN at 60 GHz band. Fast session transfer allows multi-band operation across 2.4/5/60 GHz bands [6]. Thus, IEEE 802.11ad WLAN is backward compatible by switching to legacy WLANs at 2.4/5 GHz bands. Fast session transfer also supports both simultaneous and non-simultaneous operations and also support both transparent and non-transparent fast session transfer [6]. For transparent fast session transfer, the MAC address is the same in the bands, while for non-transparent fast session transfer, the MAC addresses are different in different bands [6].

5.4.2 Beacon Interval

The basic BI of IEEE 802.11ad MAC is shown in Figure 5.2. Each BI consists of four parts [4]: a beacon time interval (BTI), A-BFT, ATI, and a data transfer time interval (DTI).

The DTI consists of SPs and CBAPs and provides transmission opportunities for devices in the IEEE 802.11ad network. A beacon frame is used to startup the BF procedure between an AP/PCP and a receiving device (non-AP/non-PCP). In this chapter, devices also mean stations. They are equivalent. To join the network, a non-AP/non-PCP device scans for a beacon sent out by the AP/PCP and proceeds with the BF process with the AP/PCP in the A-BFT. The beacons can be sent out in different directions. The beacon enables synchronization, discovery of new devices, re-establishment of lost BF link of non-AP/non-PCP devices with the AP/PCP [6], etc. Association between the non-AP/non-PCP device and the AP/PCP is then done in the ATI or CBAP. A-BFT allows BF among a number of non-AP/non-PCP devices with the AP/PCP in the same A-BFT at the same time. Thus, associations between the AP/PCP and the non-AP/non-PCP devices do not have to be associated before the non-AP/non-PCP devices start to do BF with the AP/PCP. A-BFT is slotted. Management between AP/PCP and a non-AP/non-PCP device is done via request–response frame exchanges. AP/PCP uses A-BFT to provide CBAP and SP allocations in the DTI to the non-AP/non-PCP devices. All types of frames, including application data frame, can be transmitted in CBAPs and SPs. Transmissions in the CBAPs are based on a modified IEEE 802.11 EDCA [7] operation that is tweaked for directional frames exchanges. On the other hand,

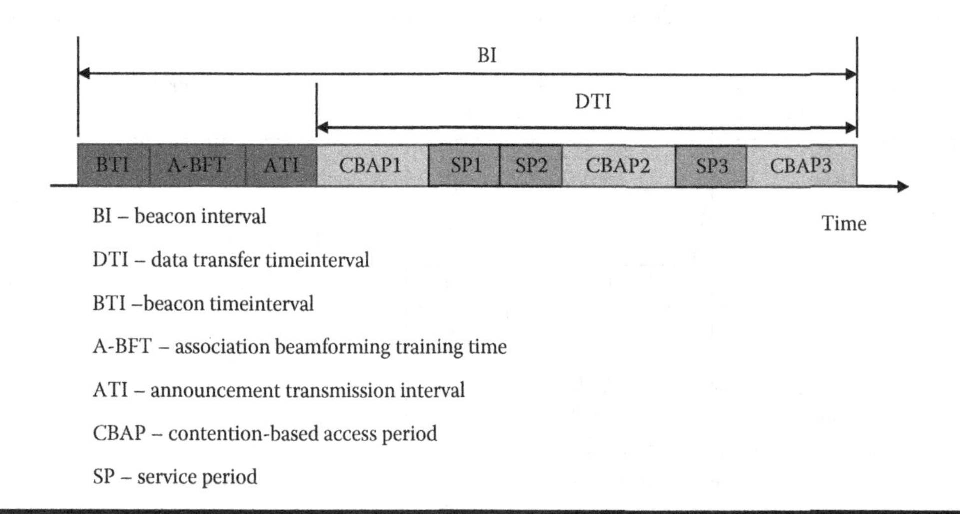

Figure 5.2 Basic beacon interval of an IEEE 802.11ad MAC.

transmissions in the SPs are scheduled and allocations of SPs are allocated to a particular two non-AP/non-PC devices or a non-AP/non-PCP device and the AP/PCP for communication in each SP. Only these specified non-AP/non-PCP devices or a non-AP/non-PCP device and the AP/PCP in the SP allocation can be transmitted in each of the SPs.

That is, an SP is used for transmission between two devices [6], including AP/PCP if necessary. The AP/PCP can also be considered as a device. SPs can be used not only for data transmissions but also for BF [6] as well as measurements for spatial sharing. BF can be done for an AP/PCP to a non-AP/non-PCP device or for a non-AP/non-PCP device to another non-AP/non-PCP device. Spatial sharing enables two or more SPs with transmissions at the same time. More on BF and spatial sharing are presented in Sections 5.4.3 and 5.4.5, respectively.

5.4.3 Beamforming

BF is a mechanism that is used by a pair of devices to determine appropriate antenna system settings for both transmission and reception. In IEEE 802.11ad, the BF training is a bidirectional training transmissions through sector sweep (SSW).

A device that initiates the BF training is referred to as the initiator, and the peer device that participated in the BF training with the initiator is referred to as the responder. Figure 5.3 shows an example of BF training procedure.

The BF training starts with a sector level sweep (SLS) and a beam refinement protocol (BRP) may follow if required. The purpose of the SLS is to select the best transmit or receive antenna pattern between the initiator and the responder that enable their communication at the control PHY rate or higher MCS. Furthermore, the BRP phase is to enable iterative refinement of the antenna weight vector (AWV) of both transmitter and receiver at both participating devices.

The SLS phase has four components: an initiator sector sweep (ISS) to train the initiator link, a responder sector sweep (RSS) to train the responder link, a SSW feedback, and a SSW acknowledgement (ACK). Normally, an initiator begins the ISS phase and a responder follows with the RSS, except when the ISS occurs in the BTI and the RSS occurs in the A-BFT. During the ISS, the initiator can select to execute either an initiator transmit sector sweep (TXSS) or an initiator receive sector sweep (RXSS) for the purpose of training the transmitter or receiver link to the responder, respectively. In a similar way, during the RSS, the responder can also execute either a responder

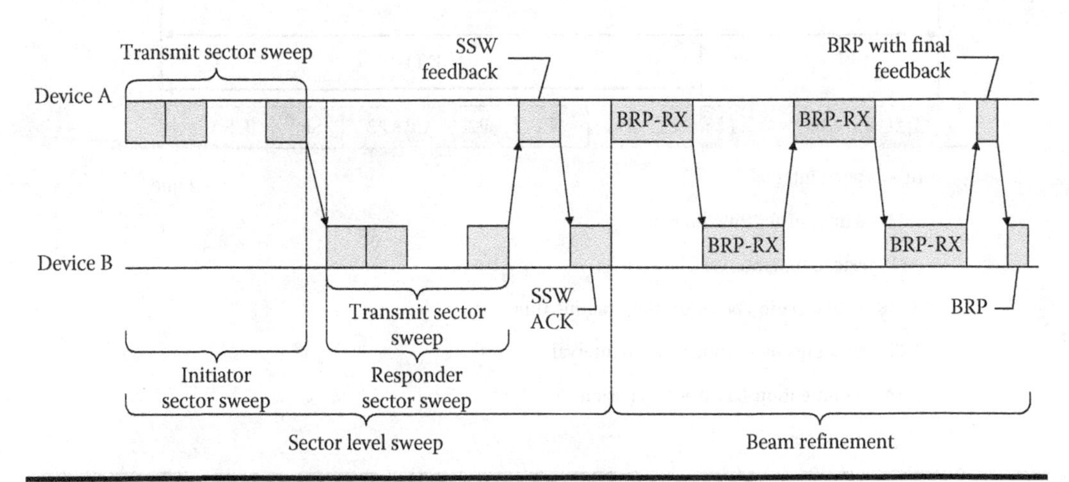

Figure 5.3 An example of beamforming training procedure.

TXSS or a responder RXSS for the purpose of training the transmitter or receiver link to the initiator, respectively, and transmit the feedback on the BF training result with the best quality obtained during the ISS. The SSW feedback occurs following each RSS. During the SSW feedback, the initiator transmits the feedback on the BF training result with the best quality obtained during the RSS to the responder. Afterward, the SSW ACK follows if present and the responder acknowledges the BF training result received from the initiator. During the SSW feedback and the SSW ACK, both the initiator and the responder use the best antenna pattern indicated by the feedback information obtained during the ISS and the RSS to transmit or receive with each other.

The BRP is a process in which a device trains its antenna array and further improves its antenna configuration using an iterative procedure. The BRP phase is composed of a BRP setup subphase, a multiple sector ID detection (MID) subphase, a beam combining (BC) subphase, a subset of the previous subphases, and one or more beam refinement transactions. The BRP setup subphase is to exchange the intent and capabilities to conduct the following subphases. The MID and BC subphases, collectively called the MIDC subphase, are used to find better initial AWVs between initiator and responder for iterative beam refinement than the best antenna pattern obtained from the SLS phase. However, either the BRP setup or the MIDC subphase or both can be skipped if both peer devices have no request for this operation. The beam refinement is a request/response-based process. Either device can request transmit or receive beam refinement training by sending a BRP frame to a peer device. For transmit beam refinement training, the requesting device sends a BRP frame containing the training fields and requests for the feedback from the responding device. The responding device replies with a BRP frame to indicate the sector with the best receive quality in the received BRP frame. For receive beam refinement training, the responding device receiving a BRP frame replies with a BRP frame containing the training fields to refine the receive beam of the requesting device. This process continues until both the requesting and responding devices have no request for trainings. After the successful completion of the BF training, BF is said to be established.

5.4.4 AP/PCP Clustering

IEEE 802.11ad AP/PCP clustering mechanism is able to improve spatial sharing and interference mitigation among co-channel networks. There are two types of clustering: the decentralized AP/PCP clustering and centralized AP/PCP clustering.

Figure 5.4 shows an example of decentralized AP/PCP cluster which consists of one synchronization-AP/synchronization-PCP (S-AP/S-PCP) and (n – 1) number of member APs/PCPs.

In every BI, a set of SPs is scheduled for the beacon frame transmissions, which are denoted by Beacon SP_n, n = 1, …, ClusterMaxMem. Here, the ClusterMaxMem refers to the maximum number of APs/PCPs that can participate in this cluster. The first Beacon SP is reserved for the S-AP/S-PCP, starting from the beginning of the S-AP/S-PCP's BI.

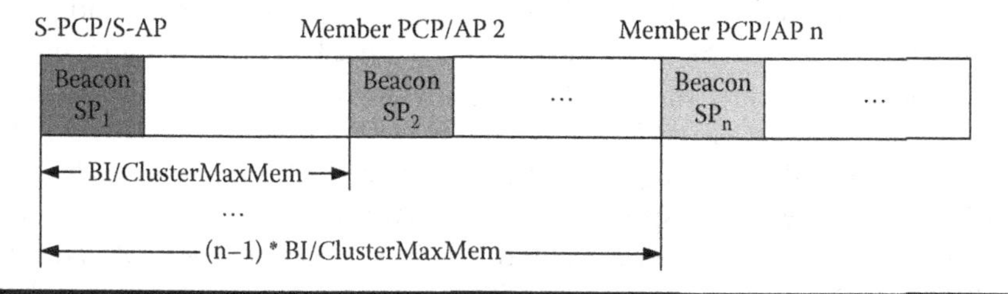

Figure 5.4 An example of decentralized AP/PCP cluster.

The following ones are staggered to one another for an offset equal to the BI length of the S-AP/PCP divided by the ClusterMaxMem. Moreover, the length of each Beacon SP is set by the S-AP/S-PCP through the cluster information contained in the beacon frame and its maximum size can be equal to BI/ClusterMaxMem.

If another AP/PCP operates on the same channel and discovers the presence of a S-AP/S-PCP within a cluster, it may find an empty Beacon SP of this cluster to become a member AP/PCP by setting the same length of BI with S-AP/S-PCP, adjusting its BI to start from the beginning of the empty Beacon SP, and transmitting the beacon frame during this Beacon SP. If no Beacon SP is empty, the AP/PCP may cease its activity on this channel and attempt operations on a different channel.

Within a cluster, an S-AP/S-PCP and a member AP/PCP should not transmit or schedule transmissions during a Beacon SP that is not its own Beacon SP. Obviously, the clustering mechanism is to properly stagger the starting time of each AP/PCP's BI as well as protect the beacon frame transmission to solve the co-channel interference problem.

To further mitigate any interference with the transmissions, an AP/PCP may receive the beacon frame during every Beacon SP except for the one occupied by itself to know the scheduling information of other APs/PCPs and then allocates its CBAPs or SPs during the following DTIs on a first-allocate-first-use basis.

On the other hand, an AP/PCP can set up a centralized AP/PCP cluster by performing the following two steps in order: configuration and verification. The configuration step is to obtain the configuration information of this centralized cluster from a centralized coordination service root (CCSR), which plays the role of the coordinator for all the S-APs within the same service set. The verification step is to monitor and find an empty channel on which the device setups a centralized AP/PCP cluster. After the successful execution of the aforementioned steps, the devices can become a S-AP by transmitting the beacon frame with the centralized cluster information configured by CCSR.

Another AP/PCP that receives the beacon frame from the S-AP can choose to join this centralized AP/PCP cluster as a member AP/PCP. However, the process of finding an empty Beacon SP is different with that in a decentralized AP/PCP cluster. In a centralized AP/PCP cluster, it not only needs to monitor the channel to find the empty Beacon SPs but also needs the help of a secondary non-AP/non-PCP device to associate with the S-AP to obtain the available Beacon SP set from it. Thus, the Beacon SP index is selected from the intersection of the empty Beacon SPs monitored by itself with the available Beacon SP set indicated by the S-AP. After that, the secondary non-AP/non-PCP device responds to the S-AP with the selected index to update the available Beacon SP set.

5.4.5 Spatial Sharing

The spatial sharing mechanism is to allow SPs belonging to different devices in the same spatial vicinity to be scheduled concurrently over the same channel with interference mitigation.

As illustrated in Figure 5.5, SP_e is an existing SP that has been allocated for communication between a pair of devices A and B, and SP_c is a candidate SP that is allocated for communication between a pair of devices C and D. Here, candidate SP refers to an SP that is to be assessed for spatial sharing with other existing SPs. We assume that the BF training has been performed between each pair. For the sake of spatial sharing, the AP/PCP transmits a request to devices C and D to measure over SP_e's allocation and also transmits a request to devices A and B to measure over SP_c's allocation. After that, each station (STA) performs the measurement and reports the result to the AP/PCP.

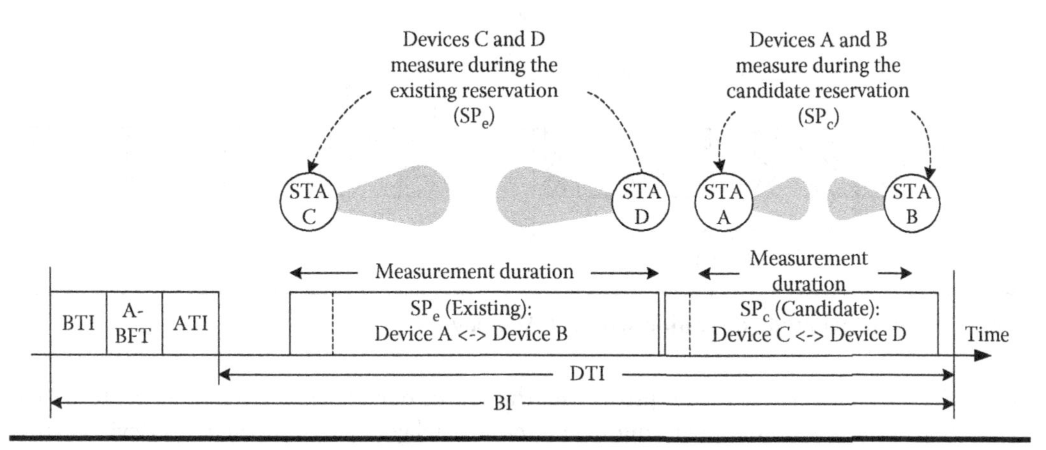

Figure 5.5 An example of spatial sharing.

The AP/PCP estimates the channel quality based on the feedback information. If the relevant interference level is lower than the predefined receiver sensitivity, AP/PCP may schedule the SP_c overlapping with the SP_e in time.

However, the spatial sharing mechanism in IEEE 802.11ad WLANs is a blind selection process based on measurement. No priori information can be provided for the AP/PCP to appropriately choose an exact existing SP for a candidate SP to assess. Thus, the efficiency of this kind of spatial sharing mechanism needs to be further improved when many candidate SPs intend for spatial sharing with the existing SPs.

5.4.6 Link Switching Relaying

In IEEE 802.11ad network, a source device can transmit frames to a destination device with the help of another relay device. The link between only the source device and the destination device is called direct link, and the link between the source device and the destination device via the relay device is called relay link. A source device, a destination device and a relay device can establish two types of relaying: link switching or link cooperating. The link switching means the transmission between the source device and the destination device via either the direct link or the relay link. However, the link cooperating adopts a different way that the transmission from the source device to the destination device is simultaneously repeated by the relay device which can possibly increase the signal quality received at the destination device.

Before link switching or link cooperating, the source device and the destination device must perform the relay selection procedure and the relay link setup (RLS) procedure. The relay selection procedure is described as follows. First, the source device sends a relay search request to the AP/PCP to initiate the discovery of the relay devices. Upon receiving the relay search request, the AP/PCP replies the source device and the destination device, respectively, with the list of all the relay devices in this network. In the following time, the AP/PCP also schedules two SPs for each relay device: one SP for BF training between the source device and the relay device and the other one for BF training between the relay device and the destination device. Except of performing the BF trainings for each relay link, the source device and the destination device also perform the BF training between each other for the direct link. Thus, the source device and the destination device can use the direct link to exchange the channel measurement information obtained in the BF trainings for each relay link and select one of them as the relay

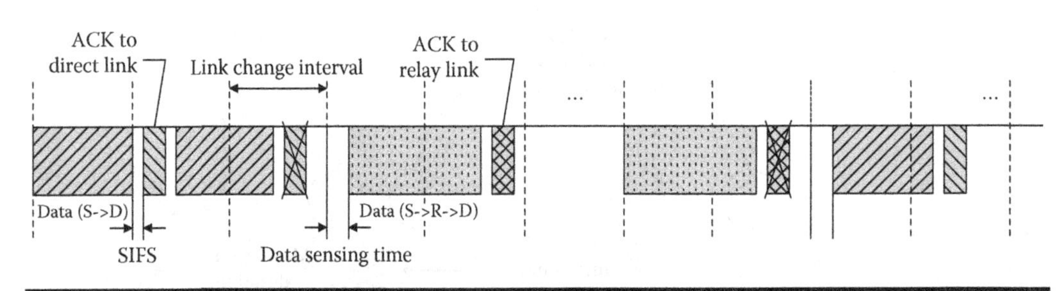

Figure 5.6 Normal mode operation with FD-AF relay.

link. Following the relay selection procedure, the source device sends a RLS request for either the link switching or link cooperating to the selected relay device and the relay device that is willing to participate in the relay link shall forward this request to the destination device. If the destination device is also willing to participate in this relay link, it shall reply a RLS response to the relay device and the relay device forwards this response to the source device. Then, the source device transmits an announcement to the AP/PCP to indicate that the RLS procedure has completed successfully.

The link switching relaying operates as follows. If the AP/PCP broadcasts an SP allocation to a pair of source device and destination device, a relay device shall check whether or not this pair of devices has successfully completed the RLS procedure with it. If the relay device is selected for the relay link between this pair of devices, it shall operate in either full-duplex amplify-and-forward (FD-AF) or half-duplex decode-and-forward (HD-DF) during that SP allocation according to its ability.

A FD-AF relay device can work under two transmission modes: normal and alternative. If the normal mode is used, a pair of devices transmits via either the direct link or the relay link until the current link becomes unavailable. As shown in Figure 5.6, if the source device decides to change the link, it will begin its transmission at the start of the following link change interval after a data sensing time period. Here, the link change interval and the data sensing time period are predefined by the source device during the previous RLS procedure. On the other hand, if the alternative mode is used, a pair of device transmits via both links and the use of each link alternates at the start of each link change interval. Note that the used transmission mode is also indicated during the RLS procedure and may be changed in the following exchanges of RLS request and response.

However, a HD-DF relay device can only work under the normal mode. The frame transmission between the source device and the destination device via the relay link is performed in two periods as shown in Figure 5.7.

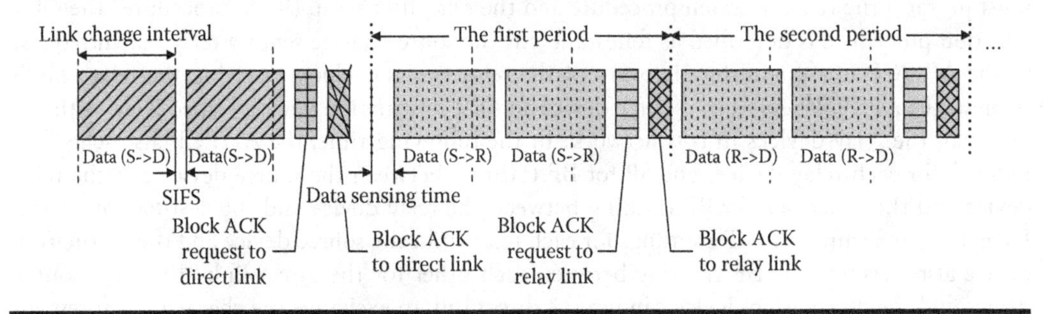

Figure 5.7 Normal mode operation with HD-DF relay.

In the first period, the source device transmits a frame to the relay device, and the relay device responses if needed. In the second period, the relay device forwards the frame received from the source device to the destination device, and the destination device responses if needed. This process is repeated as long as the relay link is used.

5.4.7 Link Cooperating Relaying

If a pair of source device and destination device intends to use the link cooperating relaying, they should not only complete the RLS procedure (defined in Section 5.4.6) but also perform the transmission time-point adjustment (TPA) procedure. Since the direct link and the relay link have different transmission delay, the TPA procedure is to eliminate the time variations when the destination device receives two duplicated frames transmitted from the source device and the relay device, respectively.

The TPA procedure is triggered by the destination device following the RLS procedure. First, the destination device requests the AP/PCP to allocate an SP for the source device and itself. Then, the AP/PCP broadcasts and indicates that the source of SP is the destination device and the destination of SP is the source device. Thus, the relay device that receives this SP allocation will perform the TPA procedure during this SP.

Within the allocated SP, the destination device first sends the TPA requests to the relay device and the source device one by one. Then, the relay device and the source device that receive the TPA request shall reply with a TPA response to the destination device after the same predefined delay interval (denoted by aDtime), respectively. Upon receiving the TPA responses, the destination device can estimate the transmission delays between itself and the relay device (denoted by T_{DR}), and between itself and the source device (denoted by T_{DS}) according to the TPA request transmitting time and the receiving time. In a similar way, the source device can also estimate the transmission delay between itself and the relay device (denoted by T_{SR}). After this procedure, the destination device sends one more TPA request to the relay device and indicates the transmission delay offset $T_{DS} - T_{DR}$. Upon receiving this TPA request, the relay device replies the TPA response after a time interval of aDtime + $T_{DS} - T_{SR}$. Then, the destination shall estimate whether or not the two-way transmission delay between itself and the relay device is equal to the value of $2 \times T_{DR} + (T_{DS} - T_{DR})$. If the estimated transmission delay meets the calculated value $2 \times T_{DR} + (T_{DS} - T_{DR})$, the TPA procedure is said to be successful. The last step is that the destination device sends a TPA report to the source device to indicate whether the last TPA is successful or not.

After completing the RLS procedure and the TPA procedure, the link cooperative relaying operates in two steps as shown in Figure 5.8. Within an allocated SP, the source device first transmits a data frame to the relay device. Then, after a predefined time (called aPtime) plus T_{SR}, the source device retransmits the same frame to the destination device. Similarly, upon receiving the data frame, the relay device also transmits the same frame to the destination device after a time interval of aPtime + $T_{DS} - T_{DR}$. Obviously, the destination device can receive the transmission from the source device and the relay device at the same time, and the total transmission delay time is equal to aPtime + $T_{SR} - T_{DS}$. This process is repeated as long as the relay link is used.

5.5 Performance Analysis of MAC Saturated Throughput

This section analyses and evaluates the saturated throughput of IEEE 802.11ad in the CBAP of the IEEE 802.11ad MAC *directionally* with *multi-rate*. The analytical approach here can also be applied to IEEE 802.15.3c MAC and other directional and multi-rate MAC protocols.

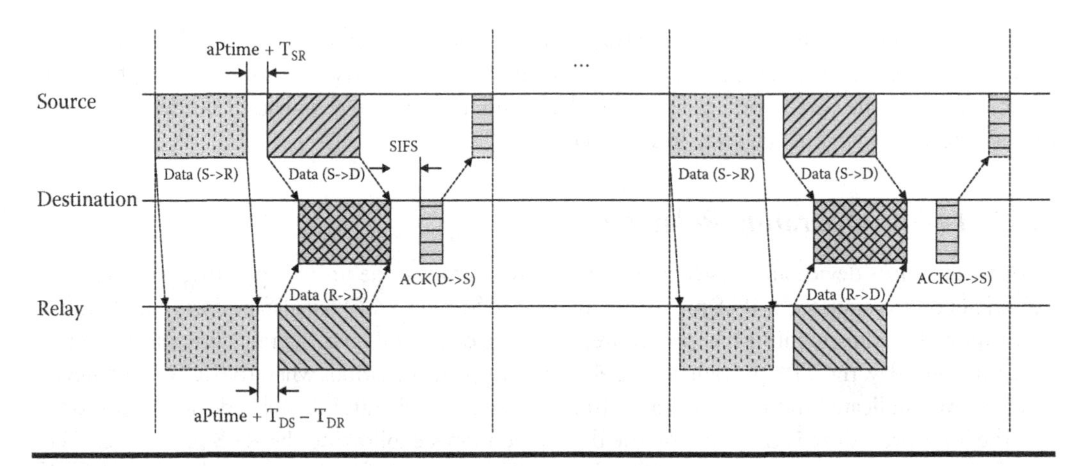

Figure 5.8 Data transmission with link cooperating relay.

In this section, an approximate analytical formulation of the saturated throughput of IEEE 802.11ad MAC in the CBAP is presented for directional access with multi-rate. The analytical framework is formulated for N non-AP/non-PCP devices, excluding the AP/PCP. The state transition diagram is modeled by a two-dimensional discrete-time MC. One dimension of the MC is for the backoff stage and the other dimension is for the value of the backoff counter. The saturated throughput is approximated by the product of a weighted ratio of the throughput of the CBAPs and the throughput of a CBAP period minus the average period necessary to transmit a request-to-send (RTS) frame, a clear-to-send (CTS) frame, a data packet, an ACK frame, and four short inter-frame space (SIFS) times before the end of the current CBAP, the probability of the number of devices in a beam sector and the number of beam sectors, the probabilities of each non-AP/non-PCP device in a particular region with respect to the AP/PCP and the weighted average period of a successful packet transmission. Packets that are transmitted prior to this period at the end of a CBAP via the RTS/CTS access mode have chances of being successfully transmitted, while packets cannot be successfully transmitted via RTS/CST access mode within this period at the end of the CBAP and are thus not transmitted. Numerical results of the saturated throughput corresponding to typical parameter values are presented. These results show the performance results of IEEE 802.11ad MAC in the CBAPs using the OFDM mode with a varying number of sectors and multi-rate scenarios of 10, 7, and 3 data rates. The agreement between the analytical and simulation results for the directional and multi-rate IEEE 802.11ad MAC is good.

5.5.1 CBAP MAC Protocol

We assume that BF has been completed. Transmission in the CBAP is done using CSMA/CA in a directional manner, similar to IEEE 802.11e with one traffic class, and there is a *period* before the end of the CBAP that must be considered before transmitting a packet. This *period* includes a RTS frame, a CTS frame, a packet transmission time, an ACK frame time, and four SIFSs. If the time at the point of decision for transmitting a RTS frame to the end of the current CBAP is less than this *period*, the RTS frame will not be send out as the whole RTS frame/CTS frame/packet/ACK frame with four SIFS will not be successfully transmitted as the remaining time to the end of the CBAP for frame and packet transmissions is too short. Note that a SIFS time is considered after the ACK frame for this *period*, rather than a DIFS time, because a new RTS frame cannot be transmitted at the end of the CBAP. On the other hand, if the time at the point of decision for transmitting a RTS frame to the end of the current CBAP is at least this *period*, the RTS frame can be send out as the whole RTS frame/CTS frame/packet/ACK frame with four SIFS and could be successfully transmitted as the remaining time to the end of the CBAP for the three frames, a packet transmission and four SIFS times is long enough, subjected that there is no RTS frame collision for the packet to be successfully transmitted via the RTS/CTS access mode. We also assume that the AP/PCP's antenna switching time is small and is ignored in our calculations. There are M CBAPs of equal sizes corresponding to M directions from the AP/PCP with devices and their corresponding data rates, R_ks, and distances, D_ks, from the AP/PCP as shown in Figure 5.9. Non-AP/non-PCP devices are assumed to be uniformly distributed in the M pie areas of a circle. The nearer a device is to the AP/PCP, the higher is its data rate.

CSMA/CA MAC: Assuming IEEE 802.11 DCF MAC as the baseline, the CSMA/CA MAC protocol works directionally in each CBAP as follows:

- If the channel is idle for more than a distributed coordination function inter-frame space (DIFS) time, a device can transmit a RTS frame immediately.
- If the channel is busy, the device will generate a random backoff period. This random backoff period is uniformly selected from zero to the current contention window size. The backoff counter will decrement by 1 if the channel is idle for each time slot and will freeze if the channel is sensed busy or if its CBAP is not active.
- The backoff counter is reactivated to count down when the channel is sensed idle for more than a DIFS or when its CBAP is active. At the initial backoff stage, the current contention window size is set at the minimum contention window size.
- If the backoff counter reaches zero, the device will attempt to transmit its RTS frame if the remaining time to the end of its active CBAP greater than or equal to the *period* mentioned

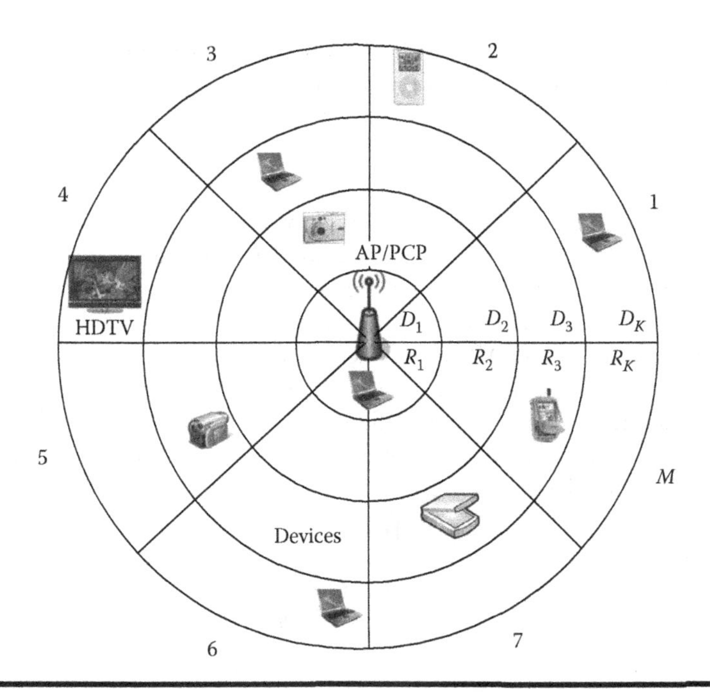

Figure 5.9 Devices are uniformly distributed in the *M* directions of the AP/PCP.

earlier. If it is successful, the destination device will reply with a CTS frame after a SIFS time, the source device will transmit the packet after a SIFS time and the destination device will reply with an ACK frame after a SIFS time, and the current contention window size is reset to the minimum contention window size. If it is not successful, it will increase the current contention window size by doubling it and adding one in the next backoff stage and a new random backoff period is selected as mentioned earlier.

- If the backoff counter reaches zero, the device will not attempt to transmit its RTS frame if the remaining time to the end of its current active CBAP is less than the *period* mentioned earlier. Instead, it will increase the current contention window size by doubling it and adding one in the next backoff stage and a new random backoff period is selected as mentioned earlier, and the counting down of the backoff counter will start after a DIFS after the next ATI. If the maximum retry limit is reached, the packet will be dropped and the next packet will start its backoff process with a minimum contention window size after a DIFS from the beginning of its next active CBAP. The backoff counter does not countdown immediately as it could reach zero but it still cannot transmit via the RTS/CTS access mode as the period at the end of the CBAP is not long enough. Thus, the countdown effort in this case is simply wasted.
- If a busy period ends within the *period* mentioned earlier, all the backoff counters will freeze until after a DIFS from the beginning of its next active CBAP.
- This process repeats itself until the RTS frame is first successfully transmitted via RTS/CTS access mode or until the maximum retry limit is reached. If the RTS frame is still not successfully transmitted and the maximum retry limit is reached, the packet is dropped.
- If a device does not receive a CTS frame within a CTS timeout period after a RTS frame is transmitted and a SIFS time, it will continue to attempt to retransmit the RTS frame first via the RTS/CTS access mode according to the backoff algorithm.

We assume that there are no errors in the RTS frame, CTS frame, packet, and ACK frame. Only collisions may happen to RTS frames if more than one device transmits at the same time slot.

5.5.2 Analytical Model

We assume that the CBAPs and SPs do not overlap for simplicity and trackability in the analysis. That is, no spatial reuse is considered. Furthermore, we assume that there are no concurrent transmissions in multiple sectors. In addition, we assume that the CBAPs are divided into equal intervals according to the different directions for simplicity and trackability too. Let N, n, and M denote the total number of non-AP/non-PCP devices, excluding the AP/PCP device, the number of non-AP/non-PCP devices in a beam sector, excluding the AP/PCP device, and the number of beam sectors, respectively. Note that the definition of n here differs from that in Section 5.4.4. Let us first consider a *beam sector*. Assuming uniform distribution of devices around the AP/PCP device, the probability of n devices in one beam sector, excluding the AP/PCP device, denoted by $\Pr(n)$, is given by

$$\Pr(n) = \binom{N}{n} q^n (1-q)^{N-n}, \quad (5.1)$$

where

$$q = \frac{1}{M}. \tag{5.2}$$

We consider a generic K number of data rates, denoted by R_ks, where $k = 1, \ldots, K$, and $R_1 > R_2 > \ldots > R_K$ and their respective coverage distances are denoted by D_k's, $D_1 < D_2 < \ldots < D_K$ as shown in Figure 5.9. The nearer a non-AP/non-PCP device is to the AP/PCP, the higher its data rate. By the same token, the farther a non-AP/non-PCP device is away from the AP/PCP, the lower its data rate. We assume that all devices are uniformly distributed in a circular coverage area of radius R, where $R = D_K$, and the AP/PCP is at the center of the circle. In real situations, the coverage area may not be a perfect circular disc and the assumption of uniform distribution of devices from the AP/PCP may not hold. Nevertheless, this assumption gives first-cut results, and simplifies and helps in the analytical model. Non-AP/non-PCP devices within circle of radius D_1 use a data rate R_1 to communicate with the AP/PCP, while non-AP/non-PCP devices within concentric circles of radii D_{k-1} and D_k use a data rate R_k, $k = 2, \ldots, K$, to communicate with the AP/PCP. Let P_k, $k = 1, \ldots, K$, denote the probability of being in each of these regions using of data rate R_k from the source non-AP/non-PCP device to the AP/PCP and it is given by

$$P_k = \begin{cases} D_k^2/D_K^2, & k = 1 \\ (D_k^2 - D_{k-1}^2)/D_K^2, & k = 2, \ldots, K \end{cases}. \tag{5.3}$$

Consider the probability of being in each of these regions using data rate R_{l_i} for the ith source non-AP/non-PCP device, denoted by P_{l_i}, where $l_i = 1, \ldots, K$, and $i = 1, \ldots, n$ or $i = 1, \ldots, N$, and is given by

$$P_{l_i} = \begin{cases} \dfrac{D_{l_i}^2}{D_K^2}, & l_i = 1, i = 1, \ldots, n \text{ or } i = 1, \ldots, N, \\ \dfrac{(D_{l_i}^2 - D_{l_i-1}^2)}{D_K^2}, & l_i = 2, \ldots, K, i = 1, \ldots, n \text{ or } i = 1, \ldots, N \end{cases} \tag{5.4}$$

Let $a(t)$ be a random process representing the backoff stage x, $x = 0, 1, \ldots, L_{retry}$, at time t, where L_{retry} is the retry limit. Let $b(t)$ be a random process representing the value of the backoff counter at time t. The value of the backoff counter $b(t)$ is uniformly chosen in the range of $(0, 1, \ldots, W_x)$, where $W_x = 2W_{x-1} + 1$ for $0 \le x \le L_{retry}$.

$$W_x = (W+1)2^x - 1, 0 \le x \le L_{retry} \tag{5.5}$$

where $W = W_0$.

Let p denote the probability that a transmitted RTS frame collides in a CBAP. Let δ denote a time slot in the backoff counter. The two-dimensional random process $\{a(t), b(t)\}$ is a discrete-time MC. Therefore, the state of each device is described by $\{x, y\}$, where x stands for the backoff stage taking values from $\{0, 1, \ldots, L_{retry}\}$ and y stands for the backoff delay having values from $\{0, 1, \ldots, W_x\}$ in time slots.

Let $b_{x,y} = \lim_{t\to\infty} \Pr[a(t) = x, b(t) = y]$ be the stationary distribution of the MC. Its non-null transition probabilities are as follows:

$$\Pr[(0,y)\,|\,(x,\ 0)] = \frac{(1-p)}{(W_0+1)},\ 0 \le y \le W_0,\ 0 \le x < L_{retry}$$

$$\Pr[(0,y)\,|\,(L_{retry},\ 0)] = \frac{1}{(W_0+1)},\ 0 \le y \le W_0,\ x = L_{retry}$$

$$\Pr[(x,y)\,|\,(x,y+1)] = 1,\ 0 \le y \le W_x - 1,\ 0 \le x \le L_{retry}$$

$$\Pr[(x,y)\,|\,(x-1,\ 0)] = \frac{p}{W_x+1},\ 0 \le y \le W_x,\ 0 \le x \le L_{retry}$$

The transition probability to each state in backoff stage 0 from backoff stage x, $0 \le x < L_{retry}$, is given in the first equation, while the transition probability to each state in backoff stage 0 from backoff stage L_{retry} is given in the second equation. The backoff counter decrementing by 1 is given in the third equation, while the transition probability to each state in backoff stage x is given in the fourth equation.

Similar to References 8 and 9, we have $b_{0,0}$ which is given by

$$b_{0,0} = \frac{2(1-p)(1-2p)}{(W_0+1)(1-p)(1-(2p)^{L_{retry}+1}) + (1-2p)(1-p^{L_{retry}+1})}. \tag{5.6}$$

A non-AP/non-PCP device only attempts to transmit its RTS frame when its backoff counter reaches zero, that is, the device is at any of the states $\{x, 0\}$, where $0 \le x \le L_{retry}$. The probability that a non-AP/non-PCP device transmits in a generic time slot, denoted by τ, is given by

$$\tau = \sum_{x=0}^{L_{retry}} b_{x,0} = \sum_{x=0}^{L_{retry}} p^x b_{0,0} = \left[(1-p^{L_{retry}+1})b_{0,0}\right] / (1-p) \tag{5.7}$$

A transmitted RTS frame collides when one or more non-AP/non-PCP devices in the beam sector transmit its/their RTS frame(s) during a time slot in a CBAP. The probability that a non-AP/non-PCP device senses the channel busy in the backoff stage, denoted by p, is given by

$$p = 1-(1-\tau)^{n-1}. \tag{5.8}$$

The probabilities τ and p can be solved numerically.

The probability that the channel is busy in a CBAP period happens when at least one non-AP/non-PCP device transmits its RTS frame during a time slot, denoted by p_b, and is given by

$$p_b = 1-(1-\tau)^n. \tag{5.9}$$

The probability of a successful RTS frame transmission in a time slot, denoted by p_s, is given by

$$p_s = n\tau(1-\tau)^{n-1}. \tag{5.10}$$

The probability of an idle channel for a time slot is $(1-p_b)$. The probability that the channel is neither idle nor successful for a time slot is $[1-(1-p_b)-p_s] = p_b - p_s$.

Let T_{BI} be the BI, T_{BTI} be the BTI, $T_{A\text{-}BFT}$ be the A-BFT, T_{ATI} be the ATI, T_{SP} be the sum of all SPs and $T_{CBAP/M}$ be a CBAP period, assuming equal CBAP periods. $T_{CBAP/M}$ is given by

$$T_{CBAP/M} = (T_{BI} - T_{BTI} - T_{A-BFT} - T_{ATI} - T_{SP})/M. \tag{5.11}$$

Let T_Q be the average time required to successfully transmit a RTS frame, a SIFS, a CTS frame, a SIFS, a contention period packet including header and payload, a SIFS, an ACK frame and a SIFS at the point of decision, weighted by P_k. If the remaining time before the end of the current active CBAP is greater than $T_{Q,k}$ (using data rate R_k) when the backoff counter is zero, the packet can be attempted to be transmitted via the RTS/CTS access mode in the remaining active CBAP period, otherwise it will not be transmitted. That is,

$$T_{Q,k} = T_{RTS} + T_{CTS} + T_H + T_{E(L),k} + T_{ACK} + 4T_{SIFS}, \tag{5.12}$$

$$T_Q = \sum_{k=1}^{K} T_{Q,k} P_k, \tag{5.13}$$

where $T_{Q,k}$, denote the time required to successfully transmit a RTS frame, a SIFS, a CTS frame, a SIFS, a contention period packet including header and payload using data rate, R_k, a SIFS, an ACK frame, and a SIFS at the point of decision. If the remaining time to the end of the CBAP at the point of decision for RTS frame transmission is shorter than $T_{Q,k}$, the packet will not be successfully transmitted via RTS/CTS access mode as the *period* is not long enough. On the other hand, if the remaining time to the end of the CBAP at the point of decision for RTS frame transmission is at least $T_{Q,k}$, the packet will have a chance of being successfully transmitted via RTS/CTS access mode as the *period* is long enough. T_H, L, $T_{E(L),k}$, T_{SIFS}, T_{ACK}, T_{RTS}, and T_{CTS} denote respectively the time to transmit the header (including preamble and frame header consisting of MAC header, PHY header, header check sequence [HCS], and Reed–Solomon parity bits), the length of the payload including frame check sum (FCS) and pad bits using data rate R_k, the SIFS time, the time to transmit an ACK, a RTS frame, and a CTS frame.

Let $T_{E(L)}$, T_{l_i}, and T_c, T denote the time to transmit a payload with length $E(L)$, the average time required to *successfully* transmit a RTS frame, a CTS frame, a CBAP packet using data rate R_{l_i} of the ith device, an ACK frame, three SIFS times and one DIFS time, and the average time that the CBAP channel has a *collision*, the average length of a slot time for n non-AP/non-PCP devices, respectively. Furthermore, let T_{DIFS} denote a DIFS time for a non-AP/non-PCP device. The fraction of CBAP in a BI, denoted by $F_{CBAP/M}$, is given by

$$F_{CBAP/M} = \frac{T_{CBAP/M}}{T_{BI}}, \tag{5.14}$$

while the fraction of CBAP minus T_Q in a BI, denoted by $F_{CBAP/M-TQ}$, is given by

$$F_{CBAP/M-TQ} = \frac{T_{CBAP/M} - T_Q}{T_{BI}}. \tag{5.15}$$

The saturated throughput, denoted by S, is given by

$$S = M\sum_{n=1}^{N} \Pr(n)\sum_{l_1=1}^{K}\sum_{l_2=1}^{K}\cdots\sum_{l_n=1}^{K} P_{l_1}P_{l_2}\ldots P_{l_n} \times \frac{p_s E(L)}{T} \times \left(\begin{array}{l} \dfrac{T_{CBAP/M}}{2T_{CBAP/M} - T_Q} F_{CBAP/M} \\ + \dfrac{T_{CBAP/M} - T_Q}{2T_{CBAP/M} - T_Q} F_{CBAP/M-TQ} \end{array} \right). \tag{5.16}$$

where

$$T = \sum_{i=1}^{n} ((1 - p_b)\delta + p_s T_{l_i} + (p_b - p_s)T_c)/n, \tag{5.17}$$

$$T_{l_i} = T_{RTS} + T_{CTS} + T_H + T_{E(L),l_i} + 3T_{SIFS} + T_{ACK} + T_{DIFS}, \tag{5.18}$$

$$T_c = T_{RTS} + T_{CTS\_timeout} + T_{DIFS}, \tag{5.19}$$

and

$$T_{CTS\_timeout} = T_{SIFS} + T_{CTS}. \tag{5.20}$$

Note that Pr(n) is the probability of n devices in one beam sector, excluding the AP/PCP device and the number of devices in a beam sector, n, varies from 1 to N. For n devices, each of them can use a data rate from R_1, R_2 to R_K with a probability of P_{l_1} for device 1, a probability of P_{l_2} for device 2 and a probability of P_{l_n} for device n. P_{l_i} is equivalent to the probability of being in each regions, P_k, using data rate R_{l_i}, except that it is for the ith device, $i = 1, \ldots, n$. That is, $l_i = k = 1, \ldots, K$. The average slot time, T, is the sum of probabilistically weighted by the average idle time, average successful packet transmission time and the average RTS frame collision time for all n devices and divided by n for the n devices, each transmitting with a data rate R_{l_i}. Note that p_s is the probability of successful RTS transmission in a time slot and $E(L)$ is the average packet length. Note that $T_{E(L),l_i}$ is equivalent to $T_{E(L),k}$, except that it is for the ith device, $i = 1, \ldots, n$, and $l_i = k = 1, \ldots, K$. That is, $T_{E(L),l_i}$ is the length of the payload including FCS and pad bits for the ith device using data rate R_{l_i}. We consider two cases for the CBAP transmissions. One of the cases is a whole CBAP being used for transmissions, while the other case is that for the CBAP minus the *period*, at the end of the CBAP, being used for transmissions. Within this *period*, transmissions using RTS/CTS access mode will not be successfully transmitted as the *period* is not long enough and thus will not be transmitted. These two cases are represented by $F_{CBAP/M}$ and $F_{CBAP/M-TQ}$, respectively. There are cases that fall between these two cases, but we focus on these two extreme cases for simplicity of analysis. Thus, this part is an approximation. Nevertheless, the simulation results in Section 5.5.3 show that this approximation is reasonable for the typical values used and the agreement between the numerical and simulation results is good. The weighted probabilities for each of these two cases are simply the probability of each case divided by the sum of the probabilities of the two cases. This is shown in the last two weighted terms for the two cases in the brackets of Equation 5.16. M is the number of beam sectors. Note that Equation 5.16 is computationally intensive for large N and K due to the nested summations depending on the number of devices in each beam sector.

5.5.3 Numerical Results

In this section, we present result for the saturated throughput of IEEE 802.11ad MAC in the CBAPs. The agreement between the numerical and simulation results is good. Here, we assume the OFDM mode of the PHY of the IEEE 802.11ad.

The parameter values used in the numerical examples for the IEEE 802.11ad MAC are based on the OFDM mode and are tabulated as in Table 5.4. The distance versus data rate is obtained from Reference 10. The RTS/CTS CSMA/CA access is assumed in this section. Furthermore, the sum of the SPs is assumed to be zero so as to assess the maximum achievable saturated throughputs for multi-rate in the directional CBAP. The sectors are also assumed to be equal sizes.

Table 5.4 Parameter Values Used for IEEE 802.11ad MAC

Symbol	*Value*	*Symbol*	*Value*
T_{BI}	1 s	W_j	{31, 63, 127, 255, 511, 1023, 2047}
T_{BTI}	1000 μs	T_H (preamble and MAC and PHY headers)	1.89 1 + 0.58 2 + 5.65198052 = 8.12498052 μs
$T_{A\text{-}BFT}$	5000 μs	$E(L)$	7920/16 × 7920 bytes (payload)
T_{ATI}	5000 μs	R_k (K = 10, 7, 3)	{1386, 1732, 2079, 2772, 3415, 4158, 4504.5, 5197.5, 6237, 6756.75}, {2772, 3415, 4158, 4504.5, 5197.5, 6237, 6756.75}, {5197.5, 6237, 6756.75} Mbps
W	31	D_k (K = 10, 7, 3)	{10, 7, 5.25, 4, 3.25, 2.5, 2, 1.75, 1.5, 0.5}, {4, 3.25, 2.5, 2, 1.75, 1.5, 0.5}, {1.75, 1.5, 0.5} m
L_{retry}	6	T_{RTS}	10.45225325 μs
δ	5 μs	T_{CTS}	10.45225325 μs
T_{SIFS}	3 μs	T_{ACK}	10.45225325 μs
T_{DIFS}	13 μs	M	1, 4, 8, 12
K	10, 7, 3		

The simulation model is implemented using the simulation kernel *smpl* [11] and *C* programs. The *C* programs use the scheme mentioned in Section 5.5.1 to handle the cases when backoff counter is zero and the remaining time before the end of the current active CBAP is greater or equal to, or less than, the *period* needed to successfully transmit a RTS frame, CTS frame, data packet, an ACK frame, and four SIFSs before the end of the current active CBAP. The analytical results for $M = \{4, 8, 12\}$ are obtained from Section 5.5.2, while that for $M = 1$ is obtained from the Appendix. Analytical results are denoted by symbols and "Anal," while simulation results are denoted by lines and "Sim."

For the results in this section, T_{SP} is set to zero. The saturated throughputs of devices for number of beam sectors, $M = \{1, 4, 8, 12\}$, payload of 7920 bytes, $R = 10$ m, and $K = 10$ data rates are shown in Figure 5.10. The range of the number of devices (non-AP/non-PCP), N, is from only 2 to 10 as we expect typical number of devices in IEEE 802.11ad WLAN to be no more than 10. Another reason is also due to the computation complexity for large N and K. The saturated throughput with $M = 1$ is higher than that for $M = 4$, which in turn is higher than that for $M = 8$, which in turn is higher than that for $M = 12$. In this region, as M increases, the saturated throughput decreases and the slope of the saturated throughput with respect to the number of devices, N, decreases. The number of devices is small from $N = 2$ to $N = 10$. The number of devices per sector is even lower as M increases. That is, the mean number of devices per sector is smaller for higher number of beam sectors, M. From Figure 5.10, even with multi-rate, a small number of beam sectors should be used when the number of devices, N, is small. Note that these data rates are below 1 Gbps at the MAC layer. This is because of the effect of multi-rate, causing large portion of the channel to be occupied by lower data rate data packets, while the smaller portions are occupied by higher data rate data packets. Another reason for achieving saturated throughputs below 1 Gbps at the MAC layer is because of the relatively low payload of 7920 bytes.

From the results, at small number of devices, the saturated throughput for omnidirection ($M = 1$) is better than that for those with a number of sectors ($M = \{4, 8, 12\}$).

Next, we consider the effect on the performance of the saturated throughputs as the coverage area and number of data rates decrease. The saturated throughputs of devices for number of beam sectors, $M = \{1, 4, 8, 12\}$, payload of 7920 bytes and devices' coverage of radius, $R = 4$ m and $K = 7$ data rates, and $R = 1.75$ m and $K = 3$ data rates are shown in Figures 5.11 and

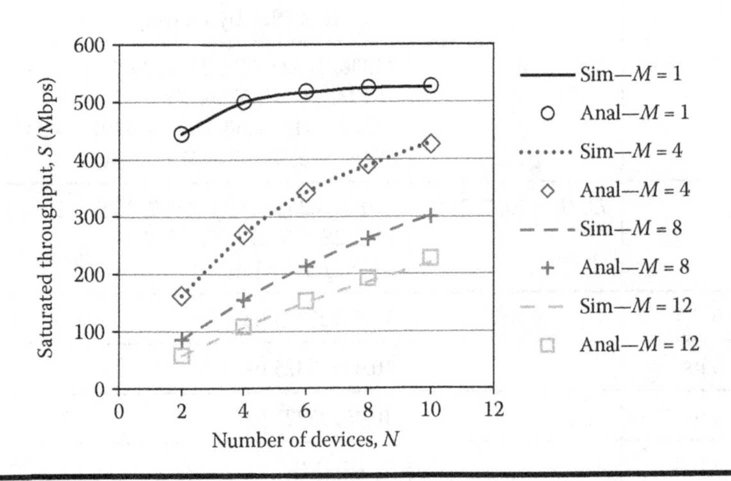

Figure 5.10 Saturated throughput for IEEE 802.11ad MAC with $E(L) = 7920$ bytes, $M = \{1, 4, 8, 12\}$, $R = 10$ m, and $K = 10$ data rates from OFDM mode.

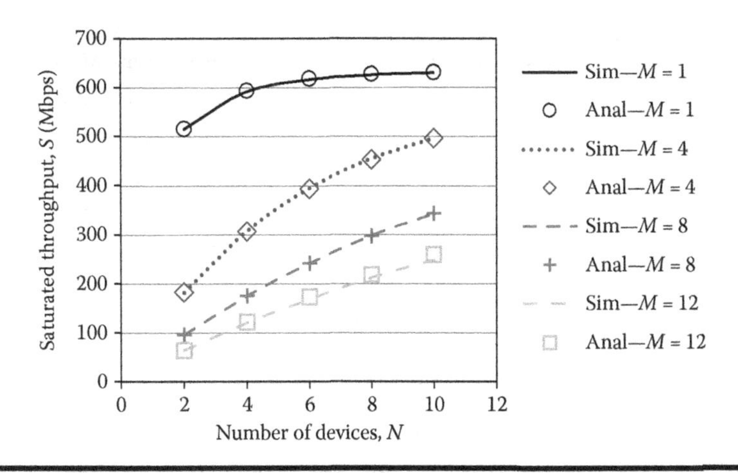

Figure 5.11 Saturated throughput for IEEE 802.11ad MAC with with $E(L) = 7920$ bytes, $M = \{1, 4, 8, 12\}$, $R = 4$ m, and $K = 7$ data rates from OFDM mode.

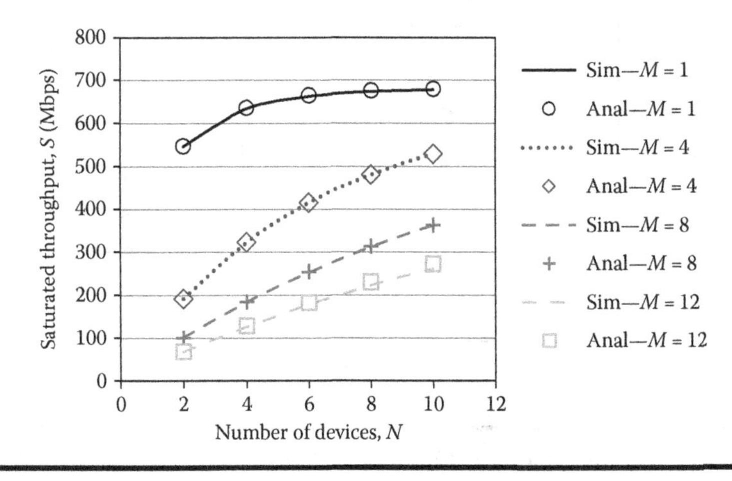

Figure 5.12 Saturated throughput for IEEE 802.11ad MAC with with $E(L) = 7920$ bytes, $M = \{1, 4, 8, 12\}$, $R = 1.75$ m, and $K = 3$ data rates from OFDM mode.

5.12, respectively. As the coverage area decreases, the saturated throughputs increase. The reason for this is that the data rates that can be used at shorter distances from the non-AP/non-PCP devices to the AP/PCP are higher. For the omnidirectional case, saturated throughput of close to 0.7 Gbps at the MAC layer can be achieved at $N = 10$. At small number of devices, the saturated throughput using omnidirectional antenna is better than that for those with a number of sectors ($M = \{4, 8, 12\}$).

Next, we evaluate the performance of the saturated throughputs as the payload is increased by 16 times. The saturated throughputs of devices for number of beam sectors, $M = \{1, 4, 8, 12\}$, payload of 16×7920 bytes and devices' coverage of radius, $R = 10$ m and $K = 10$ data rates, $R = 4$ m and $K = 7$ data rates and $R = 1.75$ m and $K = 3$ data rates are shown in Figures 5.13 through 5.15, respectively. Comparing Figures 5.10 through 5.12 with Figures 5.13 through 5.15, the saturated throughputs increase as the payload increases from 7920 to 16×7920 bytes.

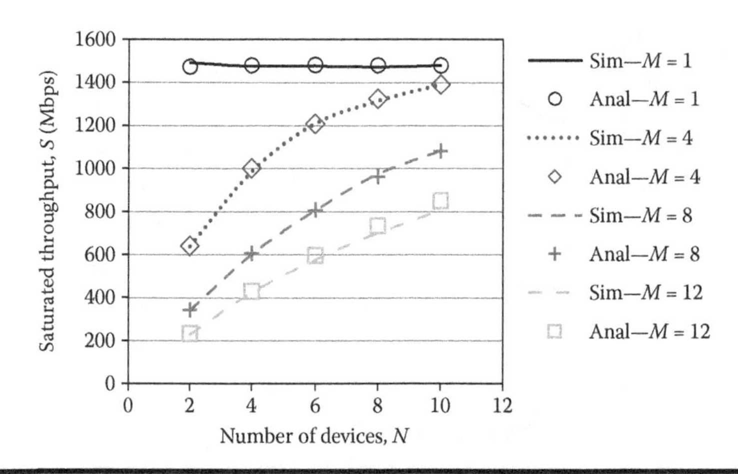

Figure 5.13 Saturated throughput for IEEE 802.11ad MAC with $E(L) = 16 \times 7920$ bytes, $M = \{1, 4, 8, 12\}$, $R = 10$ m, and $K = 10$ data rates from OFDM mode.

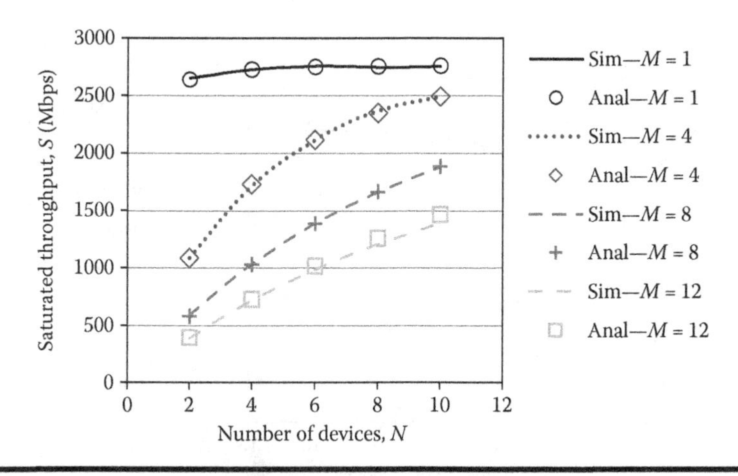

Figure 5.14 Saturated throughput for IEEE 802.11ad MAC with $E(L) = 16 \times 7920$ bytes, $M = \{1, 4, 8, 12\}$, $R = 4$ m, and $K = 7$ data rates from OFDM mode.

From Figure 5.15, the saturated throughput for the omnidirectional case can be reached up to about 4 Gbps at $N = 10$.

Finally, we look at the performance of the saturated throughput for large number of devices up to $N = 100$. The saturated throughputs of devices for number of beam sectors, $M = \{1, 4, 8, 12\}$, payload of 16×7920 bytes and devices' coverage of radius, $R = 1.75$ m and $K = 3$ data rates are shown in Figure 5.16. The analytical results are limit to up to 25 devices as the computation complexity is large for N. However, we evaluate the saturated throughputs performance for up to 100 devices by simulation.

For small number of devices ($N = 5$–25), the saturated throughput decreases as the number of sectors, M, increases. This is due to smaller CBAP size as M increases and also due to the small number of devices.

However, for large number of devices ($N = 100$), the saturated throughputs for $M = \{4, 8\}$ can be larger than that for $M = 1$. The reason for this is due to the number of sectors ($M = \{4, 8\}$) and

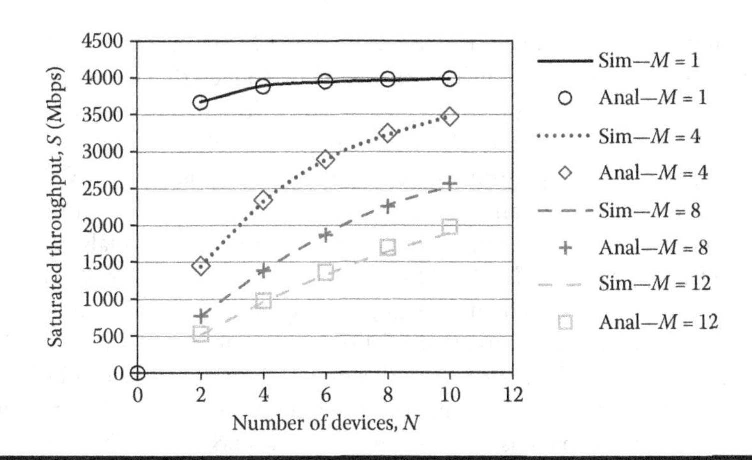

Figure 5.15 Saturated throughput for IEEE 802.11ad MAC with $E(L) = 16 \times 7920$ bytes, $M = \{1, 4, 8, 12\}$, $R = 1.75$ m, and $K = 3$ data rates from OFDM mode.

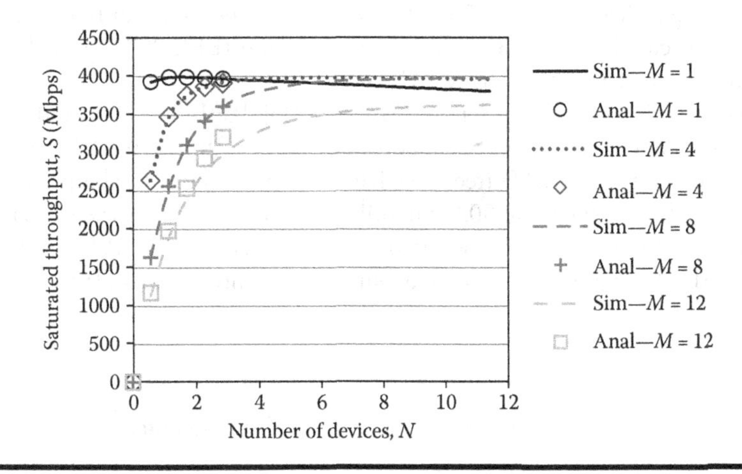

Figure 5.16 Saturated throughput for IEEE 802.11ad MAC with $E(L) = 16 \times 7920$ bytes, $M = \{1, 4, 8, 12\}$, $R = 1.75$ m, $K = 3$ data rates from OFDM mode, and $N = 5$ to 100.

smaller number of devices in each sector as compared to the omnidirectional case. The saturated throughputs for $M = \{1, 4, 8\}$ are slightly less than 4 Gbps at $N = 100$, while that for $M = 12$ is close to 3.6 Gbps. Thus, very high saturated throughputs can be achieved for a number of directional sectors in the IEEE 802.11ad wireless network.

5.6 Conclusions and Other New Technologies

IEEE 802.11ad WLAN is described in this chapter. A list of applications is provided for IEEE 802.11ad WLAN. An overview of the PHY of IEEE 802.11ad WLAN is presented and the new BI and MAC architecture is described. Fast session transfer for multi-band operation across

2.4/5/60 GHz bands is also presented. The new directional MAC protocols include BF, AP/PCP clustering, spatial sharing, link switching relaying, and link cooperating relaying. These MAC protocols are explained with detailed illustrations. An analytical formulation of the saturated throughput of IEEE 802.11ad MAC in the CBAP is also presented. Numerical examples with typical parameters are used to show the performance of the saturated throughput of CBAP transmission devices in the presence of *directivity* and *multi-rate*. The achievable saturated throughputs for IEEE 802.11ad MAC in the CBAP are presented. The agreement between the analytical and simulation results for *directional* and *multi-rate* IEEE 802.11ad MAC in the CBAP is good. From these results, at small number of devices, the saturated throughput using omnidirectional antenna is better than that for those with a number of sectors, that is, directional antennas. However, at large number of devices, the saturated throughputs of four and eight numbers of sectors can be better than that using omnidirectional antenna. These saturated throughputs can reach close to a maximum of 4 Gbps. The saturated throughput has been used to gauge the maximum achievable throughput for IEEE 802.11ad MAC in this chapter. One other traffic model not covered in this chapter includes the on/off traffic model for each device. Furthermore, the duration sizes of the CBAPs can also be allocated according to the number of devices in each sector or CBAP, instead of the assumption of equal beam sectors in this chapter. These issues are beyond the scope of this chapter.

Other interesting articles for IEEE 802.11ad WLAN can be found in References 12 through 26. These articles discuss IEEE 802.11ad MAC protocols and IEEE 802.11ad PHYs and a few of these articles discuss on IEEE 802.11ac WLAN too.

Upcoming standards to watch out for are IEEE 802.11aj, IEEE 802.11ah, and IEEE 802.11ax. IEEE 802.11aj provides enhancements to support operation in Chinese mmWave frequency bands including the 59–64 and the 45 GHz frequency bands to enable multi-gigabit per second throughput and lower power [27], while IEEE 802.11ah utilizes sub-1 GHz license-exempt bands to provide extended range for WLAN with a large group of stations or sensors [28]. IEEE 802.11ax improves the efficiency and the throughput is increased to four times the throughput of IEEE 802.11ac [29,30].

Appendix

For the omnidirectional case ($M = 1$), the saturated throughput, denoted by S, is given by

$$S = \sum_{l_1=1}^{K}\sum_{l_2=1}^{K}\cdots\sum_{l_N=1}^{K} P_{l_1}P_{l_2}\ldots P_{l_N} \times \frac{p_s E(L)}{T} \times \left(\begin{array}{l} \dfrac{T_{CBAP/M}}{2T_{CBAP/M} - T_Q} F_{CBAP/M} \\ + \dfrac{T_{CBAP/M} - T_Q}{2T_{CBAP/M} - T_Q} F_{CBAP/M-TQ} \end{array} \right), \tag{A.1}$$

and

$$T = \sum_{i=1}^{N} ((1 - p_b)\delta + p_s T_{l_i} + (p_b - p_s)T_c)/N, \tag{A.2}$$

where T denotes the average length of a slot time for N devices, rather than n devices in a beam sector, and T_{l_i} is defined in Equation 5.18. P_{l_i} is equivalent to the probability of being in each regions, P_k, using data rate R_{l_i}, except that it is for the ith device, $i = 1, \ldots, N$. That is, $l_i = k = 1, \ldots, K$.

References

1. Standard ECMA-386—High Rate Ultra Wideband PHY and MAC Standard, ECMA International, December 2005.
2. IEEE Standard 802.15.3c Standard, October 2009.
3. Standard ECMA-387—High Rate 60GHz PHY, MAC and HDMI PAL Standard, ECMA International, December 2008.
4. IEEE P802.11ad Standard Part 11: Wireless LAN Medium Access Control (MAC) and Physical Layer (PHY) Specifications Amendment 3: Enhancements for Very High Throughput in the 60GHz Band, December 2012.
5. Online, http://www.theregister.co.uk/2014/07/03/qualcomm_swallows_wilocity_promises_wigig wifi_crossover_chip/, dated 6 July 2014.
6. Cordeiro, C. 2011. Next generation multi-Gbps wireless LANs and PANs, *Tutorial, at Institute for Infocomm Research 79 p.*
7. "IEEE Standard 802.11e," September 2005.
8. Bianchi, G. 2000. Performance analysis of the IEEE 802.11 distributed coordination function. *IEEE Journal on Selected Areas in Communications*, 18(3) pp. 535–547.
9. Kong, Z. N., D. H. K. Tsang, and B. Bensaou, 2004. Performance analysis of IEEE 802.11e contention-based channel access, *Journal on Selected Areas in Communications*, 22(10) pp. 2095–2106.
10. Zhu, X., A. Doufexi, and T. Kocak, 2011. A performance evaluation of 60 GHz MIMO Systems for IEEE 802.11ad WPANs, *IEEE International Symposium on Personal, Indoor, and Mobile Radio Communications 2011*, pp. 950–954.
11. MacDougall, M. H. 1987. *Simulating Computer Systems: Techniques and Tools*, MIT Press, Cambridge, MA.
12. Perahia, E., C. Cordeiro, M. Park, and L. L. Yang, 2010. IEEE 802.11ad: Defining the next generation multi-gbps Wi-Fi, *IEEE Consumer Communications and Networking Conference* 2010, pp. 1–5.
13. Cordeiro, C., D. Akhmedov, and M. Park, 2010. IEEE 802.11ad: Introduction and performance evaluation of the first multi-Gbps WiFi technology, *ACM mmWave*, pp. 3–7.
14. Perahia, E. and M. X. Gong, 2011. Gigabit wireless LAN: An overview of IEEE 802.11ac and 802.11ad, *ACM SigMobile Mobile Computing and Communications Review* 15(3), pp. 23–33.
15. Charfi, E., L. Chaari, and L. Kamoun, 2013. PHY/MAC enhancements and QoS mechanisms for very high throughput WLANs: A survey, *IEEE Communication Surveys and Tutorials*, 15(4), pp. 1714–1735.
16. Vaughan-Nichols, S. J. 2010. Gigabit Wi-Fi is on its Way, *Computer*, 43(11), pp. 11–14.
17. Garber, L. 2012. Wi-Fi Races to a Faster Future, *Computer*, 45(3), pp. 13–16.
18. Verma, L., M. Fakharzadek, and S. Choi, 2013. WiFi on Steroids: 802.11ac and 802.11ad, *IEEE Wireless Communications Magazine* 2013, 20(6), pp. 30–35.
19. Singh, H., J. Hsu, L. Verma, S. S. Lee, and C. Ngo, 2011. Green operation of multi-band wireless LAN in 60 GHz and 2.4/5 GHz, *IEEE Consumer Communications and Networking Conference 2011*, 2011, pp. 787–792.
20. Zhu, X., A. Doufexi, and T. Kocak, 2011. Throughput and coverage performance for IEEE 802.11ad Millimeter-Wave WPANs, *IEEE Vehicular Technology Conference 2011 Spring*, pp. 1–5.
21. Zhu, X., A. Doufexi, and T. Kocak, 2011. Beamforming performance analysis for OFDM-based IEEE 802.11ad millimeter-wave WPANs, *IEEE International Workshop on Multi-Carrier Systems and Solutions 2011*, pp. 1–5.
22. Zhu, X., A. Doufexi, and T. Kocak, 2012. A performance enhancement for 60 GHz wireless indoor applications, *IEEE International Conference on Consumer Electronics 2012*, pp. 209–210

23. Zhang, C., Z. Xiao, H. Wu, L. Zeng, and D. Jin, Performance analysis of the ODFM PHY of the IEEE 802.11ad Standard, *IEEE International Conference on Computational Problem-Solving 2011*, pp. 708–713.
24. Zaaimia, M., R. Touhami, A. Hamza, and M. C. E. Yagoub, 2013. Design and performance evaluation of 802.11ad Phys in 60 GHz multipath fading channel, *IEEE 8th International Workshop on Systems, Signal Processing and Their Applications*, pp. 521–525.
25. Shevchenko, A., R. Maslennikov, and A. Maltsev, 2014. Comparative analysis of different hardware decoder architectures for IEEE 802.11ad LDPC code, *IEEE Mediterranean Electrotechnical Conference 2014*, pp. 415–420.
26. Park, M. and H. K. Pan, 2012. A spatial diversity technique for IEEE 802.11ad WLAN in 60 GHz Band, In *IEEE Communications Letters*, 16(8), pp. 1260–1262.
27. IEEE 802.11aj website Online, http://www.ieee802.org/11/Reports/tgaj_update.htm, dated 30 June 2014.
28. Wikipedia Online on IEEE 802.11ah, http://en.wikipedia.org/wiki/IEEE_802.11ah, dated 30 June 2014.
29. IEEE P802.11ac Standard Part 11: Wireless LAN Medium Access Control (MAC) and Physical Layer (PHY) Specifications Amendment 4: Enhancements for Very High Throughput for Operations in Bands below 6 GHz, December 2013.
30. Wikipedia Online on IEEE 802.11, http://en.wikipedia.org/wiki/IEEE_802.11, dated 3 July 2014.

MILLIMETER WAVE

Chapter 6

MAC Layer Protocols for Wireless Networks with Directional Antennas

Muhammad Irfan Rafique

Contents

6.1 Introduction

In the last two decades wireless local area networks (WLANs) have become increasingly popular inspiring broad range of applications and ease of installation. Now, WLANs are equally popular for household networking application and enterprise networks. In the near future, almost all of the electronic device would be equipped with wireless transceivers to exploit the benefits of WLAN. Currently, WLANs are exploited in two major topologies: infrastructure mode and adhoc mode. The Institute of Electrical and Electronics Engineers (IEEE) has defined a number standards to improve performance of classical (infrastructure-based) WLANs [1–3]. In its recent recommendations (IEEE 802.11ac) [3], IEEE has standardized beamforming technique to improve the performance of traditional WLANs.

In recent years, directional antennas have been recognized as prominent technology for beamforming transmission. They can be exploited in wireless networks to improve transmission reliability, higher spatial reuse, larger transmission range, and lower power consumption. Directional antennas radiate energy in a specific direction which yields higher SNR than omnidirectional antenna for a constant transmission power. The higher signal to noise ratio (SNR) can be transformed into higher transmission range for a constant BER at the receiver. Thus a node laying outside the transmission range of omnidirectional antenna could be reached with beamforming. For example, in the scenario shown in Figure 6.1, node A can transmit to B using beamforming, whereas link A to B is not valid for omnidirectional antenna as B lies outside the transmission range of omnidirectional antenna. The higher SNR obtained by radiating energy in a intended direction applying beamforming could also be translated to higher transmission rate for a constant transmission range and bit error rate (BER) at the receiver. Furthermore, for a constant transmission range and BER, the higher SNR due to beamforming could be applied to save transmission power. As beamforming transmits in directional mode, it hence causes interferences to only those nodes that lie in the direction of the beam. The directional transmission mode of beamforming can be utilized to enable concurrent transmissions. This is illustrated in Figure 6.2 where parallel communication is possible applying beamforming on links A to B and C to D as both pairs lie outside the transmission direction of each other. In case node A apply omnidirectional antenna for transmission, C cannot transmit at the same as time as both A and C are in interfering range with each other. Also, the directional transmission mode of beamforming improves the security of a WLAN by reducing the risk of eavesdropping.

Although, IEEE 802.11ac [3] proposes mechanisms to utilize spatial reuse feature of beamforming for concurrent transmissions, it relies on traditional channel access mechanisms [1,2] to share a common wireless medium between nodes. However, incorporation of directional antennas in wireless networks to utilize their full advantages requires new medium access control (MAC) layer protocols as traditional protocols—based on omnidirectional antenna—failed to exploit the

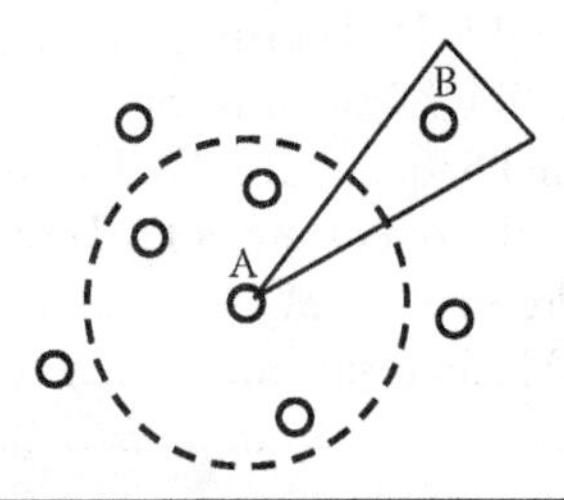

Figure 6.1 Example scenario of large transmission range.

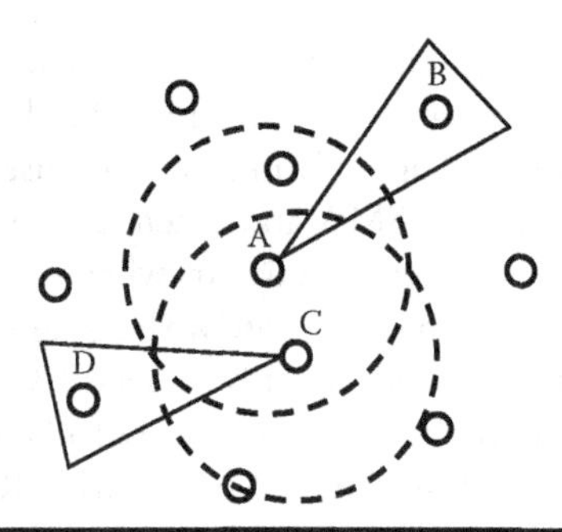

Figure 6.2 Example scenario of parallel communication.

potential benefits of directional antennas. Since changing transmission technique at the physical layer (PHY) impacts more at the MAC layer compared to the above layers, MAC layer protocols play a significant role to explore features of directional antennas in wireless networks. A number of MAC layer protocols has been proposed in the literature to apply features of directional antennas in wireless networks. This chapter enlightens the challenges raised by directional antennas at the MAC layer of wireless networks and explains MAC layer protocols for wireless networks using directional antennas. The chapter also explains conventional MAC protocols of IEEE infrastructure-based WLANs; however, its major focus is on MAC layer protocols for wireless multihop networks with directional antennas.

The rest of the chapter is organized as follows: Section 6.2 briefly explains IEEE MAC. Section 6.3 describes challenges raised by directional antennas at the MAC layer. Section 6.4 elaborates few proposed MAC protocols for wireless networks with directional antennas. Section 6.5 explains few existing link scheduling mechanisms which exploit spatial reuse feature of directional antennas. Section 6.6 summarizes the chapter.

6.2 Conventional MAC Protocols for Wireless Networks

The main responsibility of MAC protocol is to enable fair and efficient sharing of common wireless channel between nodes. In the absence of channel access protocols, multiple nodes may try to access the channel at the same time which result in conflicts that lead to less channel utilization. MAC protocol for wireless network may be classified into two major categories: contention-based and contention-free MAC. In the contention-based MAC, nodes access the channel through a random access and if a conflict between nodes is occurred, a distributed conflict resolution algorithm is used to resolve it. On the other hand, channel is allocated to each node according to a predetermined schedule in the contention-free MAC protocols.

6.2.1 Contention-Based MAC Protocols

This section briefly explains the conventional contention-based channel access mechanisms.

6.2.1.1 CSMA and CSMA/CA

The most commonly considered contention-based MAC mechanisms are the carrier sensing multiple access (CSMA) and CSMA with collision avoidance (CSMA/CA). In CSMA, node listens the channel before transmission. If another node is transmitting, that is, the channel is sensed busy, then the node defers its transmission and waits until the channel is sensed idle. CSMA successfully avoids collision with transmitting node in a carrier sense range.

CSMA/CA exploits the benefits of CSMA and extends it to reduce the collision due to the hidden node problem. The hidden node problem occurs when two nodes laying outside the carrier sense range with each other attempt to communicate with a single node. In CSMA/CA, when the channel is sensed idle, the transmitter sends a ready to send (RTS) frame to its desired receiver. Upon reception of the RTS frame, if the channel is sensed idle then the receiver replies with a clear to send (CTS) frame to the transmitter. Upon overhearing RTS/CTS frames, nodes lying in the vicinity of the transmitter and receiver, refrain from transmission and reception for the time interval mentioned in the RTS/CTS frames.

6.2.1.2 Distributed Coordinated Function

The IEEE 802.11 distributed coordinated function (DCF) is one of the CSMA-/CA-based protocols, which has lately received a great attention due to its simplicity. In IEEE 802.11 DCF, a node sends a RTS frame if the channel is sensed idle for a specific distributed interframe space (DIFS) delay. The desired receiver sends a CTS frame after a short interframe space delay (SIFS) if the channel is sensed idle. Upon reception of the RTS/CTS frames, other nodes update their network allocation vector (NAV) with time duration (given in the RTS/CTS frame), and defer their transmission for this time duration. Once the value of an NAV is set, countdown is started until new RTS/CTS/DATA frame is received. A node cannot start its transmission until its NAV is zero. The mechanism of DCF is illustrated in Figure 6.3. If a node does not receive CTS or ACK after transmitting RTS or DATA frames, respectively then the node chooses a random backoff number between 0 and its contention window size. The sender can transmit only when the channel is sensed idle for DIFS time interval following random backoff time.

6.2.1.3 Enhanced Distributed Channel Access

Enhanced Distributed Channel Access (EDCA) is proposed in Reference 1. It enhances DCF and provides distributed mechanism to access the channel using eight different user priority traffic.

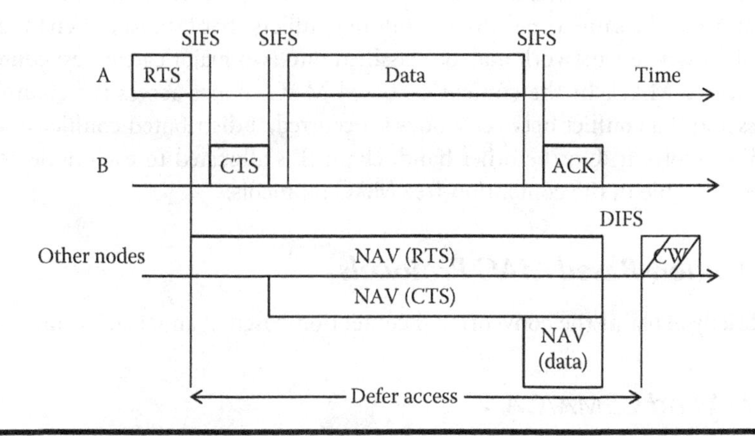

Figure 6.3 Channel reservation in distributed coordinated function medium access control (DCF MAC).

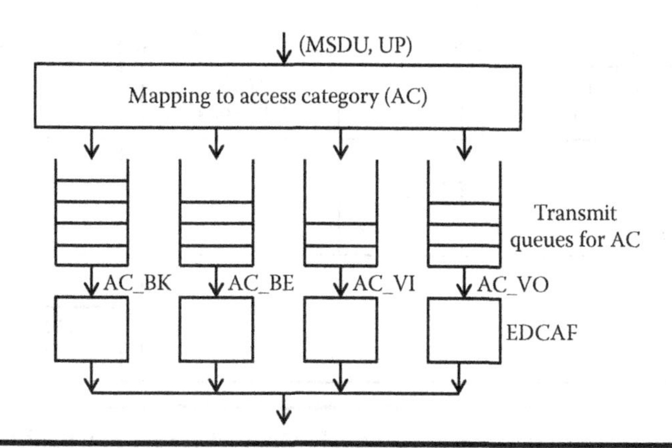

Figure 6.4 IEEE enhanced distributed channel access (EDCA) reference implementation model.

Nodes map the traffic into four access categories (AC). When data arrives, the IEEE 802.11e MAC puts packets into the corresponding queue as shown in Figure 6.4. Each queue behaves like a virtual node and contends for the channel access independently using its EDCA function (EDCAF). To prioritize channel access, the channel access parameters are set different for each queue as shown in Table 6.1. The arbitration interframe space number (AIFSN) is the time duration for which the channel must be idle before the transmission. At first, queues contend internally among themselves, and the queue with smallest backoff wins the contention. Then, the winning queue contends externally for the wireless medium. The external contention is similar to DCF. Furthermore in EDCA, an interval of time, called transmission opportunity (TXOP), is assigned to a node on the channel access. In a TXOP, a node can transmit multiple packets. If the TXOP limit is set to zero, then the node can transmit only one packet.

6.2.2 Contention-Free MAC Protocols

The mechanisms explained in this section require a central entity (like access point) in the network.

6.2.2.1 Point Coordinated Function

The IEEE 802.11 point coordinated function (PCF) is a channel access mechanism that works on infrastructure-based networks. In PCF, all nodes (stations) access media through a single access point (AP) that acts as a point coordinator (PC). In PCF-enabled basic service set (BSS), the

Table 6.1 Default EDCA Parameters

AC	*CWmin*	*CWmax*	*AIFSN*	*TXOP Limit (11b/11a)*
AC_BK	aCWmin	aCWmax	7	0/0
AC_BE	aCWmin	aCWmax	3	0/0
AC_VI	(aCWmin + 1)/2 – 1	aCWmin	2	6.016 ms/3.008 ms
AC_VO	(aCWmin + 1)/4 – 1	(aCWmin + 1)/2 – 1	2	3.264 ms/1.504 ms

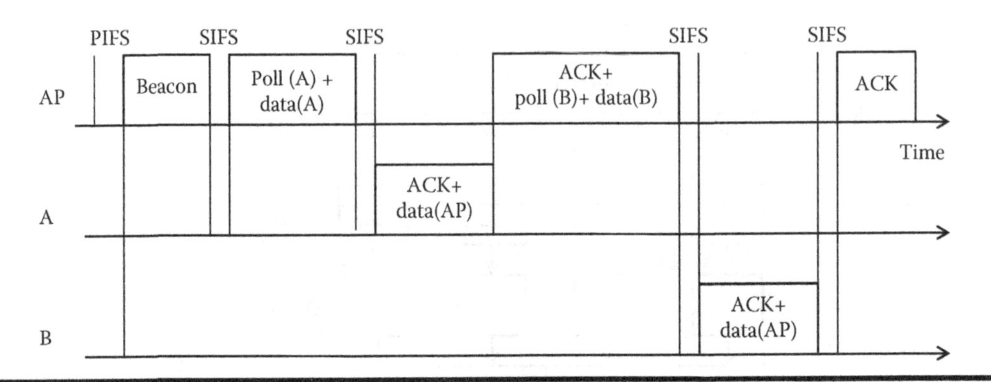

Figure 6.5 IEEE point coordinated function (PCF) channel access mechanism.

channel access time is divided into contention-free periods (CFP) and contention periods (CP). In CP, stations utilize DCF to access the channel. On the other hand in CFP, AP polls stations to transmit. The CFP is started with beacon transmission of AP as shown in Figure 6.5. These beacons contain information about the duration of CFP and CP. In CFP, only those stations are allowed to transmit which are polled by AP. Upon receiving a POLL from AP, the station starts transmission after PCF interframe space (PIFS). If a polled station does not have packet to transmit, it sends a NULL packet. To ensure that DCF-enabled stations do not interfere with PCF operation, PIFS are set to shorter than DIFS intervals.

6.2.3 HCF Controlled Channel Access

Hybrid control function (HCF) controlled channel access (HCCA) proposed in IEEE 802.11e exploits both contention-based and contention-free channel access mechanism. It uses an hybrid coordinator (HC) which runs at a access point. Similar to PCF, the interval between two beacons is divided into two periods, that is, Contention Free Period (CFP) and Contention Period (CP). However, HCCA enables AP to use CFP at any time during CP as illustrated in Figure 6.6. If the channel is sensed idle for one PIFS, then the AP can initiate frame exchange sequences and allocate TXOPs to itself or other stations even in a contention period. This kind of CFP in CP is called a controlled access phase (CAP) in 802.11e. During a CAP, the AP controls the access to the medium. During the CP, all stations use EDCA to access the channel. The second major difference with the PCF is the usage of traffic class (TC) and traffic streams (TS). In HCCA,

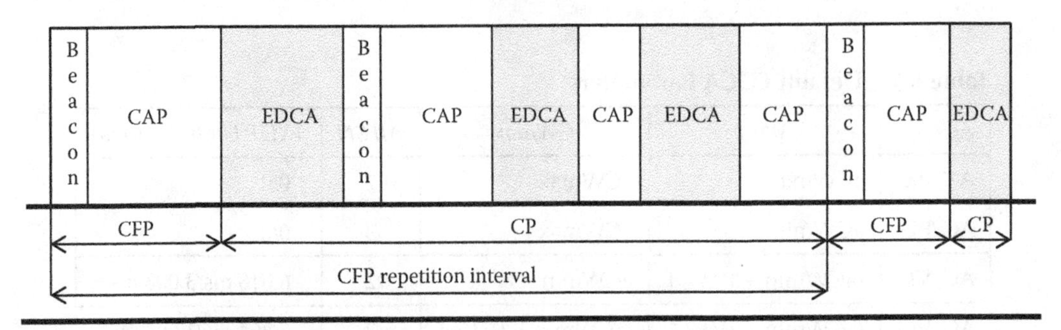

Figure 6.6 IEEE HCF controlled channel access (HCCA) channel access mechanism.

stations give information about the length of their queue for each traffic class, and the AP utilizes the information to poll stations. The AP can use this information to give priority to one station over another. Another difference is that stations are given a TXOP; thus, they may send multiple packets in a row for a given time period selected by the HC.

6.3 MAC Challenges with Directional Antennas

Conventional wireless MAC protocols overcome issues raised by hidden and exposed nodes. However, the unique features of directional antennas pose another type of challenge which should be considered in the design of MAC protocols for wireless networks with directional antennas. In this section, the major beamforming related challenges facing the MAC layer are discussed. However, Reference 4 explains these challenges in detail.

6.3.1 Deafness

In beamforming, deafness occurs when the node is not aware of ongoing directional communication at its intended receiver. Thus the node continues RTS transmission and if the failure of RTS frames exceeds a specific threshold value then the link is dropped and route discovery algorithms are started. The mechanism can also be explained with Figure 6.7. Assume that node A is engaged in directional communication with node B. Now, node C initiates directional communication with node D in parallel using benefits of beamforming. Then, node A does not hear exchange of RTS/CTS between C and D as it is deaf to node C. If node A wants to communicate with node C after finishing communication with node B, then RTS frames from node A would be lost as node C is deaf to A (due to its directional transmission with D). These unsuccessful retransmissions reduce the network capacity. As backoff contention window increases exponentially with failures of frames, these retransmissions result in lower channel utilization. The consequences of deafness may also lead to unfair channel allocation at the MAC layer. Let us assume node C has multiple frames to transmit to node D. Then, after a successful transmission to D, node C selects a backoff interval from the minimum contention window and transmits its next frame to D. On the contrary, the backoff interval for node A increases exponentially on each RTS frame failure. The probability is higher that while node A is engaged in backoff process, node C keeps transmitting. In this way, node A drops multiple packets before it gets channel access.

6.3.2 Head-of-Line Blocking

The directional transmission enables nodes to utilize the spatial reuse feature for parallel transmissions. However, besides degree of freedom, the core factor which limits the utilization of spatial

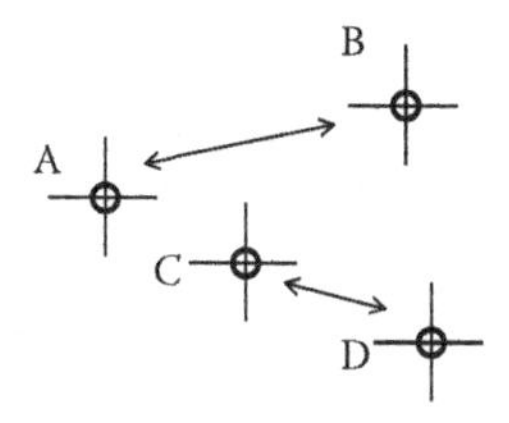

Figure 6.7 Example scenario of deafness.

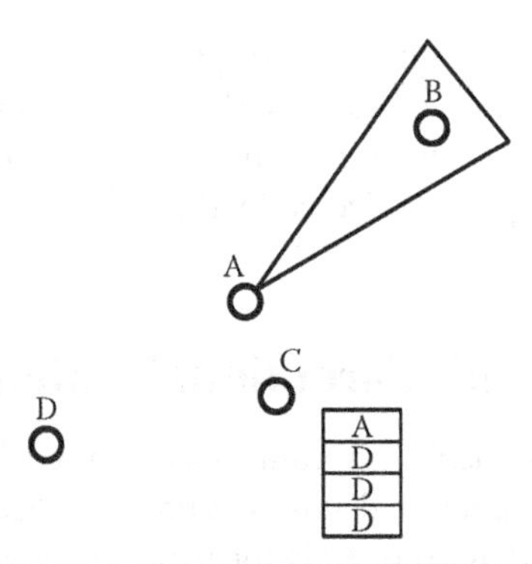

Figure 6.8 Example scenario of head-of-line (HOL).

reuse feature is the head-of-line (HOL) blocking problem. The problem was first identified in Reference 5. It occurs when the receiver of the first available packet in the first-in first-out (FIFO) queue is already engaged in directional communication, thus even in the possibility of parallel transmission for the next available packet the node has to wait. For example, in the scenario shown in Figure 6.8, if node A is communicating with B, then node C cannot communicate with D concurrently using beamforming as the first available frame in its queue is for node A.

6.3.3 MAC Layer Capture Effect

The third major issue which limits the exploitation of potential advantages beamforming is the capture effect. In traditional behavior, the MAC layer decides to accept or drop the packet depending upon the destination and the SNR. Thus irrespective of the packet destination, the packet is received by PHY layer. This does not limit the performance of omnidirectional antenna as only one transmission is expected in a transmission-sensing region. However, the behavior limits the performance of beamforming antennas as the time a node spent to receive energy of an undesired packet could be utilized for parallel transmission. In an ideal mode, to receive energy from all directions, a node listens in omnidirectional mode while upon reception of signals, the node steers beam in the direction of maximum power. Thus, irrespective of the destination of a packet, the node will be busy to receive the packet. Figure 6.9 illustrates the MAC layer capture problem. If node A is communicating to node B, and node C has packet for node D. Though with beamforming concurrent communication is possible on links A to B and C to D, however, the parallel communication is uncertain in this scenario. This is followed by the fact that when communication starts earlier between node A and node B, the node D will not hear RTS frames from node C as it is busy in packets reception of nodes A and B.

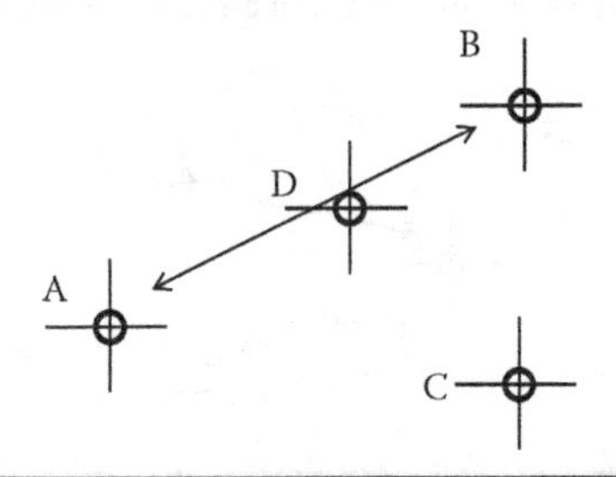

Figure 6.9 Example scenario of MAC layer capture.

6.4 MAC Protocols with Directional Antennas

A number of MAC protocols for wireless networks with directional antennas exist in the literature. These protocols can be classified into two groups: random (distributed) channel access protocols and synchronized channel access protocols. Here, only few of these protocols are explained as it is not possible to include all of these in one chapter.

6.4.1 *Random Channel Access Protocols*

A number of contention-based MAC protocols exist in the literature which modify conventional protocols to incorporate directional antennas in wireless networks. In this section, few of the existing random access protocols are explained. These protocols are subdivided according to the transmission modes of control frames (RTS/CTS) used in the channel access mechanism.

6.4.1.1 Omnidirectional RTS Transmission

The proposed MAC protocol by Nasipuri et al. in Reference 6 is one of the first approaches to utilize directional antennas in wireless ad hoc networks. It has been considered that each node is equipped with multiple directional antennas. The channel access mechanism is similar to traditional IEEE 802.11 DCF with RTS/CTS mechanism. It is assumed that a node listens ongoing transmissions on all of its antennas and applies selection diversity to use the signal from the antenna receiving maximum power of the desired signal. A channel is sensed idle when it is free in all directions or received signals have less power than that of interference threshold. The RTS/CTS frames are exchanged omnidirectionally before sending a data packet. The omnidirectional transmissions of RTS/CTS frames enable nodes to detect the appropriate antenna for transmission and reception depending on the signal strength. This mechanism is illustrated in Figure 6.10, where node A wants to send data packet to node B. Node A transmits RTS frame on all of its antenna elements as it does not know the direction of B at the start. Upon reception of the

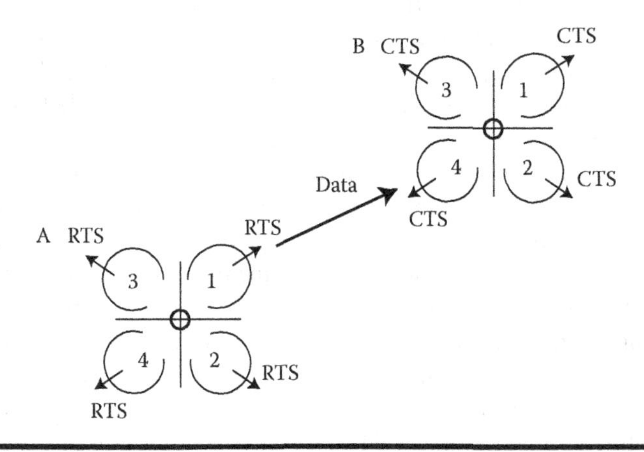

Figure 6.10 Example scenario of channel access mechanism proposed. (Adapted from Nasipuri, A. et al., 2000. In *IEEE Wireless Communications and Networking Conference (WCNC)*, Vol. 3, pp. 1214–1219.)

RTS frame from A, node B applies selective diversity and notes the antenna element at which maximum power is received (in this scenario antenna 4 is selected). Similarly, upon reception of CTS frame from the receiver, the transmitter A gets information about the suitable antenna element for further transmission. The DATA/ACK frames are transmitted directionally on appropriate (selected) antenna elements.

It is observed that the directional transmission of data frames increases SNR and improves the throughput. However, as transmission power is considered constant for both omnidirectional and directional transmission, the omnidirectional transmission of RTS/CTS frames limit the protocol to utilize potential of directional antennas for higher transmission range.

6.4.1.2 Unidirectional RTS Transmission

In Reference 7, it is assumed that nodes are equipped with global positioning system (GPS) and their location is known to each other. A node applies 802.11 DCF mechanism to gets access to the channel and when the channel is accessed, the node sends directional RTS (DRTS) including the physical position of intended receiver. The receiver transmits CTS with omnidirectional antenna (OCTS) and includes its own location information and location information of the RTS transmitter in the frame. This information helps neighboring nodes to set their NAV and to avoid interference at the nodes engaged with ongoing communication. It is considered that an explicit NAV is assigned to each directional antenna element. Upon reception of DRTS/OCTS, the neighboring nodes block the antenna (by setting their NAV) which could cause the interference at ongoing communication. The mechanism also helps to utilize spatial reuse by concurrent transmission/reception on the antennas which are not blocked. On successful exchange of DRTS/OCTS, DATA, and ACK frames are transmitted directionally.

The approach is illustrated in Figure 6.11, when a node A wants to send data packet to node B, applying location information of node B, it transmits DRTS frame utilizing appropriate antenna element (antenna 1) toward node B. The suitable antenna toward the intended receiver is chosen using the location information of the receiver. In this example, node A send RTS using antenna 1, and node B replies with omnidirectional CTS. Since transmission of CTS is done omnidirectionally which is achieved by transmitting the frame over all directional antennas, therefore node B will reply with the CTS frame when all of its antennas are free—that is, there is no ongoing communication in the neighborhood. Upon reception of CTS, node C saves information about ongoing communication between A and B and blocks its antenna (1) for the transmission to avoid the interference at node B. Now if node C wants to send data packet to D, it sends DRTS using appropriate antenna if the antenna is not blocked. Since antenna 2 is pointed toward node D and is not blocked due to the communication at link A to B, therefore C sends DRTS using antenna 2 while node D sends CTS omnidirectionally in response to DRTS. Thus parallel communication is possible at links A to B and C to D. Now if node C has a packet for E, it would not start its transmission as suitable antenna (antenna 1) to node node E is blocked due to transmission between A and B. As node A transmits RTS in the specific direction, so neighboring nodes laying in other direction are unaware of the RTS frame transmission. On the other hand, the nodes which receive omnidirectional the CTS frame, block their antenna both toward RTS and CTS transmitters, and in this scenario node C blocks antennas toward node A and B. However, the neighboring nodes of A which do not receive CTS frame, can send RTS to their desired receivers. For example, in Figure 6.11, node F is unaware of the ongoing communication between A and B, thus it can send RTS to node G on antenna 4.

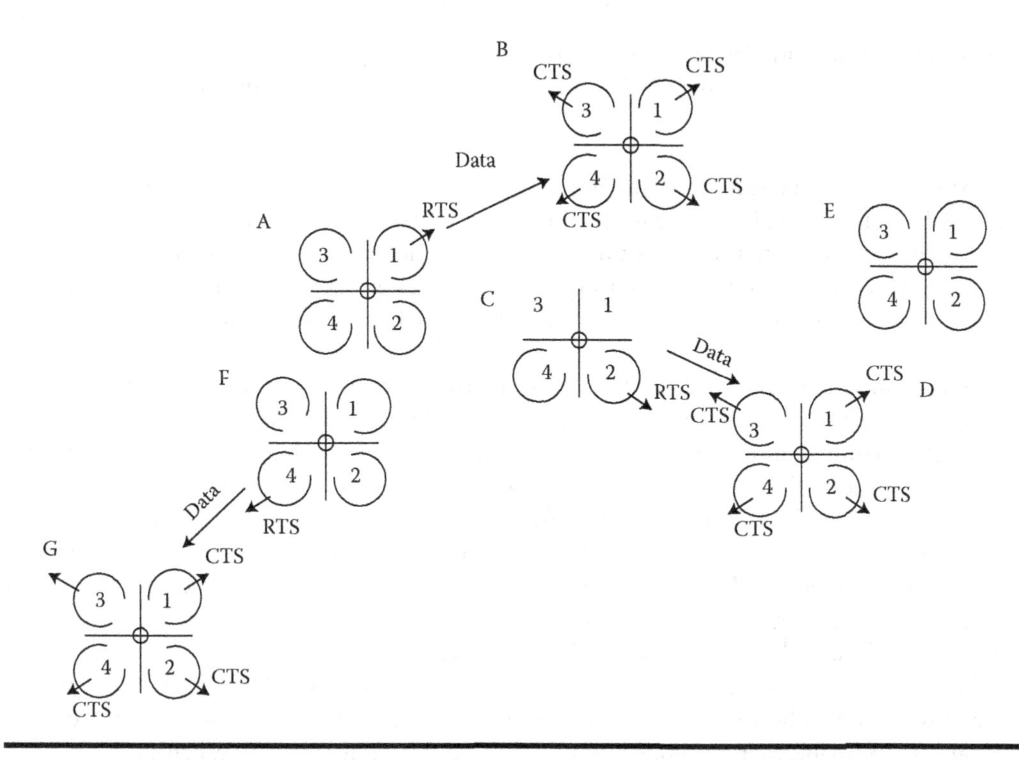

Figure 6.11 Example scenario of directional-MAC (D-MAC) P.

Figure 6.12 illustrates the scenario where nodes A and B are already engaged in communication while node C wants to send data packet to A. As A has transmitted DRTS, thus node C is unaware of ongoing communication between A and B. In this way, DRTS from node C will be lost (due to deafness). To avoid such situation, a node can send both DRTS and ORTS prior to the data transmission. The rules to select ORTS and DRTS are listed hereunder:

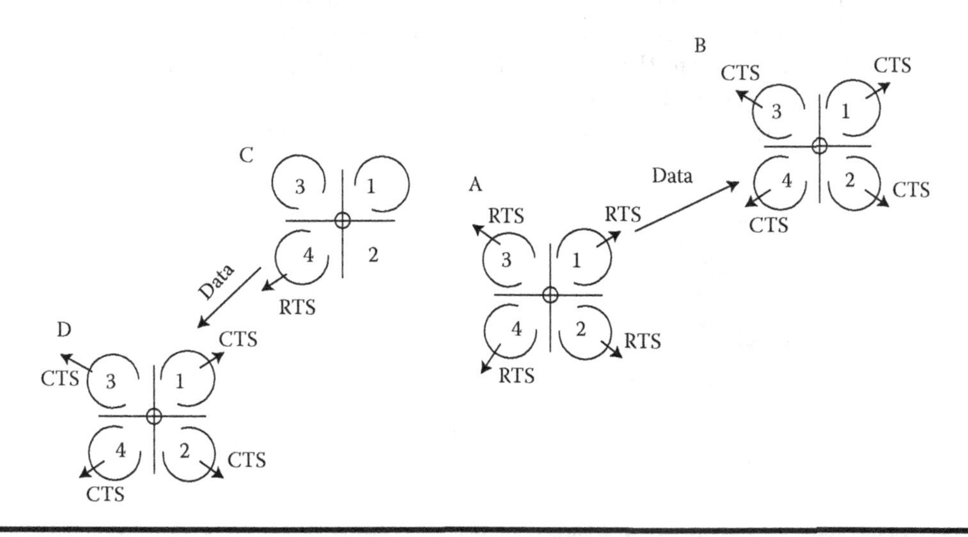

Figure 6.12 Example scenario of D-Mac deafness avoidance.

- ORTS is transmitted when all antennas are free.
- DRTS is transmitted if one of the antenna elements is blocked by ongoing transmission of neighboring nodes.

For example, now node A will send ORTS prior to start sending data packets to B. Node C will hear ORTS from A and block antenna 2 pointed to A for time duration given in the RTS frame.

D-MAC exploits spatial reuse feature of directional antennas and enables parallel communication. However, the protocol does not explore advantages of higher transmission range due to directional antennas as CTS is transmitted omnidirectionally. Further, GPS feature increases the cost of nodes.

Takai et al. propose the concept of directional virtual career sensing (DVCS) in Reference 8. It allows traditional IEEE 802.11 MAC protocol to determine direction specific channel availability. The technique minimizes the usage of additional resources like orthogonal channel for transmission of control and data frame, or external devices such as GPS, ultrasound, and compass. To incorporate DVCS in the original IEEE 802.11 MAC protocol, three fundamental capabilities are added: caching the angle of arrival (AOA), beam locking and unlocking, and use of directional network allocation vector (DNAV). DNAV is similar to NAV, however, it reserves the channel for others only in a range of directions. On hearing of a signal, a node estimates the AOA of the transmitter and stores the information in a DNAV table. When a node wants to send a data packet to other node, first of all it checks the AOA of intended receiver (from DNAV table) and applies antenna which forms beam in the direction of the receiver. If AOA information for the desired receiver is not available then the node sends RTS packet omnidirectionally. When a node receives RTS frame, it estimates appropriate beam pattern toward the transmitter and use this pattern for further communication. The transmitter also locks the beam pattern for further communication upon the reception of CTS frame, and then DATA/ACK frames are transmitted directionally using these patterns. Each node maintains a DNAV table which consists of multiple DNAVs. Each antenna element has its own DNAV as shown in Figure 6.13a. The value in DNAV indicates that the node must not transmit in the given direction for the specific time interval. When a node wants to send data to other, it checks the stored AOA information of desired receiver and looks into DNAV. If the time value in the DNAV is null or has expired than the node sends DRTS using that antenna element. In case of valid time value in the DNAV, the node defers its transmission and performs backoff similar to IEEE 802.11. For example, in Figure 6.13b, when node A wants to send a data packet to node B and if it has no AOA information about B, then it sends RTS omnidirectionally. Upon reception of RTS, neighboring nodes C and D estimate AOA of node A

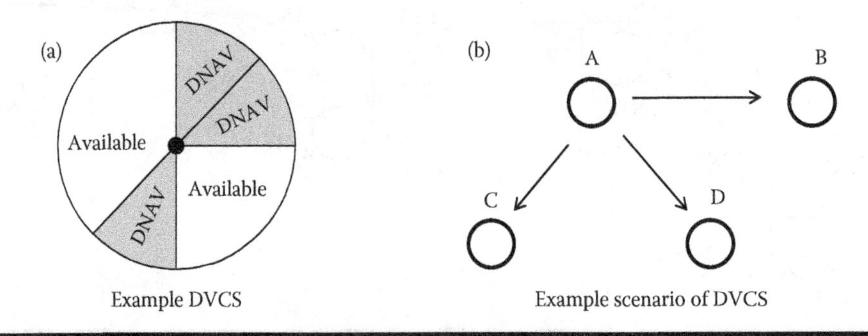

Figure 6.13 Example scenario of directional virtual career sensing (DVCS).

and update their DNAV information of the corresponding antenna elements. On receiving RTS from node A, node B locks its pattern toward node A after estimating AOA and sends (directional) CTS frame. Now if node C wants to communicate with node A, then the DNAV settings block node C to communicate until the ongoing communication between A and B is finished. However, if node C has AOA information for node D, then it can send (directional) RTS to node D. In this way, concurrent communication is possible on links A to B and C to D.

In Reference 9, the proposed multihop RTS MAC (MMAC) protocol it exploits higher transmission range advantages of beamforming and extends the concept of the MAC protocol introduced in Reference 7. It is shown that routing hops could be decreased by exploring the advantages of higher transmission range due to directional antennas, thus throughput of multihop network could be increased. It is assumed that an upper layer is aware of neighboring node and provides transceivers profiles required for the communication with other node. The neighboring nodes are classified into direction-omni (DO) neighbor and direction-direction (DD) neighbors. A node x is DO neighbor of a node y if it can receive directional transmission from node y while listening in omnidirectional mode. A node x is DD neighbor of a node y if it can receive directional transmission from node y only by steering its beam toward y. All DO neighbors are DD neighbors as well but all DD neighbors might not be DO neighbor. Furthermore, a DD neighbor of a node may be accessed using route through DO neighbors.

When a node wants to send data packet to a destination, the upper layer sends the data packet and the transceiver profile of DD to the MAC layer. The profile is comprised of routing information to next DD neighbor in path toward destination. The routing information has a list of DO nodes (laying in the path) to reach the desired DD node. Furthermore, the transceiver profile has information about the appropriate antenna element for the transmission toward the receiver (i.e., next DO neighbor in the path). In MMAC, a node listens in omnidirectional mode, and estimates direction of arrival (DOA) upon the reception of the directional transmission. Also, it sets DNAV value accordingly. The DNAV keeps the record of directions at which neighboring nodes are engaged in communication. The value in DNAV indicates that node must not transmit in a direction for the specific time interval.

Upon reception of data packet and transceiver profile, the MAC layer checks directional DNAV. If the value of DNAV is 0 and channel is sensed idle then node sends DRTS applying appropriate antenna in the direction of the receiver. For example, considering the scenario given in Figure 6.14, the MAC layer of node S receives data packet (from upper layer) and transceiver profile for next DD neighbor E. The transceiver profile includes the information for next DD neighbor and its direction, route to DD neighbor, and direction of next DO receiver. Node A uses directional information of DD, sets beam in the direction, and transmits the RTS frame to node E. The RTS frame includes its transmission time duration and number of hops (DO neighbors) to reach node E. It is possible that node E does not receive this RTS frame as it listens in the omnidirectional mode.

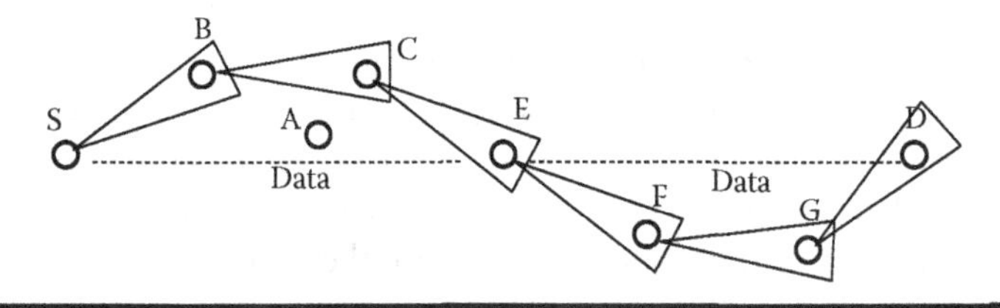

Figure 6.14 Example scenario of MMAC.

However, the transmission of the RTS frame enables potential interfering nodes to set their DNAV, for example in the scenario, node A will estimate DOA (θ) and sets its DNAV upon the reception of DRTS from node S. In addition, node A sets DNAV in the direction $(\theta + 180°)\,mod\,360$ to avoid the interference at the destination node E. The time duration value for DNAV is equal to *time required for one RTS frame transmission* × *number of hops to reach E*. If node E has already pointed its beam toward S, then node E might receive the RTS frame, and in this case E sends CTS frame. Node A remains beamformed after the transmission of RTS frame in the direction of E to receive CTS frame.

To send the RTS frame to DD neighbor via multiple hops, special RTS is constructed (called forwarding RTS). This RTS frame includes route comprising DO neighbors and the duration of subsequent DATA/ACK frames. Nodes which receive forwarding RTS frame, send to next DO neighbor according to the information given in the route. In this scenario, node S sends RTS frame of destination E to node B using information from the transceiver profile, whereas node B (using DOA information of C) forwards RTS frame to node C. The nodes which forward the RTS frame do not update their DNAV. On receiving forwarding RTS frame, node E sends CTS frame by pointing its transmission beam in the direction of node S. Upon reception of CTS frame, DATA/ACK frames are exchanged directionally. Nodes which overhear DATA/CTS frames, update their DNAV accordingly. The limitations of this protocol includes the long delay of RTS propagation and the risk of losing RTS over multiple hops. Also, the intermediate multihop paths for RTS propagation may not always be available.

Kolar et al. in Reference 5 proposed modifications to handle the HOL blocking problem. It considers directional MAC protocol similar to Reference 7 where RTS/DATA/ACK frames are transmitted directionally. To avoid HOL problem with beamforming antennas and FIFO queuing, authors propose a new greedy queuing policy. Traditional FIFO queue at the MAC layer is divided into interlinking queue (IQ) and MAC queue (MQ). The IQ is implemented with a set of FIFO queues for each priority. The MAC layer dequeues packet from IQ and enqueues to MQ. The criterion to dequeue packet is the mode of its transmission, if AOA information of the intended receiver of first packet exists then the packet is dequeued from IQ and enqueue to MQ. Otherwise, broadcast packets and unicast packet with receiver of unknown AOA are kept in IQ. The packets from IQ are transmitted only when all of the directional antennas are idle and neighboring nodes are not engaged in communication. On the other hand, a packet from MQ is dequeued which has the least wait time respecting the priorities of MQ. If MQ is empty then packets from IQ are dequeued to MQ or transmitted omnidirectionally.

In Reference 10, the authors proposed an approach called receiver initiated directional MAC (RIDMAC) to minimize the effect of deafness raised during directional transmission of DATA/ACK and control frames (RTS or CTS). In RIDMAC, each node transmits data packet and their corresponding control frames directionally, while a node listens in the omnidirectional mode. To address the issue of deafness, each node maintains a polling table. The polling table stores information about neighboring nodes which may experience deafness. The transmitter checks the destination of next data frame, and if the frame has the same destination as the current data frame then the transmitter appends the information about the size of next data frame in the header of current data frame, otherwise information is set to zero. Upon reception of a data frame, neighboring nodes update their polling tables. The additional information of the next packet in the header of the current data frame help to avoid deafness caused by a node's transmission to its intended receiver. After successful exchange of DATA/ACK frame, the transmitter and receiver check their polling table. If entries of potential deafness nodes exist then it sends RTR to one of the deafness nodes for transmission. The criteria to select potential transmitter to send RTR frame are the longest delayed node and waiting time for its

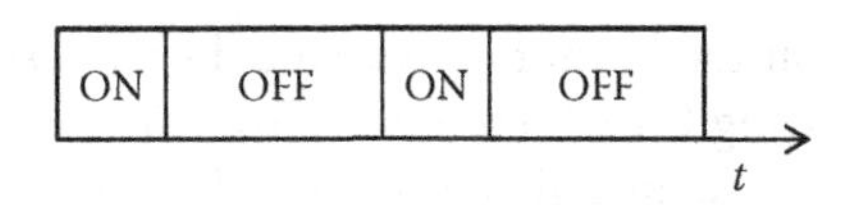

Figure 6.15 ON–OFF durations capture-aware directional MAC (CADMAC).

own pending packet for the transmission. If the waiting time of its own pending packet is longer, then the node defers polling; otherwise, node with longer delay is selected for potential transmission. Upon reception of RTR frame, the polled node transmits the data frame.

Choudhury et al. proposed capture-aware directional MAC (CADMAC) technique in Reference 11 to address the capture effect due to directional antennas. It is assumed that nodes are strictly synchronized and time is divided into cycles with each cycle subdivided into ON and OFF durations as show Figure 6.15. In an ideal mode, a node listens in omnidirectional mode during ON duration. Upon reception of a signal, a node steers beam in that direction to increase the reliability of the signal, and the MAC layer records the beam used to receive it. The received packets are characterized into productive traffic and capture traffic. In case the received packet is broadcast packet or unicast packet and destined to the node itself, then it is called productive traffic; otherwise, it is capture traffic. If capture traffic is received, then the beam applied to receive the packet is blacklisted. At the end of ON duration, the node turns off all the blacklisted beams for OFF duration and exploits rest of beams for transmission/reception. For example, in the scenario shown in Figure 6.16, assume node A is communicating with node C via node B while node D is communicating with E. Upon reception of packets from node D and E, node B turns off its antennas 3 and 2 in OFF subcycle and listens/transmits/receives at antennas 1 and 4. Similarly node E turns off its antennas 1 and 4 upon reception of packets from A and C. As a result, communication among nodes A, B, and C, does not capture E, and communication between nodes D and E does not capture B. A node utilizes single beam for transmission or reception and listens on all available turned-on beams during OFF period. DVCS is also modified to minimize the capture effect at the receiver when capture traffic is received from the same direction from which RTS was received. DVCS is enhanced in a such a way that after reception of RTS/CTS prior to the desired

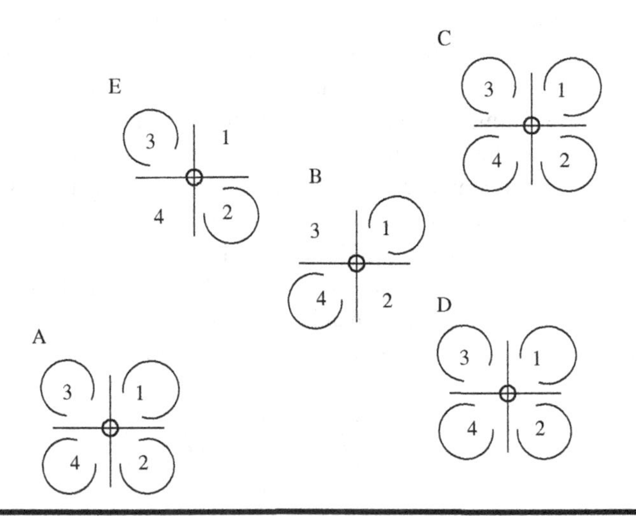

Figure 6.16 Example scenario of CADMAC.

data packet, PHY layer turns off the beam for a specific duration so that only desired packet is transmitted/received after receiving CTS/RTS frames. To reduce capture effect caused by reception of undesired RTS/CTS frames, an early destination detection mechanism is also proposed, where destination address is included in initial bits of the header.

6.4.1.3 Multidirectional Sequential RTS Transmission

The directional transmission of RTS/CTS/DATA/ACK frames successfully exploits spatial reuse and higher transmission range advantages of directional antennas. However, in mobile multihop networks it is hard to estimate the direction of a receiver prior to send a RTS frame because of rapid channel variation. This can be overcome by transmitting directional RTS in all directions. At first, Korakis et al. proposed a protocol, called circular RTS MAC (CRM) in Reference 12. In CRM, prior to sending data frames, RTS is transmitted sequentially over all antenna elements to inform neighboring nodes. The duration value in the RTS frame is decreased by a RTS transmission period after a RTS frame transmission over a directional antenna element or in a direction. In an idle mode, a node listens in the omnidirectional mode. Upon reception of the RTS frame, the desired receiver selects appropriate antenna or beam and transmits CTS frame applying that beam. The receiver waits for a specific time interval before transmission of CTS frame so that transmitter could transmit RTS in all directions. After transmission of RTS frames, a node listens in omnidirectional mode to receive CTS and selects antenna beam to transmit DATA/ACK accordingly. In Reference 13, the authors have enhanced this approach and proposed circular transmission of RTS and CTS frames to avoid interference caused by hidden nodes, the protocol being called circular RTS and CTS MAC (CRCM).

Though sequential directional transmission of RTS/CTS exploits advantages of higher transmission range, spatial reuse and decreased signal processing complexity at the transmitter and receivers, however, it increases additional delay caused by sequential transmission of RTS/CTS frames.

Gossain et al. in Reference 14 proposed a MAC protocol for directional antennas (MDA) which addresses the issue of time delay caused by CRCM [13]. The protocol proposed circular transmission of RTS and CTS frames at the same time after successful exchange of directional RTS and CTS frames in the directions of receiver and transmitter, respectively. The technique requires the AOA or directional information of neighboring nodes. This is achieved through routing process, upon the reception of PREQ/PREP frame which are transmitted using omnidirectional antenna, nodes update direction information of their neighboring nodes in directional neighbor table (DNT) and maintains it on overhearing packets at the MAC layer. The sequential transmissions of RTS/CTS at the same time decrease the delay with respect to CRCM. To avoid coverage overlap of circular transmission of RTS/CTS frames, a diametrically opposite direction (DOD) is defined where RTS/CTS are transmitted only in the direction of existing neighboring node. The DOD mechanism is illustrated in Figure 6.17. If A wants to send data packet to node B then first it estimates the antenna by which receiver (node B) reaches it. It is assumed that a node can estimate this antenna element by using neighbor information. If N_{AB} is the antenna element of A used to transmit to B then antenna element used by B to transmit A (N_{BA}) is estimated by

$$N_{BA}(N_{AB},K)=\begin{cases} N_{AB}+\dfrac{K}{2} & N_{AB}<\dfrac{K}{2} \\ N_{AB}-\dfrac{K}{2} & \text{otherwise} \end{cases}$$

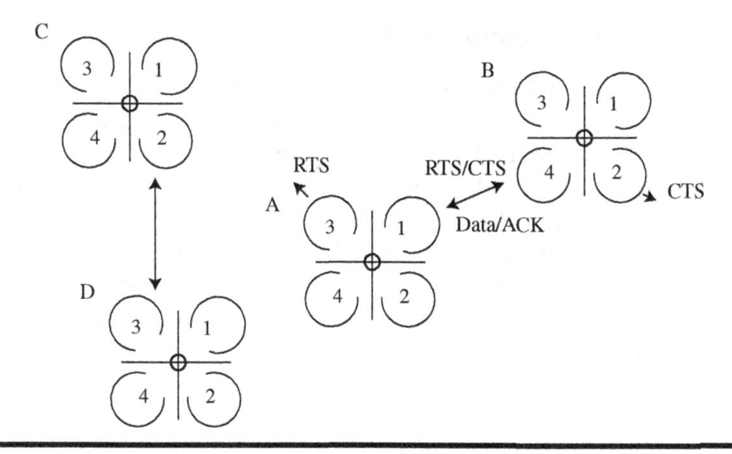

Figure 6.17 Example scenario of CRCM.

where K is an even number of antenna elements at nodes. If $K = 4$, then node A estimates that $N_{BA} = 3$ and sets DOD_{RTSEND} to 3. Therefore, node A transmits RTS on antenna 3. Similarly, B sets DOD_{CTSEND} to 2 and transmits CTS on antenna 2. The RTS/CTS frames transmitted in the directions other than receiver/transmitter are called DOD-RTS/CTS. To avoid deafness caused by DOD-RTS/CTS transmission, enhanced DNAV(EDNAV) is proposed. EDNAV differentiates between collision avoidance and deafness avoidance. EDNAV has two explicit tables, that is, DNAV table and deafness table (DT). Whenever a node has a packet to be sent over one direction, it looks in both DNAV and DT. On the contrary, upon reception of a packet the node will either modify its DNAV or its DT, not both. The DNAV is to be modified when node lays on communication path between the transmitter and the receiver. The DT is modified whenever the node receives either a DOD-RTS or DOD-CTS. EDNAV when compared to DNAV has advantages to differentiate between deafness and collision scenario. For example, in the scenario shown in Figure 6.17, nodes A and B exchange their RTS and CTS at their antennas 1 and 4, respectively. Upon successful exchange of RTS/CTS frames, node A transmits DOD-RTS frames using its antenna element 3. Upon reception of DOD-RTS transmitted from antenna 3 of A, node C updates its DT table. The entry in DT prevents deafness as node A is unable to communicate with C—similar to DNAV. When C wants to send packet to D, it checks its DNAV and DT table. If an antenna entry exists in DNAV table, then node defers its transmission using that antenna element assuming that it is lying in ongoing communication path. If an antenna entry exists only in DT table, then node can communicate to other nodes concurrently. In this scenario, EDNAV enables node C to communicate with D concurrently.

In Reference 15, Takata et al. proposed directional MAC with deafness avoidance (DMAC/DA). It reduces the access overhead caused by transmission of additional control frames to minimize deafness. Each node keeps information about the activity of its neighboring nodes. Regardless to the destination of a packet, upon reception of a packet, these information are updated. When a node wants to send a packet to neighboring receiver, after successful channel access, it sends RTS in the direction of desired receiver. The receiver of RTS sends directional CTS to the transmitter while other neighboring nodes (laying in the same direction) update their information. After successful exchange of RTS/CTS, wait to send (WTS) frames are transmitted toward potential transmitters. The potential transmitter is estimated from the information available in the neighbor table.

6.4.1.4 Multidirectional Concurrent RTS Transmission

To minimize delay of sequential transmission of RTS/CTS, Capone et al. proposed multidirectional concurrent transmission in Reference 16. It is a power controlled directional MAC (PCDMAC) protocol for wireless mesh networks. Each node broadcasts control frames at a constant power. Upon receiving the broadcast frames, nodes estimate the minimum power to access the neighboring node. When a node accesses the channel, it simultaneously transmits RTS in multiple directions with appropriate power in each direction. Similarly, the receiver transmits CTS frames in multiple direction. After successful exchange of RTS/CTS, DATA and ACK packets are transmitted directionally with the minimum required power to reduce the interference and increase the spatial reuse.

6.4.2 Synchronized Access Protocols

In Reference 17, authors proposed TDMA-based MAC algorithm for wireless ad hoc networks with directional antennas. Time is divided into frames, whereas each frame is divided into three subframes. The first subframe is used for neighbor discovery while second subframe is used to reassure connection and to make data reservation. Third subframe is used for actual data transmission. Each subframe is divided into slots and each slot within neighbor discovery and reservation subframe is divided into multiple minislots. In the neighbor discovery, a node (called A) sequentially scans for its neighbors and transmits a message with specific power level. A node (called node B) which receives the message, decides that either a connection is possible with the transmitter or not. If the received power is greater than threshold, then node B estimates the required transmission power to transmit message toward the transmitter. If the required power is greater than maximum power, then B assumes that the connection is not possible with A and abandons its transmission. Similarly, node A listens for the return transmissions and estimates if the connection is possible with node B depending upon the power level of received message from the node B. In case of a valid connection, node A sends a message to complete handshake and informs node B about the time slot for data reservation. Then in a agreed time slot, node A initiates three-way handshake to reassure connection and to make reservation for data transmission. In case node A wants to reserve slots for data transmission, it also indicates the number of desired slots. Node B computes the power level from message of node A and computes the set of free slots which are common. In case the number of common slots is less than the required by A, no reservation can be made. However, the two nodes use the directional information and may continue in the next agreed minislot in the following frames. The node B replies to node A if the number of available common slots for both nodes is equal or greater than that required by node A. It transmits reply with a newly computed power level and the set of common slots reserved for data transmission.

In Reference 18, Jakllari et al proposed a polling based medium access protocol (PMAC) in detail. It is assumed that each node is synchronized with its neighbors in time, and time is divided into contiguous frames. Furthermore, each frame is divided into three segments, that is, searching, polling, and data transfer. In a search segment, each node searches for a new neighbor by steering beam in a random direction. The search slot is further divided into four subslots. In first two subslots, a node transmits or receives a pilot tone. If it transmits in first subslot, then it receives in the second subslot. A node which receives pilot tone in the first subslot, transmits pilot tone in the same direction in the second subslot. In this way, connection is established between two neighbors. In subslots 3 and 4, the nodes that successfully exchanged pilot tones in subslots 1 and 2, exchange their lists of slots which are unused in their polling segments. This information is used (by nodes) to identify a polling slot which can be used for scheduled polling. Once nodes agree upon a polling slot, they communicate in the same polling slot frame after frame until the path is lost. In a chosen polling slot of a frame, the nodes steer beams in the same direction in which communication occurred in the previous frame. In this way, polling slots also help to re-establish connection. To avoid collisions, control frames similar to RTS/CTS are also exchanged in the polling time slots. In case of two successive collisions of these control frames, the two communicating nodes chose last common free polling slot in order to communicate with each other. For this purpose, the nodes also piggyback a list of their free polling slots at the end of each data exchange (in data segment) between them. If it is also failed then the nodes resort to a search. Each node announces (during its polling slot) the next data frame that it needs to send and its length. It also indicates the available instances (slots) in the data transfer segment for its transmission and reception from the corresponding neighbor. The data transfer is scheduled at the earliest time possible in either the current frame or in a future frame. In the data transfer segment, the schedule data transfer takes place. If the nodes were scheduled (in the polling slot) to transmit, in the data transfer segment node steers its beams toward the receiver. RTS/CTS frames are also exchanged in data transfer segment to avoid collisions.

6.5 MAC Layer Link Scheduling Mechanisms

In Reference 19, authors proposed HMAC protocol to explore spatial reuse features of switched-beam antennas and enables multiple packet transmission and reception. It also address the issue of deafness and HOL. It is assumed that each node is equipped with wide-azimuth switched-beam smart antenna comprised of an multi beam adaptive antenna (MBAA) and capable to extract AOA of incoming signal. To transmit concurrently from multiple beams, a node waits for CW-based random backoff before DIFS interval. HMAC uses node-based backoff where all beams wait for equal duration after transmission/reception or access failure. Thus, a common backoff mechanism is used for all beams. When a transmission failed on a beam, the CW gets doubled and all beams have to wait for (a duration) before transmission. Similarly, in case of successful transmission, common backoff is set. In HMAC, hybrid network allocation vector (HNAV) table is defined in which each node maintains the following information about its neighboring node:

- Specific beam to access to node
- Time duration until this neighbor is engaged in communication
- Silence time for a beam
- Transmission probability for this neighbor

To mitigate deafness issue, each node also stores information about destination of received a RTS at each beam. Furthermore, HMAC defines scheduling message (SCH) to inform other potential neighboring transmitters about the intended transmission. SCH frame consists of transmitter address and duration of intended transmission. Upon reception of SCH frame, nodes update their HNAV.

Upon reception of RTS frame which is not destined to itself, the node also updates its HNAV. When a node gets access to the channel, it sends RTS to desired receiver applying corresponding beam while SCH frame is transmitted by all other nonblock beams toward its potential transmitters. Nodes also keep the record of their previous transmitters from which they have received RTS frames (destined to themselves). A beam is blocked if the transmission/reception is going on in its

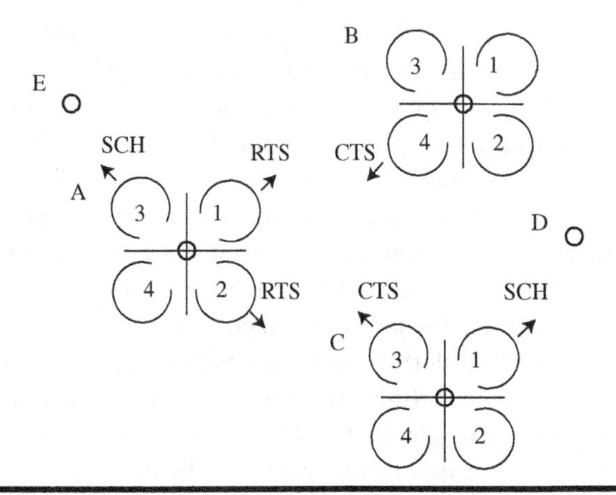

Figure 6.18 Example scenario of HMAC.

direction. If a node receives an invalid CTS (destined for other node) in the direction of intended transmission, the RTS/SCH is transmitted from all nonblocking beams. It helps to avoid interference with ongoing communication. Upon reception of RTS frame, the desired receiver sends CTS to the transmitter while SCH frames are transmitted using nonblocked beam on which RTS (destined to itself) is received. If a CTS (destined to other node) is received in the direction for intended reception, the CTS/SCH is transmitted from all nonblocking beams.

Figures 6.18 and 6.19 illustrate the operation of HMAC. Assuming that node A wants to send a packet each to nodes B and C, it sends RTS to both B and C and an SCH to E. All the neighbors of A update their HNAV from the duration in these messages. Similarly, the neighbors of B and C update their HNAV after listening to the CTS/SCH from B and C, respectively.

In Reference 20, the authors have proposed receiver-oriented multiple access (ROMA) to explore spatial reuse feature of multibeam adaptive antenna (MBAA) for transmission scheduling.

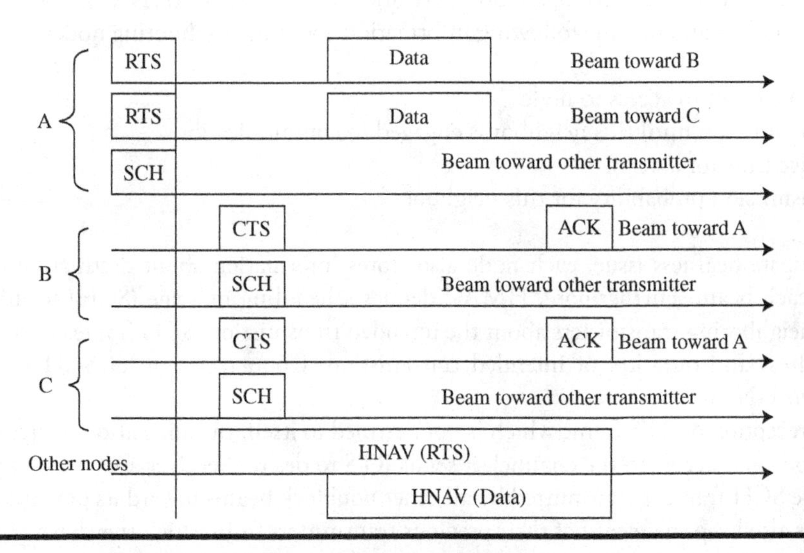

Figure 6.19 Example scenario of frame exchange in HMAC.

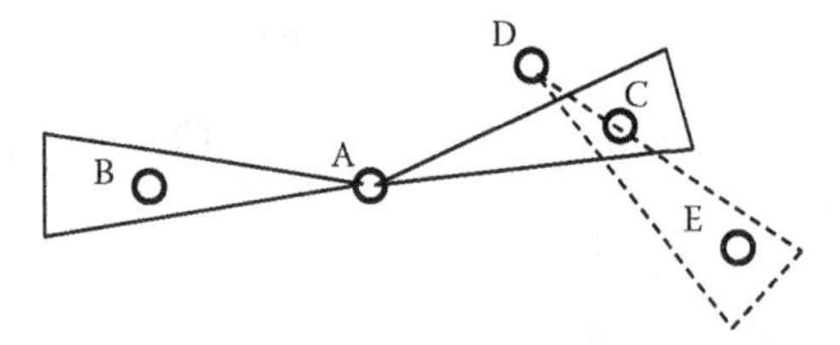

Figure 6.20 Example scenario of receiver-oriented multiple access (ROMA).

In ROMA, each node in the network is assigned a unique identification (ID) number and mounted with an MBAA antenna. It exploits multibeam forming feature of MBAA and enables receiver-oriented multiple access. It uses information of two-hop neighboring nodes and links bandwidth allocations to decide whether a node is a receiver or a transmitter. Furthermore, it also decides which corresponding links can be activated for reception or transmission during a time slot. In ROMA, all mobile nodes are synchronized and each node broadcasts information about its one-hop neighbors. At first, a node randomly separates its two-hop neighbors including itself in transmitters and receivers. If a node is transmitter or receiver then it chooses up to K of its active outgoing or incoming links for transmission or reception, respectively. In ROMA, each node and link is assigned *a priority* which is calculated using its identifiers and the current time slot. If the priority of a node is odd in a current time slot, then the node is considered as a transmitter; otherwise, it is considered as a receiver. As a result, nodes are randomly separated into two classes. It is also possible that a node and all its one-hop neighbors belongs to same class (transmitters or receivers). When a node is in reception or transmission mode, it computes up to K active incoming and outgoing links, respectively in such a way that links do not cause direct interference and raise hidden node problem.

Figure 6.20 shows the basic mechanism of ROMA in a sample network with MBAA antennas which are capable of forming up to two antenna beams. Node A can transmit concurrently to its desired receiver B and C. However, D detects hidden node problem at C incurred from A and D itself. Thus, the link D to E is not activated.

In Reference 21, a simple link activation algorithm is proposed that trade offs path length to increase the network capacity. The proposed algorithm uses spatial TDMA to take the advantage of concurrent transmissions or receptions capability in a given time slot. It reduces the number of colors required for a given graph and maximize the number of links scheduled within a time slot. The algorithm considered a graph-theory-based approach to perform link scheduling. Consider a connected graph $G = (V, E)$, where V and E denote the set of vertices and edges, respectively. At the start, algorithm selects nod with the highest degree, if more than one node have same degree then one node is selected randomly. Then, edge connected to the children of selected node is removed. After that, algorithm selects the next node with highest degree. After visiting all nodes, the algorithm calls a graph-coloring algorithm. In the end, a time slot corresponding to its assigned color is allocated to each node. The mechanism is illustrated in Figure 6.21. In normal mode, four different colors are required to assign all nodes. The concurrent transmission is possible only on nodes having same color; otherwise, it causes interference. The algorithm selects node B due to its higher clique degree. After iterating through node B's children, the algorithm removes edges connecting to children (C to F, E to F, A to D, and D to E). The resulting topology is shown in Figure 6.21b, and it only requires two colors. However, now nodes only reach each other via node B.

MUD-MAC enables multiple nodes to concurrently transmit with a single receiver. It is assumed that nodes enjoy frame level synchronization. In MUD-MAC each data packet is split

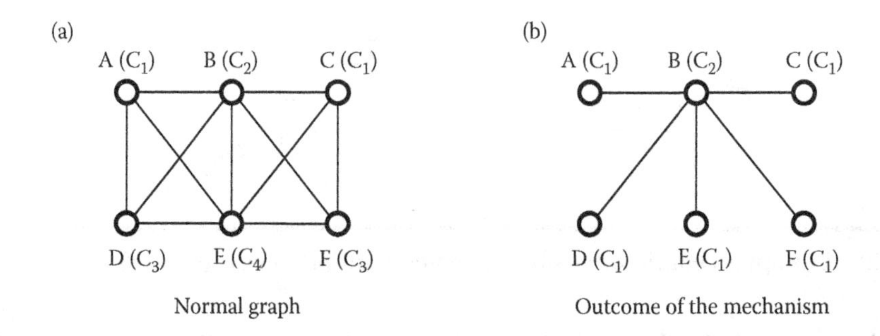

Figure 6.21 Example scenario of mechanism proposed. (Adapted from Chin, K.-W., 2008. A new link scheduling algorithm for concurrent Tx/Rx wireless mesh networks. In *IEEE ICC '08*, May, pp. 3050–3054.)

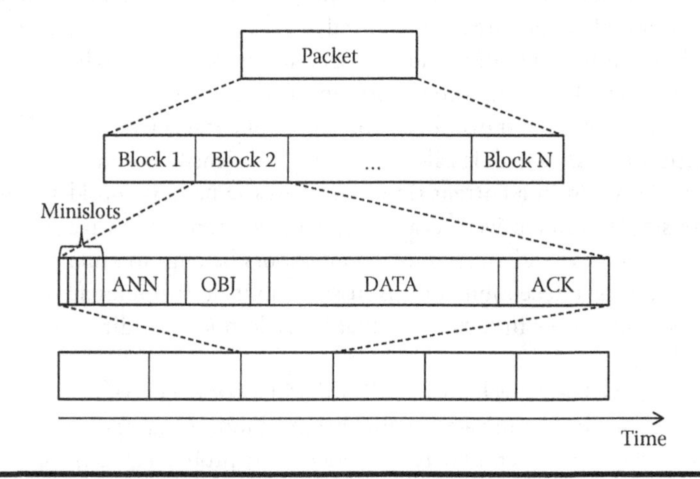

Figure 6.22 MUD-MAC frame structure.

into several blocks as shown in Figure 6.22, and each block comprises minislots, announcement message ANN, objection message OBJ, data payload DATA, acknowledgment message ACK, and interframe spacing (IFS) as a guard time to separate these messages. Prior to data transmission, a node announces for it in a ANN slot. Upon reception of the announcement, a node may send object depending on the interference level and the receiver capability. In case no object is received, the transmitter proceeds packet transmission using n successive blocks. However, only one ANN is transmitted for each packet and rest of $n - 1$ ANN slots are used by neighboring nodes for their transmissions. Contrary to data message, interference to the control messages (ANN, OBJ, and ACK) is considered as collision. To avoid collision of ANN frames, minislots are introduced. A node randomly selects minislot and starts ANN transmission unless it detects any ANN from other nodes before the chosen slot. In case a ANN frame is detected from other node before transmission, the node avoids to transmit ANN frame and goes to backoff for random time duration. Upon reception of ANN frame, node sends OBJ frame if interference streams in DATA are higher than its interference cancellation capability or ANN frame is not decodable. To transmit data frame, the transmitter uses unique signature which is generated from its MAC address by a specific routine. The neighboring nodes can also estimate signature from the MAC address of ANN transmitter. Nodes use the signature information to decode data frame. Upon successful reception of data packet, ACK is transmitted.

6.6 Summary

This chapter presents a number of the MAC layer protocols, which have been proposed to exploit the advantages of directional antennas in wireless networks. The incorporation of directional antennas in wireless networks has been extensively studied in the past few years, and a plethora of new protocols is proposed in this regard. However, there still exist several open challenges that are of practical relevance and hence need attention. The advantages of spatial reuse can be extended by applying cooperative interference cancellation where transmitting and receiving nodes exchange information regarding the nodes to which nulls are intended to be set. The exploration of such schemes in the future would be of great interest. Furthermore, the influence of link scheduling on routing can be investigated and addressed in the future.

References

1. IEEE 802.11 Part 11: Wireless LAN Medium Access Control (MAC) and Physical Layer (PHY) Specifications, Amendment 8: Medium Access Control (MAC) Quality of Service Enhancements, 2005.
2. IEEE P802.11n/D11.0, Part 11: Wireless Medium Access Control (MAC) and Physical layer (PHY) specifications, Amendment 5: Enhancement for higher throughput, June 2009.
3. Ong, E. H., J. Kneckt, O. Alanen, Z. Chang, T. Huovinen, and T. Nihtila, 2011. IEEE 802.11ac: Enhancements for very high throughput WLANs. In *IEEE 22nd International Symposium on Personal Indoor and Mobile Radio Communications (PIMRC)*, Toronto, Canada, pp. 849–853.
4. Bazan, O. and M. Jaseemuddin, 2012. A survey on MAC protocols for wireless ad hoc networks with beamforming antennas, *Communications Surveys Tutorials, IEEE*, 14(2), 216–239.
5. Kolar, V., S. Tilak, and N. Abu-Ghazaleh, 2004. Avoiding head of line blocking in directional antenna [MAC protocol], In *29th Annual IEEE International Conference on Local Computer Networks*, November, Tampa, USA, pp. 385–392.
6. Nasipuri, A., S. Ye, J. You, and R. Hiromoto, 2000. A MAC protocol for mobile ad hoc networks using directional antennas. In *IEEE Wireless Communications and Networking Conference (WCNC)*, Vol. 3, Chicago, USA, pp. 1214–1219.
7. Ko, Y.-B., V. Shankarkumar, and N. Vaidya, 2000. Medium access control protocols using directional antennas in ad hoc networks. In *IEEE INFOCOM*, Vol. 1, Tel Aviv, Israel, pp. 13–21.
8. Takai, M., J. Martin, R. Bagrodia, and A. Ren, 2002. Directional virtual carrier sensing for directional antennas in mobile ad hoc networks. In *ACM MobiHoc'02*, New York, NY, USA, pp. 183–193.
9. Choudhury, R. R., X. Yang, R. Ramanathan, and N. H. Vaidya, 2002. Using directional antennas for medium access control in ad hoc networks. In *Proceedings of the 8th Annual International Conference on Mobile Computing and Networking, ser. MobiCom '02*, New York, NY, USA, pp. 59–70.
10. Takata, M., M. Bandai, and T. Watanabe, 2006. A Receiver-Initiated Directional MAC Protocol for Handling Deafness in ad hoc Networks. In *IEEE ICC '06*, Vol. 9, Istanbul, Turkey, pp. 4089–4095.
11. Choudhury, R. R. and N. H. Vaidya, 2007. MAC-Layer capture: A problem in wireless mesh networks using beamforming antennas. In *IEEE SECON*, San Diego, USA, pp. 401–410.
12. Korakis, T., G. Jakllari, and L. Tassiulas, 2003. A MAC protocol for full exploitation of directional antennas in ad-hoc wireless networks. In *ACM MOBIHOC*, Annapolis, USA, pp. 98–107.

13. Jakllari, G., I. Broustis, T. Korakis, S. Krishnamurthy, and L. Tassiulas, 2005. Handling asymmetry in gain in directional antenna equipped ad hoc networks. In *IEEE 16th International Symposium on Personal, Indoor and Mobile Radio Communications (PIMRC)*, Vol. 2, Berlin, Germany, pp. 1284–1288.
14. Gossain, H., C. Cordeiro, and D. Agrawal, 2005. MDA: An efficient directional MAC scheme for wireless ad hoc networks. In *IEEE GLOBECOM '05*, Vol. 6, St. Louis,USA, pp. 3633–3637.
15. Takata, M., M. Bandai, and T. Watanabe, 2007. A MAC protocol with directional antennas for deafness avoidance in ad hoc networks. In *IEEE GLOBECOM '07*, November, Washington, USA, pp. 620–625.
16. Capone, A., F. Martignon, and L. Fratta, 2008. Directional MAC and routing schemes for power controlled wireless mesh networks with adaptive antennas. *Ad Hoc Netw.*, 6, 936–952.
17. Zhang, Z., 2005. Pure directional transmission and reception algorithms in wireless ad hoc networks with directional antennas. In *IEEE ICC'05*, Vol. 5, Seoul, Korea, May, pp. 3386–3390.
18. Jakllari, G., W. Luo, and S. Krishnamurthy. 2007. An integrated neighbor discovery and MAC protocol for ad hoc networks using directional antennas, *Wireless Communications, IEEE Transactions on*, 6(3), 1114–1024.
19. Jain, V., A. Gupta, and D. P. Agrawal, 2008. On-demand medium access in multihop wireless networks with multiple beam smart antennas, *IEEE Trans. Parallel Distrib. Syst.*, 19(4), 489–502.
20. Bao, L. and J. Garcia-Luna-Aceves, 2002. Transmission scheduling in ad hoc networks with directional antennas. In *Proceedings of the 8th Annual International Conference on Mobile Computing and Networking, ser. MobiCom '02*. New York, NY, USA: ACM, pp. 48–58.
21. Chin, K.-W., 2008. A new link scheduling algorithm for concurrent Tx/Rx wireless mesh networks. In *IEEE ICC '08*, May, pp. 3050–3054.

Chapter 7

Millimeter-Wave Wireless Networks: A Medium Access Control Perspective

Shiwen Mao and Theodore (Ted) S. Rappaport

Contents

Abstract

With vast spectrum availability, millimeter-wave (mm-wave) communications have been recognized as a promising technology for the next generation 5G mobile broadband and have attracted considerable interest from academia, industry, and standards bodies. In this chapter, we examine the mm-wave spectra frontier, in particular, the 60-GHz band, for multigigabit-rate wireless networks from a medium access control (MAC) perspective. We discuss the technical challenges and the state-of-the-art in 60-GHz MAC protocols. We then present three case studies of 60-GHz MAC protocol design to illustrate the importance of leveraging spatial reuse while reducing MAC control overhead. This chapter concludes with a discussion of open problems.

7.1 Introduction

With the dramatic advances in wireless communications and networking technologies, there is a rapidly growing demand for wireless data service. According to a recent study by Cisco, there has been a 1000-fold mobile data traffic growth since 2005. An acceleration of mobile data traffic growth was first observed in 2003, and the growth is predicted to be 11-fold between 2013 and 2018 [1]. In addition, mobile video, including all video data that travels over 2G, 3G, and 4G networks, is already half of the overall mobile data traffic and will be 69% of the mobile data traffic by 2018. This trend is driven by the proliferation of smartphones and the compelling need for ubiquitous access to data and video content over wireless access networks, and these enormous increases in data consumption will significantly stress the capacity of the existing wireless networks and strongly influence the design of future, that is, the fifth generation, wireless networks.

To this end, the millimeter-wave (mm-wave) band, especially the 60-GHz band with the vast unlicensed spectrum, is at the spectral frontier for high-bandwidth commercial wireless communication systems. The new applications of mm-wave communications in the 60 GHz band have gained considerable interest from academia, industry, and standards bodies. This is due to the large unlicensed bandwidth (i.e., up to 7 GHz) that is available in most parts of the world. The massive unlicensed bandwidth provides great potential to meet the surging wireless data demand. Many bandwidth-demanding new applications (e.g., high-definition (HD) video [2,3], low-cost wireless backhaul, home gaming devices, position location systems, and wireless memory devices) can be supported in mm-wave wireless networks [4].

Several standards have been or are being defined to achieve multigigabit rates in 60-GHz networks, such as WirelessHD, Wireless Gigabit Alliance (WiGig), IEEE 802.11ad, IEEE 802.15.3c, and ECMA-387. WirelessHD occurred earlier (first products were produced in 2008) and targeted for wireless personal area network (WPAN). It supports the transmission of uncompressed HD video and audio data and can be a wireless replacement of high-definition multimedia interfaces (HDMI). The WiGig alliance was formed in 2009 to enable multigigabit speeds for high-performance wireless data, display and audio applications in both WPAN and wireless local area network (WLAN), and harmonized with the IEEE 802.11ad standard body in 2013 with products shipping in that year. The IEEE 802.15.3c standard was published in 2009, specifying an mm-wave physical layer (PHY) for existing IEEE 802.15.3 WPANs. It allows high data rates over 2 Gbps as well as high coexistence with other mm-wave systems in the IEEE 802.15 family of WPANs. ECMA-387 is a standard published by ECMA International, Geneva, CH, in December 2008. It specifies the high-rate 60-GHz PHY, MAC, and HDMI Phase Alternating Line (PAL) standards for bulk data and multimedia streaming in WPANs. In August 2013, the The Federal Communications Commission (FCC) has amended rules to allow longer communication ranges for outdoor unlicensed 60-GHz point-to-point systems, in order to provide broadband service to office buildings and other commercial facilities [5].

In the 60-GHz mm-wave band, many antennas can be integrated on-chip due to the small wavelength, thus enabling flexible beam-forming and highly directional transmissions. Different antenna beam widths can be achieved by varying the number of elements used in a phased array (Chapter 4, [5]). The system architecture, including analog-mixed signal channel equalization, is ready, and the use of complementary metal oxide semiconductor (CMOS) and the increasing level of integration mean that 60 GHz is on the verge of success [4]. However, the unique 60-GHz channels also create many technical issues that are significantly different from conventional cellular networks and raise substantial technical challenges.

In this chapter, we suggest ways to exploit the 60-GHz spectra frontier for multigigabit-rate wireless networks from a medium access control (MAC) perspective. In wireless networks, MAC protocols are the "workhorse" of the protocol stack. Their efficiency is pivotal for higher-layer protocols (e.g., transmission control protocol (TCP) or video streaming) to savor the benefits of emerging PHY technologies. In the development of Wi-Fi protocols, it has been widely recognized that simply increasing the PHY data rate without reducing the control overhead will only achieve limited performance gains with respect to throughput and delay. At 60 GHz, the wavelengths are so small (5 mm in free space) so that directional antennas can be used for increased gain and for bouncing-off of reflective objects in order to overcome signal outages caused by humans and objects that block the line-of-sight (LOS) radio path. This introduces a new complexity into the MAC layer as rapid and low-overhead beam-forming must be conducted. In order to accommodate existing and emerging bandwidth-intensive applications, it is thus crucial to improve the efficiency of wireless MAC protocols, while new PHY technologies are being adopted for higher channel data rates.

In the remainder of this chapter, we first discuss the new technical challenges brought about by the unique 60-GHz channels and review proposed solutions for 60-GHz-network MAC protocols. We then examine three different approaches to the design of 60-GHz MAC protocols, that is, a directional carrier sense multiple access with collision avoidance (CSMA/CA) approach to overcome the deafness problem, an interference measurement-based approach to leverage spatial reuse, and a frame-based scheduling approach to minimize control overhead and to maximize spatial reuse. We conclude this chapter with a discussion of open problems and future perspectives.

7.2 Technical Challenges and State-of-the-Art

7.2.1 Technical Challenges

Although high data rates up to multigigabit per second can be supported, mm-wave communications in the 60-GHz band suffer severe attenuation, especially when there is not an LOS propagation path between the transmitter and the receiver. The propagation attenuation of 60-GHz signals in free space is 22 dB higher than that of 5-GHz signals, and atmospheric absorption of 60-GHz signals ranges from 15 to 30 dB/km [4]. As a result, the range of a 60-GHz link can be limited, especially over long distances. Since the free space propagation loss between isotropic antennas scales as λ^2, where λ is the wavelength, beam-forming becomes necessary to achieve directivity that scales as $1/\lambda^2$ to overcome attenuation. The 60-GHz systems have been shown viable for short-range communications [6], such as WPANs that may require high-data-rate transmissions over short distances (e.g., high-definition television (HDTV) or mass storage synchronization), while new applications such as 60-GHz mesh networks or mm-wave backhauls are emerging. Deep understanding of the propagation and characteristics of the 60-GHz channel is crucial for its success.

The highly directional links in 60-GHz systems entail a complete rethinking of network protocols originally designed for traditional CSMA-based Wi-Fi or mesh networks. The highly directional link is a double-edged sword: on the one hand, carrier sensing and node coordination are nontrivial due to the inherent deafness problem (i.e., two close nodes cannot communicate since their beams are not pointing to each other); on the other hand, the highly directional transmissions greatly reduce interference and allow spatial reuse, providing great potential to enhance network capacity.

Furthermore, as a lesson learned from Wi-Fi standards, solely pursuing higher PHY data rates does not necessarily achieve superior network throughput performance. Rather, reducing control overhead and improving protocol efficiency are the keys to transforming plenty of 60-GHz spectra for high-throughput application [7].

Although mm-wave prototype chipsets are emerging [4], their performance in a *network setting* and the design of efficient 60-GHz networking protocols remain open areas of future research. Efficient MAC scheduling is indispensable to fully harvest the multigigabit rates in PHY for high-throughput applications [8]. As discussed earlier, MAC protocols for the 60 GHz band must support transmission and reception with highly *directional* and *adaptive* beams. *To achieve high throughput performance, leveraging spatial reuse and improving protocol efficiency are key approaches for mm-wave networks.*

7.2.2 Existing Solutions

There have been considerable works on directional MAC protocols in the literature, though these past works may not be directly adopted in 60-GHz networks. This is because the unique channel properties in the mm-wave band have not been fully considered in these earlier schemes [8]. Some proposals require high control overhead or have high complexity, while others rely on omnidirectional communications for control messages, which may not be feasible for mm-wave systems that operate in the multigigabit domain with highly directional transmissions [6].

There have been several directional MAC protocols designed for 60-GHz systems, which can be roughly classified into two categories: (i) *centralized* and (ii) *distributed* MAC protocols. In centralized schemes, all data transmissions are coordinated by an access point (AP) (or piconet coordinator (PNC)). In Reference 9, the authors propose a directional CSMA/CA protocol to exploit virtual carrier sensing and rely on AP to distribute network allocation vector (NAV) information. Another centralized approach, the multihop relay directional MAC (MRDMAC) protocol, is based on the conventional AP-based single-hop MAC architecture for keeping primary connectivity [10]. Most data transmissions are via an AP with a sequential polling policy. The AP selects an intermediate node to relay the traffic, whenever the direct link is blocked. Spatial reuse, as enabled by the highly directional transmissions, is not fully exploited in these schemes. In a recent work [11], concurrent transmissions was scheduled based on the notion of exclusive regions, although the scheduling scheme uses greedy heuristics without performance guarantees.

The memory-guided directional MAC (MDMAC) [12] and directional-to-directional MAC (DtDMAC) [13] are both fully distributed schemes. MDMAC achieves coordination between highly directional transmitters and receivers in a distributed fashion, by employing memory to achieve approximate time-division multiplexed (TDM) schedules without explicit coordination or resource allocation. In DtDMAC, sending nodes cache the location information about their neighbor nodes and use such information later to determine the direction in which it should first try to send request to send messages. Both directional MAC protocols can alleviate the deafness effect without centralized control, which is highly desirable. However, there is still room for improvement to achieve high data rates, such as improving the low frame transmission efficiency and mitigating the high overhead of control messages.

7.3 Case Studies

Now, let us consider three different approaches as examples for the design of 60-GHz MAC protocols, to illustrate the importance of spatial reuse and overhead reduction. We first examine a directional CSMA/CA approach to overcome the deafness problem and then introduce an interference measurement-based approach that allows concurrent transmissions. Finally, a frame-based scheduling approach is presented to minimize control overhead and to maximize spatial reuse.

7.3.1 CSMA/CA-Based Approach

Currently, several 60-GHz standards, such as ECMA-387 and IEEE 802.15.3c, focus on using time-division multiple access (TDMA) for data communications. Because data traffic is bursty, the required medium time (i.e., the period when the medium is busy for one transmission) is often highly unpredictable. A TDMA-based MAC protocol may cause either high overhead for on-the-fly medium reservation or under-allocated or over-allocated medium time for individual users. Contention-based MAC protocols, such as CSMA/CA, work well with bursty traffic and operate robustly in unlicensed bands. However, the conventional CSMA/CA protocol does not work well with directional antennas due to impaired carrier sensing at the transmitters [9].

In Reference 9, a directional CSMA/CA-based MAC protocol is presented for 60-GHz WPANs. Consider a wireless network with one or more mobile devices (DEVs) and a PNC. Before associating with the PNC, the DEVs first perform beam-forming training for both transmission and reception such that both the transmitting and the receiving antennas can provide sufficient beam-forming gain. After beam-forming training, a DEV always beam-forms toward the PNC in its idle mode (i.e., waiting to receive), meaning that there is high receiver antenna gain. The PNC receives in the omnidirectional mode with a lower receiver antenna gain when it is idle. In the proposed protocol, because a DEV is always beam-formed toward the PNC before any data transmission or reception and the PNC coordinates the transmission within the network, the *deafness problem* is thus easily solved.

We next use a typical example where DEV-1 has data to transmit to DEV-2 to illustrate the operation of the directional CSMA/CA-based MAC protocol. In the idle mode, both DEV-1 and DEV-2 are beam-formed toward the PNC. Before DEV-1 can communicate with DEV-2 directly, it transmits a target request to send (TRTS) to the PNC. The TRTS contains three addresses: receive address (i.e., PNC), transmit address (i.e., DEV-1), and target address (i.e., DEV-2). Upon receiving the TRTS, the PNC will broadcast a target clear to send (TCTS) message in the omnidirectional mode.* This ensures that all associated nodes can receive it. The TCTS contains three addresses: receive address (i.e., DEV-1), transmit address (i.e., PNC) and target address (i.e., DEV-2). Both TRTS and TCTS indicate the duration of the transmission opportunity (TXOP) and are transmitted using a low-rate modulation and coding scheme (MCS), such that these control messages can be successfully decoded when only one end of the link has high beam-forming gain. If the PNC does not receive the TRTS from DEV-1 either due to channel error or due to a collision on TRTS, DEV-1 will not receive a TCTS after transmitting a TRTS. DEV-1 will then assume that a collision has occurred and will start an exponential back-off procedure.

Since all the DEVs point to the PNC when it is idle, the PNC can collect their location/direction information and then distribute such information to all the DEVs. After receiving the TCTS and recognizing that its MAC address is the target address, DEV-2 steers its beam toward

* Note that the omnidirectional model can be avoided by employing rapid consecutive transmissions or parallel transmissions made in many different spatial directions with high-gain antennas, to assure that links are properly made while distributing the PNC messages to all users throughout the space.

DEV-1. After receiving the TCTS, DEV-1 also steers its beam toward DEV-2. The TRTS/TCTS exchange thus sets up a TXOP in the network. Within the TXOP, DEV-1 can transmit one or more aggregated MAC protocol data units (A-MPDUs) directly to DEV-2 at a high data rate. Upon receiving an A-MPDU, DEV-2 replies with a block ACK (BA) that identifies which MPDUs in the A-MPDU have been received successfully. Other DEVs learn from the TCTS that there will be an ongoing transmission and thus set their NAV for the duration of the TXOP indicated in the TCTS.

In Reference 9, the directional CSMA/CA protocol was evaluated with analysis and simulations. It was shown that the proposed protocol achieves reasonably high MAC efficiency, especially when the optimal minimum contention window (CW_{min}) values are chosen.

7.3.2 Interference Measurement-Based Approach

The directional CSMA/CA MAC can be further enhanced by allowing concurrent transmissions, as long as they do not interfere with each other, thus achieving *spatial reuse*. The key is how the PNC can identify the set of links that can coexist with highly directional transmissions. The design of 60-GHz MAC protocols depends on the specific antennas used, the channel propagation, and the interference conditions, as well as the specific locations and environmental factors of the network. For practical and heterogeneous systems, an empirical scheme based on interference measurement is presented in Reference 14 to exploit spatial reuse.

Consider a WLAN with an AP and multiple nodes. An existing 60-GHz MAC protocol is used to schedule the directional transmissions in the network. Time is divided into a scheduling phase, where the AP collects transmission requests from the nodes, and a transmission phase, where some nodes are scheduled by the AP to transmit data. Nodes that are not scheduled for transmission are, instead, scheduled to measure interference while pointing to their target peer nodes. Such interference measurement results will also be returned to the AP in the next scheduling phase. Based on the interference measurement, the AP can classify the links into groups, where links in the same group do not interference with each other if scheduled for concurrent transmissions (as shown in Figure 7.1). After such a transient interference measurement

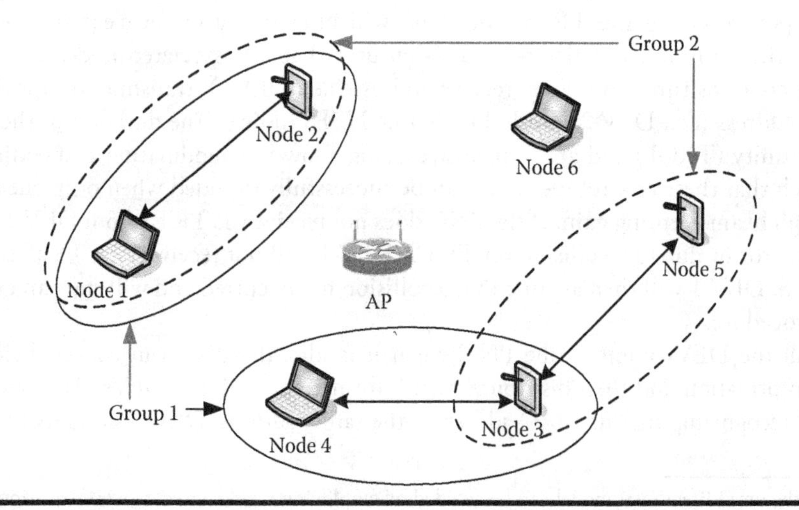

Figure 7.1 Two groups are formed based on interference measurement results.

period, noninterfering groups are formed based on particular directional beam headings of various nodes, and the AP can start to schedule concurrent transmissions for links within the same group.

The advantages of this approach are that it does not depend on any antenna propagation model and can be built on the fly in real-world in situ conditions, almost as if a real-time directional radio map were being built for the network, which is useful for deployments in cluttered environments where no theoretical model can be sufficiently accurate. The drawback is that a measurement and classification mechanism is required. In typical applications, such as an indoor WPAN, the nodes are stationary, and the measurement results could be valid unless the topology is changed (e.g., moved to a different location). In fact, the interference measurement can be continued during normal network operation, and the classification of links can be progressively refined, to make it robust for network and environment changes.

7.3.3 Frame-Based Scheduling

The third approach is focused on *efficient scheduling* and *overhead reduction* in mm-wave networks [7]. Consider an mm-wave WPAN or WLAN consisting of an AP (or PNC) and multiple devices (DEV) (see Figure 7.2). With a suitable bootstrapping mechanism [15], the nodes are aware of each other's locations (received from the PNC) and always point their beams to the PNC when it is idle. The PNC collects traffic demands from the DEVs and computes schedules to enable concurrent directional transmissions in the WPAN.

To amortize control overhead, *frame-based scheduling* is adopted, where network time is partitioned into a sequence of nonoverlapping, variable-length intervals, termed *frames*. Each frame consists of (i) a *scheduling phase*, where the PNC collects traffic demands from DEVs and computes and disseminates a transmission schedule, and (ii) a *transmission phase*, where DEVs start concurrent transmissions following the schedule. Packets arriving during the current frame will be stored at the nodes in the *virtual queues* (one for each neighbor) and then scheduled to be transmitted in the next frame. When it is idle, all DEVs point their beams to the PNC. In the scheduling phase, the PNC polls each DEV with directional transmissions to collect traffic demands from the DEVs, computes a schedule to serve the traffic demands, and then transmits the schedule to each of the DEVs (also with directional transmissions). A schedule consists of (i) a sequence of topologies, each indicating how the DEVs are paired to form directional links, and (ii) a corresponding sequence of time intervals, each indicating how long each topology should sustain. During the transmission phase, the PNC and DEVs pair with each other and start transmitting packets for a number of time slots, as specified in the schedule. We term this approach *frame-based directional MAC* (FDMAC) [7].

The design objective of FDMAC is to achieve the dual goal of spatial reuse and efficiency, that is, (i) to leverage collision-free concurrent transmissions to fully exploit spatial reuse and (ii) to amortize control overhead over a long sequence of data transmissions for high efficiency. The latter has been long recognized as highly effective for wireless network systems, such as gated/exhaustive service in polling systems, frame aggregation in Wi-Fi and 802.15.3c mm-wave WPANs, and limited-k and gated-service-based MAC for WLANs [7].

Let there be N DEVs in the WPAN and one DEV, denoted as DEV-1, that also serves as the PNC. The FDMAC operation is illustrated in Figure 7.2. In the beginning of the scheduling phase, the PNC polls the DEVs using adaptive beam antennas for their virtual queue backlogs and assembles the traffic demand matrix $\mathbf{D}$. Each element d_{ij} of $\mathbf{D}$ is the number of backlogged

(a)

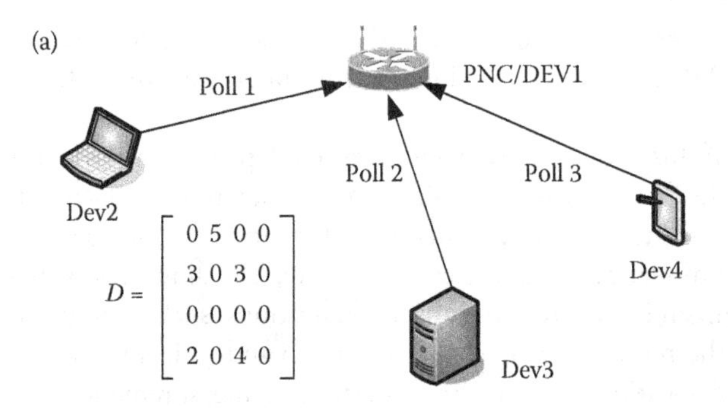

(b)

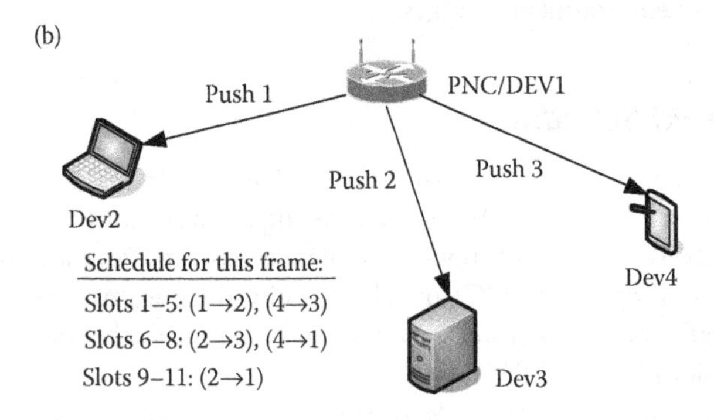

(c)

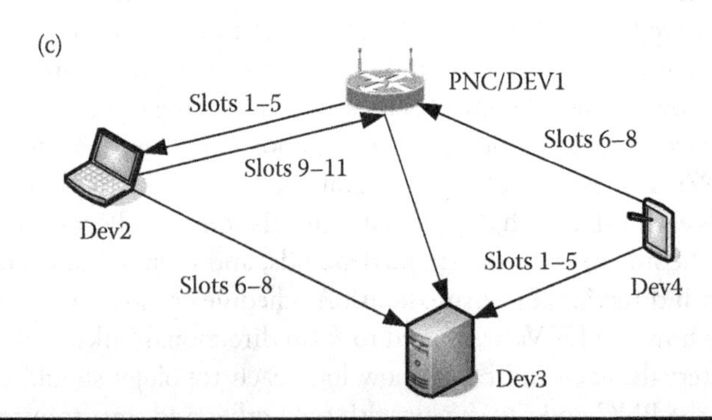

Figure 7.2 Operation of the frame-based scheduling directional MAC protocol (FDMAC). (a) PNC polls traffic demands from the DEVs to construct the traffic demand matrix *D*. (b) PNC computes a schedule for demand *D* and pushes the schedule to the DEVs. (c) DEVs start concurrent transmissions following the schedule received from the PNC. (From Son, I.-K. et al., 2012. On frame-based scheduling for directional mmwave WPANs. In *Proc. IEEE INFORCOM'12*, Orlando, FL, March, pp. 2149–2157. Copyright 2012 IEEE.)

packets from DEV i to j. The PNC then computes a schedule to serve the traffic demand **D**. The schedule consists of multiple elements, denoted by matrix $\mathbf{A}^k = \left[a_{ij}^k\right]_{N\times N}$ (where $a_{ij}^k = 1$ indicates a directional link from DEV i to j; $a_{ij}^k = 0$ indicates no transmission from DEV i to j). Each element $\mathbf{A}^k$ pairs some DEVs for δ^k time slots. Written in the matrix form, the traffic demand **D** and its transmission schedule **S** for the example in Figure 7.2 are as follows:

$$\mathbf{D} = \begin{bmatrix} 0 & 5 & 0 & 0 \\ 3 & 0 & 3 & 0 \\ 0 & 0 & 0 & 0 \\ 2 & 0 & 4 & 0 \end{bmatrix}, \tag{7.1}$$

$$\mathbf{D} \prec \mathrm{S} = \delta^1 \mathbf{A}^1 + \delta^2 \mathbf{A}^2 + \delta^3 \mathbf{A}^3$$

$$= 5 \begin{bmatrix} 0 & 1 & 0 & 0 \\ 0 & 0 & 0 & 0 \\ 0 & 0 & 0 & 0 \\ 0 & 0 & 1 & 0 \end{bmatrix} + 3 \begin{bmatrix} 0 & 0 & 0 & 0 \\ 0 & 0 & 1 & 0 \\ 0 & 0 & 0 & 0 \\ 1 & 0 & 0 & 0 \end{bmatrix} + 3 \begin{bmatrix} 0 & 0 & 0 & 0 \\ 1 & 0 & 0 & 0 \\ 0 & 0 & 0 & 0 \\ 0 & 0 & 0 & 0 \end{bmatrix}, \tag{7.2}$$

With this schedule, the traffic demand **D** will be served with $5 + 3 + 3 = 1$ time slots. The transmission schedule is then transmitted to all the nodes.

In the transmission phase, the nodes pair with each other following the transmission schedule and start directional, concurrent transmissions, as shown in Figure 7.2c. It takes FDMAC 11 time slots to serve traffic demand **D**, a considerable reduction comparing to the conventional sequential approach that transmits one packet at a time via the PNC (which takes 24 time slots). The overhead (i.e., the scheduling phase) will be amortized by the long transmissions in the frame, which is analogous to *gated service* in polling systems that can achieve 100% throughput. During the current frame period, new packet arrivals to the nodes are backlogged in the corresponding virtual queues and will be scheduled for transmission in the next frame.

To account for the potential cochannel interference (CCI), the directional antenna gain can be modeled as $h_{ij}\Gamma(\theta)$, where h_{ij} is the maximum gain for link i, j, and θ is the angle offset from the peak gain direction. When DEVs i and j point their beams to each other with transmit power P_t, the received power is $|h_{ij}|^2 P_t$ when the angle between their beams is $\theta(i, j)$, the received power is $|h_{ij}|^2 \Gamma(\theta(i,j))^2$. For a scheduled link indicated by $a_{ij}^k = 1$ in a scheduling matrix $\mathbf{A}^k$, the signal-to-interference plus noise ratio (SINR) at the receiver DEV j can be written as in Equation (7.3),

$$\mathrm{SINR}_{i,j}^k = \frac{|h_{ij}|^2 P_t}{\sum_{m \in \{m | a_{mn}^k = 1, m \neq i, j\}} |h_{mj}|^2 \Gamma(\theta(m,j))\Gamma(\theta(j,m))P_t + \sigma^2}, \tag{7.3}$$
$$\forall i, j \in \{i, j \mid a_{ij}^k = 1\}, \forall k,$$

where P_t is the transmitter power, h_{ij} is the channel gain from DEV i to DEV j, and σ^2 is the noise power.

The scheduling algorithm decomposes the traffic demand **D** into K matrices, each being an *adjacency matrix* $\mathbf{A}^k$ describing a set of links that can be scheduled concurrently, and a duration δ^k describing how long the set of links will last. A feasible schedule should satisfy $\mathbf{D} \prec \mathrm{S}$, where $\mathrm{S} = \mathrm{S} = \delta^1 \mathbf{A}^1 + \cdots + \delta^K \mathbf{A}^K$. For a given traffic demand **D**, the optimal schedule should clear the backlog **D** using the minimum number of time slots, which implies maximum concurrent transmissions. The optimal frame-based scheduling problem can be formulated as a mixed integer nonlinear programming (MINLP) problem as follows.

$$\text{minimize:} \quad \sum_{k=1}^{K} \delta^k$$

$$\text{subject to:} \qquad a_{ij}^k \in \begin{cases} \{0,1\}, & \text{if } d_{ij} > 0 \\ \{0\}, & \text{otherwise,} \end{cases} \quad \forall\, i,j,k$$

$$\backslash\backslash \quad \textit{binary variable indicating link schedules}$$

$$\sum_{k=1}^{K} (\delta^k \cdot a_{ij}^k) \begin{cases} \geq d_{ij}, & \text{if } d_{ij} > 0 \\ = 0, & \text{otherwise,} \end{cases} \quad \forall\, i,j$$

$$\backslash\backslash \quad \textit{all backlogs should be served by end of the frame}$$

$$\sum_{j=1}^{N} (a_{ij}^k + a_{ji}^k) \leq 1, \forall\, i,k$$

$$\backslash\backslash \quad \textit{half}-\textit{duplex operation of the links}$$

$$\text{SINR}_{i,j}^k \geq \xi, \forall\, i,j \in \{i,j \mid a_{ij}^k = 1\}, \forall\, k. \tag{7.4}$$

$$\backslash\backslash \quad \textit{CCI should be upper bounded}$$

In Reference 7, we propose a graph coloring-based scheduling algorithm, termed a *greedy coloring* (GC) algorithm that can compute near-optimal schedules with respect to the total transmission time with low complexity. We compare the GC-based FDMAC with two state-of-the-art 60-GHz MAC protocols, that is, MRDMAC [10] and MDMAC [12], while relaxing the SINR constraint (7.4) according to the "pseudo-wired" property of 60-GHz links (i.e., assuming that all links are good with adequate SNR and not interfering with each other due to the narrow beam widths) [16].

The simulation study clearly demonstrates the value of spatial reuse and control overhead reduction. The delay and throughput performance of the three MAC protocols are shown in Figure 7.3, for different traffic loads under uniform on–off bursty traffic. The maximum throughput of MRDMAC is under 100% and its average delay diverges when the offered load is between 0.55 and 0.6, since it cannot support simultaneous transmissions between many nodes that use directional antennas. With MDMAC, the nodes operate in the ad hoc mode with concurrent transmissions scheduled distributedly. MDMAC achieves higher throughput than MRDMAC, and as shown in Figure 7.3, the delay curve diverges when the offered load exceeds 3.05 (i.e., 305% of the maximum capacity of a single-directional DEV to DEV communication). FDMAC, in addition to maximizing spatial reuse as in MDMAC, also minimizes the associated control overhead with the frame-based approach. FDMAC can significantly reduce the average delay and support a heavier traffic load. As shown in Figure 7.3, FDMAC keeps the system stable until the

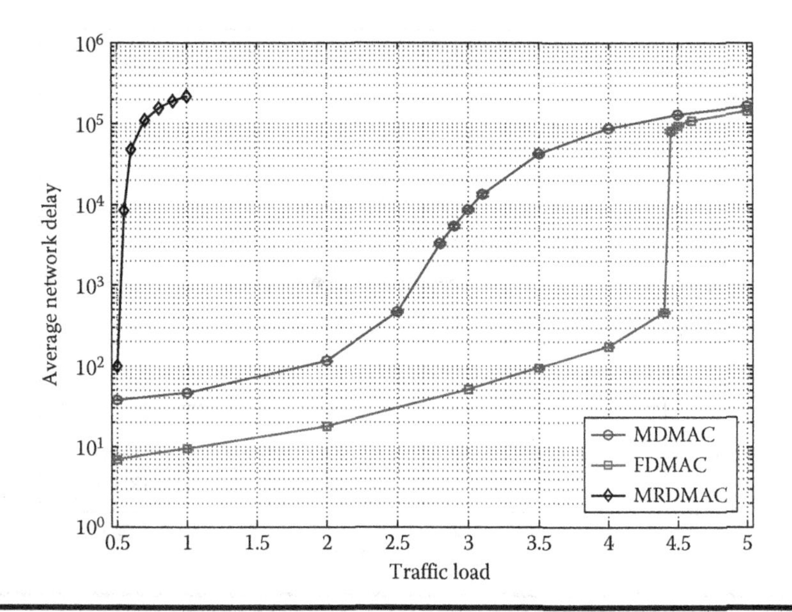

Figure 7.3 Delay under uniform on–off bursty traffic. (From Son, I.-K. et al., 2012. On frame-based scheduling for directional mmwave WPANs. In *Proc. IEEE INFORCOM'12*, Orlando, FL, March, pp. 2149–2157. Copyright 2012 IEEE.)

offered load reaches 4.5 (i.e., 450% of the maximum capacity of a single-directional DEV to DEV communication), which is close to the upper bound $\lfloor N/2 \rfloor = 5.0$.*

We also examine the fairness performance of the three protocols. In Figure 7.4, the fairness indices, as defined in Reference 17, of the three schemes under uniform on–off bursty traffic are plotted. We find both FDMAC and MDMAC achieve better fairness performance than MRDMAC, while FDMAC outperforms MDMAC when the offered load is higher than 3.0 (Figure 7.4).

It can be clearly seen that letting DEVs with directional antennas that use the channel for an extended period of time does not hurt the fairness performance, while achieving considerable throughput and delay gains. As more DEVs become active in the WPAN, the overall throughput will continue to increase since more concurrent transmissions can be scheduled. We conjecture that the throughput will eventually reach its maximum as the network scales in nodes, when the network becomes spatially saturated and the interference constraint (7.4) becomes the limiting factor.

7.4 Conclusion and Outlook

In this chapter, we examined the challenging problem of MAC protocol design for 60-GHz wireless networks. We discussed the technical challenges and reviewed existing proposals. As case studies, we presented three different 60-GHz MAC protocol designs to illustrate the importance of spatial reuse and overhead reduction.

* The upper bound is achieved when all the DEVs are paired for concurrent transmissions. There are at most $\lfloor N/2 \rfloor$ pairs for a N node WPAN.

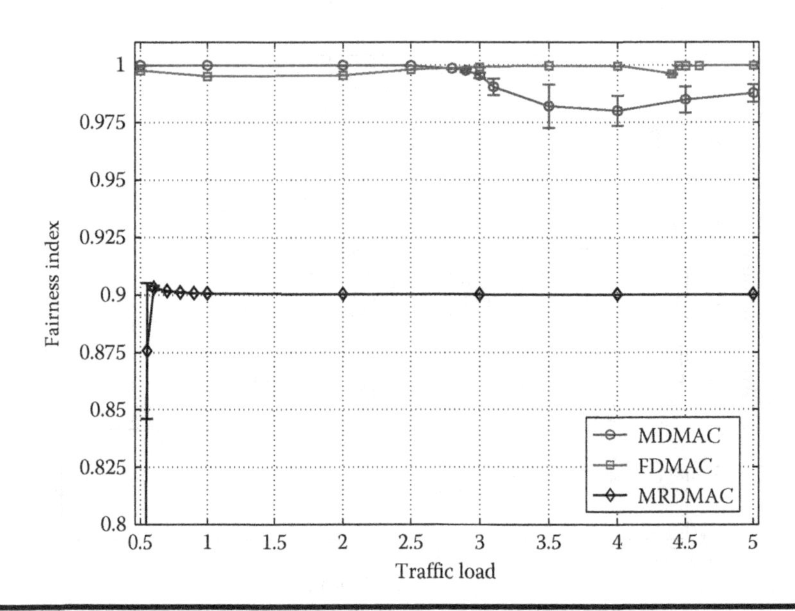

Figure 7.4 Fairness under uniform on–off bursty traffic. (From Son, I.-K. et al., 2012. On frame-based scheduling for directional mmwave WPANs. In *Proc. IEEE INFORCOM'12*, Orlando, FL, March, pp. 2149–2157. Copyright 2012 IEEE.)

Although there have been many existing works in this highly important and high potential problem area, there are many interesting problems that call for more research efforts. We now briefly discuss these open problems.

- Bootstrapping: In a distributed environment, researchers must figure out how to best discover the dynamic set of nodes with highly directional transmissions. How should a network elect a PNC for the group of DEVs? Such a neighbor discovery/bootstrapping scheme is the basis of an effective MAC protocol. A general and effective approach that performs robustly over a wide range of adaptive antenna configurations and propagation environments will be highly desirable.
- Range extension and multihop networks: Researchers must determine how to effectively extend the range and coverage of 60-GHz networks? Relays have been shown effective in 4G cellular networks. How can we best incorporate relays in 60-GHz networks? How can we best establish a multihop 60-GHz mesh network? In such a mesh network, how frequent should the beams be steered (i.e., a semistatic topology or a dynamic topology)?
- Interfacing with wireline networks: Given the multigigabit rate of 60-GHz networks, the AP to the wireline network may become the bottleneck. It is worth studying how to interface with wireline networks. Would it be helpful to exploit the locality of traffic?
- Validation and performance benchmark: Any proposed 60-GHz MAC protocols should be rigorously validated. There is a compelling need for a deep understanding of the propagation characteristics and accurate and efficient models to be incorporated into simulators. Testbeds that allow flexible programming and testing of 60-GHz techniques are also in a great need.
- Coexistence and interoperability: As more 60-GHz (or 5G wireless) network protocols are standardized, it is important to investigate the coexistence and interoperability of heterogeneous wireless networks operating in the same mm-wave band.

The introduction of mm-wave wireless networks that exploit directional and adaptive beamforming antennas opens up new horizons and opportunities, never before used in consumer wireless networks. We have attempted to show some of the potentials and challenges that remain to be solved as networks take on the added dimensionality of spatial processing for vastly improved throughput as compared to today's wireless networks. Applications, after all, are the most important reason we care about this field, and all the previous research questions will help ensure that future applications will work as required as new MAC and PHYs are developed.

Acknowledgments

This work is supported in part by the US National Science Foundation (NSF) through grants CNS-1320664 and CNS-1320472 and through the NSF I/UCRC Broadband Wireless Access & Applications Center (BWAC) site at Auburn University. Any opinions, findings, and conclusions or recommendations expressed in this chapter are those of the authors and do not necessarily reflect the views of the foundation.

References

1. Cisco, 2014. Cisco visual networking index: Global mobile data traffic forecast update, 2013–2018, February, [online] Available: http://www.cisco.com/c/en/us/solutions/collateral/service-provider/visual-networking-index-vni/white_paper_c11-520862.html.
2. He, Z. and S. Mao, 2013. Multiple Description Coding for uncompressed video streaming over 60GHz networks. In *Proc. First ACM Workshop on Cognitive Radio Architectures for Broadband (CRAB 2013)*, in conjunction with *ACM MobiCom 2013*, October, Miami, FL, pp. 61–68.
3. He, Z. and S. Mao, 2014. Adaptive multiple description coding and transmission of uncompressed video over 60 GHz networks, *ACM Mobile Computing and Communications Review (MC2R)*, 18(1), pp. 14–24.
4. Rappaport, T. S., J. N. Murdock, and F. Gutierrez, 2011. State of the art in 60 GHz integrated circuits and systems for wireless communications, *Proceedings of the IEEE*, 99(8), 1390–1436.
5. Rappaport, T. S., R. W. Heath, R. D. Jr., and J. Murdock, 2014. *Millimeter Wave Wireless Communications*. Upper Saddle River, NJ: Pearson Prentice Hall.
6. Geng, S., J. Kivinen, X. Zhao, and P. Vainikainen, 2009. Millimeter-wave propagation channel characterization for short-range wireless communications, *IEEE Trans. Veh. Tech.*, 58(1), 3–13.
7. Son, I.-K., S. Mao, M. X. Gong, and Y. Li, 2012. On frame-based scheduling for directional mmwave WPANs. In *Proc. IEEE INFORCOM'12*, Orlando, FL, March, pp. 2149–2157.
8. An, X., 2010. *Medium access control and network layer design for 60 GHz wireless personal area networks.* PhD dissertation. Technische Universiteit Delft.
9. Gong, M. X., R. J. Stacey, D. Akhmetov, and S. Mao, 2010. Performance analysis of a directional CSMA/CA protocol for mmwave wireless PANs. In *Proc. IEEE WCNC'10*, Sydney, Australia, April, pp. 1–6.
10. Singh, S., F. Ziliotto, U. Madhow, E. M. Belding, and M. Rodwell, 2009. Blockage and directivity in 60 GHz wireless personal area networks: From cross-layer model to multihop MAC design, *IEEE J. Sel. Areas Commun.*, 27(8), 1400–1413.
11. Cai, L. X., L. Cai, X. Shen, and J. W. Mark, 2010. REX: A Randomized EXclusive region based scheduling scheme for mmWave WPANs with directional antenna, *IEEE Trans. Wireless Commun*, 9(1), 113–121.
12. Singh, S., R. Mudumbai, and U. Madhow, 2010. Distributed coordination with deaf neighbors: Efficient medium access for 60 GHz mesh networks. In *Proc. IEEE INFOCOM*, San Diego, CA, March, pp. 1–9.

13. Shihab, E., L. Cai, and J. Pan, 2009. A distributed asynchronous directional–to–directional MAC protocol for wireless ad hoc networks, *IEEE Trans. Veh. Tech.*, 58(9), 5124–5134.
14. Gong, M. X., D. Akhmetov, R. Want, and S. Mao, 2010. Directional CSMA/CA protocol with spatial reuse for mmWave wireless networks. In *Proc. IEEE GLOBECOM'10*, Miami, FL, December, pp. 1–5.
15. Ning, J., T. S. Kim, S. V. Krishnamurthy, and C. Cordeiro, 2011. Directional neighbor discovery in 60 Ghz indoor wireless networks, *Elsevier Performance Evaluation Journal*, 68(9), 897–915.
16. Mudumbai, R., S. Singh, and U. Madhow, 2009. Medium access control for 60 GHz outdoor mesh networks with highly directional links. In *Proc. IEEE INFOCOM 2009 (Mini Conf.)*, Rio de Janeiro, Brazil, April, pp. 2871–2875.
17. Jain, R., A. Durresi, and G. Babic, 1999. Throughput fairness index: An explanation. In *ATM Forum/99-0045*, February.

Chapter 8

Directional MAC Protocols for 60 GHz Millimeter Wave WLANs

Kishor Chandra and R. Venkatesha Prasad

Contents

Abstract

In this chapter, we study the Medium Access Control (MAC) protocols proposed for directional communication in 60 GHz frequency bands. We include the MAC protocols specified in standards such as ECMA-387, IEEE 802.15.3c and IEEE 802.11ad, and MAC protocols proposed in literature as well. We will discuss the various 60 GHz MAC features, such as beamforming protocols, discovery and association mechanisms, relaying, and fallback options in case of unavailability of fragile 60 GHz links for each of the standard MAC protocols. We also present a scheme to select the appropriate access point beamwidth for IEEE 802.11ad Wireless Local Area Networks (WLANs) employing Carrier Sense Multiple Access using Collision Avoidance (CSMA/CA) mechanism for channel access. Finally, we conclude with the challenges to be solved for robust WLAN connectivity at 60 GHz frequency band.

8.1 Introduction

In recent years, a huge surge has been witnessed in data traffic due to widespread use of the Internet and the ever increasing popularity of applications such as video streaming and video gaming. To resolve the scarcity of spectrum at lower frequency bands, new frequency bands are being explored at higher frequencies such as millimeter wave bands (30–300 GHz). In 60 GHz frequency band, 5–9 GHz of bandwidth is available worldwide for unlicensed use. Figure 8.1 shows the available bandwidth in different countries. Due to large available bandwidth, 60 GHz band has emerged as a promising candidate for multi-Gb/s wireless connectivity at short distances of 10–20 m. Several standards such as IEEE 802.15.3c [1], ECMA-387 [2], and IEEE 802.11ad [3] have come up with Physical (PHY) and MAC specifications for short-range 60 GHz communications. However, 60 GHz band transmission significantly differs with the 2.4/5 GHz transmissions, which requires novel approaches for medium access and network management.

8.1.1 Signal Propagation Characteristics in 60 GHz Frequency Band

60 GHz signal transmission possesses some unique characteristics as compared to 2.4 GHz and 5 GHz signals. The wave propagation in 60 GHz is significantly different from that in the lower frequency bands. The most important feature of 60 GHz wireless propagation is its high path loss. As free space path loss is proportional to the square of the career frequency, path loss at 60 GHz is more than 20 dB worse compared to that of 5 GHz.

The received power $P_r(d)$ at a distance d from an isotropic antenna is given as

$$P_r(d) = G_t G_r P_t \left(\frac{\lambda}{4\pi}\right)^2 \left(\frac{1}{d}\right)^{\alpha} \tag{8.1}$$

where G_t and G_r are transmit and receive antenna gains, respectively, P_t is transmitted power, λ is the career wavelength, and α is the path loss coefficient. For 60 GHz frequency band λ is much lower than that of 2.4 GHz, which results in very high free space path loss.

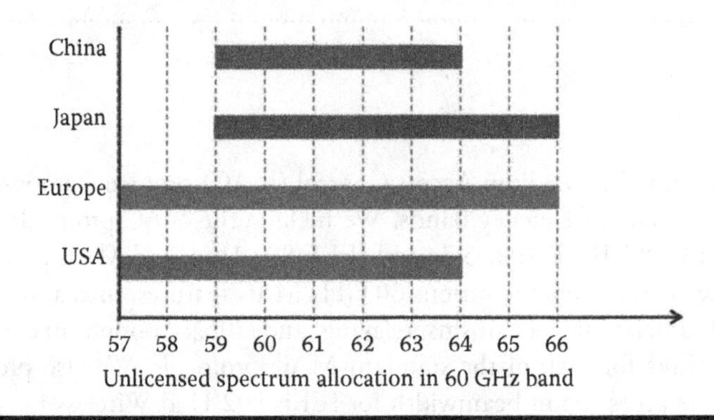

Figure 8.1 Frequency allocation in different countries.

High oxygen absorption (10–15 dB/km) is another problem at 60 GHz frequency band, but it is prominent in outdoor environments at a distance of more than 100 m, and does not have a significant role in indoor environment [4].

Another important characteristic is its limited ability to diffract around the obstacles and inability to penetrate the walls, furniture, and other objects. Because of this, it is suitable mainly for line of sight (LOS) communication [5].

To overcome the high path loss in 60 GHz communication, directional antennas are used to constrain signal power in a particular direction of interest. Interestingly, due to the smaller wavelength at 60 GHz, compact directional antenna arrays can be easily implemented. Directional antennas overcome the high path loss and provide high spatial multiplexing capability [6].

8.1.2 Directional Antennas

An isotropical antenna radiates the transmit power equally in all directions, while a directional antenna confines the transmit power in a particular direction. This results in increased gain for the antenna, which is highly desired in 60 GHz band to overcome the high free space path loss. Figure 8.2 shows an example of the transmit pattern of a directional antenna in which a majority of the signal power is restricted to one direction, referred to as "main lobe," and the remaining small fraction of signal power is leaked through the "sidelobe." The gain of a directional antenna in a given direction is indication of relative power in that direction compared to that of an omni-directional antenna. The unit of antenna gain is dBi and its value for an omni-directional antenna equals zero dBi. A directional antenna is reciprocal in its transmission and reception characteristics, which means that it exhibits both transmission gain and reception gain. If $D(\theta, \phi)$ is the directivity of an antenna, then its gain $G(\theta, \phi)$ is given by [7] as below

$$G(\theta,\phi) = \varepsilon_L(1-|\Gamma|^2)D(\theta,\phi) \tag{8.2}$$

where ε is the efficiency factor of antenna and Γ is the reflection coefficient.

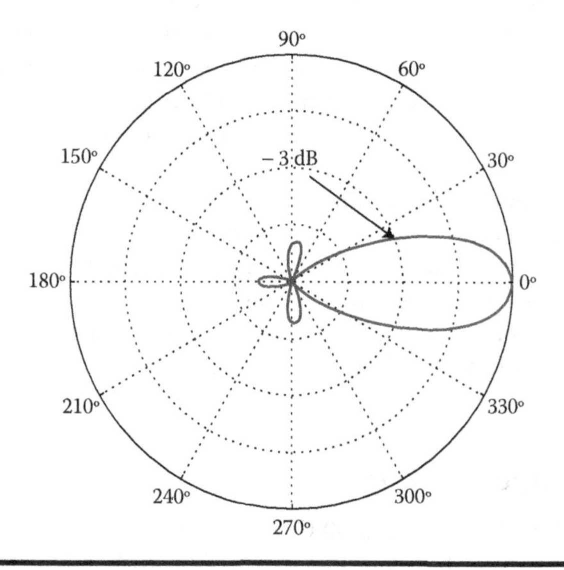

Figure 8.2 Directional antenna transmit pattern.

Apart from the directivity function $D(\theta, \phi)$, beamwidth is an important parameter of directional antennas, and is usually calculated as the half-power beamwidth—which is the angle between the half-power (–3 dB) points of the antenna main lobe. The directional antenna beamwidth defines the angular coverage area of transmission. The smaller the beamwidth of an antenna, the more the power will be concentrated within the angular coverage area of a directional antenna—implying an inverse relation between the antenna gain and beamwidth.

Real world antenna patterns are usually complex and require many details to accurately represent them in a mathematical model. To avoid this complexity, simplified antenna models such as cone plus circle, Gaussian main lobe antennas, have been widely used in literature.

IEEE 802.11ad standard [3] for 60 GHz communication also uses a simplified antenna model in which main lobe gain is characterized by a circularly symmetric Gaussian function represented as

$$G_{\theta\,\mathrm{dB}} = G_{0\,\mathrm{dB}} - 12\left(\frac{\theta}{\theta_{-3\,\mathrm{dB}}}\right)^2 \tag{8.3}$$

where G_0 is the maximum gain of antenna main lobe which depends on the half-power beamwidth $\theta_{-3\,\mathrm{dB}}$. The relation between G_0 and $\theta_{-3\,\mathrm{dB}}$ is given by

$$G_0 = \left(\frac{1.6162}{\sin\theta_{-3\,\mathrm{dB}}/2}\right)^2 \tag{8.4}$$

The main lobe beamwidth is generally determined by the –20 dB power level relative to the maximum gain value. It can be calculated from Equation 8.3.

Figure 8.3 shows the main lobe patterns for different values of $\theta_{-3\mathrm{dB}}$. It is evident that the lower the half-power beamwidth $\theta_{-3\mathrm{dB}}$, the higher will be the maximum antenna gain.

Beamforming antennas can be broadly classified into two categories: (i) switched beam antennas and (ii) beamsteering antennas. Switched beam antennas use predefined fixed sectors while beamsteering antennas can steer the beam in any desired direction. Switched beam antennas are relatively simple but resulting antenna gain can be less if desired transmit direction does not coincide with the peak of antenna sector main lobe. On the other hand, steered beam antenna can provide maximum gain, but additional cost in terms of complex signal processing and power consumption is involved.

Although employing directional antennas can overcome the high path loss at 60 GHz band, it complicates the medium access mechanism due to transmission and reception in selective directions. Therefore, MAC protocols proposed in 60 standards are hybrid in nature employing contention-based access as well as fixed access mechanisms. In the following sections, we present the comparison of MAC protocols proposed in 60 GHz standards namely, ECMA-387 [2], IEEE 802.15.3c [1], and IEEE 802.11ad [3], based on our work done in Reference 8.

8.2 ECMA-387 Specifications

ECMA-387 published by ETSI, defines 60 GHz Wireless Personal Area Networks (WPANs) operating over four channels with a separation of 2.16 GHz within the frequency bands between 57.24–65.880 GHz. ECMA-387 specifies two types of devices, *viz.*, device Type A and device Type

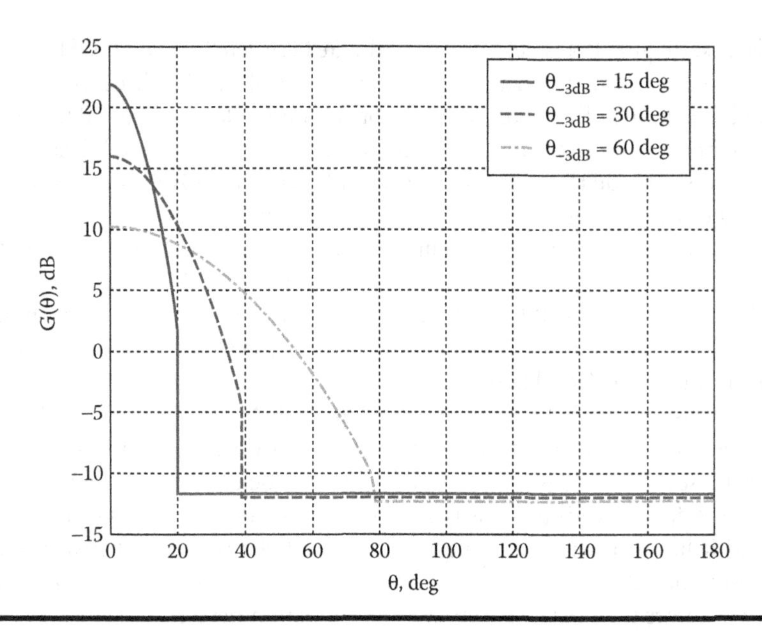

Figure 8.3 Antenna gain pattern for different values of θ_{-3dB}.

B. Type A device is expected to support high data rates (up to 6.350 Gb/s), multilevel Quality of Service (QoS), robust multipath performance, and adaptive antenna arrays capable of beamforming and beamsteering. Type B device aims to be simple, low power, low cost, and targeted to be suitable for handheld devices supporting data rate up to 3.175 Gb/s. Channel bonding of adjacent channels is facilitated to increase the data rates of both the device types. The Type A PHY includes two general transmission schemes, namely Single Carrier Block Transmission (SCBT), also known as Single Carrier with Cyclic Prefix, and Orthogonal Frequency Division Multiplexing (OFDM). The advantage of the SCBT mode over the OFDM is that it lowers the peak to average power ratio (PAPR) and, hence, is preferred. On the other hand, Type B minimizes the complexity and power consumption of the receiver, and may not support antenna training for beamforming.

ECMA-387 provides a decentralized MAC protocol for both the device types, enabling coexistence, interoperability, QoS provisions, and spatial reuse. This standard supports NoAck, ImmAck, and BlockAck policies. Primarily, the ECMA-387 standard supports a completely decentralized operation where each station sends its beacon over the discovery channel. There are two kind of devices: (i) that which can send beacons and (ii) that which cannot send beacons. Coordination among beacon-capable devices is fully distributed, while, in the case of beacon-sending and non-beacon devices existing together, a beacon-sending device works as a controller. It also defines the protocol adaptation layer (PAL), which interacts with the MAC layer through the multiplexing sublayer to support different applications. To ensure communication among heterogeneous devices, a separate discovery channel is reserved. ECMA-387 provides two power management modes in which a device can operate: active and in hibernation. Devices in active mode transmit and receive beacons in every superframe. Devices in hibernation mode hibernate for multiple superframes and do not transmit or receive in those superframes. In addition, this standard provides facilities to support devices that sleep for portions of each superframe in order to save power.

To provide a better WPAN experience, ECMA-387 provides a low rate 2.4 GHz control channel called out of band (OOB) control channel to support the unstable 60 GHz channel. A MAC convergence layer see Figure 8.4 is defined to coordinate between 2.4 and 60 GHz channels, and to support device discovery, synchronization, association control, service discovery, 60 GHz channel reservation, and scheduling. There are two OOB operation modes, namely, ad hoc and infrastructure mode. In ad hoc mode there is no controller and each device sends OOB beacons periodically, while in infrastructure mode the controller periodically sends the OOB beacons to which devices respond with association requests over the 2.4 GHz OOB channel and form the network.

An important function of the OOB control channel is to report the loss of 60 GHz link to the transmitter if adequate signal power is not received over 60 GHz data link. The transmitter and receiver then switch over to 60 GHz discover channel and restart the beamforming procedure to relinquish the 60 GHz data link. Devices can also use the OOB control channel to transmit acknowledgment frames for 60 GHz data.

Apart from the use of the OOB control channel for WPAN management, ECMA-387 has laid provisions for using intermediate devices as amplify and forward relays if LOS connection is not feasible between a source and destination device due to blockage or bad channel conditions. In this case the source device needs to identify a relay device through which it can reach the destination device, and it has to reserve the timeslots on agreed channel pairs between source-to-relay and relay-to-destination devices. Thus, using relay nodes, alternative paths can be found in case of nonavailability of direct path between source and destination devices.

8.3 IEEE 802.15.3c Specifications

IEEE 802.15.3c was the first standard proposed by IEEE for 60 GHz WPAN services. IEEE 802.15.3c defines three millimeter wave (mmWave)-based PHY named as single carrier (SC) PHY, high-speed interface (HSI) PHY, and audiovisual (AV) PHY, respectively. The SC PHY mode, also known as the office desktop model, is designed to support low cost, low complexity, while maintaining relatively high data rate to support high-performance applications with data rate in

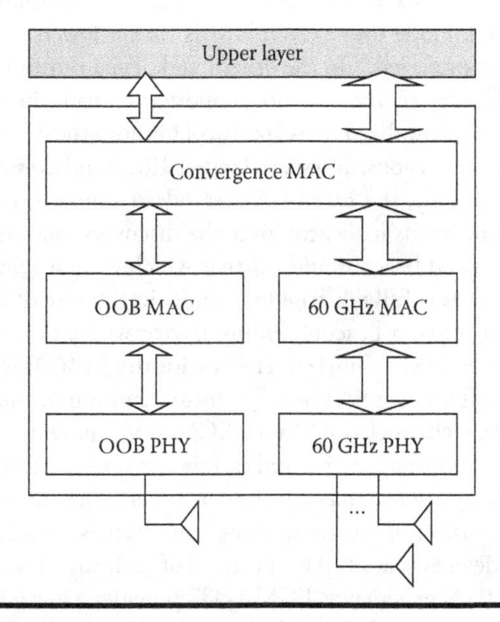

Figure 8.4 OOB control channel used by ECMA-387.

excess of 3 and 5 Gb/s, respectively. HSI PHY is designed for devices with low latency, bidirectional high-speed data, and using OFDM, which is suitable for the conference ad hoc user model with base rate for data at 1.54 Gb/s, and highest up to 5.77 Gb/s. AV PHY is designed for a typical audio–video consumer electronics usage model. For these applications two different sub-PHY modes are defined: high data rate PHY (HRP) for video transmission and low data rate PHY (LRP) for control signal. Both modes use OFDM. The data rate for LRP is 2.5 to 10.2 Mb/s and for HRP it is 0.952 to 3.807 Gb/s. Common mode signaling (CMS) with data rate of 25.2 Mb/s is supported by all the three PHYs for control and management frame transmissions.

The operating area of IEEE 802.15.3c is typically around a radius of 10 m. This standard proposes a completely centralized network architecture where one device assumes the role of piconet coordinator (PNC) of the piconet. The piconet either operates in omni mode or in quasi-omni mode in which directional communication is supported.

IEEE 802.15.3c employs a hybrid MAC protocol which uses both the contention-based access and fixed time division multiple access or TDMA-based medium access mechanisms. Timing in IEEE 802.15.3c is based on the superframe (SF). The superframe consists of three parts: beacon, contention access period (CAP), and channel time allocation period (CTAP). The beacon is used to communicate the timing allocations and management information for the piconet. CAP is used to communicate commands and asynchronous data if it is present in the superframe. The channel access mechanism used in CAP period is CSMA/CA. CTAP is used for isochronous data transmission and channel time allocation in CTAP is purely TDMA based which is allocated during CAP period. It is guaranteed that no other Devices (DEVs) will compete for the channel during the indicated time duration of the CTA allotted to a DEV. In order to ensure reliable frame transmissions, IEEE 802.15.3c provides three acknowledgment mechanisms, namely, ImmAck, BlockACk, and DelayACK. It also supports NoACK mode when acknowledgment is not sought after frame transmission.

To support directional communication, IEEE 802.15.3c provides a three-tier beamforming mechanism (see Figure 8.5). The widest beamwidth level is called quasi-omni (QO). Each QO level can have several sectors having narrow beamwidths. Further, each sector can be divided into very fine beams and called beam level. During beacon and CAP, QO level beamwidth is used for broadcasting management information and channel contention by devices, respectively. During CTAP periods, device pairs can further narrow down their beamwidths up to sector levels or high-resolution beam levels. Beamforming mechanism is used to select the best transmit receive beam pairs at each level. Further, during data transmission, devices use special training packets to track best beam pairs in order to maintain the link quality. Since special training packets are used, it is called *out-packet training.*

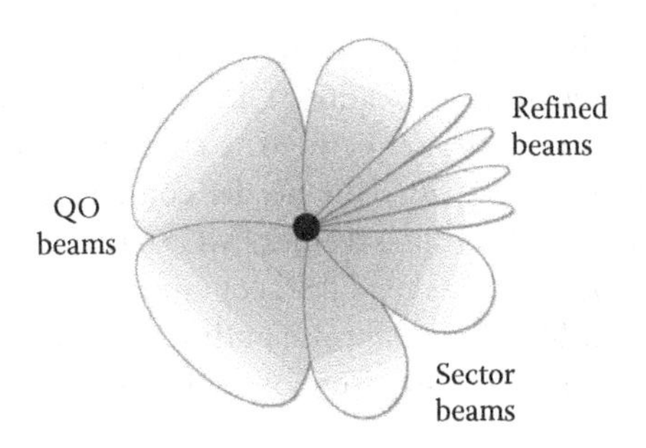

Figure 8.5 Different beam levels in IEEE 802.15.3c and IEEE 802.11ad.

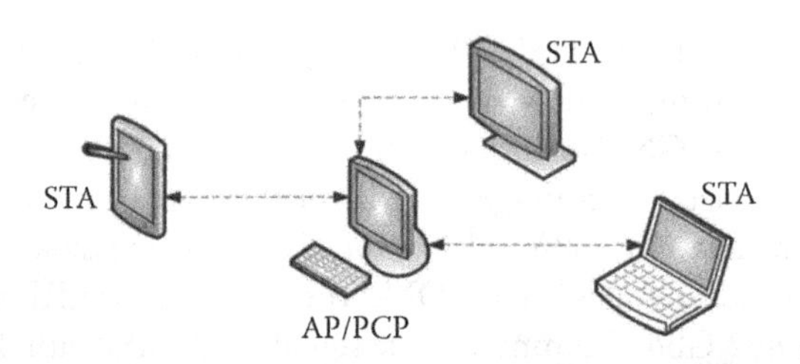

Figure 8.6 Example of an IEEE 802.11ad architecture.

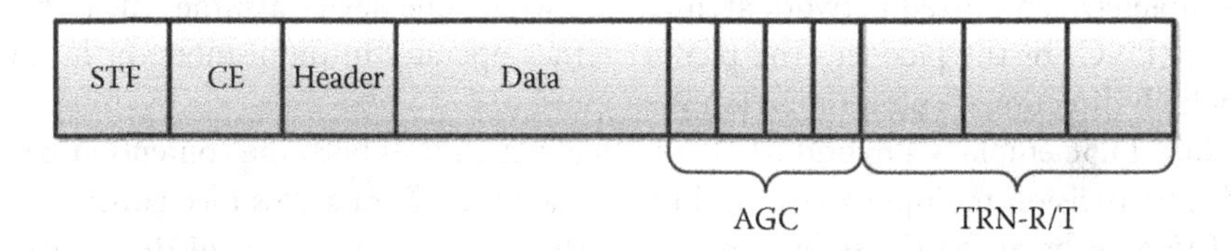

Figure 8.7 Structure of an IEEE 802.11ad packet.

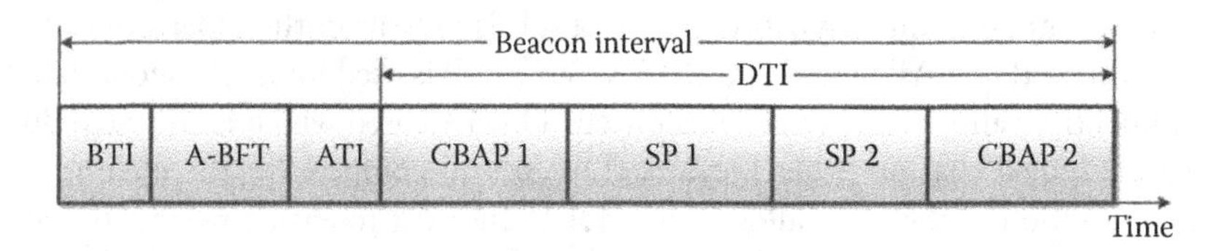

Figure 8.8 A superframe of the MAC layer of the IEEE 802.11ad protocol.

8.4 IEEE 802.11ad Specifications

The IEEE 802.11ad amendment requires the mobile stations (STAs)* to communicate independently of each other; therefore, it uses a personal basic service set (PBSS). To assign basic timing to the STAs, one STA is required to be the PBSS central point (PCP), as shown in Figure 8.6.

The 802.11ad protocol has defined three different packet structures: control PHY, single carrier (SC) (with low-power SC) PHY, and OFDM PHY [9]. Control PHY operates at the lowest data rate, but uses the highest coding gain, such that it can be used for low signal-to-interference-plus-noise ratio (SINR) situations. It is used before a beamformed link is set up or for control frame transmissions. SC PHY is designed for low-power and low complexity transceivers. The last is OFDM PHY, which can achieve the highest data rate.

A packet in the PHY layer has a structure as shown in Figure 8.7. The first two fields are the short training field (STF) and channel estimation (CE) fields. They help with signal acquisition, automatic gain control (AGC) training, predicting the characteristics of the channel, frequency offset estimation, and synchronization [10]. The header contains general information about the packet, such as the modulations and coding schemes (MCS), the size of the packet and also if the optional training fields are appended. The data field contains the MAC header and MAC data.

The access methods used in 802.11ad comprise both CSMA/CA and TDMA [9]. A frame is referred to as a beacon interval (BI). The structure of such a shown in Figure 8.8.

* In this chapter we interchangeably use term device (DEV) or station (STA) that refer to the user equipments such as mobile phones, tablets and laptops, etc.)

The BI consists of multiple parts. The first part is the beacon transmission interval (BTI), in which the PCP/AP transmits one or more beacons in different directions. STAs willing to join the PBSS, can be trained in the association beamforming training (A-BFT) stage of the BI. During announcement time (AT), the PCP/AP can transmit information to the STAs in a request/response fashion. The main data transmission part is the data transmission interval (DTI), in which two periods are present. The contention-based access period (CBAP) and service period (SP) allow any frame exchange, including data transmissions, where CSMA/CA is used in CBAPs, and TDMA is used in SPs. It is possible to use any combination in the number and order of SPs and CBAPs in the DTI [9]. IEEE 802.11ad also provides dynamic channel allocation in which PCP/AP polls STAs either during CBAP or SP periods, and grants channel access. During CBAP, EDCA mechanism can be used by an STA for prioritized channel access.

One of the main advantages of IEEE 802.11ad with respect to other protocols is that it has the capability to switch between 2.4/5 bands and 60 GHz band transmissions. This is called fast session transfer (FST), and it allows seamless connectivity. This is a major cornerstone for 802.11ad since the link quality can quickly degrade in a 60 GHz network due to movement or blockage. FST can operate in both transparent and nontransparent mode. The MAC address is the same in both bands if the STAs are in transparent operation and different in nontransparent operation. FST also supports both simultaneous and nonsimultaneous operation. However, frequent switching from 60 GHz band to 2.4/5 GHz band can be annoying for users. An important role for the MAC layer is the seamless connection it should offer. In order to achieve this, FST needs to perform efficiently.

In order for devices to communicate at a high data rate, the 802.11ad protocol employs beamforming. The beamforming setup consists of three phases similar to IEEE 802.15.3c. The first phase is the sector level sweep (SLS). Its purpose is to allow communication between two STAs. SLS is followed by the beam refinement phase (BRP) in which STAs narrow down there beams. The different level beams can be seen in Figure 8.5. The last phase is the beamtracking (bt) phase and it is done to further track the beams/channel.

8.5 Comparative Analysis

In this section we compare the various features of ECMA-387, IEEE 801.15.3c, and IEEE 802.11ad as follows.

Device Discovery: In ECMA-387, device discovery is achieved using beacon and polling frames. For same type of devices, device discovery is done using beacon frames employing the CSMA/CA protocol. On the other hand, heterogeneous device discovery is done on a master–slave basis, using a polling protocol where the Type A device works as a master and the Type B device as a slave. In IEEE 802.15.3c and IEEE 802.11ad, PNC and PCP/AP periodically send beacon frames in different QO directions. Once the STAs detect beacons, association requests are sent during association CAP and A-BFT periods by IEEE 802.15.3c and IEEE 802.11ad STAs using CSMA/CA protocol, respectively.

Medium Access: ECMA-387 provides a distributed MAC mechanism in which, following the device discovery phase, device pairs reserve the channel for data transmission without intervention of any coordinator. On the other hand, IEEE 802.15.3c provides hybrid channel access mechanism in which devices use CSMA/CA or TDMA-based channel access in CAP and CTAP durations, respectively. To reserve the TDMA slots, it uses CSMA/CA during CAP periods. Thus, the CAP period is used for data transmissions as well as for CTAP reservations. IEEE 802.11ad also provides a hybrid channel access similar to IEEE 802.15.3c. CSMA/CA-based data transmission is done during CBAP periods but the reservation of TDMA slots (called SPs) is done using polling by PCP/AP during ATI period. Further, IEEE 802.11ad also has a provision for dynamic channel access—in which PCP/AP can dynamically poll STAs during CBAP or SP durations for fast channel access.

Beamforming: For antenna training and tracking, ECMA-387 used special frames called TRN frames to determine the appropriate antenna weight vectors. Open-loop and closed-loop training and tracking mechanisms are given. In closed-loop training, transmit antenna derives its weight vectors based on the feedback provided by receiver antenna while in open-loop training there is no provision for feedback and the same training weights are used of transmission and reception. IEEE 802.15.3c provides a three-level antenna training mechanism using beam codebooks [12], namely: (i) best QO pattern training; (ii) best sector level training; and (iii) best beam pair training. During this procedure, it also uses special training frames. IEEE 802.11ad also uses a similar three-level beamforming mechanism. However, it does not use special frames; rather, data frames are used and, hence, this is called *in-packet training*. On the other hand, the ECMA-387 and IEEE 802.15.3c training mechanism are called *out-packet training*.

Network Architecture: Primarily, ECMA-387 supports a completely distributed architecture without any controller (if only Type A devices are present). If Type B devices are also present, then one of the Type A device acts as a coordinator and network operates on master–slave basis. On the other hand, IEEE 802.15.3c proposes completely centralized network architecture in which PNC coordinates communications among device pairs. Similarly, IEEE 802.11ad PBSS is centrally coordinated by PCP/AP. However, peer-to-peer communication is supported by both IEEE 802.15.3c and IEEE 802.11ad.

Relay and Fallback Option: Since 60 GHz links are highly susceptible to link blockage due to channel variations—because of obstacles or misalignment of antenna beams, it is desirable to have alternate means to reclaim the lost links between devices. Also, device discovery and association becomes difficult due to directional communication at 60 GHz. ECMA-387 has an option of using 2.4 GHz OOB signaling for WPAN management. OOB is used for device discovery and association. Apart from OOB signaling it also proposes use of intermediate devices as relay if a 60 GHz LOS link is broken to discover the alternate 60 GHz path. IEEE 802.11ad also provides support for relays at 60 GHz. Apart from relay support, it also provides fast session transfer mechanism to switch over 2.4 or 5 GHz channel. On the contrary, IEEE 802.15.3c does not mention either support for relay or fallback option to lower frequency to relinquish the lost/blocked links. We have summarized all these comparison in Table 8.1.

8.6 Enhancing the 60 GHz MAC: Adaptive Beamwidth Allocation Mechanism

As explained in the earlier sections, all MAC protocols proposed in 60 GHz standard have a structural similarity, that is, hybrid medium access mechanism employing CSMA/CA and TDMA-based fixed access. During CSMA/CA period, devices contend for the channel time, while TDMA slots are allocated on a fixed schedule basis. In TDMA periods transmitter and receiver beamwidth will have great impact on communication range and link throughput. This is

Table 8.1 Comparison on the Basis of Various Mechanisms

Options	*ECMA-387*	*IEEE 802.15.3c*	*IEEE 802.11ad*
Network Architecture	*Distributed*	*Centralized*	*Centralized*
Medium access	CSMA/CA and TDMA	CSMA/CA and TDMA	CSMA/CA, TDMA, Polling
Dynamic Channel Access	No	No	Yes, PCP/AP can dynamically poll STAs during CBAP
Backward Compatibility	No	No	Yes, back compatible to IEEE 802.11 b/g/n/ac
Relay	Yes	No	Yes
Fallback to 2.4 GHz	No	No	Yes, Fast session transfer mechanism if 60 GHz link is not available
WPAN Management	Provision of 2.4 GHz control plane	PNC operating over 60 GHz	PCP/AP operating over 60 GHz

Source: Adapted from IEEE 802.15.3c working group, tgc3, 2009; High rate 60 GHz PHY, MAC and HDMI PALs, *Standard ECMA-387,* December 2010; Draft standard for information technology—telecommunications and information exchange between systems—local and metropolitan area networks—Specific requirements—Part 11: Wireless lan medium access control (mac) and physical layer (phy) specifications—Amendment 4: Enhancements for very high throughput in the 60 GHz band, *IEEE P802.11adTM/D9.0,* July 2012; IEEE standard for information technology–telecommunications and information exchange between systems–local and metropolitan area networks–specific requirements—part 11: Wireless lan medium access control (mac) and physical layer (phy) specifications amendment 3: Enhancements for very high throughput in the 60 GHz band, *IEEE Std 802.11ad-2012 (Amendment to IEEE Std 802.11-2012, as amended by IEEE Std 802.11ae-2012 and IEEE Std 802.11aa-2012),* pp. 1–628, 2012; Cordeiro, C., D. Akhmetov, and M. Park, 2010. *Proceedings of the 2010 ACM International Workshop on mmWave Communications: From Circuits to Networks, ser. mmCom '10.* New York, NY. ACM, pp. 3–8.).

rather straightforward, as antenna gains have a direct relation with beamwidth of transmitter and receiver. Hence, using the narrow beams would always be a better choice for the better link quality. However, performance of CSMA/CA MAC protocol is not that straightforward. In CSMA/CA employing directional antennas, beamwidth of WLAN PCP/APs plays an important role because a device can access the channel only if PCP/AP is listening to the sector device is placed. On the other hand, collisions among devices contending for channel time also limits the channel utilization. We proposed a scheme [13] to determine the appropriate PCP/AP beamwidth, as follows.

Figure 8.9 shows an IEEE 802.11ad PBSS where PCP is in the center of the circle and STAs are distributed around the area covered by the PCP. The typical radius of a PBSS is about 10–20 m. CBAP uses CSMA/CA for medium access. The different QO levels of PCP/AP are switched in a round robin fashion. As shown in Figure 8.9, only those STAs that are within the beamwidth of the current PCP/AP QO level will contend for the channel during the current CBAP duration. Let

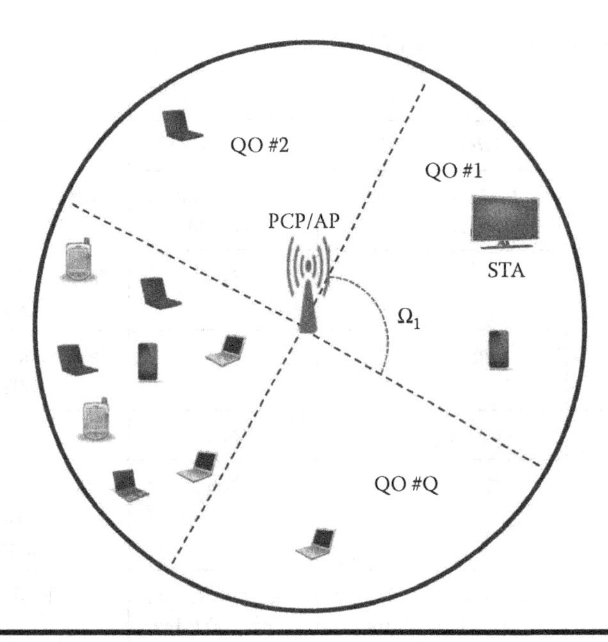

Figure 8.9 IEEE 802.11ad system model.

the beamwidth of PCP/AP for k^{th} QO level having n_k STAs be denoted by Ω_k, where, $1 \le k \le Q$ and Q is the maximum number of QO levels.

Let U_k be the channel utilization in k^{th} QO level, which is defined as the fraction of time that the channel is used to transmit payload successfully. Assuming that all the QO patterns are allotted CBAPs in one BI, overall channel utilization during BI (assuming there are no SPs) can be given by

$$U = \frac{1}{Q} \sum_{k=1}^{k=Q} U_k. \tag{8.5}$$

For each QO direction, channel utilization U_k is calculated using Bianchi's model, which employ Markov chains to calculate the transmission and collision probabilities [14]. To improve the channel utilization, several solutions are proposed, such as controlling minimum and maximum window size and retransmission limits to reduce the collision probability, which require that all the STAs adapt these parameters changes [15–17]. However, IEEE 802.11ad employing directional antennas provides one extra degree of freedom (i.e., beamwidth) and can control the number of simultaneously contending STAs by restricting the channel access in some spatial directions while allowing it in others. This is the motivation behind adaptive beamwidth allocation, and a novel algorithm is proposed in the next section.

IEEE 802.11ad/802.15.3c does not provide any mechanism for selecting the beamwidth of individual QO levels. It only defines them as QO levels [18]. Expression for average channel utilization [14] during CBAB indicates that the fraction of total time channel is being used for data transmission, which largely depends on the conditional collision probability. Consequently, using the equal beamwidth QO levels can lead to very high collision probability in the densely populated regions or underutilization of channel in the QO levels having less number of STAs. To address this problem, we propose an algorithm for appropriate beamwidth selection for each QO level which tries to maximize CSMA/CA channel utilization.

Let beamwidth Ω_k of kth QO level vary from Ω_{min}* to Ω_{max}. The value of Ω_{min} is decided by the capability of antenna array to narrow down its beamwidth to the least possible value (highest possible beam resolution) and Ω_{max} is limited by the intended maximum distance the PCP/AP has to cover. Here, for the sake of simplicity, we consider a perfect conical antenna model for which antenna gain $G(\Omega)$ for a beamwidth of Ω can be given by $G(\Omega) = (2\pi/\Omega)$, where, G is the gain of omni-directional antenna. Thus, antenna beamwidth will decide the maximum distance it could cover for a particular MCS and transmitted power. Hence, the value of Ω_{max} depends on the data rate and intended distance to be covered by the transmitter.

To determine the beamwidth of QO levels, we start with the minimum antenna beamwidth Ω_{min} for the first sector, and keep increasing the beamwidth by $\Delta\Omega$ until the throughput reaches its maximum value. With increasing beamwidth, first CSMA/CA throughput increases because initially channel utilization increases due to the increase in number of devices, but, when we keep increasing the beamwidth, after a certain beamwidth, CSMA/CA throughput starts decreasing due to the increase in collisions. The detailed algorithm is described in Algorithm 2.

- ***Step 0*** Initialization (see line 1):
  During the association process (A-BFT duration), STAs discover the PCP/AP and train their beams with PCP/AP using SLS procedure. During this process, PCP/AP collects the angle information (i.e., angle w.r.t the position of PCP/AP) of all the STAs which are associated with the PCP/AP. Let A represent the set of angular information β_j ($j = 1, 2, \ldots, n$) of all the n STAs associated with the PCP/AP.
- ***Step 1*** Evaluation for the minimum beamwidth Ω_{min} (see line 3):
  Starting with a beamwidth Ω_{min}, PCP/AP calculates the number of devices in this beam area and CSMA/CA throughput. Rename this beamwidth as Ω_p (i.e., past value) and corresponding throughput as U_p.
- ***Step 2*** Increment by the differential beamwidth $\Delta\Omega$ (see line 4):
  Take Ω_p and U_p as reference values. Then, increase the beamwidth by $\Delta\Omega$. New beamwidth is denoted as Ω_n (since, $\Omega_n = \Omega_p + \Delta\Omega$, more devices are likely to be included in the increased beam area). Calculate the CSMA/CA throughput denoted as U_n.
- ***Step 3*** Determine the appropriate QO beamwidth (see lines 5–9):
  Compare U_n and Ω_n with U_p and Ω_{max}, respectively. If $U_n \geq U_p$ and $\Omega_n \leq \Omega_{max}$, set $U_p = U_n$, $\Omega_p = \Omega_n$ and go to Step 3. Otherwise, select $\Omega_k = \Omega_p$ as the optimum beamwidth for sector k (initially, k = 1).
- ***Step 4*** Iteration to include all the STAs (see lines 2 and 10).

After deciding the beamwidth for k^{th} sector, go to Step 2. Take the end point of previous sector as the starting point for the next sector. Repeat the same procedure to decide the beamwidth for $k + 1^{th}$ sector. Do the same procedure until all the devices are included or the complete area around the PCP/AP is traversed.

Algorithm 1 Adaptive QO beamwidth selection:

1. **initialize** A := $\{\beta_i | 1 \leq i \leq n\}$, Ω_{min}, $\Delta\Omega$, and Ω_{max};
2. **while** $A \neq \emptyset$ **do**

---

* Ω_{min} is the minimum beamwidth.

3. $\Omega_p \leftarrow \Omega_{\min}$ and calculate U_p;
4. $\Omega_n \leftarrow \Omega_p + \Delta\Omega$ and calculate U_n;
5. **if** ($U_n \geq U_p$ and $\Omega_n \leq \Omega_{\max}$) **then**
6. $\Omega_p \leftarrow \Omega_n$, $U_p \leftarrow U_n$, and go to step 4;
7. **else**
8. $\Omega_k \leftarrow \Omega_p$;
9. **end if**
10. go to step 2 and repeat the procedure for $(k+1)^{th}$ sector;
11. **end while**
12. **return** Ω, $n_k := \{\Omega_k \mid 1 \leq k \leq TotalSector\}$;

We assume 60 GHz PCP/AP placed at the centre in the ceiling of room. We simulate a conference room environment (radius = 10 m) for different number of STAs, as shown in Figure 8.9. Beamwidth of all the STAs is assumed to be equal and taken as 60°. Location of an STA is identified by the Euclidean distance and angle from the PCP/AP. The Euclidean distances of STAs from the PCP/AP are uniformly distributed in the range [1,10 m], while angles are generated from a Gaussian distribution with mean equal to 180° and a standard deviation of 90°. Thus, device distribution ensures uneven distribution of devices (some regions are densely packed while some are not). All the parameters' values are taken from the IEEE 802.11ad [3] and listed in Table 8.2. For the RTS, DMG CTS, and ACK frames, DMG control PHY (27.5 Mbps) is used. Data frames are transmitted using mandatory MCS4 (1.15 Gbps). Both the minimum beamwidth $\Omega_{\min}$ and differential beamwidth $\Delta\Omega$ are assumed equal to 20°. To calculate the minimum required CBAP duration, total number of requests per QO level are assumed equal to the number of STAs in that QO level. For fixed beamwidth, per QO level beamwidth is equal to 90° while for adaptive beamwidth, each QO level has a different beamwidth and decided using Algorithm 1.

Figure 8.10 shows the CBAP channel utilization for fixed and adaptive beamwidth QO levels. It is clear that the adaptive beamwidth approach performs better than the fixed beamwidth allocation. The reason behind this is that the adaptive beamwidth approach always tries to accommodate STAs in such a manner that maximum channel time is utilized. If the number of STAs is increased, throughput of fixed beamwidth approach deteriorates, but adaptive beamwidth approach is able to maintain a fair throughput. This is because when the number of STAs is increased, fixed beamwidth QO levels suffer increased collisions due to more number of STAs, while the adaptive beamwidth approach is able to reduce the beamwidth in heavily populated regions, and thus restrict the number of STAs under the each QO level. This minimizes the collisions and maintains a steady throughput which is 20%–30% more than the fixed beamwidth sectors.

8.7 Further Challenges for Robust WLAN Connectivity at 60 GHz

In the preceding sections we compared various aspects of three standards proposed for short-range multi-Gb/s communications at 60 GHz frequency bands. Various schemes proposed by these standards were discussed. However, to realize robust multi-Gb/s WLAN connectivity at 60 GHz frequency bands like that of 2.4 GHz bands is still a challenge. The main issues are: (i) severe blockage of signals due to obstacles and (ii) link outage due to mobility while using

Table 8.2 CBAP Analysis Parameters

RTS	20 Octets
DMG CTS	26 Octets
ACK	14 Octets
SIFS	2.5 μs
RIFS	9 μs
DIFS	13.5 μs
CCA Detect Time	4 μs
Minimum window size W_0	8
Maximum backoff stage (m)	3
Retry limit (H)	5
Data size	1024 octets
Transmit power P_t	10 dBm
Omni antenna gain G	1 (linear)
Fading loss X_σ	2 dB
Receiver sensitivity (MC4)	–64 dBm
Receiver sensitivity (MCS0)	–78 dBm
Link margin	20 dB
Path loss exponent α	2

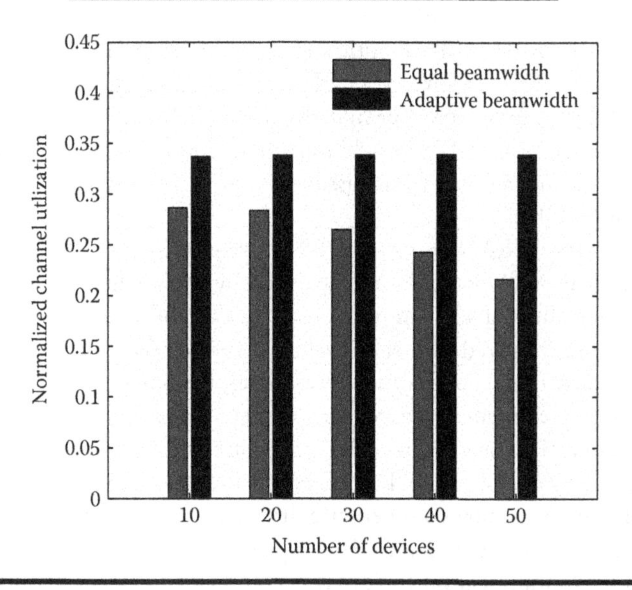

Figure 8.10 CBAP performance comparison for adaptive and fixed beamwidth sectors.

directional antennas. IEEE 802.11ad has already emerged as the most favored 60 GHz standard among device manufacturers. Hence it is desirable to further strengthen IEEE 802.11ad so that a reliable and robust WLAN service similar to WiFi can be delivered at 60 GHz frequency bands.

To tackle the link blockage, there is a provision for relay devices so that the alternate path can be used. However, further research is required in this domain. A cooperative MAC protocol using intermediate STAs as relay nodes is proposed [19]. The cooperating relaying enhances the performance of IEEE 802.11ad and extends the communication range. Further Kim et al. have proposed a scheme for relay selection in IEEE 802.11ad multihop network while maximizing the video quality [20]. The relay selection depends on the video quality achieved in a multihop IEEE 802.11ad network. Relays play an important role if LOS connection is blocked or if the source and destination STAs are far apart. For seamless user experience, intelligent relay selection mechanisms are required so that smooth link transition can be facilitated to users without any interruption in service delivery. Spatial diversity is used to combat the human induced shadowing, and it is shown that desired link quality can be achieved by combining multiple streams pointing in slightly different directions from each other [21].

Beamforming for the initial link setup, and beamtracking to retain the communication link between moving devices are important for better user experience. A novel beam searching algorithm is proposed [22], which fastens the code book searching procedure specified [12]. The angle-of-arrival-based approach is proposed to select the secondary beam if the best beam is blocked due to obstacles [23]. Xueli et al. [24] have proposed a learning-based beam switching algorithm if LOS path is blocked and proved that the learning-based approach is better than the instantaneous decision-based approach. It is extremely important for moving device pairs to keep their beam aligned to achieve desired link quality. Utilizing motion sensors embedded in the user device to track the user movement and thus switch the antenna beams in the appropriate direction is also a promising technique for beamtracking. Exploiting data gathered from motion sensors to reorient the beam direction is an idea less explored, and requires more investigation [25,26]. At present, literature on beamtraining and tracking is limited and lacks measurement-based studies in deployments, and thus requires more efforts.

IEEE 802.11ad also provides fallback option using fast session transfer to 2.4/5 GHz channel. However, if multiple links fallback on 2.4/5 GHz simultaneously, interference can limit the data transmission capabilities of each link. Specifically, if PCP/AP is involved in frequent switching from 60 to 2.4 GHz, then other STAs which are able to communicate at 60 GHz have to suffer unnecessarily. This is because a frame transmission at 2.4/5 GHz would take about 10 times more channel time than at 60 GHz.

Further, to realize the WLAN concept at 60 GHz, multiple PCP/APs need to be installed to cover the indoor areas as different rooms separated by walls require their own PCP/AP. This makes network management a challenging task. In case of mobility, frequent association–disassociation events are triggered. Thus, device discovery or AP discovery becomes an important challenge due to directional communication. Therefore, management of multi-Gb/s 60 GHz WLAN is an important challenge and requires novel approaches. Using 2.4/5 GHz channel for transmission of control information—with occasional fallback option for data transmission when 60 GHz link is not available—can be a potential alternative to provide seamless multi-Gb/s WLAN coverage. Figure 8.11 shows a schematic drawing of frame transmissions using 2.4 GHz as control plane and 60 GHz as data plane. This type of network architecture with opportunistic fallback of data plane communication to 2.4 GHz could be a viable option for seamless WLAN experience. Mandke et al. [27] have discussed the motivation for a dual band WLAN operating

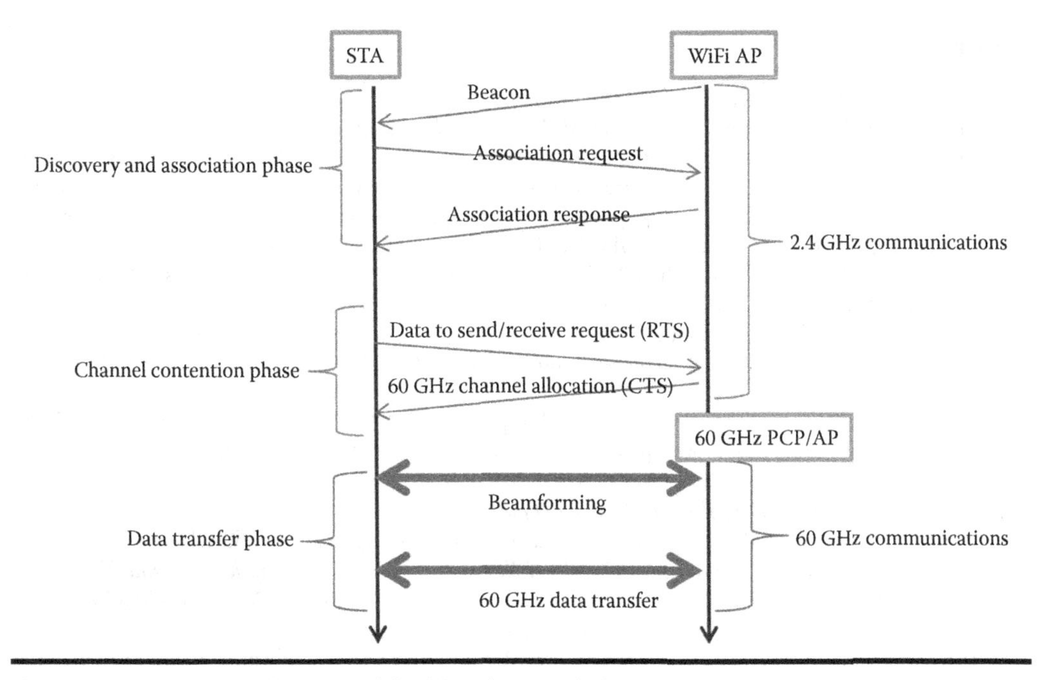

Figure 8.11 Sequence diagram of dual band transmission.

simultaneously over 2.4 and 60 GHz. Classification of traffic over the 2.4 and 60 GHz frequency band is discussed. A 2.4 GHz assisted 60 GHz neighbor discovery and association mechanism are proposed [28], and it is shown that with the help of 2.4 GHz transmission, 60 GHz device discovery procedure can be accelerated. However, simultaneous operation over 2.4/5 GHz frequency band (for control/management information transmission) and 60 GHz frequency band (for data transmission) is not explored much and requires novel approaches for network architecture and MAC protocol design. We see further investigations that may lead to amendments to IEEE 802.11ad standard.

8.8 Summary

In this chapter, we studied MAC protocols proposed for 60 GHz short-range communications. Since IEEE 802.11ad has emerged as the most popular standard for short-range multi-Gb/s communication at 60 GHz frequency band, we discussed in detail about the research challenges and deployment issues pertaining to IEEE 802.11ad. We explored the issues related to efficient channel access such as directional MAC protocol and its dependency on PCP/AP beamwidth, device discovery mechanism, and beamforming procedure. We further elaborated the challenges arising in network management and handover due to closely spaced PCP/APs which are of paramount importance in order to provide robust WLAN connectivity at 60 GHz. In this context we presented importance of relaying, fallback options to 2.4/5 GHz band and out of band control plane. We believe that a WiFi-like WLAN connectivity at 60 GHz frequency band would be an integral part of the 5G era, which would require tackling challenges at many levels.

References

1. IEEE 802.15.3*c* working group, tgc3, Report 2009.
2. High rate 60 GHz PHY, MAC and HDMI PALs, *Standard ECMA-387*, December 2010.
3. Draft standard for information technology—telecommunications and information exchange between systems—local and metropolitan area networks—Specific requirements—Part 11: Wireless lan medium access control (mac) and physical layer (phy) specifications—Amendment 4: Enhancements for very high throughput in the 60 GHz band, *IEEE P802.11adTM/D9.0*, July 2012.
4. Smulders, P. 2002. Exploiting the 60 GHz band for local wireless multimedia access: Prospects and future directions, *IEEE Communications Magazine*, 40(1), 140–147.
5. Jacob, M., S. Priebe, R. Dickhoff, T. Kleine-Ostmann, T. Schrader, and T. Kurner, 2012. Diffraction in mm and sub-mm wave indoor propagation channels, *IEEE Transactions on Microwave Theory and Techniques*, 60(3), 833–844.
6. Singh, S., F. Ziliotto, U. Madhow, E. M. Belding, and M. J. W. Rodwell. Millimeter wave WPAN: Cross-layer modeling and multihop architecture, In *Proc. IEEE INFOCOM*, May 2007, Alaska, USA, pp. 2336–2340.
7. Mailloux, R. J. 2005. *Phased Array Antenna Handbook*. Boston, USA.
8. Chandra, K., A. W. Doff, Z. Cao, R. V. Prasad, and I. G. Niemegeers, 2015. 60 GHz MAC standardization: Progress and way forward. In *2015 IEEE 12th Consumer Communications and Networking Conference (CCNC): CCNC 2015 Workshops—IEEE Standards (CCNC 2015—CCNC 2015 Workshops—IEEE Standards)*, January, Las Vegas.
9. IEEE standard for information technology–telecommunications and information exchange between systems–local and metropolitan area networks–specific requirements—part 11: Wireless LAN medium access control (MAC) and physical layer (PHY) specifications amendment 3: Enhancements for very high throughput in the 60 GHz band, *IEEE Std 802.11ad-2012 (Amendment to IEEE Std 802.11-2012, as amended by IEEE Std 802.11ae-2012 and IEEE Std 802.11aa-2012)*, pp. 1–628, 2012.
10. Shankar, N., D. Dash, H. El Madi, and G. Gopalakrishnan, 2012. WiGig and IEEE 802.11ad - For multi-gigabyte-per-second WPAN and WLAN, *ArXiv e-prints*, November 2012.
11. Cordeiro, C., D. Akhmetov, and M. Park, 2010. Ieee 802.11ad: Introduction and performance evaluation of the first multi-Gbps wifi technology. In *Proceedings of the 2010 ACM International Workshop on mmWave Communications: From Circuits to Networks, ser. mmCom '10*. New York, NY. ACM, pp. 3–8.
12. Wang, J. et al. 2009. Beam codebook based beamforming protocol for multi-Gbps millimeter-wave wpan systems. *IEEE Journal on Selected Areas in Communications,* 27(8), 1390–1399.
13. Chandra, K., R. V. Prasad, I. G. Niemegeers, and A. R. Biswas, 2014. Adaptive beamwidth selection for contention based access periods in millimeter wave WLANs. In *Consumer Communications and Networking Conference (CCNC), 2014 IEEE 11th*. IEEE, January 10–13, 2014, Las Vegas, NV, USA, pp. 458–464.
14. Bianchi, G., 2000. Performance analysis of the IEEE 802.11 distributed coordination function. *IEEE Journal on Selected Areas in Communications*, 18, 535–547.
15. Gannoune, L. and S. Robert, 2004. Dynamic tuning of the contention window minimum (cw min) for enhanced service differentiation in IEEE 802.11 wireless ad-hoc networks. In *15th IEEE International Symposium on Personal, Indoor and Mobile Radio Communications, 2004*, September 5–8, 2004, Barcelona, Spain, Vol. 1, pp. 311–317.
16. Qiao D. and S. Choi, 2001. Goodput enhancement of IEEE 802.11 a wireless LAN via link adaptation. In *Communications, 2001. ICC 2001. IEEE International Conference on*, June 11–14, 2001, Helsinki, Finland, Vol. 7, pp. 1995–2000.
17. del Prado Pavon, J. and S. Choi, Link adaptation strategy for IEEE 802.11 WLAN via received signal strength measurement. In *Communications, 2003. ICC'03. IEEE International Conference on*, May 11–15, 2003, Anchorage, Alaska,USA, Vol. 2, pp. 1108–1113.
18. Wang, J. et al. 2009. Beam codebook based beamforming protocol for multi-Gbps millimeter-wave WPAN systems. *IEEE Journal on Selected Areas in Communications*, 27(8), 1390–1399.

19. Chen, Q. J. Tang, D. Wong, X. Peng, and Y. Zhang, 2013. Directional cooperative MAC protocol design and performance analysis for ieee 802.11ad WLANs, *IEEE Transactions on Vehicular Technology*, 62(6), 2667–2677.
20. Kim, J. Y. Tian, S. Mangold, and A. Molisch, 2013. Quality-aware coding and relaying for 60 GHz real-time wireless video broadcasting. In *2013 IEEE International Conference on Communications (ICC)*, June, Budapest, Hungary, pp. 5148–5152.
21. Xiao, Z. 2013. Suboptimal spatial diversity scheme for 60 GHz millimeter-wave WLAN, *Communications Letters, IEEE*, 17(9), 1790–1793.
22. Li, B., Z. Zhou, W. Zou, X. Sun, and G. Du, 2013. On the efficient beam-forming training for 60 GHZ wireless personal area networks, *IEEE Transactions on Wireless Communications*, 12(3), 504–515.
23. Tsang, Y. and A. Poon, 2011. Successive AoA estimation: Revealing the second path for 60 GHz communication system. In *2011 49th Annual Allerton Conference on Communication, Control, and Computing (Allerton)*, September, University of Illinois Monticello, IL, USA, pp. 508–515.
24. An, X., C.-S. Sum, R. Prasad, J. Wang, Z. Lan, J. Wang, R. Hekmat, H. Harada, and I. Niemegeers, 2009. Beam switching support to resolve link-blockage problem in 60 GHz WPANs. In *2009 IEEE 20th International Symposium on Personal, Indoor and Mobile Radio Communications*, September 13–16, 2009, Tokyo, Japan, pp. 390–394.
25. Shim, D.-S., C.-K. Yang, J. H. Kim, J. P. Han, and Y. S. Cho, 2014. Application of motion sensors for beam-tracking of mobile stations in mm wave communication systems, *Sensors*, 14(10), 19622–19638. [Online]. Available: http://www.mdpi.com/1424-8220/14/10/19622.
26. Doff, A. W., K. Chandra, and R. V. Prasad, 2015. Sensor assisted movement identification and prediction for beamformed 60 GHz links. In *2015 IEEE 12th Consumer Communications and Networking Conference (CCNC) (CCNC 2015)*, January 09–12, 2015, pp. 648–653.
27. Mandke, K. and S. M. Nettles, 2010. A dual-band architecture for multi-Gbps communication in 60 GHz multi-hop networks. In *Proceedings of the 2010 ACM International Workshop on mmWave Communications: From Circuits to Networks, ser. mmCom '10.* New York, NY. ACM, pp. 9–14. [Online]. Available: http://doi.acm.org/10.1145/1859964.1859969
28. Park, H., Y. Kim, T. Song, and S. Pack, 2014. Multi-band directional neighbor discovery in self-organized mm wave ad-hoc networks, *IEEE Transactions on Vehicular Technology*, Vol. 99, pp. 1–1.

Chapter 9

Performance Improvements of mm-Wave Wireless Personal Area Networks Using Beamforming and Beamswitching

Seokhyun Yoon

Contents

9.1 Introduction

Wireless personal area network (WPAN) standards, such as IEEE 802.15.3c and 802.11ad, provide Gbps short range wireless transmission over 60 GHz band, which should be implemented with very low hardware cost and power consumption [1]. The primary reason for using the 60 GHz band is that a massive amount of spectrum is available in this band to support such a high data rate transmission [2]. However, this is not the only advantage of using the 60 GHz band. As its wavelength is only several millimeters long, multiple antenna technologies, such as beamforming [3–5] or spatial multiplexing/diversity [6–9], can be easily implemented on small area of tiny portable devices. The question is which multiple antennae technique is most suitable for wireless PAN, for example, having lowest hardware cost and power consumption for tiny devices.

In general, multiple antennae techniques can be classified into three types, namely, beamforming/smart antennas [3–5], spatial multiplexing [6,7], and spatial diversity [8,9], according to the usage of the spatial domains. The spatial diversity techniques utilize the so-called space–time coding to improve the diversity order and the outage of the communication channel, especially for multimedia traffic. These can be easily implemented with only a marginal increase in hardware and overhead costs. However, it cannot fully utilize many antennas that can be mounted on a small area of tiny devices. The spatial multiplexing is a technique used primarily to improve the data rate by transmitting different data through each of the antenna ports. By utilizing the channel side information at the receiver/transmitter, the matrix channel can be converted into a multiple of parallel channels, resulting in throughput scaling with the number of antennas. This may be a good choice for WPAN but the main disadvantage of spatial multiplexing is the high hardware cost because each antenna must have separate Radio Frequency (RF) and baseband processors.

Beamforming is a traditional multiple antenna technique of which the primary objective is for interference mitigation and increasing the range by utilizing the directivity of the signal transmission and reception. From a spectral efficiency point of view, the beamforming technique is inferior to spatial multiplexing since it utilizes only one channel (path) among a multiple of parallel spatial channels. While, spatial multiplexing utilizes all the spatial channels to improve the throughput, which increases with the number of "resolvable paths." On the other hand, beamforming can be implemented at a much lower hardware cost than spatial multiplexing techniques since every antenna elements can share the same RF and baseband processor. Most of all, however, it was shown [10,11] that beamforming is the optimum choice under certain conditions, when there exists a strong LOS (line of sight) path and the perfect channel side information is not available at the transmitter. Overall, taking the hardware cost into account, beamforming is the most suitable technology for WPAN, and here the detailed issues around the application of beamforming techniques to mm-wave wireless PAN are covered.

9.2 Channel and Signal Models

9.2.1 Channel Model

The channel model to describe the beamforming technique is a modified version of those developed by the IEEE 802.15.3c task group [12–14], where the channel gain, the time delay, and the angle of arrival (AoA) are defined for each path (or subpath). In addition to these parameters, the angle of departure (AoD) will also be added to each path in the proposed model, making it possible to examine the performance with the beamforming at the transmitter side too.

According to References 12 and 14, the 2-dimensional channel impulse response is defined as follows:

$$h(t,\theta_A,\theta_D) = \underbrace{a_{0,0}\delta(t-\tau_{0,0})\delta(\theta_A-\theta_{0,0,AoA})\delta(\theta_D-\theta_{0,0,AoD})}_{LOS} + \sum_{l=1}^{L}\sum_{k=0}^{K_l-1} a_{k,l}\delta(t-\tau_{k,l})\delta(\theta_A-\theta_{k,l,AoA})\delta(\theta_D-\theta_{k,l,AoD}) \quad (9.1)$$

where L is the number of paths, K_l is the number of subpaths for the lth path, $a_{k,l}=|a_{k,l}|\exp(j\phi_{k,l}),\ k=1,2,\ldots,K_l,\ l=1,2,\ldots,L$ is the complex channel gain for the kth subpath of the lth path with $\phi_{k,l}\sim U(0,2\pi)$, $\tau_{k,l},\ k=1,2,\ldots,K_l,\ l=1,2,\ldots,L$ is the time delay, $\theta_{k,l,AoA}$, and $\theta_{k,l,AoD}$, $k=1,2,\ldots,K_l,\ l=1,2,\ldots,L$ are the angle of arrival and the angle of departure, respectively. All these channel parameters are random, and their distributions for some typical environment settings can be found in Reference 13. Although the real channel is three-dimensional, where, typically, a two-dimensional planar antenna array is used in practice, we focus on a two-dimensional channel and one-dimensional antenna array for analytical simplicity. The extension to a three-dimensional space will will be briefly introduced later.

Using the model in Equation 9.1, the complex channel impulse response, between the uth antenna element of the transmitter and the sth antenna element of the receiver, is given by

$$h_{u,s}(t) = \sum_{l=0}^{L}\sum_{k=0}^{K_l-1}\Phi_{u,s}(\theta_{k,l,AoA},\theta_{k,l,AoD})\cdot a_{k,l}\delta(t-\tau_{k,l}) \quad (9.2)$$

where the number of subpaths for LOS component $K_0 = 1$ and

$$\Phi_{s,u}(\theta_{AoA},\theta_{AoD}) = \exp\left(j\eta\cdot d_{t,s}\sin\theta_{AoD}\right)\cdot\exp\left(j\eta\cdot d_{r,u}\sin\theta_{AoA}\right) \quad (9.3)$$

η is the wave number defined as $\eta \equiv 2\pi/\lambda = 2\pi f_c/c$, where λ is the carrier wavelength in meters, $d_{t,s}$ is the distance in meters from the transmitter antenna s to the reference ($s = 0$), and $d_{r,u}$ is that from the receiver antenna u to the reference ($u = 0$). Assuming half wavelength antenna spacing, $d_{t,s}=d_{r,s}=(\lambda/2)\cdot s,\ s=0,1,\ldots,M_t-1 \text{ or } M_r-1$.

802.15.3c and 802.11ad support both the single carrier transmission (SC) (with frequency domain equalization at the receiver) and the orthogonal frequency division multiplexing (OFDM) as their physical layer technology [1], of which our main focus here is on the latter. As OFDM-based transmission is being considered, it is necessary to derive the channel matrix for each subcarrier. To do so, the discrete time channel impulse response (DT-CIR), $h_{u,s}[n]$ is firstly defined, which is obtained by first convolving $h_{u,s}(t)$ with the "chip" pulse shaping filter, $c(t)$, and then resampling it. The channel gain of the mth subcarrier, $H_{u,s}[m]$, is then given by the N-point DFT of $h_{u,s}[n]$ (N can be regarded as the number of subcarriers in the ODFM system). Finally, denoting the number of antennas at the transmitter and the receiver as M_r and M_t, respectively, the $M_r \times M_t$ channel matrix H_m is defined for the mth subcarrier, with its (u, s) th element given by $H_{u,s}[m]$. For analytical convenience, it is assumed that $E|H_{u,s}[m]|^2=1$ by using the normalization condition

$$\sum_{0}^{N-1} E\,|h_{u,s}[n]|^2 = 1 \quad (9.4)$$

Denoting the beamsteering vector of the transmitter and the receiver as **w** and **c**, respectively, of which $\mathbf{c}^H\mathbf{c} = 1$ and $\mathbf{w}^H\mathbf{w} = 1$ (typically, each component of **c** and **w** is a complex number with a magnitude of $1/\sqrt{M_t}$ satisfying the above condition), let $\mathbf{y}_m$ be the $M_r \times 1$ received signal vector for the *m*th subcarrier. This is given by

$$\mathbf{y}_m = \mathbf{H}_m\mathbf{w} \cdot x_m + \mathbf{n}_m \tag{9.5}$$

where $\mathbf{n}_m$ is the $M_r \times 1$ noise vector of which the covariance matrix is given by $E[\mathbf{n}_m\mathbf{n}_m^H] = \sigma^2\mathbf{I}$. Note that, with $\mathbf{w}^H\mathbf{w} = 1$, the total transmission power is normalized to 1, the same as that of the single antenna transmission, $E|x_m|^2 = 1$, even if using the M_t transmit antennae. The decision variable of the mth subcarrier is then given by

$$r_m = H_{\mathbf{c},\mathbf{w}}[m] \cdot x_m + \mathbf{c}^H\mathbf{n}_m \tag{9.6}$$

where $H_{\mathbf{c},\mathbf{w}}[m]$ is the effective channel gain for the *m*th subcarrier after antenna filtering, given by

$$H_{\mathbf{c},\mathbf{w}}[m] = \mathbf{c}^H\mathbf{H}_m\mathbf{w} \tag{9.7}$$

and the variance of the effective noise $\mathbf{c}^H\mathbf{n}_m$ by

$$E|\mathbf{c}^H\mathbf{n}_m|^2 = \sigma^2.$$

The SNR of the *m*th subcarrier is then given by

$$SNR_{m,BF}(\mathbf{c},\mathbf{w}) = \frac{|H_{\mathbf{c},\mathbf{w}}[m]|^2}{\sigma^2} \tag{9.8}$$

Since σ^2 is assumed to be constant, the denominator is ignored in the optimization, which will be discussed later. Note that, from Equation 9.8, the SNR for the single antenna system is given by $SNR_{m,\text{single}} = |H_{0,0}[m]|^2/\sigma^2$, where $H_{0,0}[m]$ is the channel gain of the *m*th subcarrier between the first (0th) transmit antenna and the first (0th) receive antenna.

The effective channel gain, $H_{\mathbf{c},\mathbf{w}}[m] = \mathbf{c}^H\mathbf{H}_m\mathbf{w}$, can also be formulated in a different way by defining the function $G(\mathbf{w},\theta)$ of a given beamsteering vector $\mathbf{w} = [w_1, w_2, \ldots, w_M]^T$ at the receiver (or transmitter) and the angle of arrival (or departure) θ as

$$G(\mathbf{w},\theta) = \sum_{m=0}^{M-1} w_m \cdot \exp(j\eta \cdot d_m \sin\theta) \tag{9.9}$$

The beam pattern is defined by $|G(\mathbf{w},\theta)|$ and the effective channel impulse response between transmitter and the receiver can also be represented by

$$h_{\mathbf{c},\mathbf{w}}(t) = \sum_{s=0}^{M_t-1}\sum_{u=0}^{M_r-1} c_u \cdot h_{u,s}(t) \cdot w_s = \int_{-\pi}^{\pi}\int_{-\pi}^{\pi} G(\mathbf{c},\theta_D) \cdot G(\mathbf{w},\theta_A) \cdot h(t,\theta_A,\theta_D)\, d\theta_A d\theta_D \tag{9.10}$$

$$= \sum_{l=0}^{L} \sum_{k=0}^{K_l - 1} G(\mathbf{c}, \theta_{k,l,AoD}) \cdot G(\mathbf{w}, \theta_{k,l,AoA}) \cdot a_{k,l} \delta(t - \tau_{k,l})$$

The effective DT-CIR, $h_{\mathbf{c},\mathbf{w}}[n]$, is obtained by first convolving $h_{\mathbf{c},\mathbf{w}}[t]$ with the chip pulse shaping filter, $c(t)$, and then sampling it. The channel gain of the mth subcarrier, $H_{\mathbf{c},\mathbf{w}}[m]$, is then given by the N-point DFT of $h_{\mathbf{c},\mathbf{w}}[n]$.

9.2.2 *System Configurations: Front-End versus Back-End Beamforming*

When applying beamforming techniques to an OFDM system, two different system configurations, namely the front-end and the back-end beamforming, can be considered. In the front-end beamforming, shown in Figure 9.1a, the transmitter beamsteering vector is applied after the OFDM modulation, and the receiver beamsteering vector is applied before the OFDM demodulation. This type of configuration is a conventional beamforming system, where a bank of phase shifters are inserted at RF front-end and all the subcarriers share the same beamsteering vector pair. Alternatively, in the back-end beamforming, shown in Figure 9.1b, the transmitter beam steering vector is applied before OFDM modulation, and the receiver beamsteering vector is applied after OFDM demodulation. In this case, the beamsteering is performed in the digital domain and can be optimized for each subcarrier. Compared to front-end configuration, however, the hardware is much more complex since it requires RF and baseband modem separately for each antenna.

9.3 Optimum Beamforming and Beamswitching

To implement the beamforming technique on an OFDM-based system, two generic types, namely, the optimum per-subcarrier beamforming and beamswitching can be considered. In the first type, the beamforming is performed and optimized on a per-subcarrier basis for which the beamforming must be implemented with the back-end configuration. While, in the second type, the beamforming can be performed in a block fashion and be implemented with the front-end configuration, requiring less complex hardware at the cost of performance degradation.

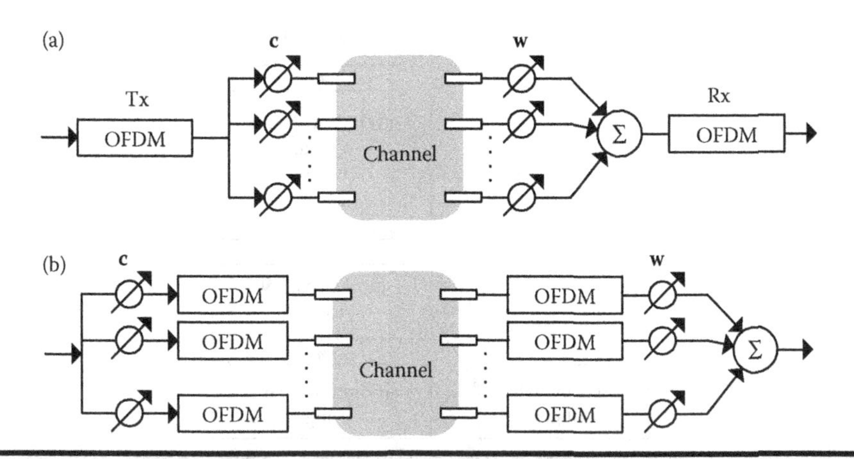

Figure 9.1 Two system configurations (a) front-end beamforming; (b) back-end beamforming.

9.3.1 Optimum Beamforming

With the term "optimum beamforming," it is implicitly assumed that the beamsteering vectors **c** and **w** are adjusted based on a certain optimality criterion for given channel matrix. For example, the beamsteering vector can be adjusted to maximize the SNR separately for each subcarrier, that is,

$$\max_{\mathbf{c},\mathbf{w}} \frac{|\mathbf{c}^H \mathbf{H}_m \mathbf{w}|^2}{M_r M_t \sigma^2} \tag{9.11}$$

where the maximum can be achieved by singular value decomposition (SVD) of the channel matrix assuming the perfect channel estimation at the receiver. That is, the channel matrix $\mathbf{H}_m$ for the mth subcarrier can be decomposed into

$$\mathbf{H}_m = \mathbf{U}_m \mathbf{D}_m \mathbf{V}_m^H \tag{9.12}$$

where $\mathbf{U}_m$ and $\mathbf{V}_m$ are unitary matrices of size $M_r \times M_r$ and $M_t \times M_t$, respectively, and $\mathbf{D}_m$ is $M_r \times M_t$ matrix of which the (j, j)th element denoted by $\lambda_{m,j}$ for $j = 1, 2, \ldots, \min(M_r, M_t)$ is a singular value (a non-negative real number) and all other elements are zero. It is assumed that the singular values are ordered such that $\lambda_{m,1} = \lambda_{m,\max} \geq \lambda_{m,2} \geq \ldots$. Hence, by setting $\mathbf{c} = \mathbf{u}_{m,1}$ and $\mathbf{w} = \mathbf{v}_{m,1}$, the first column vectors of $\mathbf{U}_m$ and $\mathbf{V}_m$ respectively, the maximum SNR is obtained as

$$\max_{\mathbf{c},\mathbf{w}} \frac{|\mathbf{c}^H \mathbf{H}_m \mathbf{w}|^2}{\tilde{\sigma}^2} = \frac{|\lambda_{m,1}|^2}{\sigma^2} \equiv \gamma_m \tag{9.13}$$

Figure 9.2 shows a typical procedure to implement the optimum beamforming, which has the following requirement:

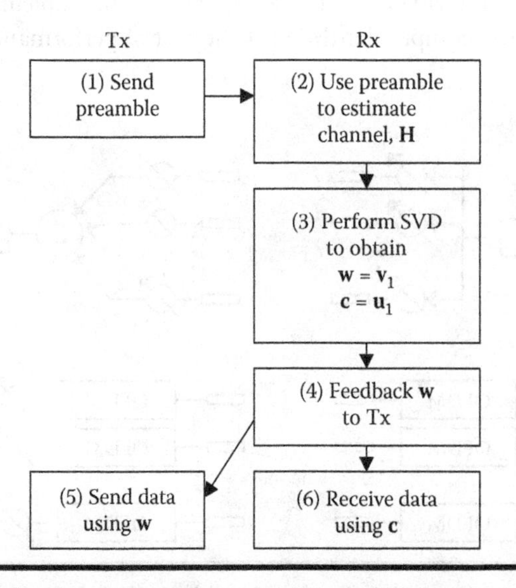

Figure 9.2 Protocols for implementing the optimal beamforming.

1. It requires channel estimation at the receiver by using, for example, preambles, which, in turn, requires the back-end configuration in Figure 9.1b having high hardware cost.
2. (If the channel between transmitter and receiver does not hold reciprocity) The beamsteering vector obtained at the receiver must be fed back to the transmitter so that the transmitter can adjust the values of the beamsteering vectors.

In practice, the channel side information feedback from the receiver to the transmitter is a burdensome task since the complex valued beamsteering vector itself has a large amount of information and must be fed back for each subcarrier. Although one can devise an adaptive protocol to iteratively search the optimal beamsteering vector with marginal protocol overhead, it still may require continuous transmission of the signal, either pilot or data, and sometimes, continuous feedback as well, which is not a realistic scenario in packet data communications, such as WPANs.

A straightforward figure of merit would be the average SNR, and the one that is achieved with the per-subcarrier beamforming is given by

$$\gamma_{ave,opt} = \frac{1}{N}\sum_{m=0}^{N-1}\gamma_m \tag{9.14}$$

Sometimes, however, an "effective value" would be more informative than the average value. Specifically, for a given set of maximum SNRs $\{\gamma_m;\ m = 0,1,\ldots,N-1\}$ with N being the number of subcarriers, our focus is on the effective SNR defined by the error rate performance approximations in References 16 and 17

$$p_{error} \approx \exp(-\beta\cdot\gamma_{eff,opt}) = \frac{1}{N}\sum_{m=0}^{N-1}\exp(-\beta\cdot\gamma_m) \tag{9.15}$$

where β is a parameter depending on the modulation and coding scheme used. In the literature, such an effective SNR mapping is known as the exponential effective SNR mapping (EESM) [16,17], where one can find some example values of β. Another useful performance criterion is the spectral efficiency bound defined for given SNRs $\{\gamma_m;\ m = 0, 1,\ldots, N-1\}$ by

$$R_{opt} = \frac{1}{N}\sum_{m=0}^{N-1}\log_2(1+\gamma_m) \tag{9.16}$$

9.3.2 Beamswitching

To ease the receiver feedback in the per-subcarrier optimum beamforming, one can use beamswitching where, instead of performing SVD and feeding back the beamsteering vector itself, the transmitter and receiver shares a set of predefined beamsteering vectors, namely the "beam codebook," and the receiver feeds back only the indices of the beams for the transmitter to use for data transmission. Figure 9.3a shows a typical process for the codebook-based beamswitching. The receiver first estimates the channel, $\mathbf{H}$, and selects the beamsteering vector pair $(\mathbf{w}, \mathbf{c})$ such that

$$\max_{\mathbf{c},\mathbf{w}\in C}|\mathbf{c}^T\mathbf{H}\mathbf{w}|^2 \tag{9.17}$$

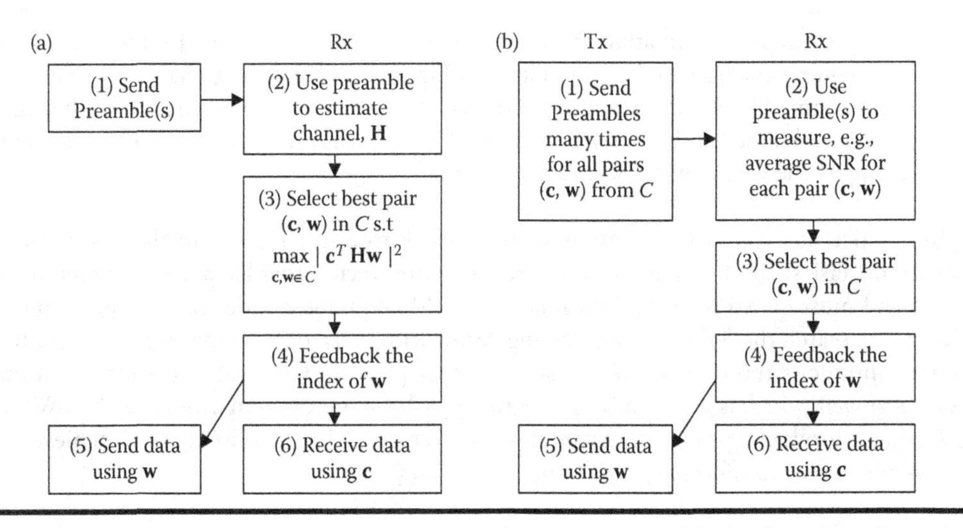

Figure 9.3 Protocols for implementing beamswitching (a) on top of the back-end configuration; (b) on top of the front-end configuration.

where C is the beam codebook. The receiver then feeds back the index of $\mathbf{w}$ to the transmitter. In this scheme, the amount of feedback information is reduced quite a bit, while it still requires the estimation of channel matrix $\mathbf{H}$ at the receiver using the back-end configuration in Figure 9.1b as the estimation of channel matrix cannot be performed with front-end configuration.

For low complexity implementation, (for example, by using the front-end configuration in Figure 9.1a) one can also resort to the protocol in Figure 9.3b. Here, before sending data packets, the transmitter sends preambles many times with different beamsteering vector pairs, with which the receiver measures some selection criteria (for example the effective SNR) and selects the best one for both the transmitter and the receiver. The receiver then notifies the transmitter which beamsteering vector should be used at the transmitter. Once the beamsteering vectors are set at both sides, data packets are transmitted. Note that with the bank of the phase shifters at RF front-end, the beamsteering vectors for the transmitter and the receiver are applied the same to all the subcarriers and, due to this reason, we call it "block beamswitching."

Since the block beamswitching use the same beamsteering vector pair for all the subcarrier, we also need to select the pair based on a collective measure for all the subcarriers. Corresponding to the three criteria in Equations 9.14 through 9.16, they can be represented as, for the average SNR criteria,

$$\gamma_{ave,beam-switch} = \max_{\mathbf{c},\mathbf{w}\in C}\left(\frac{1}{N\sigma^2}\sum_{m=0}^{N-1}|\mathbf{c}^H\mathbf{H}_m\mathbf{w}|^2\right) \tag{9.18}$$

or, for the effective SNR criteria, as

$$\gamma_{eff,beam-switch} = \max_{\mathbf{c},\mathbf{w}\in C}\left(-\frac{1}{\beta}\right)\log\left(\frac{1}{N}\sum_{m=0}^{N-1}\exp\left(-\frac{\beta|\mathbf{c}^H\mathbf{H}_m\mathbf{w}|^2}{\tilde{\sigma}^2}\right)\right) \tag{9.19}$$

or, for the spectral efficiency criterion, as

$$R_{beam-switch} = \max_{\mathbf{c},\mathbf{w}\in C}\left(\frac{1}{N}\sum_{m=0}^{N-1}\log_2\left(1+\frac{|\mathbf{c}^H\mathbf{H}_m\mathbf{w}|^2}{\tilde{\sigma}^2}\right)\right) \tag{9.20}$$

A drawback in the block beamswitching is that the preambles must be transmitted $K_t \times K_r$ times, where K_t and K_r are the number of beams used at the transmitter and the receiver, respectively. Even with a moderate codebook size, the protocol overhead required for sending preamble many times can be very high. This issue will be discussed in the next subsection, along with beam codebook design.

9.3.3 Beam Codebook and Beam Search in Block Beamswitching

Beam codebook can be defined by a $M \times K$ matrix, $\mathbf{W}^{(M,K)}$, where M and K are the number of antennae and beams, respectively. For example, in IEEE 802.15.3c WPAN, the (m, k)th element of $\mathbf{W}^{(M,K)}$ is given by

$$w_{m,k}^{(M,K)} = j^{f_{m,k}}, \quad \text{with} \quad f_{m,k} = \text{round}\left(\frac{m.\text{mod}(k+K/2,K)}{K/4}\right) \tag{9.21}$$

It is more convenient to use the exponent matrix, $\mathbf{F}^{(M,K)}$, of which the (m, k)th element is $f_{m,k}$. The followings are two examples of beam codebook with $M = 2$ and $M = 4$, of which the beam patterns are depicted in Figures 9.4 and 9.5, respectively.

$$\mathbf{F}^{(2,4)} = \begin{bmatrix} 0 & 0 & 0 & 0 \\ 2 & 3 & 0 & 1 \end{bmatrix} \tag{9.22}$$

$$\mathbf{F}^{(4,8)} = \begin{bmatrix} 0 & 0 & 0 & 0 & 0 & 0 & 0 & 0 \\ 2 & 3 & 3 & 0 & 0 & 1 & 1 & 2 \\ 0 & 1 & 2 & 3 & 0 & 1 & 2 & 3 \\ 2 & 0 & 1 & 3 & 0 & 2 & 3 & 1 \end{bmatrix} \tag{9.23}$$

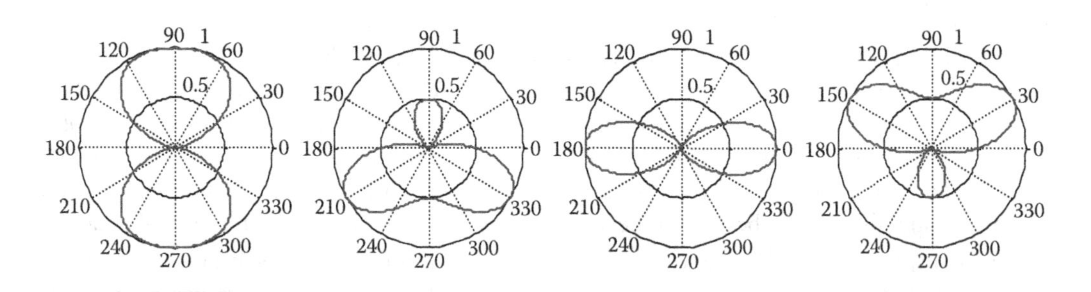

Figure 9.4 An example set of beams with 2 antenna elements (polar format).

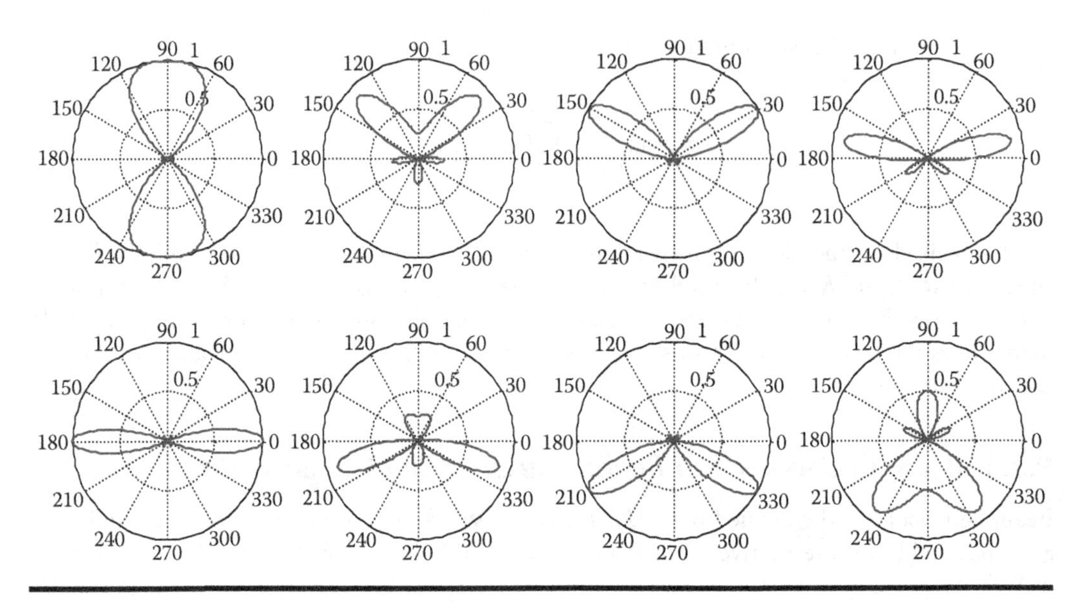

Figure 9.5 An example set of beams with 4 antenna elements (polar format).

The two examples in Equations 9.22 and 9.23 are for one-dimensional array creating a two-dimensional beam pattern. The same codebook can be used for two-dimensional array creating three-dimensional beam patterns. Let $\mathbf{w}_{3D,v,h}^{(M,K)}$ be the beam steering matrix with a two-dimensional $M \times M$ antenna array which has vertical beam pattern corresponding to $\mathbf{w}_v^{(M,K)}$ and horizontal one corresponding to $\mathbf{w}_h^{(M,K)}$. Then, it is given by

$$\mathbf{w}_{3D,v,h}^{(M,K)} = \mathbf{w}_v^{(M,K)} \cdot [\mathbf{w}_h^{(M,K)}]^T \tag{9.24}$$

Figure 9.6a and 9.6b shows the 16 three-dimensional beam patterns created by using the 4 two-dimensional patterns in Figure 9.4.

Two-step beam search [15]: As mentioned, when an analog phase shifter is used at the front-end, the beamswitching requires $K_t \times K_r$ times of preamble transmissions, which may incur considerable protocol overheads. Such overheads can be reduced by employing a multistep beam search [15]. That is, at first, wider beams are used with a smaller codebook size. Then, once the best beam has been found, the beamwidth is reduced by using more antenna elements, and then the search is carried out again. But, now, only the region corresponding to the best beam chosen in the previous step can be searched. In this way, the number of preamble transmissions can be greatly reduced.

Current WPAN protocol [15] supports a two-step beam search where the sector beam and the normal beam are defined. In the first step, sector beams are used, of which the beam is set wider than that of the normal beam by using only a subset of antenna elements, for example, 2 out of 6. Once the best sector beam indices for both the transmitter and the receiver are found, the receiver notifies it to the transmitter and then the second-step beam search is performed only within the region corresponding to the best sector beams. To show the reduction in the number of preamble transmissions, let $K_t^{(1)} = K_t^{(\text{sector})}$ and $K_r^{(1)} = K_r^{(\text{sector})}$ be the number of sector beams (sector beam codebook size) of the transmitter and the receiver, respectively, and $K_t^{(\text{beam})}$ and $K_r^{(\text{beam})}$ be the number of normal beams (normal beam codebook size) of the transmitter and the receiver, respectively.

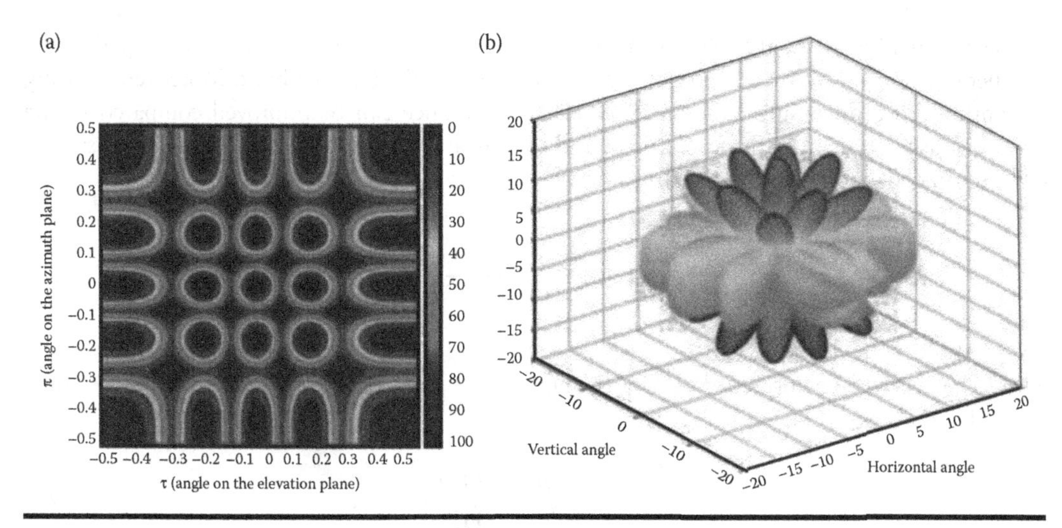

Figure 9.6 (a) Sixteen three-dimensional beams using 2 × 2 antenna array. (b) The vertical and horizontal beam patterns were created by those in Figure 9.4.

Usually, the number of normal beams corresponding to one sector are $K_t^{(2)} = K_t^{(\text{beam})}/K_t^{(\text{sector})}$ and $K_r^{(2)} = K_r^{(\text{beam})}/K_r^{(\text{sector})}$; both are integers larger than 1. Then, the total number of preamble transmissions in the two-step beam search is $K_t^{(1)} \times K_r^{(1)} + K_t^{(2)} \times K_r^{(2)}$. While, in the one-shot search, it would be $K_t^{(\text{beam})} \times K_r^{(\text{beam})}$. As an example, let $K_t^{(\text{sector})} = K_r^{(\text{sector})}$ and $K_t^{(\text{beam})} = K_r^{(\text{beam})}$. Then, in the two-step search, the number of preamble transmissions is $16 + 9 = 25$, while, in the one-shot search, it is 144, almost six times of that of the two-step search.

9.3.4 Hybrid Beamforming and Beamswitching

In terms of performance, the per-subcarrier beamforming will perform much better than the block beamswitching. However, in the former, the reverse link overhead can be very high since the transmitter beamsteering vectors, or their indices, for all the subcarriers must be fed back to the transmitter. If hardware cost is not a big problem, one can consider the use of a hybrid configuration [18], that is, *hybrid (per-subcarrier) beamforming and (block) beamswitching*, as shown in Figure 9.7. In the hybrid configuration, the transmitter beamsteering vector is applied after OFDM modulation and shared by all the subcarriers for block beamswitching, while the receiver beamsteering vector is applied after the OFDM demodulation so that they can be optimized for each subcarrier. The rationale for this is to compromise the performance and the protocol level overhead. By employing

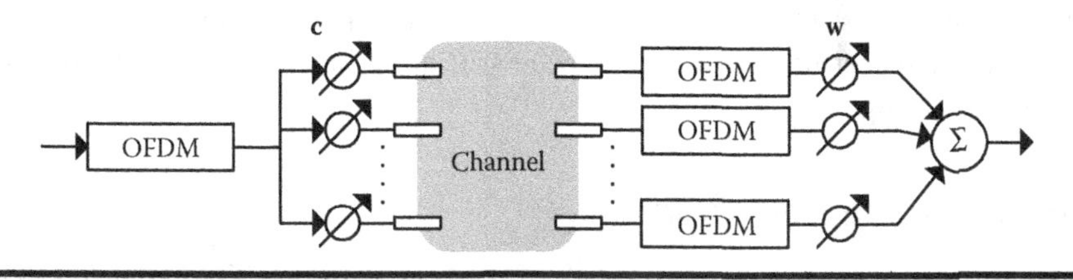

Figure 9.7 Hybrid beamforming and beamswitching.

the beamswitching at the transmitter, the number of preamble transmissions is only equal to the number of transmit antennae, and only one beam index needs to be fed back. Moreover, by using the optimum beamforming at the receiver, the performance can be improved compared to the block beamswitching at both sides. The hybrid optimization can be summarized as follows: for a given beam codebook C and the channel matrices $\{\mathbf{H}_m;\ m = 0,1,\ldots, N-1\}$,

$$\gamma_{eff,hybrid} = \max_{\mathbf{c}\in C}\left(-\frac{1}{\beta}\right)\log\left(\frac{1}{N}\sum_{m=0}^{N-1}\exp\left(-\frac{\beta}{\tilde{\sigma}^2}\cdot|\mathbf{c}^H\mathbf{H}_m\mathbf{w}_{m,opt}|^2\right)\right) \tag{9.25}$$

where the receiver beamsteering vector is obtained for the transmitter beamsteering vector $\mathbf{c}$ selected from C as

$$\mathbf{w}_{m,opt} = \sqrt{M_t}\cdot\frac{\mathbf{H}_m^H\mathbf{c}}{\left\|\mathbf{H}_m^H\mathbf{c}\right\|} \tag{9.26}$$

It is obvious from the fact that, if $\mathbf{c}$ is used for the beam steering vector at the transmitter, the effective channel vector becomes $\mathbf{H}_m^H\mathbf{c}$ and, by using Schwartz's inequality, the optimum receive beamsteering vector, for a given transmit beamsteering vector $\mathbf{c}$, is simply given by the vector matched to the effective channel vector. Note that, unlike the optimum beamforming, the singular value decomposition or the estimation of $\mathbf{H}_m$ is not needed. The hybrid beamforming and beamswitching can also be formulated under the maximum spectral efficiency criterion, that is,

$$R_{hybrid} = \max_{\mathbf{c}\in C}\left(\frac{1}{N}\sum_{m=0}^{N-1}\log_2\left(1+\frac{|\mathbf{c}^H\mathbf{H}_m\mathbf{w}_{m,opt}|^2}{\tilde{\sigma}^2}\right)\right) \tag{9.27}$$

Since $\log(1 + x)$ is a monotonically increasing function of x, the optimum receiver beamsteering vector is given the same by Equation 9.26.

When compared to the block beamswitching, the hybrid scheme requires M_r OFDM demodulator, as shown in Figure 9.2. However, the transmitter is required to send preambles only K_t times, which are typically far less than $K_t \times K_r$, and the performance can be better than the block beamswitching employing the front-end configuration. When compared to the optimum beamforming, the hardware complexity is similar. However, noting that singular value decomposition is not required, the computational complexity and the feedback overhead through the uplink might be far less than that of the optimum per-subcarrier beamforming. Of course, when the channel coherence bandwidth is large, the number of singular value decompositions, and the feedback overhead, can be reduced by grouping adjacent subcarriers, that is, by defining sub-bands. However, the feedback overhead required for the receiver to report the transmitter beamsteering vector seems to be still quite burdensome in the optimum per-subcarrier beamforming.

9.3.5 Beamforming Gain Bound

A fundamental performance measure in beamforming and beamswitching systems is the effective SNR gain over the single antenna system, which can be defined as

$$G_{BF} \equiv \frac{E[\gamma_{eff,BF}]}{E[\gamma_{eff,single-ant}]} \tag{9.28}$$

where $\gamma_{eff,BF}$ is the effective SNR with the beamforming of choice, given by one of Equations 9.15, 9.19, or 9.25, and $\gamma_{eff,single-ant}$ is the effective SNR of the single antenna system which is given by

$$\gamma_{eff,single-ant} = \left(-\frac{1}{\beta}\right)\log\left(\frac{1}{N}\sum_{m=0}^{N-1}\exp\left(-\frac{\beta}{\sigma^2}\cdot|H_{0,0}[m]|^2\right)\right) \tag{9.29}$$

where, without loss of generality, the single antenna channel was represented by the channel gain between the first transmit and the first receive antenna, indexed by 0. Since $H_{ij}[m]$s are assumed to be identically distributed random variables, the average performance is the same regardless of a specific choice of transmitter–receiver antenna pair.

Single path gain bound: It is interesting to consider the special case where only the one path exists, such that

$$h(t,\theta_A,\theta_D) = a_{0,0}\delta(t-\tau_{0,0})\delta(\theta_A-\theta_{0,0,AoA})\delta(\theta_D-\theta_{0,0,AoD}) \tag{9.30}.$$

with $a_{0,0}=1$. This is a frequency flat, Additive White Gaussian Noise (AWGN) channel and, hence, the channel matrix is simply given by

$$\mathbf{H}_m = \mathbf{H}_{single\ path} \equiv \mathbf{uv}^H \quad \text{for all } m \tag{9.31}$$

where the kth component of the vector $\mathbf{u}$ (of size $M_t \times 1$) and $\mathbf{v}$ (of size $M_r \times 1$) are given, respectively, by $u_k = \exp(j\eta\cdot d_{t,k}\sin\theta_{AoD})/\sqrt{M_r}$ and $v_k = \exp(-j\eta\cdot d_{r,k}\sin\theta_{AoA})/\sqrt{M_t}$. Note also that $\mathbf{H}_m$ has only one non-zero singular value, that is, $\lambda_{m,1}=1$ and $\lambda_{m,i}=0$ for $i\neq 0$. Let $\gamma_{m,single}$ be the SNR of the single antenna system for the mth subcarrier given by $\gamma_{m,single}=1/\sigma^2$ and $\gamma_{m,opt}$ be the SNR under the optimum beamforming given by Equation 9.13. The beamforming gain is the same for all the subcarriers and is given by

$$G_{single\ path} \equiv \frac{\gamma_{m,opt}}{\gamma_{m,single}} = M_r M_t \tag{9.32}$$

where the maximum is achieved with $\mathbf{c}=\mathbf{u}$ and $\mathbf{w}=\mathbf{v}$. In fact, $G_{single\ path}$ is the maximally achievable beamforming gain, which can be shown as follows. Let $\mathbf{H}_m$ be an arbitrary $M_r \times M_t$ channel matrix for the mth sub-carrier of which the singular values are denoted by $\lambda_{m,j}$ for $j=1,2,\ldots,$ $M\equiv\min(M_r, M_t)$ and ordered such that $\lambda_{m,1}=\lambda_{m,\max}\geq\lambda_{m,2}\geq\ldots$. Note that

$$\lambda_{m,\max}^2 \leq \sum_{i=1}^{M}\lambda_i^2 = tr[\mathbf{H}_m\mathbf{H}_m^H] = \sum_{u=1}^{M_r}\sum_{s=1}^{M_t}|H_{u,s}[m]|^2 \tag{9.33}$$

By taking expectation on both sides and from the assumption that $E|H_{u,s}[m]|^2 = 1$ from Equation 9.4, then the following is given

$$E[\lambda_{m,\max}^2] \leq M_r M_t \tag{9.34}$$

where the equality holds if and only if $\lambda_{m,1} = \lambda_{m,\max}$ and $\lambda_{m,i} = 0$ for $i \neq 0$, that is, single path case. This suggests that the practical beamforming gain in Equation 9.28 is less than or, at most, equal to the single path beamforming gain, that is,

$$G_{BF} \leq G_{single\ path} \tag{9.35}$$

9.4 Simulation and Numerical Results

To give an insight into the performance gain that can be obtained using the beamforming and beamswitching in a WPAN environment, we provide some numerical results based on the simulation parameters used in Reference 18, where, using the channel models developed within the IEEE 802.15.3 standardization body, the effective SNR gain and the spectral efficiency bound were evaluated for the three schemes, that is, the optimum per-subcarrier beamforming with back-end configuration, the block beamswitching with front-end configuration, and the hybrid beamforming and beamswitching. Various channel models have been developed by the IEEE 802.15.3 standardization body, and are publicly available. From these, two channel models have been used [18], CM.1 and CM.2 [13], which were developed based on channel measurements in typical home environments with multiple furnished rooms and their size comparable to a small office room. Specifically, CM.1 has an LOS component while CM.2 does not. The reader may refer to Reference 13 for the detailed statistical characterization for the channel parameters of the model in Equation 9.1. Although the angle of departure has not been defined in Reference 13, the same statistics as that of the angle of arrival have been used for the angle of departure.

Regarding the system parameters, they assumed 60 GHz carrier frequency, 1 GHz signal bandwidth and half wave length antenna spacing in one-dimensional array. For the OFDM parameters, the number of subcarriers was set to 512 and the cyclic prefix length to 64 samples. The parameter β used in the exponential effective SNR mapping was set to 2, which is a typical value for QPSK signaling [16,17].

Figures 9.8 and 9.9 show for the channel model CM.1 and CM.2, respectively, the effective SNR gain as a function of the number of antennae $M_t = M_r = M = 1, 2, \ldots, 6$, the same for both the transmitter and the receiver. For the block beamswitching and the hybrid scheme, the number of beams K was set to $K_t = K_r = M$. In both figures, the optimum per-subcarrier beamforming is shown to be the best, the hybrid beamforming and beamswitching is the next, and the beamswitching (beamswitch) is the worst. The performance differences between these schemes are more noticeable for CM.2, where no LOS component exists, thus verifying the rationale of the hybrid scheme.

In fact, the performance can be slightly improved by increasing the number of beams for a fixed number of antennae, which, however, means that the number of preamble transmissions must also be increased. Figure 9.10 shows the effective SNR gain with $K_t = K_r = M$, $2M$ and $3M$. Although the performance improvement in block beamswitching is higher than that of the hybrid scheme, it is still 2 dB better than the block beamswitching.

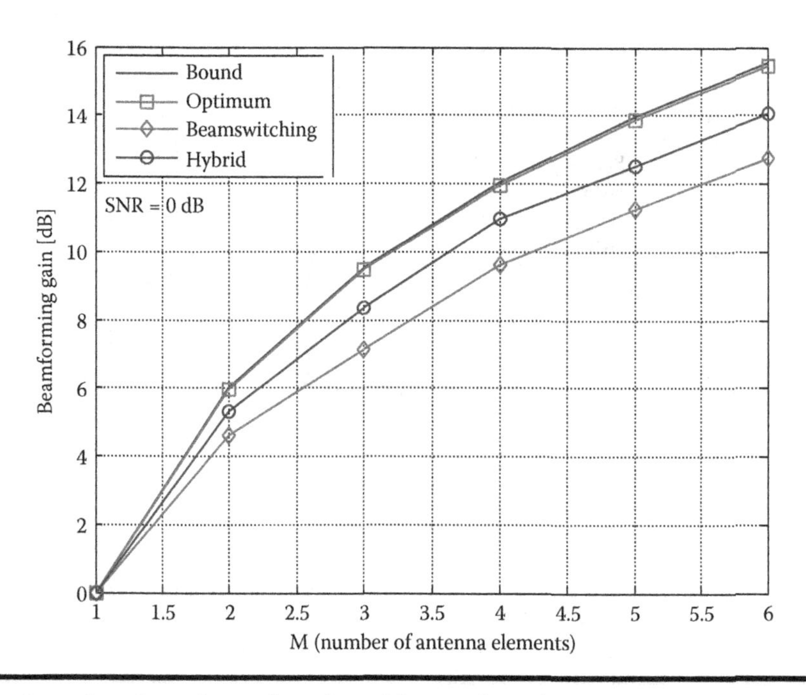

Figure 9.8 Beamforming gain as a function of the number of antennas, $M_t = M_r = M$; $K_t = K_r = M$, Channel Model: CM1.4 (with LOS), $SNR_{single} = 0$ dB.

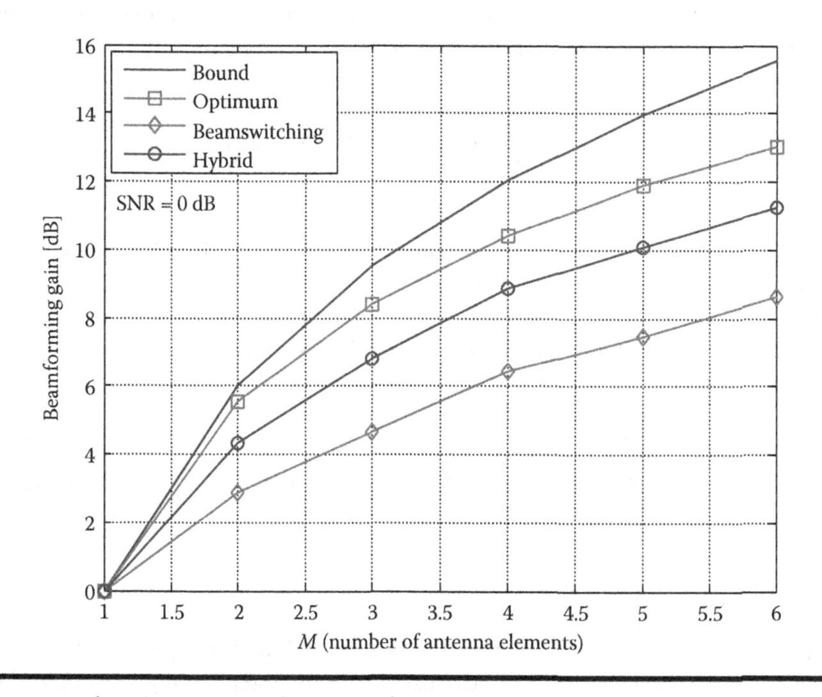

Figure 9.9 Beamforming gain as a function of the number of antennas, $M_t = M_r = M$; $K_t = K_r = M$, Channel Model: 2.4 (No LOS), $SNR_{single} = 0$ dB.

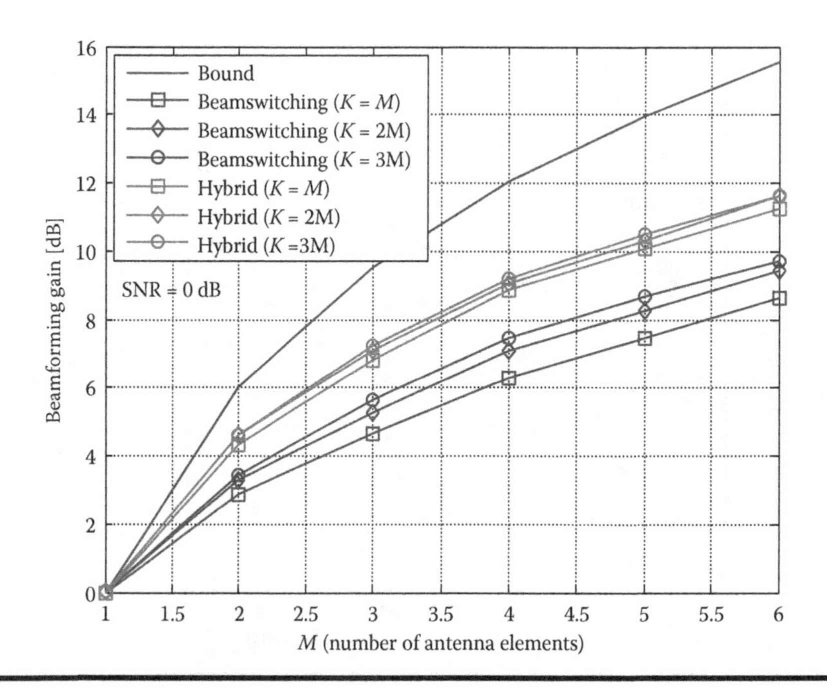

Figure 9.10 Beamforming gain as a function of the number of antennas, $M_t = M_r = M$; $K_t = K_r = M$, 2M and 3M, Channel Model: 2.4 (No LOS), $SNR_{single} = 0$ dB.

It is also interesting to compare the spectral efficiencies (under the maximum spectral efficiency criterion given by Equations 9.16, 9.18, and 9.21) with the open loop channel capacity (i.e., with the receiver channel side information only) given by

$$C = \frac{1}{N}\sum_{m=0}^{N-1} \log_2 \det(\mathbf{I} + \tilde{\sigma}^{-2} \cdot \mathbf{H}_m \mathbf{H}_m^H) \qquad (9.36)$$

Figures 9.11 and 9.12 show the channel model CM.1 and CM.2, respectively, and a comparison of the spectral efficiency bounds and the channel capacity as a function of the number of antennae $M = 1, 2, \ldots, 6$, the same for both the transmitter and the receiver. The average SNR is set to 0 dB for both the channel model, CM.1 and CM.2. Note that the channel capacity of CM.2 is much higher than that of CM.1 because CM.2, without LOS, is a richer scattering environment and the spatial multiplexing is more beneficial than the beamforming. Nevertheless, when comparing the hybrid scheme with that of the optimum per-subcarrier beamforming and the block beamswitching, it can provide good performance at a reasonable hardware and overhead cost.

9.5 Concluding Remarks

In this chapter, we considered beamforming and beamswitching for application to 60 GHz wireless personal area networks (WPAN). We discussed various system configurations and optimization criteria for OFDM-based WPAN and covered practical issues such as beam codebook and

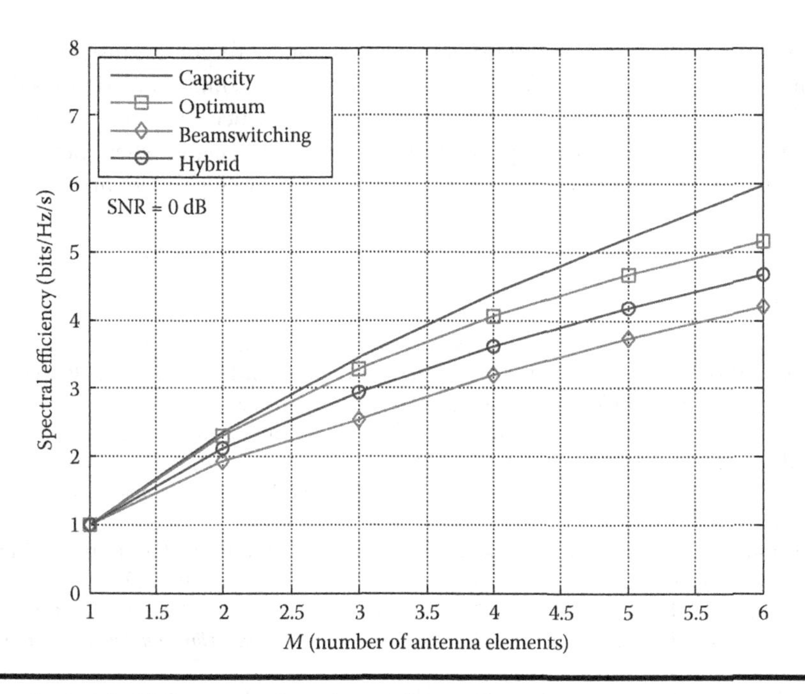

Figure 9.11 Spectral efficiency as a function of the number of antennas, $M_t = M_r = M$; $K_t = K_r = M$, 2M and 3M, Channel Model: 1.4 (with LOS), $SNR_{single} = 0$ dB.

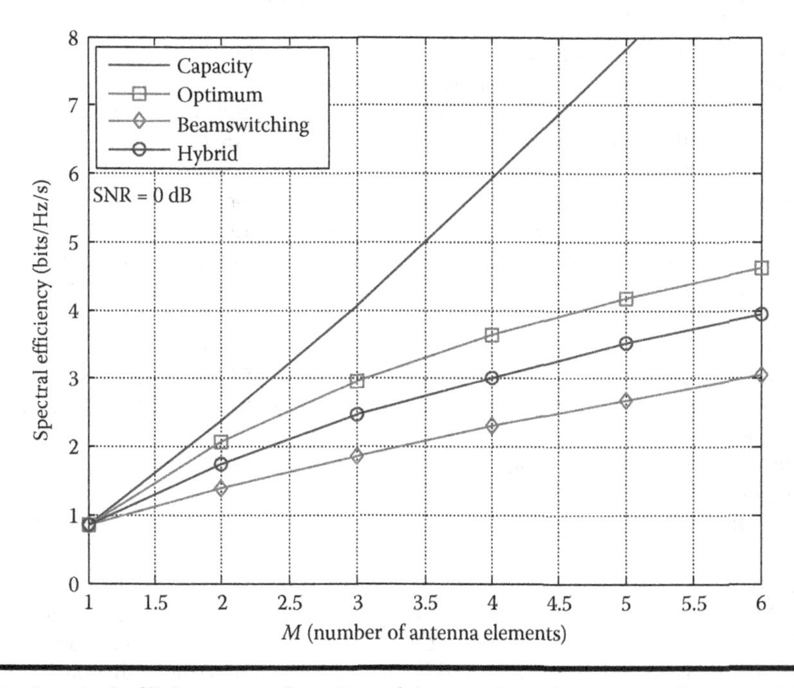

Figure 9.12 Spectral efficiency as a function of the number of antennas elements, $M_t = M_r = M$; $K_t = K_r = M$, 2M and 3M, Channel Model: 2.4 (No LOS), $SNR_{single} = 0$ dB.

beam search protocols for low-cost beamswitching techniques. To provide insight into the performance gain in typical WPAN environment, we also provided some simulation results based on the channel models with and without LOS component [18], which show that a considerable gain can be obtained by using beamforming and beamswitching. Especially, when there exists an LOS path, the SNR gain is shown to be quite close to the beamforming gain bound.

References

1. Lei, M., C. S. Choi, R. Funada, H. Harada, and S. Kato, 2007. Throughput comparison of multi-Gbps WPAN (IEEE 802.15.3c) PHY Layer Designs under Non-Linear 60-GHz Power Amplifier. In *Proceedings of PIMRC* 2007, September, pp. 1–5, Athens, Greece.
2. Smulder, P., 2002. Exploiting the 60GHz band for local wireless multimedia access: Prospects and future directions. In *IEEE Communications Magazine*, 40(1), 140–147.
3. Winters, J. H., 1998. Smart antennas for wireless systems. In *IEEE Personal Communications*, 5(1), pp. 23–27.
4. Budsabathon, M., Y. Hara, and S. Hara, 2004. Optimum beamforming for Pre-FFT OFDM adaptive antenna array. In *IEEE Transactions on Vehicular Technology*, 53(4), 945–955.
5. Zhou, S. and G. B. Giannakis, 2002. Optimal transmitter eigen-beamforming and space time block coding based on channel mean feedback. In *IEEE Transactions on Signal Processing*, 50(10), 2599–2613.
6. Foschini, G. J., 1996. Layered space-time architecture for wireless communication in a fading environment when using multi-element antennas. In *Bell Labs Technical Journal*, 1(2), 41–59.
7. Wolniansky, P. W., G. J. Foschini, G. D. Golden, and R. A. Valenzuela, 1998. V-BAST: An architecture for realizing very high data rate over the rich-scattering wireless channel. In *Proceedings of URSI Intl Symposium on Signal, Systems and Electronics Conference*, September 29–October 2, 1998, Palazzo Dei Congressi, Pisa, Italy, pp. 295–300.
8. Alamouti, S. M., 1998. A simple transmitter diversity scheme for wireless communications. In *IEEE Journal on Selected Areas in Communications* 16, 1451–1458.
9. Tarokh, V., N. Seshadri, and A. R. Calderbank, 1998. Space–time codes for high data rate wireless communication: Performance analysis and code construction. In *IEEE Transactions on Information Theory*, 44(2), 744–765.
10. Jafar, S. A. and A. Goldsmith, 2001. On optimality of beamforming for multiple antenna systems with imperfect feedback. In *Proceedings of ISIT 2001*, Washington, DC, pp. 321–321.
11. Jafar, S. A., S. Vishwanath, and A. Goldsmith, 2001. Channel capacity and beamforming for multiple transmit and receive antennas with covariance feedback. In *Proceedings of ICC* 2001, Helsinki, Finland, pp. 2266–2270.
12. Liu, C., E. Skafidas, T. S. Pollock, and R. J. Evans, 2006. Angle of arrival extended S-V model for the 60 GHz wireless desktop channel. In *Proceedings of the IEEE 17th International Symposium on PIMRC (Personal, Indoor and Mobile Radio Communications), PIMRC 2006*, September 11–14, 2006, Helsiniki, Finland. IEEE 2006.
13. IEEE P802.15. 2007. Working Group, IEEE 802.15-07-0584-00: IEEE 802.15.3*c* Channel modeling sub-committee Report, March 2007.
14. Foerster, J. R., M. Pendergrass, and A.F. Molisch, 2003. Channel model for ultrawideband personal area networks. In *IEEE Wireless Communications*, 10(6), 14–21.
15. IEEE P802.15 Working Group, IEEE P802.15-08-0355-00-003c, May 2008.
16. Lampe, M., H. Rohling, and J. Eichinger, 2002. PER-prediction for link adaptation in OFDM systems. In *Proceedings of the 7th International OFDM Workshop*, September, Hamburg, Germany.
17. Blankenship, Y., P. Sartori, B. Classon, and K. Baum, 2004. Link error prediction methods for multicarrier systems. In *Proceedings of VTC* 2004, Fall, Los Angeles.
18. Yoon, S., T. Jeon, and W. Lee, 2009. Hybrid beam-forming and beam-switching for OFDM based wireless personal area networks. *IEEE Journal on Selected Areas in Communications*, 27(8), 1425–1432.

Chapter 10

Applications of Directional Networking in Military Systems

Latha Kant, Ritu Chadha, and John Lee

Contents

Abstract

The use of directional links for mobile ad hoc networking for the military offers a number of benefits over traditional omni-directional links. One of the drawbacks of wireless networking using omni-directional links is the interference caused to all wireless receivers within range of the wireless transmitter. In contrast, the use of directional beams allows the realization of point-to-point wireless links, thereby significantly reducing interference. An immediate benefit of this is the ability to simultaneously operate multiple links at the same frequency in the same geographical region, thereby increasing overall network capacity, which is a scarce resource in wireless networks. mmW links in particular have the ability to provide long reach and high capacity (ranging from 1 Gbps at 40 Km to 10 Mbps at 60 Km under good atmospheric conditions). Another significant benefit of using highly directional mmW links for networking mobile nodes is the Low Probability of Intercept (LPI)/Low Probability of Detection (LPD) provided by these links, rendering them invaluable in contested environments.

However, the benefits of directional networks are not one-sided. The very same directional property of mmW links that provides high bandwidths, low interference, and excellent LPI/LPD properties, also presents significant challenges for networking. The challenges stem from the need for maintaining synchronization of the highly directional line-of-sight, point-to-point, mmW links in a MANET environment. This is very difficult in an environment that is typically characterized by mobile nodes and rapidly changing network conditions. The challenges are further exacerbated in an airborne environment (e.g., a network of UAVs) due to the significantly higher speeds of airborne platforms as compared to ground tactical mobile platforms.

The above challenges underscore the need for enhancements to existing mobile ad hoc networking solutions to make directional links a viable option for networking. The remainder of this chapter provides a description of some of the required enhancements that we have identified. We show that the implementation of these enhancements provides the critically needed networking capabilities needed to make directional networks a reality for the military.

10.1 Introduction

This chapter discusses the application of directional network links for mobile ad hoc networking. We describe how directional networking can be applied and managed to realize high-bandwidth mobile ad hoc networks (MANETs) for the military. The contents of this chapter are organized as follows.

We begin by describing the architecture of a directional network that provides a resilient wireless backbone network using unmanned aerial vehicles (UAVs) in Section 10.2. This architecture is used as a reference to illustrate the algorithms introduced in the next few sections. In Section 10.3, we provide an overview of algorithms for topology control for a directional network. In Section 10.4, we describe the use of a proactive route injection mechanism to improve the performance of a popular proactive link–state routing protocol (Optimized Link State Routing, or OLSR) in a high-speed directional network. Finally, in Section 10.5, we address the problem of intermittent link connectivity that often arises in directional networks, and describe solutions to this problem using alternate routing mechanisms and caching techniques. In each of these three Sections (10.3 through 10.5), we provide simulation results that illustrate the benefits of using our algorithms for directional networking.

In most of our discussion in this chapter, we refer to our experience with the use of millimeter wave (mmW) technology for directional networking; however, the technologies described are widely applicable to any type of directional wireless links.

10.2 Directional Networking Architecture for Airborne Networks Using mmW Links

In this section, we describe a basic networking architecture using directional (mmW) links that will form the basis of our discussion in the remainder of this chapter. We begin by providing a brief description of a networking pod that can be housed on UAVs in Section 10.2.1. This is followed by a description of a 20-node mmW network simulation model in Section 10.2.2.

10.2.1 Networking Pod

Figure 10.1 shows a high-level schematic of a networking pod that be carried on a variety of airborne platforms.

As shown in Figure 10.1, each pod houses a commercial router and Single Board Computer (SBC). To exploit the high link speeds in the order of gigabits per second offered by mmW links, the router shown in Figure 10.1 is a Gigabit Router. The SBC shown in Figure 10.1 is used to host the networking functions that we will describe in Sections 10.3 through 10.5. The pod shown in Figure 10.1 has the following five interfaces associated with the Gigabit Router:

1. Two high-speed directional mmW interfaces associated with mmW radios in the front (**F**) and rear (**R**),
2. Two lower speed omni-directional radios used for discovery (**D**) access (**A**), and
3. A wired Gigabit Ethernet Interface (**G**) to communicate between pods when more than one pod is housed on an aircraft.

Also shown in Figure 10.1 are the following modules:

- **Discovery Radio** (DR), which provides an omni-directional channel for network control functions and enables nodes to communicate with one another over a separate multihop ad hoc network in the absence of a directional mmW link.
- **WAP** (Wireless Access Point), which is used to connect to external wireless networks and devices using technologies such as Long-Term Evolution (LTE) or Wi-Fi, and provides access to the mmW backbone network.
- **MMW Modem**, which is used to create directional MMW radio links.
- **MMW Controller**, which sets the MMW (**F** and **R**) link characteristics (e.g., rate and direction) and returns link status information.
- **GPS-IMU** (Global Positioning System-Inertial Measurement Unit), which provides pod location and orientation information.

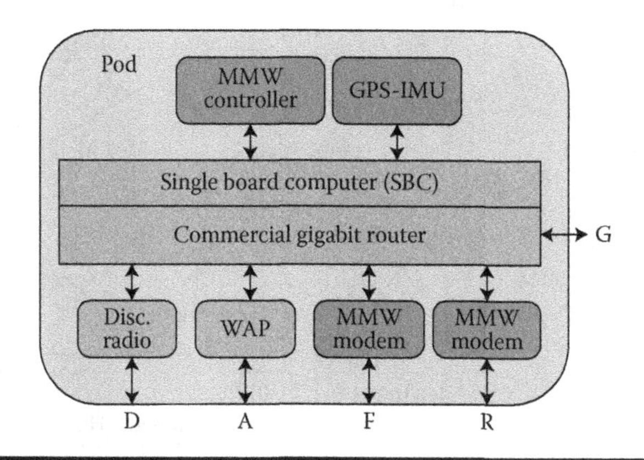

Figure 10.1 Networking pod.

10.2.2 mmW Networking Model

Figure 10.2 shows the schematic of a 20-node mmW network that we modeled using the discrete event simulator QualNet [1]. The UAVs were modeled with all the elements of the pod described in Section 10.2.1 and employed a variety of flight orbits that included periodic, aperiodic, and random trajectories. We provide below a description of the network model and traffic that were used to evaluate the performance of the topology control and routing mechanisms described in Sections 10.3 through 10.5.

Figure 10.2 provides a schematic of a 20-node network consisting of 15 UAVs (N01–N15) represented via dark circles interconnecting five ground nodes (G01–G05) spread over geographically distant locations with different network topologies, as follows. The segment of the network to the far right, including N12–N14 and G02–G03, models a tactical theatre where connectivity is sparse (tree-like topology) and link speeds are relatively lower (1–10 Mbps). The segment to the far left that includes N01–N04 and G04 models a command post characterized by a more stable ring-like topology with link speeds ranging from 100 Mbps to 1 Gbps. The segment in the center of Figure 10.2 is characterized by a mesh topology with link speeds ranging from 10 Mbps to 1 Gbps, and serves the purpose of interconnecting the tactical segment with the remote reach-back segment. The UAVs and ground nodes are interconnected via E-band mmW links that have link speeds ranging from 10 Mbps to 1 Gbps, over distances ranging from 40 Km to 60 Km under good atmospheric conditions.

The simulated UAVs employ a variety of flight orbits, as outlined below:

i. UAVs N12–N15, which appear on the far right in Figure 10.2, fly random orbits (since they model deployment in a tactical theater-like environment)
ii. UAVs N01–N04, which appear at the far left in Figure 10.2, fly a periodic pattern (elliptical and circular orbits)
iii. N05–N11, which appear towards the center in Figure 10.2, employ a mix of regular and pseudo-random (lawn-mower) flight orbits, to provide reach-back connectivity from the far left (tactical segment) to the far right (surveillance segment) of the network

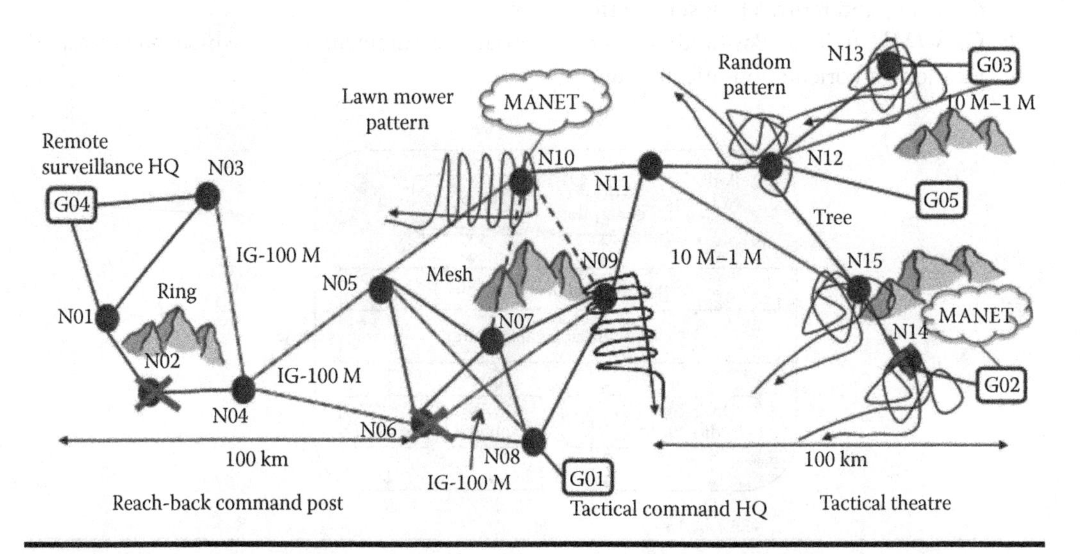

Figure 10.2 20-node mmW network scenario.

Multiple scenarios and traffic mixes were simulated. A typical scenario was run for 30 minutes (1800 s). The simulated traffic mix used in the performance evaluations captured in Figures 10.3, 10.5, and 10.6 supplied a total offered load of ~1 Tbit, and was a mixture of

a. Bulk high-bandwidth traffic (representative of Intelligence, Surveillance, and Reconnaissance (ISR) image transmissions) between air nodes and ground nodes G01, G04, and G05, categorized as medium and low priority.
b. High-priority flows simulating an aggregated MANET backhaul between nodes G02 and G03.

More specifically, the traffic of type (b) was short-lived (~3 s duration) bursty high-priority traffic that offered a low aggregate load (~200 kb/s), representative of mission-critical information, while traffic of type (a) was relatively long-lived (400–600 s) high-bandwidth (Mbps to Gbps) UDP traffic streams, representative of imagery data, that were medium and low priority. Additionally, a large FTP/TCP session involving a 6.2 GB file transfer was also simulated between two remote ground nodes (G01 and G05) and categorized as type (a).

In the sections that follow, we describe algorithms for establishing and maintaining robust network connectivity (Section 10.3), for setting up resilient communication paths (Section 10.4), and for providing alternate routing and caching to improve traffic delivery characteristics (Section 10.5) for directional networks.

10.3 Topology Control in mmW Networks

In this section, we describe topology control algorithms that leverage UAV trajectory information (e.g., based on the mission plan) to create suitable network topologies using mmW links. More specifically, we provide an overview of a distributed proactive topology control (DPTC) mechanism in Section 10.3.1, followed by a description of the details of the DPTC mechanism in Sections 10.3.2 and 10.3.3. Section 10.3.3 provides performance results using the simulated mmW network that was described in Section 10.2.2.

Broadly speaking, the term topology control (TC) in mobile ad hoc networking refers to the creation and maintenance of a network topology by using a subset of available wireless links to interconnect the nodes. Several distributed TC mechanisms for MANETs exist today [2–4]; however, they are generally not applicable to high-speed directional networks because they create and maintain network topologies based on information about which links are *currently* available, using predominantly *reactive* mechanisms. This is generally not a problem in an environment where link availability changes slowly over time, as is the case with omni-directional links. However, directional links are much more brittle and likely to disappear in a highly mobile tactical environment because of their directional nature: a directional link provides point-to-point connectivity between two nodes that have their antennas pointed towards each other, and this link is lost if the antennas get misaligned. The net result is that a tactical mobile network with directional links will suffer significant outages if mechanisms are not put in place to address this problem.

To overcome the above problem, we developed an approach that relies on *proactive* topology control algorithms to predict and proactively plan the network topology. Since military operations have a mission planning phase that involves UAV flight path and orbit planning, we leverage this information to provide proactive network topology control. We achieve this by building the network using links that are not only currently available, but also are likely to be available in the near future, based on the mission plan. It is easy to see that such an approach will result in greater network availability as the UAVs move. The next section describes the details of the approach.

10.3.1 Distributed Proactive Topology Control

We describe a distributed proactive topology control algorithm in this section that is a fully distributed mechanism and is invoked as soon as the UAVs have reached their orbit positions over an area of deployment (AoD), that is, after the launch has taken place and the UAVs have reached the AoD over which a communications network has to be formed. More specifically, once the UAVs have reached their designated locations in the AoD, each UAV uses information about the planned flight orbit, heading and speed of each of the UAVs that have been deployed to compute *feasible links* and their *future availability* over a time horizon T_h. This list of feasible links is then pruned to cater to specific mission objectives to result in a set of *selected links* that in turn go on to form the communications network topology, as explained later in this subsection. To permit distributed operations and to maintain a globally consistent topology, time horizons are synchronized using GPS. Furthermore, all the UAVs use the same T_h and trigger their topology computations at the beginning of each T_h. Since all the UAVs execute the same link selection (LS) algorithm to construct the network topology, and since the link selection algorithm does not have randomness embedded within it, this will enable a globally consistent topology which is computed in a fully distributed manner.

Additionally, if there arise situations that call for deviations from the preplanned flight orbits for one or more of the UAVs (which can occur frequently in a contested environment), the DPTC will make use of the control channel provided by the on-board omni-directional discovery radios (DR) (see Figure 10.1) to broadcast the new/changed UAV flight orbits. Since the discovery radios allow the realization of a multihop ad hoc network, the information regarding changes in the affected UAVs' flight orbits will be disseminated via the discovery radios' control channel throughout the network. All of the UAVs in the network will then use these new flight orbits to compute the new topology. Thus, such a mechanism will continue to ensure a globally consistent topology in spite of unanticipated orbit changes while retaining the fully distributed nature of the computation.

In order to compute *feasible* and *selected* links, each UAV does the following. First, it computes a list of feasible links that can be established among the UAVs in the area of deployment, over a certain time horizon. We refer to this list as the feasible link list (FLL). The time horizon (T_h) is a configurable interval of time based on the UAV speeds. The set of feasible links as well as the achievable data rates between a pair of UAVs are computed based on radio characteristics (e.g., Transmit (Tx) Power, Receiver (Rx) Sensitivity, Tx/Rx antenna gains, directional antennae look angles, etc.), and distance (as obtained from GPS) between the two UAVs for that time horizon, T_h. More specifically, the FLL contains information about a set of links over a time window T_h with timing information that indicates when each of the links will be available within that time window, along with the associated link rate for each link.* Computation of the FLL is described in Section 10.3.2.

Once the FLLs are generated, the next step is to determine a subset of the feasible links that will need to be established to realize network connectivity. We describe a Link Selection algorithm

* Note: For a given Tx power, Rx sensitivity, Tx/Rx antenna gains, the achievable data rates are a function of the inter-UAV distance.

in Section 10.3.3 that makes use of the FLL to select links based not just on their current availability, but also based on their future availability over the time horizon T_h. Further, based on mission requirements, the Link Selection algorithm takes into account the following information when computing the subset of selected links:

- Link longevity, that is, the length of time for which a link is predicted to be available.
- Link capacity, that is, the projected data rate.
- Maximum allowed platform degree, which is a configured parameter per platform for the mission, and constrains the number of allowed links per platform.

The algorithm performs appropriate tradeoffs to recommend an appropriate set of candidate links to form the topology. The output of the Link Selection algorithm will be a subset of the FLL. We refer to the links selected by the LS algorithm as the set of "selected links," or the Selected Link List (SLL), which will form the network topology.

Finally, once the Link Selection algorithm on a UAV identifies the subset of links to be established, it issues a trigger to the local MMW controller on the UAV to start the Pointing–Acquisition–Tracking (PAT) process to establish the selected mmW links.

10.3.2 Feasible Link List Computation

The Feasible Link List (FLL), as its name indicates, is a list of possible links that can be established in the network. The FLL computation is performed for every time horizon (T_h) at the beginning of that time horizon. More specifically, each UAV computes the FLL using the flight orbit and heading information obtained initially from mission planning, and subsequently via DR control channel broadcast messages whenever changes to the planned orbits take place. The FLL computations are essentially link closure computations between UAVs. They depend on the following parameters:

a. The transmit power (Tx), Receiver Sensitivity (Rs), and transmit/receive antenna gains of the UAVs
b. The path loss characteristics
c. The distance between the UAVs

Since mmW links are point-to-point links between two specific UAVs, the computation of the FLL for the network involves computing all feasible links between each pair of UAVs. Further, since each UAV has knowledge of all the other UAVs in the network along with their heading and orbits, the FLL computation at each node involves computing the set of links that can be formed between

i. The UAV and its neighbors
ii. Every other UAV pair in the network

using the canonical link closure expressions summarized below.

10.3.2.1 Link Closure Computation

We perform the following two-step computation in order to determine whether or not a link can be formed between two UAVs "A" and "B."

Step 1: Compute received power via the following expression:

$$P_B^{\text{Rx}} = P_A^{\text{Tx}} + G_A^{\text{Tx}} + G_B^{\text{Rx}} - L_{\text{FSPL}} - L_M \quad (10.1)$$

where

P_B^{Rx} = Received power (dBm) at node B
P_A^{Tx} = Transmitted Power (dBm) from node A
G_A^{Tx} = Transmitter Antenna Gain (dBi)
G_B^{Rx} = Receiver Antenna Gain (dBi)
L_{FSPL} = Free space path loss (dB)
L_{M} = Loss miscellaneous (e.g., fading, Tx/Rx coax loss, etc.) (dB)

Note: Unless otherwise specified, L_M will be assumed to be zero.
Also,

$$L_{\text{FSPL}} = \left(\frac{4\pi df^2}{c} \right)$$

where

d = distance separating the two nodes A and B
f = operating frequency of the link
c = speed of light

In units of decibels (dB), L_{FSPL} can be rewritten as follows:

$$L_{\text{FSPL}}(\text{dB}) = 20\log_{10}(d) + 20\log_{10}(f) + 32.44\,\text{dB} \quad (10.2)$$

where

d is the distance between A and B in km
f is the frequency of the link in MHz

Step 2: Determine link closure, that is, whether or not the receiving node can hear the sending node. This is done by evaluating (3) below. More specifically, the link between nodes A and B can be closed if

$$P_B^{\text{Rx}} - R_B^{\text{sen}} \geq \text{Margin} \quad (10.3)$$

where *Margin* is either provided as an input parameter that specifies the link margin (the difference between the sensitivity of the receiver and the actual received power) or is set to 0. In Equation 10.3,

P_B^{Rx} = Received power (dBm) at node B as computed from Equation 10.1
R_B^{sen} = Receiver sensitivity (dBm) at node B, which is provided as an input parameter

Note that since the UAVs will be in motion over the time horizon (T_h), the FLL computations will be discretized within the time–horizon using a configurable sampling interval (Q) within the T_h.

The output of the feasible link computation will contain information about a set of links over a time window T_h with information about the start- and end-times that each of the links are available within that T_h, along with the associated link rate for each link.

10.3.3 Selected Link List Computation

Once the FLL has been computed as described in the previous section, the next step is to select a subset of the FLL as the links that will be used to form the network topology. This problem can be formulated as a multiobjective optimization problem, for which several approaches exist in the literature. However, we argue that due to the highly stochastic nature of the mmW links and the underlying environment, algorithms that compute an optimal solution (i.e., an optimal topology) are not necessary, and are probably overkill for the problem at hand. Since arriving at optimal solutions requires an exhaustive search of the state space, computing optimal solutions would be wasteful of precious computation resources on the on-board computer, since the state of the network is expected to change very frequently. In contrast, solutions based on greedy heuristics are more attractive, because they can be obtained more efficiently and provide solutions that are close to optimal.

However, with greedy solutions, there does exist a possibility of not being able to generate a connected topology. To this end, our Link Selection algorithm applies a hybrid technique where we first apply a greedy-heuristics-based approach, and then, in case a connected network is not obtained with the greedy heuristics, we apply an enhanced minimum spanning tree algorithm to ensure that the computed network topology is fully connected. The Link Section algorithm is described next.

Link Selection Algorithm: Recall that the Link Selection algorithm is responsible for generating the Selected Link List (SLL) from the Feasible Link List (FLL). It is executed at the beginning of each time horizon T_h immediately following the FLL computation by performing the following steps:

Step 0: *Utility Calculation*: On the basis of the information in the FLL, the algorithm first computes a link score, or link utility, for each link i. A link with a higher link rate and/or higher predicted longevity is assigned a higher link utility. The basic link utility is computed using the formula

$$U_i = R_i \cdot T_i \tag{10.4}$$

In the above equation, R_i is the link rate and T_i is the predicted link longevity over the next time period. This utility function can be interpreted as the expected volume of data that can be transported over the link in the next time period. When the time window/horizon T_h encompasses Q time periods, the aggregate basic link utility function is given by

$$U_i = \sum_{j=1}^{Q} R_{i,j} T_{i,j} \tag{10.5}$$

where $R_{i,j}$ is the link rate and $T_{i,j}$ is the predicted link longevity for link i during time period j.

One of the issues with using the above utility function to drive the selection of links is that small changes in link rate, longevity, or confidence could lead to different utility values in different time periods, leading to an oscillating set of links and an unstable topology. It is therefore useful to make the system exhibit some hysteresis in order to provide a disincentive to make changes in the set of selected links across time periods. We therefore enhance the basic utility function as follows:

$$U_i = H_i \sum_{j=1}^{Q} R_{i,j} T_{i,j} \tag{10.6}$$

where H_i is a hysteresis multiplier that is set as follows. If a link i is not part of the initial topology, H_i is initialized to 1. During the next time period, if link i is not selected, H_i remains set to 1. If link i is selected in that time period, H_i is set to 2. The rationale for doing this is that once a link has been selected, we want to establish a disincentive for removing this link from the network topology. This is accomplished by increasing the utility of that link by doubling H_i, thereby doubling the utility of the link. This discourages frequent link switching in different periods, unless there is a substantial benefit in doing so.

Next, we modify the utility function in order to encourage the selection of links that improve the density of network connectivity. We do this by incorporating information about the degrees* of the source and destination of link i into the utility function as follows:

$$U_i = \frac{H_i \sum_{j=1}^{Q} R_{i,j} T_{i,j}}{1 + \alpha(d_{s(i)} + d_{d(i)})} \tag{10.7}$$

In Equation 10.7, we divide the numerator by a function of the platform degree at the source $s(i)$ and destination $d(i)$ of link i, to encourage connecting to platforms that have lower degrees, thereby promoting the density of the network connectivity. Equation 10.7 also includes a positive integer constant α to influence degree vs. rate tradeoff. The parameter α is set at the start of the mission and is the same across all platforms. A higher value of α means that the utility of a link will decay more rapidly as the degrees of the platforms at either end of the link increase, since α acts as a multiplier of these platforms' degrees in the denominator of the utility function shown in Equation 10.7.

Step 1: *Greedy Algorithm*: As its name indicates, the algorithm selects links in a greedy fashion, making the locally optimal choice at each stage of the algorithm. The lowest link identifier with the highest utility in the FLL is chosen first, and placed in the SLL after checking that it does not violate the maximum allowed platform degree for the two platforms at either end of the link. The resulting SLL contains all links thus selected to form a network. We note that while this greedy algorithm has low computational complexity, it is not guaranteed to always lead to a connected network due to the fact that locally optimal choices at each selection do not necessarily lead to a connected network topology. Hence the algorithm proceeds to Step #2 as described next.

Step 2: *Enhanced Spanning Tree*: In this step, the algorithm checks to see if the links in the SLL form a connected network topology. In a dense network scenario, with many potential links of high utility, the above greedy phase of the algorithm will, with a high likelihood, form a network topology in which all the UAVs are connected. If it does, the algorithm terminates. If not, the algorithm starts afresh with the original FLL, and an empty SLL. The Enhanced Spanning Tree phase consists of a Euclidean Spanning Tree phase, and an Enhancement phase, as follows:

Step 2a: In the Euclidean Spanning Tree phase, the algorithm first computes a Euclidean Minimum Spanning Tree among the platforms, based on their physical distances (based on GPS

* The degree of a platform is defined as the number of active links to or from that platform.

information). This Euclidean Minimum Spanning Tree is computed so that it satisfies the platform maximum degree constraints.

Step 2b: In the Enhancement phase, extra links are added to the SLL, for robustness and increased potential network performance. The algorithm greedily picks links from the remaining links on the FLL that have the highest link utility and do not violate maximum allowed platform degree constraints for the platforms at both ends of the link. This process continues until no further links are available on the FLL.

10.3.4 Performance Results

In this section, we compare the performance of a simulated directional airborne network with and without our distributed proactive topology control (DPTC) algorithms in place. In the baseline scenario, links are selected based only on knowledge of the currently available set of links. We compare this baseline scenario to one where we apply the DPTC algorithms described in the preceding sections for link selection. The measure of performance that we use is the Packet Delivery Ratio, or PDR, which is defined as the ratio of packets received to packets sent. While a variety of performance metrics exist, we have used PDR for illustrative purposes since it also serves as a proxy for other metrics in terms of qualitative network performance.

Figure 10.3 shows the performance comparison of the baseline scenario compared to one where we apply the DPTC algorithms, making use of flight orbit information to compute future link availabilities and selecting links based both on current and future availability over a time horizon T_h. A value of 60 s was used for T_h in our experiments. The traffic mix used for the simulations is the same as that described in Section 10.2.2, which consisted of (a) long-lived (400–700 s) medium-priority bulk traffic between G01, G02 and G04 via N01–N11, and (b) bursty short-lived (~3 s) high-priority traffic between the MANET segments (G02 and G03) via N12–N15.

The entire scenario lasted for 30 minutes and standard OLSR was used at the IP layer to establish communications paths. Line of sight links are established when possible by the link selection algorithm in both the baseline case as well as the case with DPTC based on aircraft geometries, constrained by any terrain obstructions (e.g., the mountainous regions depicted in Figure 10.2). Every time a link is established, a parameterized delay is introduced to account for potential physical layer delays on the acquisition of a new directional radio link due to the time taken by

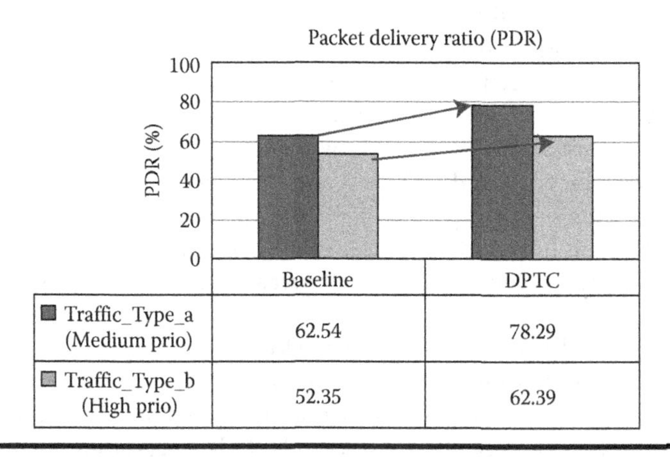

Figure 10.3 PDR performance without and with DPTC.

the Pointing–Acquisition–Tracking (PAT) process. For the simulation results presented in Figure 10.3, a PAT delay of 10 s was used.

Figure 10.3 shows the PDR for two different types of traffic, medium-priority and high-priority, for the baseline case and the case where DPTC is used. Note that there is a visible improvement in PDR for both types of traffic when DPTC is used. The improved PDR performance is due to the fact that there are more frequent network partitions (i.e., breaks in network connectivity) in the baseline case than with DPTC. Additionally, it is interesting to note that the high-priority traffic experiences worse performance than medium-priority traffic. This is due to the fact that high-priority traffic, which is exchanged between G02 and G03, is sent over a sparse tree topology in that part of the network segment.

Finally, while the use of DPTC provides improved PDR performance, we note that there is still significant room for improvement. This is because of the shortcomings of standard routing protocols when used in high-speed directional networks, which is discussed next in Section 10.4.

10.4 Proactive Route Injection Mechanism

Much work has been done in the area of routing for mobile ad hoc networks. The best-known routing protocol for MANETs is OLSR (Optimized Link State Routing) [5], which implements a version of OSPF (Open Shortest Path First) [6] optimized for MANETs. As a consequence, OLSR is a widely used routing mechanism in military MANETs. OLSR provides mechanisms to detect neighbors via HELLO and TC (Topology Control) broadcast messages, and to disseminate link state information throughout the network so that nodes can update and maintain their routing tables. However, a fundamental problem with OLSR is its convergence time, that is, the time taken to update and propagate routing changes when routes change. Based on our experience, we have observed that typical OLSR convergence times in tactical MANETs are in the order of several seconds. While such delays can be tolerated when routes change infrequently, they cause substantial disruption in the network when links are rapidly changing. In addition, when links offer high capacities in the gigabit/second range, outages of even a few seconds will result in the loss of gigabits of data. In order to address these issues, we propose a *Proactive Route Injection Mechanism* (*PRIM*) to make OLSR more suitable for use in directional networking environments.

Our proposed enhancement is to proactively inject routes into OLSR routing tables based on the output of the DPTC mechanism described in Section 10.3, thereby eliminating outages that arise due to routing convergence delays. The basic approach is to leverage knowledge about the links selected by DPTC for specific time windows, along with information about when a link is to be established at the radio layer (e.g., when a PAT trigger is issued by LS), to compute the new paths and inject them into the routers' forwarding table, so that these routes are available without having to wait for the OLSR HELLO and TC messaging to disseminate information about the new links. Thus, once the physical layer sets up a new link, the router will be ready to forward packets across this link, with no delays. OLSR will eventually discover the newly added links when its routing process converges, and will overwrite the paths that were injected into the routing table.

The PRIM process works as follows. Using knowledge about the selected link list (SLL) computed by the DPTC algorithm for a given time horizon, PRIM computes routing paths using these links, using the same path computation algorithm used by the OLSR process executing on the router. This can be viewed, in concept, as a second OLSR process running in the background, which is triggered when there is about to be a change in links. PRIM then injects these routes into the routers' forwarding tables immediately following the establishment of new links, so that the packets can be immediately sent over the newly established links. In the meantime, the primary OLSR process running on the platform eventually discovers the new links and computes forwarding paths using these links, and overwrites paths injected by PRIM. Figure 10.4 provides a summary of PRIM operation alongside the standard OLSR process running on the platform.

PRIM offers several advantages. First, it eliminates the packet losses that result due to routing convergence delays, thereby providing improved network performance by eliminating or minimizing outages resulting from new links being established and old links being destroyed or becoming unavailable. Next, the above is achieved without introducing additional overheads to achieve the increased performance. Third, and very importantly, it is compatible with the existing and popularly used proactive routing protocol (OLSR), thereby providing wide applicability to deployed platforms.

Next, we provide an evaluation of the performance of PRIM using the 20-node simulated mmW UAV network described in Section 10.2.2. In particular, to highlight the advantages obtained by using both DPTC and PRIM, Figure 10.5 shows PDR performance for the same traffic mix described in Section 10.2.2 with a PAT delay of 10 s for the following cases:

i. A baseline case that uses standard proactive routing (OLSR) and link selection using only currently active links (first column of results in Figure 10.5).
ii. An intermediate case that uses DPTC for performing link selection with standard OLSR (second column of results in Figure 10.5).
iii. A third case that uses DPTC for performing link selection and PRIM for proactive route injection (third column of results in Figure 10.5).

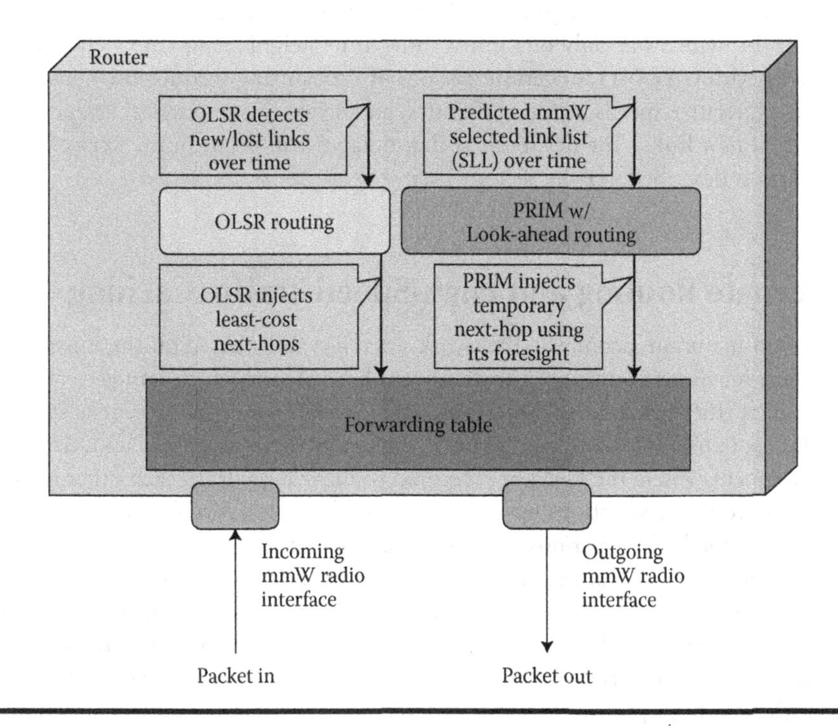

Figure 10.4 Summary of proactive routing injection mechanism.

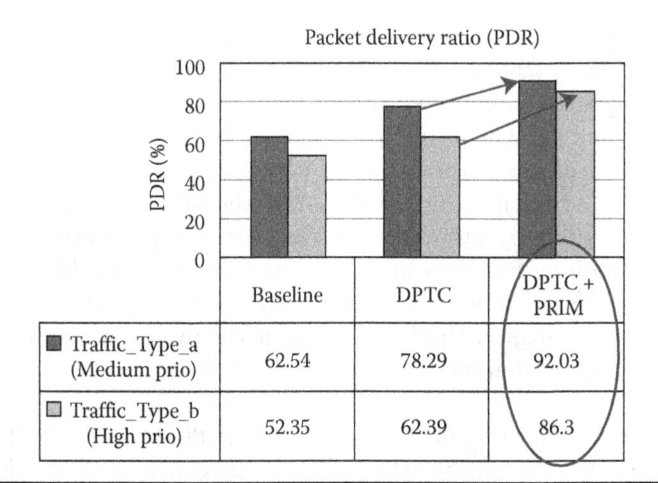

Figure 10.5 PDR performance results with the addition of PRIM.

Note that the first two cases are the same ones that were shown in Figure 10.3. We observe a significant improvement in the overall PDR for both types of traffic with the use of PRIM (92% vs. 78% for medium-priority traffic, and 86% vs. 62% for high-priority traffic). As explained while discussing Figure 10.3, the reason the high-priority traffic experiences worse performance than medium-priority traffic is because of the sparse topology available to the network in that part of the scenario. In fact, when the network topology is sparse, there arise situations where no links can be established across platforms. This happens in the network segment to the far right (tactical theatre portion of the network in Figure 10.2) where high-priority traffic is predominant. This is because UAV N14, which has only one mmW link to its neighboring UAV N15, may not have any connectivity with UAV N15 for certain periods of time during the mission, depending on the aircraft geometry, relative speeds between aircrafts, and delays due to physical layer constraints on the acquisition of new links. The net result is that outages will occur in the network, calling for other techniques as described next in Section 10.5.

10.5 Alternate Routing and High-Speed Traffic Caching

Despite efforts to maintain a connected network, establish links based on the mission plan, and proactively inject routes into routing tables, there may be occasions where connectivity is disrupted for various reasons. These include link intermittencies caused by, for example, atmospheric attenuation (e.g., clouds, rain), UAV banking and body frame obstructions, enemy jamming, or mission mobility requirements where the UAVs are required to move away from each other to accomplish their tasks. There could be situations where a UAV could become completely isolated (e.g., all its mmW links get jammed). Such periods of disconnectivity could result in the loss of large amounts of data that can no longer be transmitted across platforms. First, we propose an alternate routing and high-speed traffic caching approach to handle such situations. Briefly, alternate routing via the discovery radios' (DRs) omni-directional control channel can provide a way to deliver mission-critical traffic even when directional links are unavailable. Second, selective caching enables local storage of a subset of high-priority traffic on a UAV, so that transmission can be resumed when links become available. In such situations, it is inevitable that some traffic will have to be dropped;

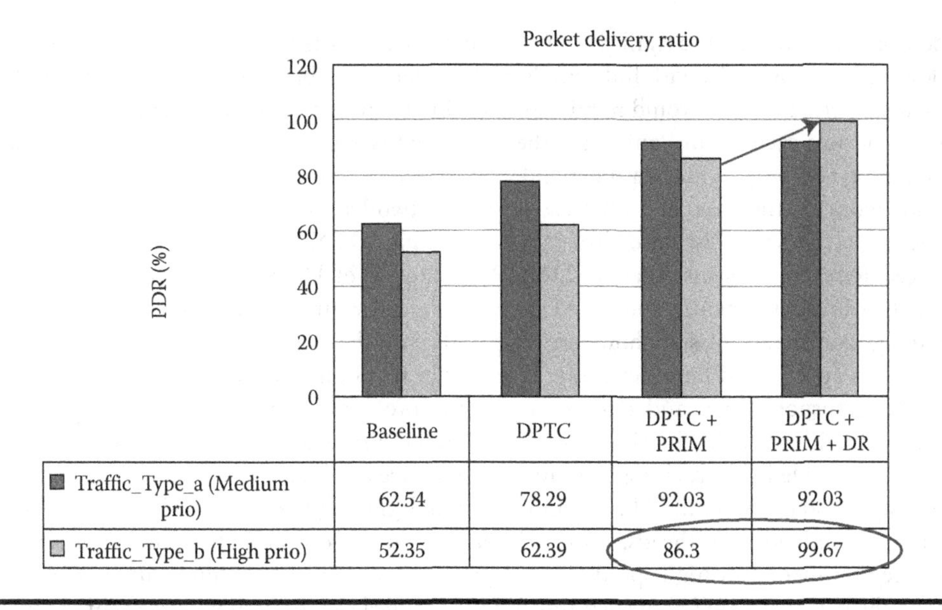

	Baseline	DPTC	DPTC + PRIM	DPTC + PRIM + DR
Traffic_Type_a (Medium prio)	62.54	78.29	92.03	92.03
Traffic_Type_b (High prio)	52.35	62.39	86.3	99.67

Figure 10.6 PDR performance results with the addition of discovery radio routing.

the approach is to drop the low-priority traffic during mmW link outages. In the remainder of this section, we describe both of the above capabilities.

10.5.1 Alternate Routing for Traffic Delivery

In the absence of mmW links, we leverage the on-board discovery radios' control channels to route the subset of traffic that is mission-critical. The discovery radio control channel is typically a low-bandwidth long-range omni-directional channel used by the on-board IMU and GPS. In order to not overload this control channel, we use it to transmit only mission-critical messages. When an outgoing packet does not have an mmW interface/link to be routed over, it is diverted to the discovery radio interface if its DSCP [7] marking indicates that it is mission-critical traffic. This enables delivery of mission-critical traffic over the control channel. Handling of other, non-mission-critical traffic during such situations is described in Section 10.5.2.

In Figure 10.6, we show the added performance benefits of incorporating discovery radio routing together with PRIM to provide sustained performance for mission-critical traffic. In particular, Figure 10.6 shows four sets of results, with the results in the first three columns being the same as in Figure 10.5. The results in the fourth column show PDR numbers when alternate routing over the discovery radio is used.

10.5.2 High-Speed Selective Traffic Caching

As mentioned in Section 10.5.1, due to bandwidth limitations of the control channel, it is not feasible to use it for anything but a small volume of high-priority mission-critical traffic. However, most military missions have other high-priority traffic that may not be mission-critical, but that needs to be delivered in a reliable fashion even if some delay is incurred while sending it. In order to accomplish this, we propose the use of a selective caching mechanism that makes use of temporary storage for storing such traffic locally on a platform when the platform has no connectivity to the destination platform. The traffic is stored until connectivity is restored, and is then delivered to the destination. Obviously, with link speeds in the order of gigabits per second, caching traffic for a prolonged period of time would require an inordinate amount of storage capacity. We therefore perform selective caching and only store the traffic that is appropriately marked, while dropping all remaining outgoing traffic, until network connectivity is restored.

Prior work in the area of traffic caching in networks with intermittent link availability includes research on the Disruption-Tolerant Networking (DTN) paradigm developed by Defense Advanced Research Projects Agency (DARPA). The goal of DTN is to provide the ability to deliver data in intermittently connected networks by providing services, including in-network data storage and retransmission; however, the paradigm suffers from several drawbacks that make it difficult to apply to military networks. First, DTN relies on a Bundle Layer [8,9] that resides above the Transport Layer and requires an overlay protocol header, which therefore makes it incompatible with the HAIPE (High Assurance Internet Protocol Encryptor) architecture. While HAIPE-compatible DTN caching is beginning to receive attention [10,11], design efforts are still at the research stage. Second, DTN requires sending signaling messages in the overlay network, resulting in bandwidth overheads. Third, DTN introduces delays due to its need for signaling, which are detrimental in high-speed networks such as those utilizing mmW links. Figure 10.7 shows a schematic of the DTN mechanism and the bundle layer, for temporarily caching information at an intermediate node between a sender and a receiver.

Our high-speed selective traffic caching (HSTC) mechanism overcomes the drawbacks associated with DTN, and works as follows. When an IP packet arrives at a router, the router determines whether or not a route exists to the packet destination. If a route exists, the packet is forwarded to the outgoing router interface. If no path exists, the packet's DSCP [7] is checked. If the DSCP indicates high-priority, non-mission-critical traffic, it is diverted to a local cache that is located at the network layer; else it is discarded. No signaling is required across nodes, as packets are simply cached locally when they cannot be transmitted.

HSTC exploits unused memory on the router to cache outgoing traffic temporarily. Doing so provides a two-fold advantage in terms of speed:

i. The IP packet structure is maintained intact, thus eliminating any processing that might be required to extract the packet payload, store it, and create a new IP packet for transmission when connectivity is restored at a later point in time.
ii. Packets are saved locally within the router, rather than being moved out of the router and being brought back in when connectivity is restored. Thus, whenever connectivity to the destination is re-established, the cached packets can be immediately forwarded to their destination. Figure 10.8 provides a schematic of the HSTC mechanism as instantiated on a UAV.

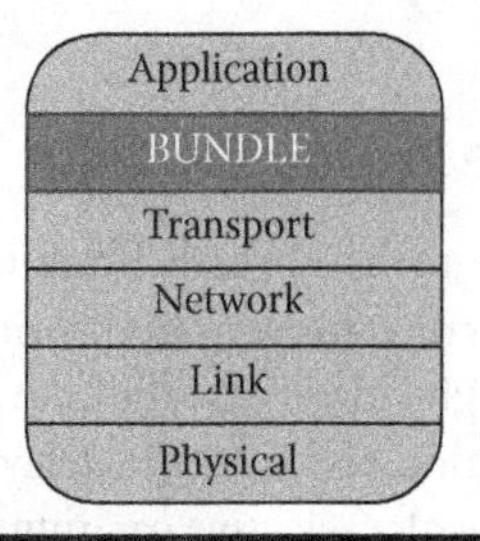

Figure 10.7 Schematic of DTN bundle layer.

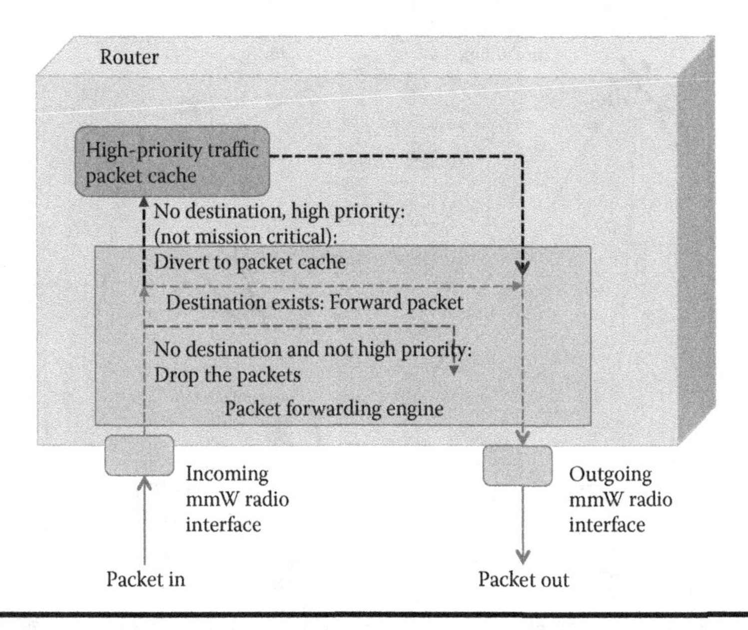

Figure 10.8 Schematic of high-speed selective traffic caching mechanism.

A significant benefit of HSTC is that it provides compatibility with HAIPE. This is because caching decisions are made solely on the basis of route availability and the DSCP marking of an IP packet. The DSCP marking of an IP packet is copied over to the encapsulating IP header upon encryption, making it observable for decision-making on the encrypted (or "black") side of the network on all transit routers. HSTC is straightforward to use and configure; DSCP packet marking is available on any IP router, and packet marking can be configured prior to the start of a mission and dynamically changed during a mission to reflect the priorities of different types of traffic for specific missions.

Figure 10.9 provides a schematic of how HSTC can be used in a HAIPE-compliant manner in an mmW network.

Figure 10.9 illustrates a scenario with five nodes: three UAVs (UAV1–UAV3) and two Ground Nodes (Gnd 1 and Gnd2) interconnected via mmW links. We consider sample application flows initiated from Gnd 1 destined to Gnd2 via the three UAVs.

On the top in Figure 10.9 are the UAVs with their flight orbits. On the bottom of the figure we show the portions of the protocol stack that the application traffic has to traverse, starting at the source (Gnd1), through the three intermediate UAVs (used as transit nodes) and finally at the destination (Gnd2). Also shown in Figure 10.9 are the HSTC instances on each of the five nodes, and how they function on each node. In the example scenario in Figure 10.9, the link between UAV2 and UAV3 suffers from fluctuations. To tide over the fluctuations, the HSTC mechanism on UAV2 rapidly caches the high-priority traffic locally and delivers it when the link comes back up. Note that none of the transit packets have to traverse any of the upper layers of the stack beyond the IP layer, thus ensuring HAIPE compatibility.

In addition to the above benefits, another benefit of HSTC is that it does not require special hardware, as packets can be cached on the local router. Based on our prior experience, we have observed that sufficient memory, in the range of ~1 GB, can be harnessed on standard router configurations. For example, we have demonstrated our HSTC mechanism using standard routers

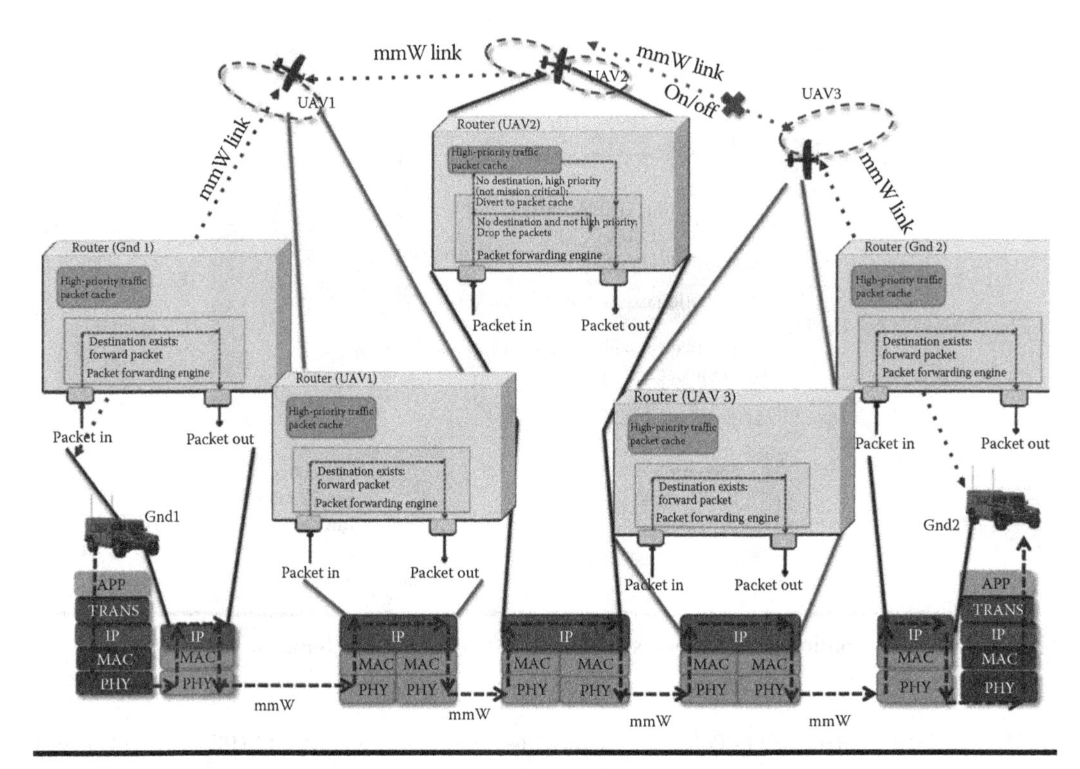

Figure 10.9 Schematic of HSTC and application traffic flows in an mmW network.

(Lanner FW-8760) on a 5-node testbed consisting of emulated 1 Gb/s directional mmW links [12]. While we were able to harness 946 MB of unused memory, we found that memory sizes of under 600 MB were sufficient to yield good performance for high-priority traffic, even with 1 Gb/s links as described next.

Figure 10.10 shows our testbed scenario consisting of five nodes employing standard routers (Lanner FS 8760) as follows: U_1, U_2, U_3 were used to represent three UAVs and G_1 and G_2 represented two remote ground nodes, interconnected using emulated 1 Gb/s mmW links. Also shown in Figure 10.10 are three sample trajectories that were used for purposes of illustration; one UA (U_1) flew a spindle-shaped trajectory, another UA (U_2) flew a figure-eight trajectory, and the third UA (U_3) flew an elliptical flight trajectory.

The traffic used in our experiment consisted of

a. Imaging traffic (representing traffic from on-board ISR cameras) generated by the UAVs and sent to the two remote ground nodes (shown via thin dashed double arrows).
b. Exchange of short messages between the remote ground nodes (denoting short confirmatory messaging regarding the course of action to take based on the images transmitted by the UAVs, shown via a thick dashed stream in the figure).
c. Random background traffic across all nodes (random traffic is not shown in Figure 10.10).

Of the above three types of traffic, (a) and (b) were higher priority traffic, and were candidates for HSTC. The mission duration in this study was 1800 s. Link intermittencies to denote the effects of jamming, UA banking, and body frame obstructions, were captured by varying the connectivity between the nodes during the course of the mission.

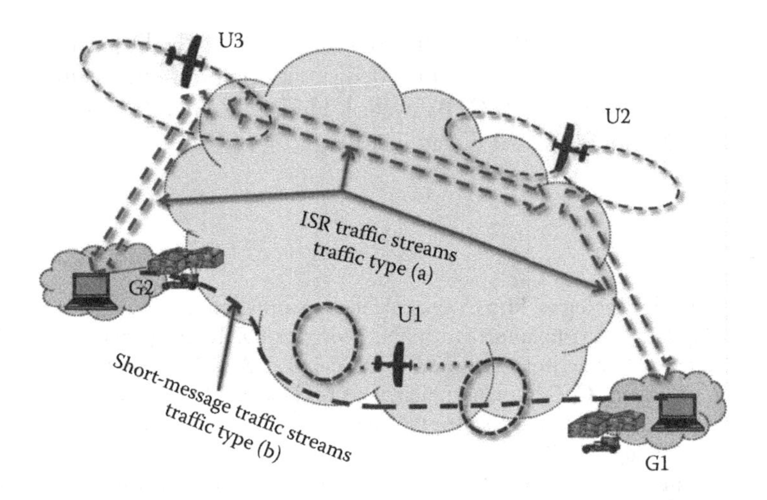

Figure 10.10 Schematic of testbed scenario.

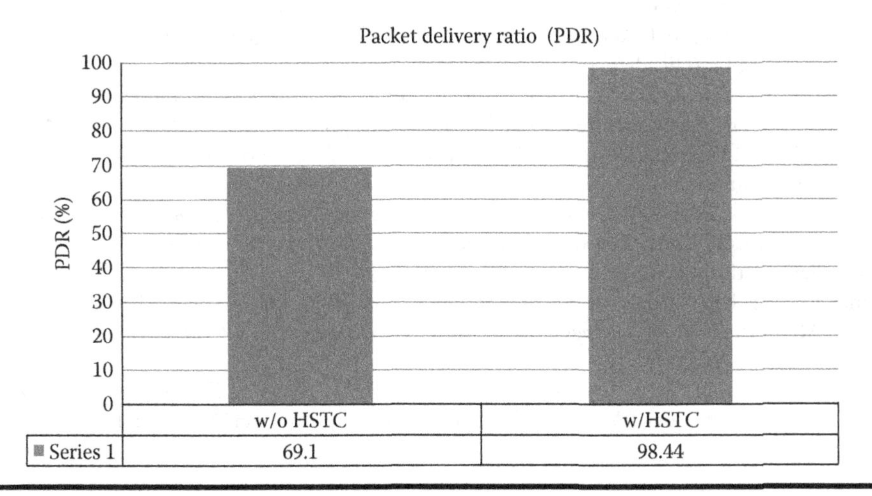

Figure 10.11 Packet delivery ratio with and without HSTC.

We collected PDR statistics and compared performance without and with HSTC, as shown in Figure 10.11. The figure shows PDR performance for traffic (a) and (b) with and without HSTC. We observe that the use of HSTC provides a substantial performance improvement; the PDR with HSTC is 98.44%, whereas it is only 69.1% without HSTC. The use of HSTC results in a ~30% improvement in the PDR for mission-critical traffic for this scenario.

Acknowledgments

The material discussed in this chapter is based upon work supported by the Defense Advanced Research Projects Agency under Contract No. HR0011-13-C-0019. The views, opinions, and/or findings contained in this article/presentation are those of the author(s)/presenter(s) and should not be interpreted as representing the official views or policies of the Department of Defense or the U.S. Government. The authors also acknowledge with gratitude the teamwork and helpful discussions with: T. K. Woodward, T. C. Banwell, J. Hodge, Richard C. Lau, A. J. McAuley, H. Kim, G.-K. Kim, K, Manousakis, and E. van den Berg, from ACS.

References

1. Scalable Network Technologies, http://web.scalable-networks.com/content/qualnet.
2. Gelal, E., G. Jakllari, S. Krishnamurthy, and N. Young. 2009. Topology management in directional antenna-equipped ad hoc networks. In *IEEE Transactions on Mobile Computing*, 8, 590–605.
3. Huang, Z., C.-C. Shen, C. Srisathapornphat, and C. Jaikaeo, 2002. Topology control for ad hoc networks with directional antennas. In *Proceedings of Computer Communications and Networks Conference*, October, pp. 16–21.
4. Jia, X., D. Li, and D. Du, 2004. QoS topology control in ad hoc wireless networks. In *IEEE INFOCOM*, pp. 1264–1272.
5. Clausen, T. and P. Jacquet, 2003. Optimized link state routing protocol (OLSR). In *IETF RFC* 3626, October.
6. Moy, J., 1998. Open shortest path first. In *IETF RFC* 2328, April.
7. Nichols, K., S. Blake, F. Baker, and D. Black, 1998. Definition of the differentiated services (DS) field in the IPv4 and IPv6 headers. In *IETF RFC* 2474, December.
8. Cerf, V., S. Burleigh, A. Hooke, L. Torgerson, R. Durst, K. Scott, K. Fall, and H. Weiss. 2007. Delay-tolerant networking architecture. In *IETF RFC* 4838, April.
9. Fall, K., 2008. DTN: An architectural retrospective. In *IEEE Journal on Selected Areas in Communications*, 26(5), June.
10. Perloff, M., *Cross Layer Cache and Queue Management for Resilient Tactical Networks (CLCQMRTN)*. Scientific Systems Inc. http://sbirsource.com/sbir/awards/63260-cross-layer-cache-and-queue-management-for-resilient-tactical-networks-clcqmrtn#
11. Elmasry, G., J. Lee, M. Jain, S. Snyder, and J. Santos, 2009. ECN-based MBAC algorithm for use over HAIPE. In *IEEE Proceedings of IEEE Military Communications Conference (MILCOM)*, Boston, MA, November 2013.
12. Woodward, T. et al., 2013. *Tactical Overlay RF Networks Addressing Data Operations (TORNADO): Architecture Report*. DARPA Mobile Hotspots Program, Contract Number: HR0011-13-C-0019, December 2013.

MIMO

IV

Chapter 11

Design and Implementation of Directional Antenna-Based LOS–MIMO System for Gbps Wireless Backhaul

Xiang Chen, Xiujun Zhang, Shidong Zhou,
Ming Zhao, and Jing Wang

Contents

Abstract

The radio access network (RAN) is evolving to accommodate rapid growth in data services and steady growth in voice traffic and to prepare for the next generation of services. Wireless backhaul is that part of the network that carries voice and data traffic in the RAN from the mobile base station to the mobile operators' core network. Since most infrastructures were initially designed for second-generation (2G) networks to carry voice traffic, most consist of time division multiplexing (TDM) and point-to-point microwave links. As more 3G and 4G services are offered, there is a need of high spectrum-efficient transmission technologies over microwave bands to support increasing backhaul capacity for 3G/4G networks. In this chapter, a directional antenna-based multiple-input multiple-output (MIMO) solution is proposed to guarantee full multiplexing gain for the line-of-sight (LOS) wireless backhaul environments at higher microwave bands. The antenna array design rule for such LOS–MIMO systems with one- or two-dimensional directional antenna elements in LOS scenarios are derived, and the strict perpendicular constraint is released in the two-dimensional case. The minimum antenna array area and the performance sensitivity to the area error are also obtained to guide the practical system design. Then, a demo MIMO-orthogonal frequency division multiplexing (OFDM) system with the designed square antenna array at 15 GHz carrier is implemented on a Gigabit Ethernet (GE) switch based software defined radio (SDR) platform, which combines the hardware accelerating units (HAUs) with the general-purpose processors (GPPs). The field evaluation results show that the system throughput and spectrum efficiency are greater than 1 Gbps and 15 bps/Hz, respectively. To the best of our knowledge, it is the first time to demonstrate the Gigabits per second LOS–MIMO–OFDM system at such microwave bands in the world, which can be a successful design example for the next generation wireless backhaul.

11.1 Introduction

With the rapid development of modern cellular network and the explosion of mobile Internet, the next generation cellular network is defined to provide higher user data rate than before, which requires the capacity of the backhaul network to improve accordingly, including the wireless backhaul in cases where it is hard to find the fiber. Traditional wireless backhaul is point to point with Single Input Single Output (SISO) antenna at microwave frequency. Cross polarization interference cancelation (XPIC) is often used to double the capacity with multiplexing gain of 2, but still not enough to catch up with next generation backhaul.

There are some possible solutions to the problem. One is to use wider bandwidth at even higher frequency band [1]. Though the throughput of such system can be times of traditional microwave system, the modulation order is often limited due to the enlargement of phase noise. In Reference 2, only Quadrature Phase Shift Keying (QPSK) is used for 60 GHz frequency, which results in low-spectrum efficiency. Another way is to increase the multiplexing gain by MIMO technology including antenna polarization [3,4]. Theoretically, the capacity can be increased unlimitedly with enough antennas when the channel state information is known at the transmitter, but it should be noted that the dimension of the antenna array will be a big problem for large antenna arrays. Some previous work [5–7] used these two technologies together, MIMO at millimeter wave frequency, to get both the multiplexing gain and wider bandwidth, but the spectrum efficiency is still low compared with MIMO system at microwave frequency.

In wireless backhaul, the channel is quasi-static LOS. Some previous works have also utilized antenna array design for LOS–MIMO. However, Reference 8 analyzed the optimal condition for uniform linear array (ULA) and a 40 GHz LOS–MIMO system is presented in Reference 9 using ULA. The antenna arrays are extended to square and rectangle in References 10 and 11, but the two axis for the array should be perpendicular. This chapter gives a more general rule to achieve optimal channel matrix (or maximum channel capacity) for one- and two-dimensional antenna arrays without perpendicular requirement. Further, the minimum area of antenna arrays is derived for optimal LOS–MIMO channel design based on our optimal rule. Further on, the impact of the array area error on channel capacity and channel matrix's condition number is analyzed to give guidelines for practical system design.

Based on the derived optimal condition of antenna array for LOS–MIMO, we design a high capacity fixed radio system using square antenna array symmetrical at both transmitter (T_x) and receiver (R_x). MIMO–OFDM—one of the most widely used technologies—is used as the basic transmission scheme, and the relatively more severe phase noise at microwave band is considered in channel estimation.

Usually, high capacity MIMO system needs complex signal processing which prefer a flexible, adaptable, programmable, and powerful SDR platform [12]. GPP-based platforms [13–15] have great adaptability and programmability, but the throughput is often limited due to the lack of real-time signal processing capacity. In contrast, SDR platforms based on HAUs, for example, field programmable gate arrays (FPGAs), can provide strong ability for real-time signal processing but with relatively poor programmability. In this chapter, a GE switch-based SDR platform for Gigabits per second radio system is presented, which takes the advantages from both HAUs and GPPs to achieve good real-time processing performance with good adaptability and flexibility. Further, the HAU resources can be flexibly scheduled by choosing MIMO–OFDM as transmission scheme and GE as platform interface.

The field evaluation verifies our design rule for LOS–MIMO antenna array, with the performance of up to Gigabits per second throughput and 15 bps/Hz spectrum efficiency. And the spectrum efficiency can be doubled to 30 bps/Hz by cross polarized antenna array without dimension increasing.

The remainder of this chapter is organized as follows. We provide the antenna array design rule for LOS–MIMO and sensitivity analysis in Section 11.2. The MIMO–OFDM based LOS–MIMO Gigabits per second radio system is presented in Section 11.3. A GE switch-based SDR platform and the implementation of the proposed Gigabits per second system are given in Section 11.4. Field evaluation results are shown in Section 11.5. Finally, conclusions and discussions for future work are drawn in Section 11.6.

11.2 Antenna Array Design

11.2.1 Channel Model and Capacity for LOS–MIMO

Consider a quasi-static LOS–MIMO channel with both antenna arrays at T_x and R_x side with the origins at (0, 0, 0) and $\mathbf{R}_0 = (0, 0, L)$ as shown in Figures 11.1 and 11.2. The coordinates for N T_x antennas and M R_x antennas are $\mathbf{t}_i = (t_{i,x}, t_{i,y}, t_{i,z})$, $i = 1, \ldots, N$ and $\mathbf{r}_i + R_0 = (r_{i,x}, r_{i,y}, r_{i,z} + L)$, $i = 1, \ldots, M$, where L is the distance between T_x and R_x. The distance for antenna pair (i, j) is $d_{ij} = \|\mathbf{t}_i - \mathbf{r}_j - \mathbf{R}_0\|$. It is reasonable to assume the distance between intraarray elements is much smaller than that between T_x and R_x array, so we have $d_{ij} \approx L$.

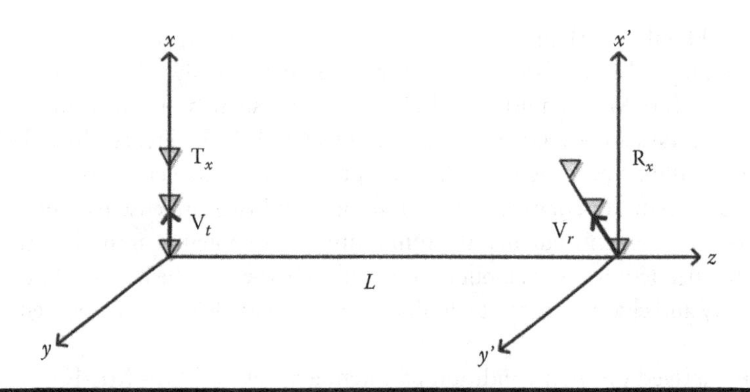

Figure 11.1 Linear antenna array with equal spacing.

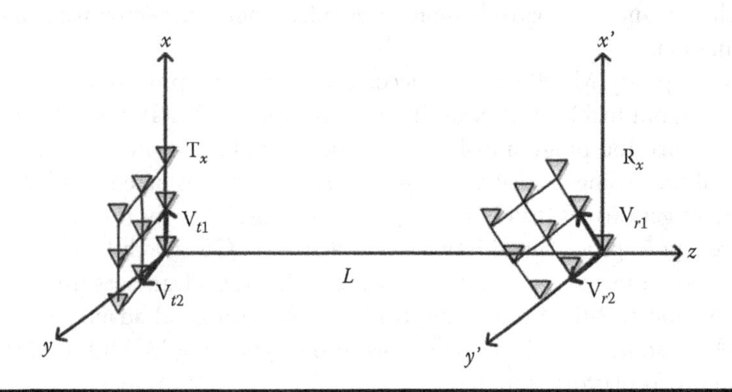

Figure 11.2 Two-dimensional antenna array.

When pure LOS is considered, the channel from the i_{th} T_x antenna to the j_{th} R_x one can be written as

$$H_{ij} = D(d_{ij})\exp\left(-j\frac{2\pi}{\lambda}d_{ij}\right) \approx D(L)\exp\left(-j\frac{2\pi}{\lambda}d_{ij}\right), \tag{11.1}$$

where $D(L)$ is the attenuation factor at transmission distance L. For a certain distance L, $D(L)$ is a constant for all antenna pairs, so it is omitted for the following capacity analysis.

Assuming $M \geq N$, the MIMO capacity under flat fading can be expressed as Reference 16

$$C = \sum_{i=1}^{N} \log_2(1 + \lambda_i SNR), \tag{11.2}$$

where $\lambda_i = eig(\mathbf{H}^H \mathbf{H})$ is the eigenvalue of $\mathbf{H}^H \mathbf{H}$.

To optimize the channel capacity, the optimal channel matrix property is studied here. Let $\mathbf{W} = \mathbf{H}^H \mathbf{H}$, then:

$$W_{kn} = \sum_{m=1}^{M} H_{mk}^{H} H_{mm} = \sum_{m=1}^{M} \exp\left(j\frac{2\pi}{\lambda}(d_{mk} - d_{mn}) \right)$$

$$= \sum_{m=1}^{M} \exp\left(j\frac{2\pi}{\lambda}\frac{d_{mk}^2 - d_{mn}^2}{d_{mk} + d_{mn}} \right),$$

$$= \sum_{m=1}^{M} \exp\left(j\frac{2\pi}{\lambda}\frac{\langle \mathbf{t}_k, \mathbf{t}_k \rangle - \langle \mathbf{t}_n, \mathbf{t}_n \rangle + 2\langle \mathbf{r}_m + \mathbf{R}_0, \mathbf{t}_n - \mathbf{t}_k \rangle}{d_{mk} + d_{mn}} \right) \tag{11.3}$$

where $\langle \cdot,\cdot \rangle$ denotes for the inner product of two vectors. Assuming T_x antenna array is in $x-y$ plane as depicted in Figures 11.1 and 11.2, $t_{i,z} = 0$, and $d_{ij} = L$, so we have

$$W_{kn} \approx \exp\left(j\pi\frac{\langle \mathbf{t}_k, \mathbf{t}_k \rangle - \langle \mathbf{t}_n, \mathbf{t}_n \rangle}{\lambda L} \right) \sum_{m=1}^{m} \exp\left(j2\pi\frac{\langle \mathbf{r}_m, \mathbf{t}_n - \mathbf{t}_k \rangle}{\lambda L} \right). \tag{11.4}$$

The conditions for the channel to achieve its maximum are as

$$C = C_{\max} \Leftrightarrow \forall k \neq n, W_{kn} = 0 \Leftrightarrow \forall k \neq n, \sum_{m=1}^{M} \exp\left(j\pi\frac{\langle \mathbf{r}_m, \mathbf{t}_n - \mathbf{t}_k \rangle}{\lambda L} \right) = 0. \tag{11.5}$$

Equation 11.5 shows that, to make the T_x and R_x array be parallel and perpendicular to the transmission direction (axis *z*) can reduce the size of antenna array, but it is not a necessary condition to achieve optimal capacity.

11.2.2 One-Dimensional Linear Array

The one-dimensional linear array case is shown in Figure 11.1 with equal spacing at both T_x and R_x side, where $\mathbf{v}_t(\mathbf{v}_r)$ is the $T_x(R_X)$ vector with amplitude of $T_x(R_x)$ array spacing and orientation of $T_x(R_x)$ array direction. Then, we have $\mathbf{t}_n - \mathbf{t}_k = (n-k)\mathbf{v}_t$, and Equation 11.5 is reduced to:

$$\forall n \in \{\pm 1, \ldots, \pm N-1\}, \sum_{m=1}^{M} \exp\left(j2\pi n\frac{\langle \mathbf{r}_m, \mathbf{v}_t \rangle}{\lambda L} \right) = 0 \tag{11.6}$$

One sufficient condition for Equation 11.6 can be found as:

$$\frac{\langle \mathbf{r}_m, \mathbf{v}_t \rangle}{\lambda L} = \frac{(m-1)p}{N}, \tag{11.7}$$

where $\mathbf{gcd}(p, N) = 1, M = kN, p, k \in \mathbb{Z}^+$

With the validity of Equation 11.7, the sum in Equation 11.6 is the sum of all the $(N/\mathbf{gcd}(n,N))$-th complex root of 1, which should be 0. Here, $\mathbf{gcd}(\cdot,\cdot)$ denotes for the greatest common divisor for two positive integers.

Now, let us consider the most compact linear array design with MIMO multiplexing gain of N. Let $p = 1$ and $M = N$, then we get $\langle \mathbf{v}_r, \mathbf{v}_t \rangle = \lambda L/N$. It is obvious to set the two arrays parallel to make the minimum antenna separation product (ASP) [4] $\|\mathbf{v}_r\|.\|\mathbf{v}_t\| = \lambda L/N$. Let R_t and R_r are the lengths of antenna arrays at T_x and R_x, then we have the optimal condition:

$$R_t R_r = \frac{(N-1)^2}{N} \lambda L \tag{11.8}$$

In case of equal size at T_x and R_x, from Equation 11.8 we get the optimal array length as:

$$R_t = R_r = (N-1)\sqrt{\frac{\lambda L}{N}} \tag{11.9}$$

11.2.3 Two-Dimensional Array

The two-dimensional antenna array case is shown in Figure 11.2, where T_x array is placed at $x-y$ plane with $N_1 \times N_2$ elements and R_x array at $x'-y'$ plane with $M_1 \times M_2$ elements. And the array elements are renumbered as:

$$\begin{aligned} \mathbf{t}_{n_1,n_1} &= (n_1-1)\mathbf{v}_{t_1} + (n_2-1)\mathbf{v}_{t_2} \\ \mathbf{r}_{m_1,m_2} &= (m_1-1)\mathbf{v}_{r_1} + (m_2-1)\mathbf{v}_{r_2}, \end{aligned} \tag{11.10}$$

where $\mathbf{v}_{t_1}$, $\mathbf{v}_{t_2}$ and $\mathbf{v}_{r_1}$, $\mathbf{v}_{r_2}$ are the two-dimensional extensions of one-dimensional $\mathbf{v}_t$, $\mathbf{v}_r$ with the similar definition. Note that $\mathbf{t}_{n_1,n_2} - \mathbf{t}_{k_1,k_2} = (n_1-k_1)\mathbf{v}_{t_1} + (n_2-k_2)\mathbf{v}_{t_2}$, Equation 11.5 can be rewritten as:

$$\begin{aligned} &\forall n_i \in \{0,\pm 1,\ldots,\pm N_i - 1\}, i = 1,2 \quad \text{and} \quad n_1 n_2 \neq 0, \\ &\sum_{m_1=0}^{M_1-1} \exp\left(j2\pi n_1 m_1 \frac{\langle \mathbf{v}_{t_1}, \mathbf{v}_{r_1}\rangle}{\lambda L}\right) \exp\left(j2\pi n_2 m_1 \frac{\langle \mathbf{v}_{t-1}, \mathbf{v}_{r_2}\rangle}{\lambda L}\right). \\ &\times \sum_{m_2=0}^{M_2-1} \exp\left(j2\pi n_1 m_2 \frac{\mathbf{v}_{t_2}, \mathbf{v}_{r_1}}{\lambda L}\right) \exp\left(j2\pi n_2 m_2 \frac{\mathbf{v}_{t_2}, \mathbf{v}_{r_2}}{\lambda L}\right) = 0 \end{aligned} \tag{11.11}$$

Set $n_1 = 0$, $n_2 = 0$ respectively, then we get the necessary conditions:

$$\begin{aligned} &\forall n_i \in \{1,\ldots,N_i - 1\}, i = 1,2 \\ &\sum_{m_1=0}^{M_i-1} \exp\left(j2\pi n_i m_1 \frac{\langle \mathbf{v}_{t_1}, \mathbf{v}_{r_i}\rangle}{\lambda L}\right) \sum_{m_2=0}^{M_2-1} \exp\left(j2\pi n_i m_2 \frac{\langle \mathbf{v}_{t_2}, \mathbf{v}_{r_i}\rangle}{\lambda L}\right) = 0. \end{aligned} \tag{11.12}$$

Table 11.1 Minimum Array Area for Different Configurations

Antenna array	2×2	2×3	3×3	2×5	2×7	3×5	3×7	2×11	5×5
Minimum area (m²)	20	32.7	53.3	50.6	64.1	82.6	104.7	85.3	128

When $\langle \mathbf{v}_{t_1}, \mathbf{v}_{r_2}\rangle = \langle \mathbf{v}_{t_2}, \mathbf{v}_{r_1}\rangle = 0$, $\langle \mathbf{v}_{t_2}, \mathbf{v}_{r_2}\rangle\langle \mathbf{v}_{t_1}, \mathbf{v}_{r_1}\rangle \neq 0$ (or $\langle \mathbf{v}_{t_2}, \mathbf{v}_{r_2}\rangle = \langle \mathbf{v}_{t_1}, \mathbf{v}_{r_1}\rangle = 0$, $\langle \mathbf{v}_{t_1}, \mathbf{v}_{r_1}\rangle \langle \mathbf{v}_{t_2}, \mathbf{v}_{r_1}\rangle \neq 0$), Equation 11.12 can be further simplified to

$$\forall n_i \in \{1,\ldots,N_i-1\}, \sum_{m_i=0}^{M_i-1} \exp\left(j2\pi n_i m_i \frac{\langle \mathbf{v}_{ti}, \mathbf{v}_{ri}\rangle}{\lambda L} \right) = 0, \quad i = 1,2. \tag{11.13}$$

Equation 11.13 is similar to the one-dimensional case. The one sufficient condition for two-dimensional array case can be concluded as

1. The first dimension of T_x/R_x and second dimension of T_x/R_x satisfy optimal linear array condition, respectively.
2. The first dimension of T_x is orthogonal to the second dimension of R_x, and so as the second dimension of T_x and the first of R_x.

Considering the most compact array design and equal number of elements at T_x and R_x, $p_i = 1$, $M_i = N_i$, we observe that the array area satisfies:

$$\frac{S_t S_r}{\lambda L} = \frac{(N_1-1)^2(N_2-1)^2}{N_1 N_2}, \tag{11.14}$$

where S_t and S_r are the array area of T_x and R_x, respectively. In case of symmetrical array at T_x and R_x, the array area yields to $S_t = S_r = \left((N_1-1)(N_2-1)/\sqrt{N_1 N_2}\right)\lambda L$.

From Equation 11.14, the multiplexing gain is $S_t S_r/(\lambda L)^2$ at large N_i, which is in accordance with the result in Reference 10. The result here is more general than in Reference 10 in that there is no perpendicular requirement for the two-dimensional array itself, so it leaves the flexible shape choice, from any parallelogram, to the system designer.

As a practical example, Table 11.1 shows the minimum array area for some typical array configurations when symmetrical array is used. The frequency and transmission distance are 15 GHz and 2 km, respectively.

11.2.4 Sensitivity Analysis for Array Area

From previous sections we know the optimal parameters, but it is reasonable to encounter some deviations from installation and manufacturing in practice, especially for the antenna array area. It is necessary to do the sensitivity analysis to see the robustness of the derived rule. For high capacity fixed wireless access application, we generally focus on high Signal to Noise Ratio (SNR) regime. Different SNR conditions are compared in the sensitivity analysis of array area.

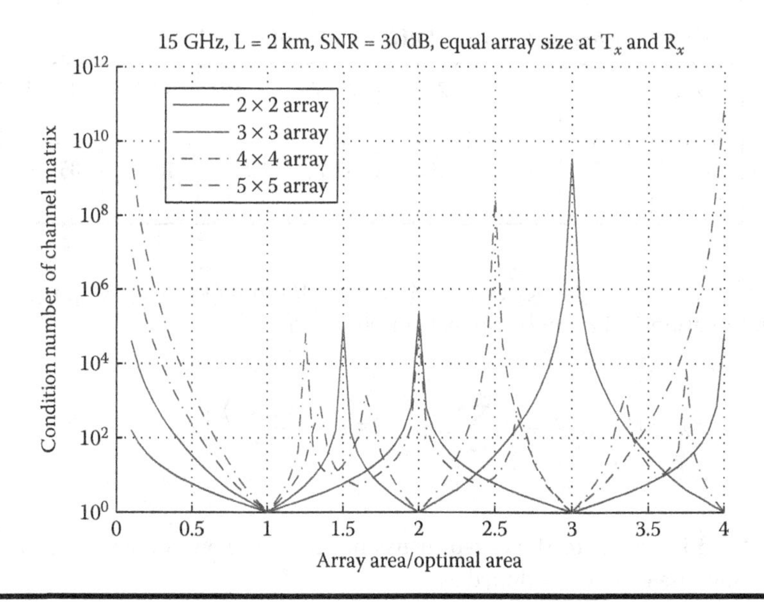

Figure 11.3 Channel condition number vs. antenna array area.

As for performance measurement, the condition number of channel matrix and channel capacity are adopted. The condition number of channel matrix **H** is denoted as cond(**H**), which is the ratio of the maximum and minimum singular values of **H**. In LOS–MIMO case, the optimal channel corresponds to cond(**H**) = 1, and large condition number leads to lower-channel capacity.

The simulation results of sensitivity analysis for condition number and capacity due to array area deviation are shown in Figures 11.3 and 11.4 under the conditions of 15 GHz frequency

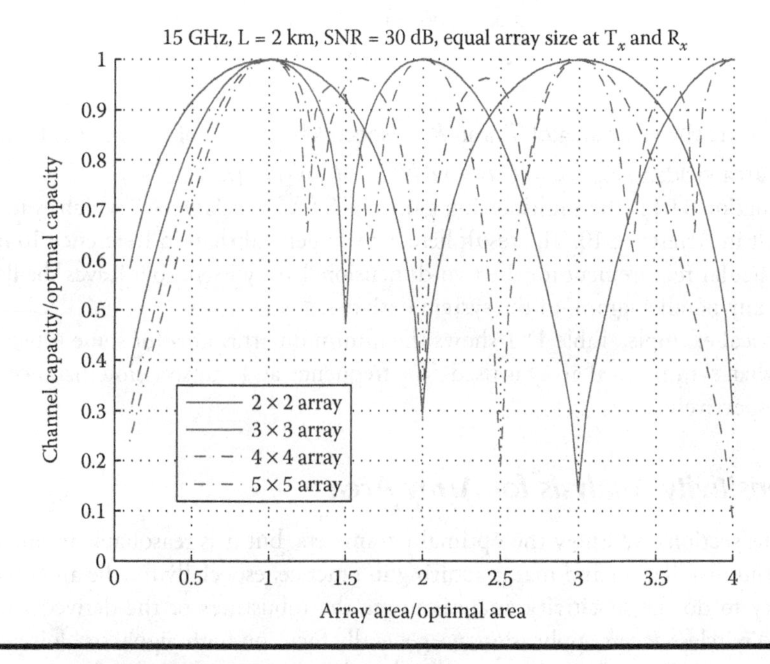

Figure 11.4 Channel capacity vs. antenna array area.

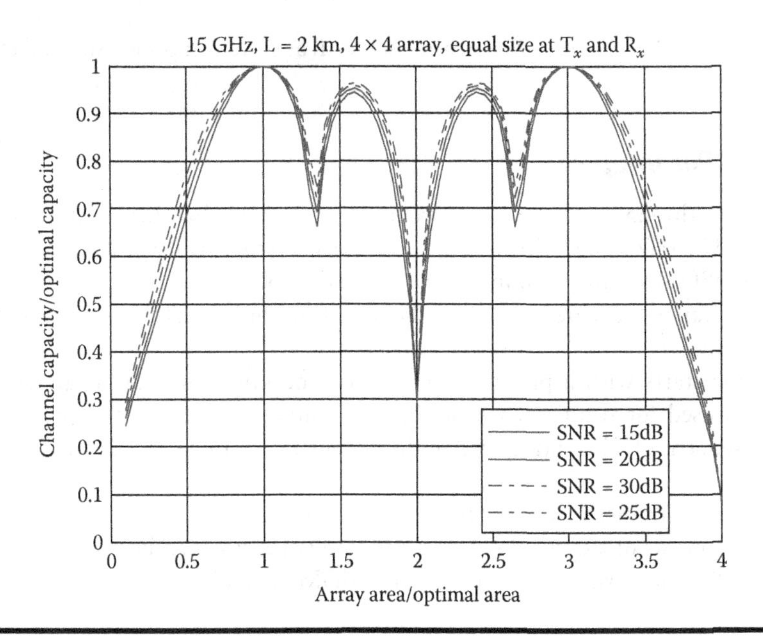

Figure 11.5 Channel capacity vs. antenna array area.

band, $L = 2$ km, $SNR = 30$ dB and equal to array size at T_x and R_x. The ratio of actual area to the optimal minimum area is used as the analyzing parameter and different array configuration are considered. In Figure 11.5, the same performance items are shown with different SNR values from 15 to 30 dB for a typical 4×4 array.

The results show that the condition number and capacity are insensitive to area deviation for most array configurations, but the performance degradation increases with more antenna elements. Take 90% of optimal capacity as a threshold, the tolerable ranges of area deviation ratio for 2×2 and 5×5 arrays are [0.46, 1.54] and [0.71, 1.2], respectively at $SNR = 30$ dB. From Figure 11.5, it can be seen that the area tolerance ranges are very close for different SNRs, which means that the proposed design rule is also not sensitive to SNR. Another interesting finding is that for 2×2 and 4×4 array, the optimal capacity can also be achieved when actual area is thrice the minimum optimal ones, and for 3×3 and 5×5, twice the minimum optimal areas are still optimal. These are due to the relatively prime property of p and N in these cases as in Equation 11.7.

11.3 MIMO–OFDM-Based LOS–MIMO Gbps System

MIMO systems provide an additional spatial dimension for wireless communications and yield a degree-of-freedom gain. These additional degrees of freedom lead to obvious capacity increase: for a system with N transmit and receive antennas respectively, the capacity increase is proportional to N. Meanwhile, OFDM modulates the information on parallel subcarriers in the frequency domain. It has a distinct advantage in antifading ability than traditional single-carrier technologies.

Therefore, MIMO–OFDM [17,18], which combines MIMO and OFDM technologies, is considered as the physical layer scheme in our Gigabits per second radio system, due to its distinct advantages in many aspects of system performance such as system capacity, spectral efficiency, and antifading ability.

11.3.1 Main Parameters

To achieve 1 Gbps throughput, the system occupies 2 × 33 MHz bandwidth at 15 GHz frequency band. For each 33 MHz bandwidth, a 4 × 4 MIMO subsystem is used. The parameters of each 4 × 4 MIMO–OFDM set are given in Tables 11.2 and 11.3.

The system's frame structure is shown in Figure 11.6. One superframe consists of 1 preamble and 32 data frames.

A superframe starts with a preamble, which has the same duration as an OFDM symbol (6.64 μs) and is used for timing synchronization, channel estimation, and phase noise correction. The Chu sequence [19] is introduced in the system as preamble for its constant modulus and cyclostationarity.

There are 32 data frames in 1 super frame, and each of them includes 8 OFDM symbols. An OFDM symbol consists of a cyclic prefix and 256 points Inverse Fast Fourier Transform (IFFT). Further, 168 of 256 subcarriers are used for each OFDM symbol.

Table 11.2 General Link Level Parameters

Parameters	*Value*
T_x/R_x antennas	4 × 4
Carrier frequency (GHz)	14.417/14.483
Bandwidth (MHz)	33 for each set
Sample rate (Msps)	40.96
Channel coding	(7680,6400) QC-LDPC
Modulation	64 QAM
Channel condition	Indoor(50 m LOS environments)

Table 11.3 OFDM Parameters

Parameters	*Value*
FFT size	256
Modulated sub-carriers	168
Subcarrier bandwidth (KHz)	160
OFDM symbol length (μs)	6.64
CP length (μs)	0.39
OFDM symbols per data frame	8
Pilot/data OFDM symbols	1/64

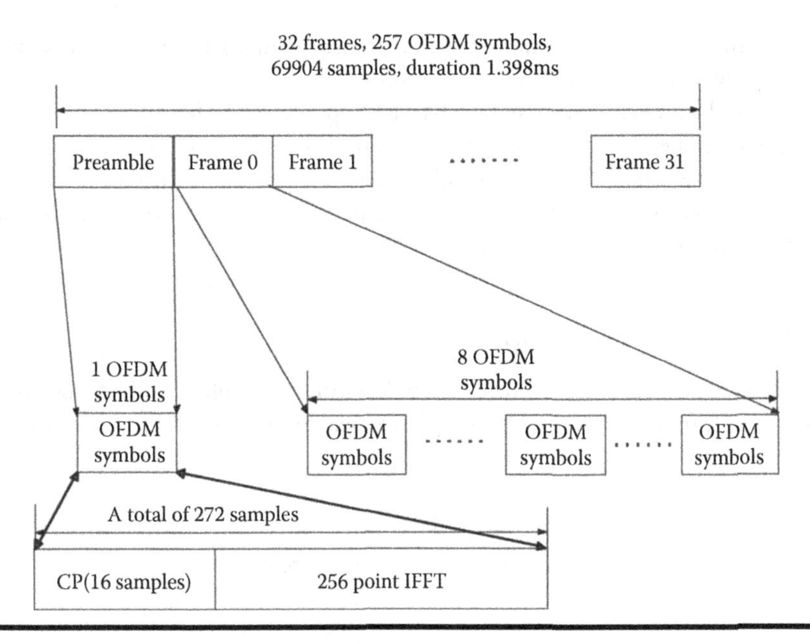

Figure 11.6 Frame structure of one superframe.

11.3.2 Algorithm Design

11.3.2.1 Preamble

To facilitate quick synchronization, the data packet is preceded with a known sequence (the preamble). The preamble is carefully designed to provide enough information for a good packet detection, phase noise estimation, and channel estimation. The Chu sequence is introduced in the system as preamble for its constant modulus and cyclostationarity. It is given by Reference 19

$$p(k)=\begin{cases} e^{j\pi(M/N_p)k^2}, & N_p \text{ is even} \\ e^{j\pi(M/N_p)k(k+1)}, & N_p \text{ is odd} \end{cases} \quad k=0,1,\ldots,N_p-1, \tag{11.15}$$

Where N_p is the length of the sequence and M, the sequence index, is relatively prime to N_p.

To estimate the MIMO channel, it is important that the subchannels from the different T_x antennas to every R_x antenna can be uniquely identified. To achieve that, the preambles on the different T_x antennas should be orthogonal. Besides, to perform the phase noise correction, a few subcarriers should be reserved in OFDM symbols not only in preamble frames but also the data frames, with which the receiver could estimate the phase rotation between the data frames and preamble frames.

11.3.2.2 Channel Estimation

The channel response can be estimated using the known training symbols within the preamble. When the timing is recovered correctly, we know which received samples correspond to the training part for channel estimation. More precisely, we know exactly which part of the received preamble is sent by transmit antenna, m. The four transmit antennas share the subcarriers in preamble orthogonally by means of frequency division.

With the preamble, we can estimate the channel response values $\mathbf{h}_m[k_1], \ldots, \mathbf{h}m[k_n]$, where m is transmit antenna number, and k denotes the location of the subcarriers sent by transmit antenna m. As only one channel value in every four subcarriers can be obtained directly from the training symbols, the values on the other subcarriers are estimated by a linear interpolation scheme based on minimum-mean-square-error (MMSE) criteria.

11.3.2.3 Phase Noise Correction

As mentioned in Section 11.3.2.1, the receiver could estimate the phase rotation between the data frames and preamble frames with the reserved subcarriers in each OFDM symbol. The phase noise matrix is denoted by

$$\Phi = \begin{bmatrix} rr_1t_1 & r_1t_2 & \cdots & r_1t_M \\ r_2t_1 & r_2t_2 & \cdots & r_2t_M \\ \vdots & \vdots & \ddots & \vdots \\ r_Nt_1 & r_Nt_2 & \cdots & r_Nt_M \end{bmatrix}, \tag{11.16}$$

where $r_i = e^{\varphi_i}$ and $t_i = e^{\theta_i}$, where φ_i and θ_i denote the phase noise on ith receive and transmit antenna, respectively. The transmit phase noise vector is defined as $\overline{\theta} = [1\, t_2/t_1 \ldots t_M/t_1]$. $\overline{\theta}$ can be obtained by

$$\overline{\theta} = \frac{1}{N}\begin{bmatrix} 1 \\ 1 \\ \cdot \\ \cdot \\ 1 \end{bmatrix}^T \begin{bmatrix} \begin{bmatrix} 1 \\ 1 \\ 1 \\ 1 \\ 1 \end{bmatrix} & \Phi_2./\Phi_1 & \Phi_3./\Phi_1 & \ldots & \Phi_M./\Phi_1 \end{bmatrix}, \tag{11.17}$$

where Φ_i is the ith column of the matrix Φ. Similarly, the receive phase noise vector is defined as $\overline{\varphi} = [r_1t_1\; r_2t_1 \ldots r_Nt_1]$, which can be estimated by

$$\overline{\varphi} = \frac{1}{M}\begin{bmatrix} \Phi_0^r./\overline{\theta} \\ \Phi_1^r./\overline{\theta} \\ \cdot \\ \cdot \\ \Phi_N^r./\overline{\theta} \end{bmatrix}\begin{bmatrix} 1 \\ 1 \\ \cdot \\ \cdot \\ 1 \end{bmatrix}, \tag{11.18}$$

where the Φ_i^r is the ith row of Φ. Once the $\overline{\theta}$ and $\overline{\varphi}$ are obtained, the channel matrix for current OFDM symbol is given by

$$\overline{\mathrm{H}} = \mathrm{diag}(\overline{\varphi})\mathrm{H}\mathrm{diag}(\overline{\theta}). \tag{11.19}$$

11.4 GE-Based SDR Platform and Implementation of the Gbps System

11.4.1 Platform Architecture

According to the OFDM theory, the whole transmission band of OFDM signals can be divided into several subbands, which can be processed independently. The multiband capability of the OFDM makes the design of broadband system much more flexible. GE interfaces are involved to connect multiple processing units and RF front ends. Based on GE switching, we establish a high-throughput pipeline for data stream, as well as a low-latency path for the control signals switched between each unit in our SDR platform. A distributed processing strategy is adopted to get high efficiency parallel computing ability and to reduce the requirements of computation capacity for each processing unit. The platform takes the advantages from both HAUs and GPPs to achieve good real-time processing performance with good adaptability and flexibility. The detailed architecture of the GE switch-based SDR platform is illustrated in Figure 11.7.

This platform can be divided into the following four function blocks: data services processing block, real-time processing block, radio frequency (RF) preprocessing block, and GPP processing block. Taking the transmitter side, for example, we can show the data switching process on the platform as follows.

The user data stream is firstly transformed into fixed-length packets by data service processing block. With GE switching, data packets are distributed to different subbands processing units and recollected after frequency-domain processing (e.g., channel coding, modulation, subcarrier mapping, and precoding) in each real-time processing block. Then the data packets in the frequency domain are distributed to each of the RF preprocessing block (performing IFFT, inserting of Cyclic Prefix (CP), etc.) through GE switching. Finally, after time-domain processing the data streams are transmitted to the RF front ends.

The basic functional partition for the receiver side is similar to the above transmitter side. However, most complex computation processes, such as channel estimation, timing and frequency synchronization, matrix decompositions for MIMO detecting and precoding, have been allocated

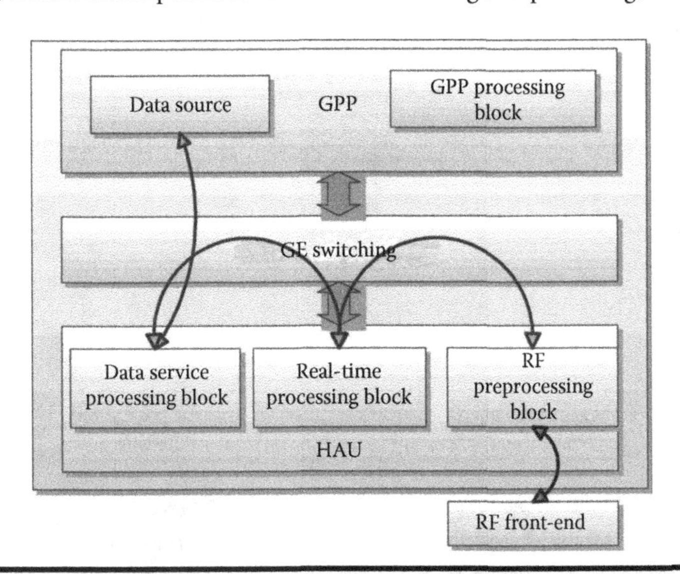

Figure 11.7 GE-based SDR platform.

to GPP processing block. The left function for the receiver is only channel decoding module in real-time processing block for HAUs.

On the other hand, GE interface is a powerful debugging tool for system developers. It is convenient to get data samples from any in–out port in this platform by devices supporting GE.

11.4.2 Implementation of the Gbps System on a GE SDR Platform

Based on the above antenna array design and MIMO–OFDM scheme for LOS–MIMO, the Gigabits per second system is implemented on the GE switch-based SDR platform, utilizing its flexibility, programmability, and real-time processing ability. Though the traffic data processing and RF front end are important modules of the system, only the transmitter and receiver path are discussed in detail.

The transmitter structure and the process partition are shown in Figure 11.8.

User application data first goes through the data service processing block and is multiplexed into constant bit rate stream and mapped to the standard GE packet for switching.

Then the data stream is distributed to four real-time processing units in real-time processing block, corresponding to four subbands, where the data stream goes through a Quasi-Cyclic Low-Density Parity-Check (QC-LDPC) encoder and is mapped to 64 Quadrature Amplitude Modulation (QAM) symbols. After that, a multiantenna multiplexing module splits the symbols into four independent parallel data streams, and each one corresponding to a transmitting antenna. Similarly, the data stream is passed to RF preprocessing block in the form of GE MAC packets.

In each RF preprocessing unit corresponding to each transmitting antenna, data samples in same frame are collected from every subband, to calculate 256-point IFFT. Then, the CP is inserted to every OFDM symbol. Finally, a preamble which includes training sequences is added to every superframe before the final T_X signal is up converted to the RF front end and transmitted.

The GPP processing block can configure the parameters for all the HAU blocks.

Compared with the transmitter, GPP processing block plays a more important role in the receiver. Although the real-time processing capacity of GPPs cannot match the needs of Gigabits

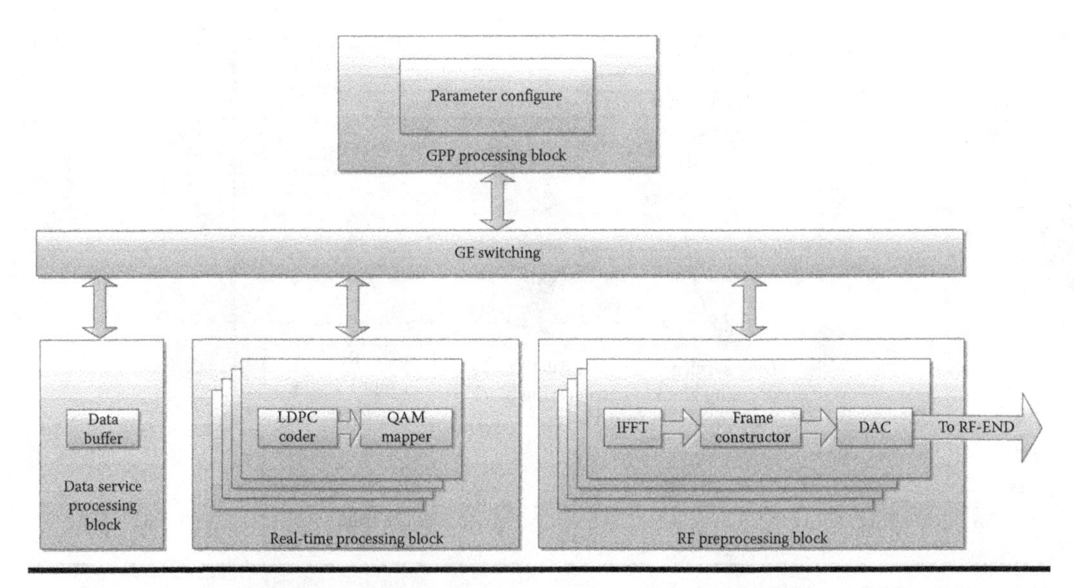

Figure 11.8 Transmitter structure.

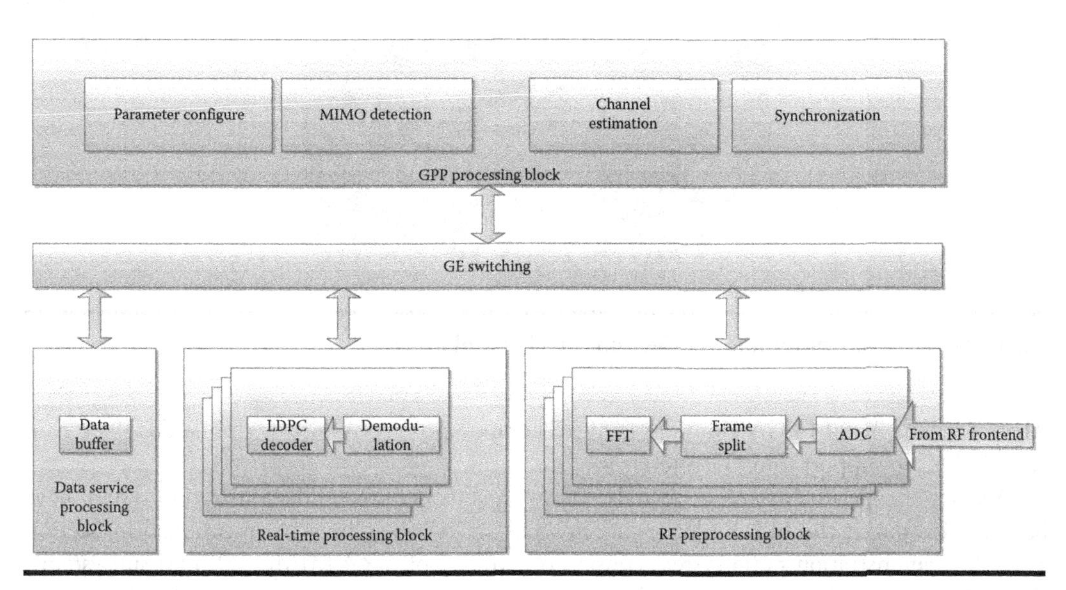

Figure 11.9 Receiver structure.

per second transmission, the algorithms working on low-rate samples such as channel estimation for slow-fading channels, timing and frequency synchronization, matrix decompositions for MIMO detecting and precoding, could be implemented on GPPs for their great flexibility and programmability. The receiver structure and the process partition are shown in Figure 11.9.

First, four receive antennas are used for each set of MIMO–OFDM system and low-rate samples of received data go through the GE interface from RF preprocessing block to GPP processing block, where a correlation method in time domain is used to achieve time synchronization and frequency offset estimation. Thereafter, GPP processing block informs the results to RF preprocessing block. In doing so, RF preprocessing units for every antenna can identify from the sampled sequences the starting points of *S*/*P* conversion of OFDM data symbols. Then, frequency-domain OFDM symbols can be obtained by (1) correcting frequency offset, (2) removing the CP, and executing FFT algorithms.

The frequency-domain OFDM symbol in the preamble is sent to GPP processing block, where channel estimation is taking place. Then, the MIMO-detecting matrix and phase noise for every subcarrier can be estimated in frequency domain. The result is transmitted to real-time processing block. In order to meet the first generation (1G) transmission data rate, four real-time processing units are used. On each unit, data samples of each subband in the frequency domain perform phase noise correction and MIMO detection, then QC-LDPC decoding algorithms are executed. Finally the user application data packet is recovered from decoded data block in data service processing block.

11.5 Field Evaluation of LOS–MIMO Gbps System

In this section, we include the field evaluation results of the end-to-end Gigabits per second transmission performance. In our MIMO–OFDM demonstration system over the GE switch-based SDR platform, a personal computer (PC) with dual-core central processing unit (CPU) operating at 1.6 GHz is used as the GPP at each side of the transmitter and receiver. For each HAU, two

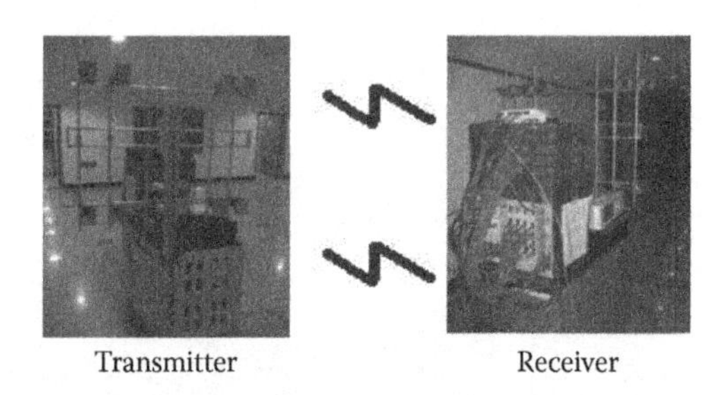

Figure 11.10 Transmitter and receiver on the GE SR platform.

Altera Cyclone 3C80 FPGA chips are used for computation, combined with one GE PHY chip (BCM5464) to support four GE interfaces.

A series of experiments and field trials of our Gigabits per second demonstration system have been carried out at the FIT Building in Tsinghua University, China. As mentioned in Section 11.3, the demonstration radio system consists of two sets of 4 × 4 MIMO–OFDM subsystems. Figure 11.10 shows the transmitter and receiver, respectively.

The field tests were done in indoor environments with the transmission distance over 50 m. The SNR and Bit Error Rate (BER) performance for different locations are depicted in Table 11.4, where the T_X and R_X logo are coordinates for T_X and R_X positions, respectively, along the corridor. In the in-building corridor scenario, there exists visible two-path effect due to reflection of the walls at each end of the corridor. The second path delay and relative power are given in Table 11.4, which suffer some variation for different locations. The second path delay is decided mainly by the length of the corridor and the relative power is below 20 dBc for most cases. The results show that the proposed and implemented system works well under such two-path scenario.

The maximum achievable information bit rates are calculated as 1.171 Gbps = 2 (two sets of subsystems) × 32(frames per superframe) × 8(OFDM symbols per frame) × 160(subcarriers per OFDM symbols) × 6(index for 64-QAM modulation) × 4(numbers of transmitting antennas) × 5/6(LDPC coding rate)/0.001398(superframe duration). Table 11.5 shows the field test results of the peak data transmission throughput of the Gigabits per second platform by different methods, wherein Netperf is a PC-based network benchmark tool. We also use self-developed

Table 11.4 SNR/BER for Different Locations

Position	*T_x Logo*	*R_x Logo*	*Path Number*	*Second-Path Delay (ns)*	*Second-Path Power (dBc)*	*SNR (dB)*	*BER*
0	15.1.3	0.4.3	2	391	− 23	23.4	5e-4
1	15.0.3	0.4.3	1	–	–	26.5	0
2	14.5.3	0.5.3	2	439	− 26	26.5	0
3	14.5.3	0.5.2	2	415	− 24.8	23.8	5e-6
4	14.5.3	0.5.4	2	342	− 25.1	22	1e-4
5	14.5.4	0.5.4	2	415	− 20.4	20	5e-4
6	14.5.4	0.5.3	2	366	− 25.2	26	0

Table 11.5 Throughput Test Results

Application	*Throughput (Mbps)*
Netperf	860
UDP	880
TCP	760

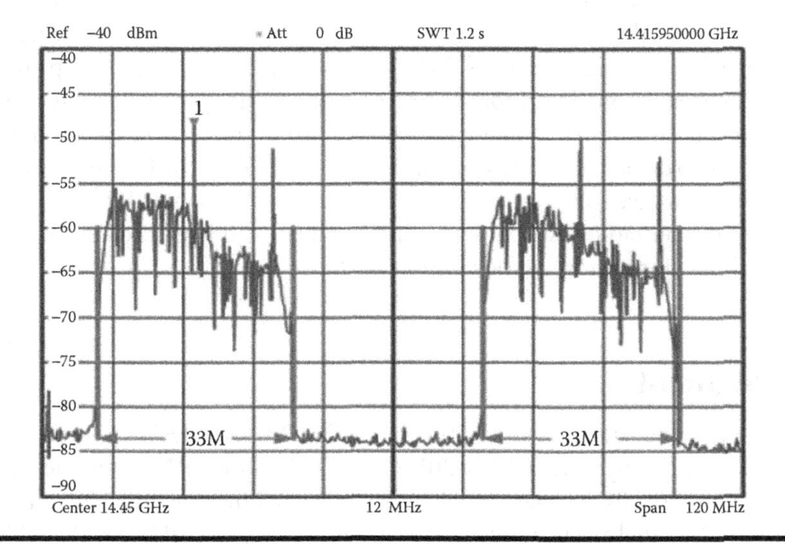

Figure 11.11 Illustration of spectral utilization.

programs to test the system throughput with Transport Control Protocol (TCP) and User Datagram Protocol (UDP) protocols.

Although the Gigabits per second demonstration system can provide over 1 Gbps data throughput at the wireless physical layer, the achieved throughput of 1 Gbps is limited due to the limitation of GE data interface itself, the performance of PC, and the overhead of Ethernet/Internet protocols.

In the above demonstration system, the initial form of carrier aggregation (proposed and standardized in long-term evolution-advanced specifications) is also considered to support deployment bandwidth even up to more than 100 MHz. Here, we just show a simple noncontiguous case with 2 × 33 MHz component carriers. The system center carrier frequency is set to 16 GHz and the total transmission bandwidth is 66 MHz. Figure 11.11 presents the spectral utilization of the Gigabits per second wireless transmission system.

From Figure 11.11 we can see that each set of 4 × 4 MIMO–OFDM subsystem occupies a bandwidth of 33 MHz. Referring to the peak data throughput, the peak spectral efficiency is higher than 15 bps/Hz.

11.6 Conclusions

In this chapter, a directional antenna-based MIMO solution is proposed to guarantee full multiplexing gain for the LOS wireless backhaul environments at higher microwave bands. The optimal design rule and sensitivity analysis for one- or two-dimensional antenna array give theoretical support for LOS–MIMO system design, and it eases the choice of array shape by moving the limitation of perpendicular arrangement from earlier work.

Under the guide of derived antenna array design rule, a MIMO–OFDM-based high capacity radio system for LOS channel at 15 GHz frequency band is proposed with symmetrical square array at both T_x and R_x. A GE switch-based SDR platform is designed to implement and verify the proposed system. Due to the good scalability and flexibility of GE interface, this platform has both the real-time processing capability in HAU and adaptability and programmability in GPP, and hence is suitable for the development of future wireless communication systems.

On the GE SDR platform, the LOS–MIMO system yields throughput up to 1 Gbps and spectral efficiency over 15 bps/Hz through field evaluation. The performance is supposed to be doubled with cross polarization of antennas at the same array area. To the best of our knowledge, this is the first time to demonstrate the Gigabits per second LOS–MIMO system with directional antennas at such microwave bands in the world. The proposed LOS–MIMO antenna array design and Gigabits per second radio system are promising solutions to fixed broadband wireless access and wireless backhaul for next generation networks.

Acknowledgments

This work is supported in part by China's 863 Project (No.2014AA01A701, 2006AA01Z282); the National Basic Research Program of China (2012CB316002); National Natural Science Foundation of China under Grant 61201192; the Key Grant Project of Chinese Ministry of Education under Grant 313005; the National Science and Technology Major Project (2011ZX03004-004); the open research fund of National Mobile Communications Research Laboratory; Southeast University under Grant 2012D02; and the International Cooperation Program (2012DFG12010).

References

1. Guo, N., R. C. Qiu, S. S. Mo, and K. Takahashi, 2003. 60-GHz millimeter-wave radio: Principle, technology, and new Results, *EURASIP Journal on Wireless Communications and Networking*, 2007(1), 41–48.
2. Makri, R. et al. 2011. Next generation millimeter wave backhaul radio: Overall system design for GbE 60GHz PtP wireless radio of high CMOS integration. In *2011 18th IEEE International Conference on Electronics, Circuits and Systems (ICECS)*, December, Beirut, pp. 338–341.
3. Mecklenbrauker, C. F., M. Matthaiou, and M. Viberg, 2011. Eigenbeam transmission over line-of-sight MIMO channels for fixed microwave links. In *2011 International ITG Workshop on Smart Antennas (WSA)*, February, Aachen, Germany, pp. 1–5.
4. Shafi, M., M. Zhang, A. L. Moustakas, P. J. Smith, A. F. Molisch, F. Tufvesson, and S. H. Simon, 2003. Polarized MIMO channels in 3-D: Models, measurements and mutual information, *IEEE JSAC*, 24(3), 514–527.
5. Torkildson, E., U. Madhow, and M. Rodwell, 2011. Indoor millimeter wave MIMO: Feasibility and performance, *IEEE Transaction on Wireless Communications*, 10(12), 4150–4160.
6. Zhang, H., S. Venkateswaran, and U. Madhow, 2010. Channel modeling and MIMO capacity for outdoor millimeter wave links. In *2010 IEEE Wireless Communications and Networking Conference (WCNC)*, April, Sydney, Australia, pp. 1–6.

7. Sheldon, C., E. Torkildson, M. Seo, C. P. Yue, U. Madhow, and M. Rodwell, 2008. "Rodwell," a 60 GHz line-of-sight 2x2 MIMO link operating at 1.2 Gbps. *IEEE International AP-S Symposium*, July, San Diego, USA, pp. 1–4.
8. Bohagen, F., P. Orten, and G. E. Oien, 2007. Design of optimal high-rank line-of-sight MIMO channels, *IEEE Transactions on Wireless Communications*, 6(4), 1420–1425.
9. Bohagen, F., P. Orten, and G. E. Oien, 2005. Modeling and analysis of a 40 GHz MIMO system for fixed wireless access. In *Proceedings of 61st IEEE Vehicular Technology Conference 2005-Spring*, May 15–17, 2005, Stockholm, Sweden, Vol. 3, pp. 1691–1695.
10. Larsson, P., E. Res, and S. Kista, 2005. Lattice array receiver and sender for spatially orthonormal MIMO communication. In *IEEE VTC*, May, Stockholm, Sweden, pp. 192–196.
11. Liu, L. et al. 2007. Characterization of line-of-sight MIMO channel for fixed wireless communications, *IEEE Antenna and Wireless Propagation Letters*, 6, 36–39.
12. Mitola, J. 1995. The software radio architecture, *IEEE Communications Magazine*, 26–38.
13. Tan, K. et al., 2009. Sora: High performance software radio using general purpose multi-core processors. In *6th USENIX Symposium on Networked Systems Design and Implementation 2009*, April 22–24, 2009, Boston, Massachusetts, USA, pp. 99–107, USENIX.
14. Chen, J., Q. Wang, Z. Zhu, and Y. Lin, 2009. An efficient software radio framework for WiMAX physical layer on cell multicore platform. In *IEEE ICC 2009*, June 14–18, Dresden, Germany,, pp. 1–4.
15. Huili Guo et al. 2009. High performance turbo decoder on CELL BE for WiMAX system. In *IEEE WCSP09*, November 13–15, 2009, Nanjing, China, pp. 1–5.
16. Telatar, I.E. 1999. Capacity of multi-antenna Gaussian channels, *European Transactions on Telecommunications,* 10, 585–595.
17. Kyung Won Park, 2005. An MIMO-OFDM technique for high-speed mobile channels, *IEEE Communications Letters*, 9(7), 604–606.
18. Yang, H. W. 2005. A road to future broadband wireless access: MIMO-OFDM-based air interface, *IEEE Communications Magazine*, 43(1), 53–60.
19. Chu, D. C. 1972. Polyphase codes with good periodic correlation properties, *IEEE Transactions on Information Theory*, 18, 531C532.

Chapter 12

MIMO and Cooperation in Cognitive Radio-Based Wireless Networks: State-of-the-Art and Perspectives

Abdelaali Chaoub

Contents

12.1 Introduction

Recently, cooperation and multiple-input-multiple-output (MIMO) technologies are emerging as new paradigms in cognitive radio (CR)-based wireless networks, as spatial diversity helps to deliver the secondary stream between cognitive users over longer distances with larger coverage and faster data rates.

This chapter summarizes and analyzes the benefits of adopting spatial diversity techniques in the context of CR and surveys the relevant and recent literature in this research field. In particular, a special emphasis is given to overlay schemes while taking advantage of cooperation and MIMO capabilities. In the proposed framework, a single noncognitive (primary) user shares knowledge of its signal messages with a set of cognitive (secondary) users. Thus, cognitive devices may enhance the noncognitive transmission and negotiate to rent special spectrum rights instead of vying for opportunistic access. More specifically, cognitive devices overhear the messages sent by the noncognitive source and use this information to eliminate the interference generated by the primary communication at the cognitive receivers as well as to improve the performance of the primary communication through relaying the accumulated messages to the primary receiver. The latter property allows a simultaneous access to spectrum for both cognitive and noncognitive users provided that the overall transmit power and the antenna degree of freedom (DoF) of the cognitive user can fairly cover the needs of its own communication and its relaying operation as well. To prevent multiuser interference, the Time Division Multiple Access (TDMA) access technique has been exploited to ensure that signals impinging from different secondary links do not disturb the output signal at the intended primary receiver and do not mutually interfere. It is apparent that the designed scheme considers many dimensions namely temporal, frequency, and spatial domains to improve spectrum utilization and avoid experiencing harmful interference. To support the analytical model, the effectiveness of the proposed MIMO cooperative CR framework is demonstrated by a comparative study with the classical cooperative single-input-single-output (SISO) CR scheme. The test case demonstrates that despite radiating at high secondary power levels, the proposed MIMO framework can markedly increase the primary link throughput up to 5% compared to the classical SISO scheme, while the achieved throughput gain is 23% for the secondary link. The resulting observations point out the complexity of the dynamics between different peers in CR networks and outline the necessity of describing the available trade-offs to find an optimal balance between the interference induced on the primary signal and the added reliability and flexibility it enjoys. The obtained throughput gain at the primary link must override the cost in terms of time leased to cognitive communications.

The remainder of this chapter is structured as follows: Section 12.2 describes the background and the motivations of this research. Further, a concise overview of spectrum-sharing paradigms and spatial diversity in CR networks is presented. Later, we have consolidated several contributions made on this research area into a logical structure. Section 12.3 raises some interesting issues related to the spatial DoF in the context of CR networks. Our goal is to conduct a performance comparison between different antenna configurations with user cooperation enabled between primary and secondary networks. The formal expression of the average throughput is computed for each scheme, taking into account the interference with the neighboring primary link and among the secondary links themselves. Being a limiting factor and a major bottleneck in multiuser multiantenna systems, interference between the coexisting secondary links has been handled by using a traditional TDMA multiplexing, whereby each cognitive pair is assigned a fixed time slot to avoid being disrupted by the surrounding secondary transmitters. Alongside, a basic transmission scenario in CR-based wireless networks is considered and results of extensive simulations are plotted to prove the effectiveness of the proposed TDMA-based cooperative MIMO model compared to the traditional cooperative single-antenna setup. Throughout this section, interesting observations are made along with some insightful comments to build a solid foundation for current results and potential improvements. Finally, Section 12.4 concludes the chapter.

12.2 A Short Introduction to CR

During the past few decades, bandwidth-consuming applications over wireless networks have grown greatly along with the tremendous proliferation of smartphones and the growing demand for high-quality mobile services, including high-speed internet and clear voice calls. This recent trend has led to a surge in demand for spectrum, and hence the need for new paradigms to overcome the scarcity of spectrum resources. The cognition in spectrum sharing and monitoring [1] brings a paradigm shift as a well-regarded agile technology to meet the expectations of the new era of services.

The licensed spectrum is not complete neither in time nor in the spatial domain. Therefore, it was attractive to explore what potential existed in gaps between active radio channels. These gaps appear as blank spaces in spectrum measurements and are referred to as "spectrum holes." This opportunity has been seized and the concept of CR has emerged to revolutionize the classical spectrum allocation strategies toward more efficient and optimal sharing of frequency resources. CR scrutinizes the surrounding environment and draws a global view of the spectrum usage so that some devices, called secondary users (SUs), can be scheduled to transmit on the vacant channels [2].

Nowadays, CR is gaining further momentum worldwide. On the 21 November 2013, European Telecommunications Standards Institute (ETSI) Future Mobile Summit [3] has cited a number of key recommendations and concluding messages to forecast what could be the roadmap of the beyond-fourth generation (4G). The unlicensed access to spectrum was one of the raised points. The concept of CR is among the promising technologies subject of research and investigations in the context of next generation networks [4].

12.2.1 CR: How the Spectrum is Shared

To enable spectrum sharing without causing harmful interference to existing traffics, cognitive users should possess a minimum of information about their nearby noncognitive users. Depending on the knowledge that is needed to coexist with the primary network, CR approaches fall into three classes, namely, interweave, underlay, and overlay [5].

The interweave-based coexistence approach has been proposed with the objective of enabling devices to occupy the spectrum rooms that have been left vacant by noncognitive users. The surrounding environment should be observed to be able to predict the state of each portion of the frequency spectrum, portions of spectrum considered as being underutilized may be accessed by secondary users as long as the primary activity remains idle. In order to facilitate the coexistence of both primary and secondary traffics within the same network in an opportunistic transmission mode, spectrum opportunities should be actively identified and monitored. Cognitive users may conduct sensing operations permanently and reliably, and different dimensions need to be explored to find the abundant spectrum gaps. Legacy sensing algorithms monitor and supervise the spectrum through three conventional dimensions: frequency, time, and space. However, other degrees of freedom such as the used code and the angle of arrival may be examined. Another alternative mechanism to identify the vacant gaps is the use of a geographic coordination through a central database, which is either a substitute or a complementary solution to selfish spectrum sensing.

In underlay systems, simultaneous cognitive and noncognitive transmissions are allowed as long as the interference level at the primary user (PU) side remains confined within the interference limit imposed to cognitive devices. In recent literature, many advanced signal processing techniques have proven to be very efficient for interference avoidance, among which we find the beamforming and the spread spectrum techniques to be excellent. Beamforming consists of exploiting the superposition concept of waves to guide the signal toward a specific receiver using multiple antennas. More importantly, in spectrum-sharing contexts constructive or destructive interference is provoked at the intended cognitive receiver to lessen the interference caused to noncognitive users while focusing the signal energy toward the desired direction. Using the spread spectrum technique, the secondary signal is multiplied by a spreading code to obtain a weaker signal with wider band. The resulting spread signal causes lower interference level to noncognitive users. The original secondary signal is recovered at the receiver side by simply multiplying the input signal with the same spreading code. The spread spectrum technique is also useful for alleviating the interference caused by the primary signals to the secondary ones. Another common solution could be limiting the power of the secondary signal to keep the interference level at the primary side bounded, albeit restricting the secondary transmissions to short range communications.

Using the overlay approach, cognitive users are able to track the PU messages or codebooks to be able to either null out the interference caused by the primary signal at the secondary receiving end or to strengthen the primary signal through serving as a relay for the licensed traffic. Unlike the underlay scheme, there is no interference temperature constraint enforced on the secondary signal power. To coexist with the licensed network without any interference, secondary users seek the best compromise between the interference induced and the improvement brought to the primary signal to achieve a stagnant signal-to-noise ratio (SNR).

Hybrid schemes using a combination of the aforementioned paradigms [6] have a great potential to improve the efficiency of spectrum sharing. The benefit of such a scheme is that it allows secondary users to maximize their transmission rate whenever a spectrum opportunity is detected.

12.2.2 Spatial Diversity in CR Networks

Since the incumbent users are the holder of spectrum license, cognitive devices are constrained to operate with low transmit powers to guarantee an interference-free environment for licensed traffics. One often advocated solution to deal with such power-constrained and error-prone transmissions is the integration of the spatial dimension to create transmit and receive diversity. Recurrent techniques for achieving this spatial diversity are, typically, cooperation and MIMO designs. Spatial diversity improves the rate and speed of communication systems without increasing the occupied bandwidth, thus preserving the precious spectral resources.

Recent literature has proven an increasing interest in using cooperative diversity [7]. The basic idea of cooperative communications is to recruit some intermediate nodes called relays each with one antenna to relay the signal of the source in a many-to-one fashion, thus forming a virtual antenna array (VAA) or a virtual MIMO. The use of cooperation for content delivery can enhance tremendously the performance of CR networks. However, cooperative schemes still need to overcome some major challenges [8] related to security vulnerabilities, resource control and management, synchronization protocols, implementation issues, delay overheads, and the number of nodes available and ready for cooperation.

MIMO and multiuser MIMO technologies are among the most powerful spatial diversity methods [9]. These techniques endow wireless devices with the capability of simultaneous transmission of multiple independent streams using antenna diversity and allow opening up the frequency bands to concurrent operating users in a noninterfering mode via interference mitigation techniques to lessen the interference caused to noncognitive users.

Given its great potential, the integration of multiple antennas in recent systems has allowed for higher data rates, less transmission errors, and increased communication ranges. Most recent systems such as WiFi, WiMax, and the 3rd Generation Partnership Project (3GPP) standards have been equipped with MIMO antennas. For instance, MIMO has been introduced in WLAN standards IEEE 802.11n and 802.11ac leading to a significant raise of the data rate up to 450 and 3.5 Gbps, respectively. WiMax standard IEEE 802.16e has also incorporated MIMO features to extend the cell coverage and allow the delivery of rich contents. Further, 3GPP is another international organism that has implemented the concept of MIMO in High Speed Packet Access+ (HSPA+), thereby increasing data rates up to 28 Mbps in release 7 and 168 Mbps in release 10. Henceforth, it is widely expected that the MIMO technology will shape the future of wireless and mobile devices, as evidenced by the current technological trend toward manufacturing less cumbersome antennas. MIMO features are rapidly evolving from traditional wireless networks to emerging cognitive radio networks (CRNs) with the aim of providing resilience to signal loss and channel fading.

For the purpose of improving the spatial diversity gain, embracing both cooperation and MIMO antennas in CRNs is an attractive research direction and a promising foundation for reliable license-exempt systems. More explicitly, the throughput of MIMO wireless networks can be significantly raised while increasing the number of antennas [10] but deploying more antennas comes at the expense of devices' size and cost. An appealing idea is to create a kind of virtual multiantenna array (VMAA, in analogy with VAA) by incorporating a set of intermediate nodes equipped with MIMO capabilities. However, MIMO-based cooperation and its potential to tackle the peculiar characteristics of CR environments is a pertinent issue to be investigated under different spectrum-sharing strategies.

12.2.3 A Literature Survey of MIMO-Assisted CRNs

In recent literature, the topic of SISO- and MIMO-based CRNs has raised a flurry of research studies among academics.

In this regard, Reference 11 surveys the use of game theoretical techniques in CR contexts and provides an overview of recent literature on this timely research area. Different game theory concepts have been presented and their applications for spectrum sharing have been debated and discussed deeply. In this regard, Reference 11 provides a detailed analysis that is very important for a complete understanding of game theory-based CR frameworks.

Other contributions like Reference 12 have proposed a cooperative single-antenna transmission scheme and based on rateless coding for interweave CR, the authors have analyzed the average end-to-end throughput and presented a low-complexity algorithm to search for the optimal number of relaying nodes. The same issue has been addressed differently in Reference 13 using the CR for virtual unlicensed spectrum (CORVUS) approach; however, the proposed cooperative scheme has the property that no coordination is needed among the collaborative relays, and therefore numerical simulations have been performed in terms of the average transmission time and the achieved throughput.

Simeone et al. have considered in Reference 14 an overlay scheme using a resource-exchange approach, wherein a licensed user leases a part of its bandwidth to some single-antenna unlicensed terminals and in exchange the incoming users are recruited to assist the licensed communication and relay the primary stream from source to destination. The licensed network seeks to maximize its expected utility in terms of either average transmission rate or outage probability while optimizing the time splitting, whereas each secondary user in turn maximizes its transmission rate through implementing a distributed power control mechanism. The paper has adopted a Stackelberg game and a Nash equilibrium to resolve this multidimensional problem.

The same model has been addressed in Reference 15, but the payment made by each secondary user is determined on the basis of the access time it has acquired. As a result, the utility function to be maximized from the PU viewpoint depends on both the transmission rate and the revenue it collects, while the utility function of the secondary user is related to its transmission rate and its charge (payment).

The use of MIMO schemes in CRNs was discussed in greater depth in References 16–18.

In Reference 16, Hua et al. have advocated the concept of MIMO in overlay scenarios and exploited the spatial diversity to enable the provision of the concurrent secondary services while simultaneously relaying the primary traffic over the same frequency channel using space–time multiple access. Cognitive devices leverage the beamforming technique to control the direction of their own as well as their relayed signals for proactive interference avoidance.

A competitive optimality criterion based on a noncooperative game approach has been successfully applied in Reference 17 to a decentralized MIMO CR design to converge to a Nash equilibrium where every secondary user is willing to unilaterally maximize its information rate provided that the interference it generates is totally canceled or at least remains bounded. The proposed MIMO framework is suitable for both underlay and interweave systems.

An adaptive cooperative MIMO CR strategy using the underlay approach is proposed in Reference 18. Each secondary user willing to convey a stream will recruit one of the remaining secondary nodes as an amplify-and-forward relay to avoid disturbing the straight link between the primary source and destination. The overall design is transformed into a joint power allocation and relay selection problem. Furthermore, a less complex scheme has been proposed, causing a negligible performance degradation with reduced complexity.

The goal of this chapter is to inspect the impact of cooperative MIMO technology on the capacity of CR systems in terms of basic theoretical concepts as well as possible architectures and practical aspects relevant to the implementation of spatial diversity in CR contexts. Our main focus is on overlay schemes due to their flexibility as they do not need to adhere to any predefined interference level constraints.

12.3 Impact of MIMO Technology on the Capacity of Overlay CR Systems

12.3.1 General Analysis

In this section, we examine a CRN collocated with a primary link. Here, time is divided into frames of fixed duration, T. Consider the practical scenario in which a common noncognitive source PT_x is sending a single stream to a common noncognitive destination PR_x during one time frame. Simultaneously, K secondary transmitter–receiver pairs $ST_x^i - SR_x^i (i \in \{1, \ldots, K\})$ operate on the same wireless channel, i.e., in the range of the ongoing primary communication. All the links between different nodes are supposed to be Gaussian and each link suffer from a Rayleigh fading process. Denote the channel gain from the primary transmitter to the primary receiver as h_{pp}. $h_{\text{TT},i}$ is the channel gain (or the vector of channel gains for the case of multiple antennas) between the primary transmitter and the ith secondary transmitter and $h_{\text{SP},i}$ is the channel gain (or the vector of channel gains) between the ith secondary transmitter and the primary receiver. The channel gain of the secondary link $ST_x^i - SR_x^i$ is denoted as $h_{\text{SS},i}$ and the channel gain (or the vector of channel gains) between the secondary transmitter j and the secondary receiver i is denoted as $h_{\text{SS},ij}$ with $i \neq j$ (Figure 12.1). We assume a frequency-flat block Rayleigh fading, which means that the channel gain is invariant during each frame but changes from one frame to the other. We also assume that noise processes over various links are zero mean and have identical variance, N_0.

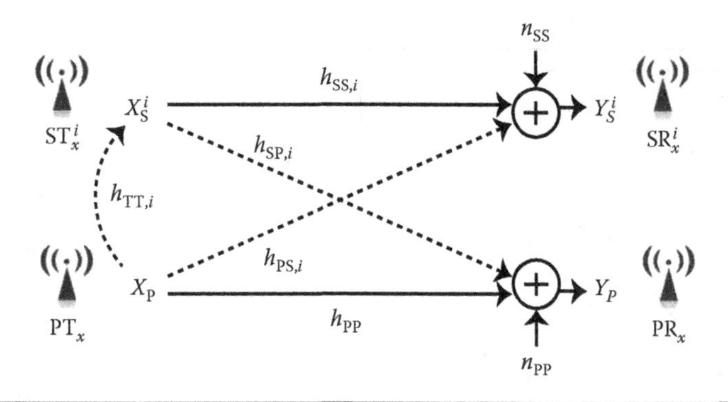

Figure 12.1 Concurrent cognitive and noncognitive transmissions.

In what follows, we suppose that the fading conditions are slowly varying so that the receiver can have an accurate estimate of the state of its associated physical resources and may disseminate these measurements to the transmitter using a dedicated control channel. Hence, a perfect knowledge of the channel state information (CSI) is made available at the transmitter side. This is a common assumption in prior works on the same issue. Besides, we assume a half-duplex constraint and suppose that each antenna can ensure the transmission as well as the reception of radio signals (i.e., transceiver).

We use the notation |.| for the absolute value in the case of a scalar and the Frobenius norm in the case of a vector.

12.3.2 Cooperative Single-Antenna Schemes

For the sake of simplicity, we start by the standard competitive access model. Popular overlay schemes proposed in research literature allow multiple coexisting single-antenna cognitive and noncognitive nodes to share the same infrastructure by using time-slitting protocol. Secondary users are allowed to rent some portions of frequency bands for a negotiated power cost and following a predetermined time schedule subject to an explicit agreement with the license holders. Concretely, the time frame structure is decomposed into three blocks according to two parameters α and β ($0 < \alpha, \beta < 1$) as illustrated in Figure 12.2. During the first phase, the noncognitive source keeps broadcasting the initial stream until it receives a sufficient number of positive acknowledgments returned from cognitive relays that have successfully received and decoded the transmitted data. These decode-and-forward relays will collaborate in the second phase transmission through the use of a distributed space time coding (DSTC) to help accumulating enough data at the noncognitive destination to recover the original message. The third phase is granted to the cognitive devices for secondary content delivery as a reward for their cooperation and assistance (Figure 12.2). A robust approach to select the appropriate set of relays is to sort the paths connecting the primary source and the available secondary transmitters according to the channel gain and select those with the K-highest values (Figure 12.3).

Based on the model described above, the primary source PT_x transmits continuously a given stream X_P until the K-relaying nodes $(ST_x^i)_{i\in\{1,\ldots,K\}}$ are able to successfully decode the transmitted message. Then, the decoding relays perform parallel transmissions of the same stream over the second hop to the primary destination. Obviously, the final signal Y_P is dominated by the worst hop among the first two stages. During the third phase, each cognitive user ST_x^i conveys its private

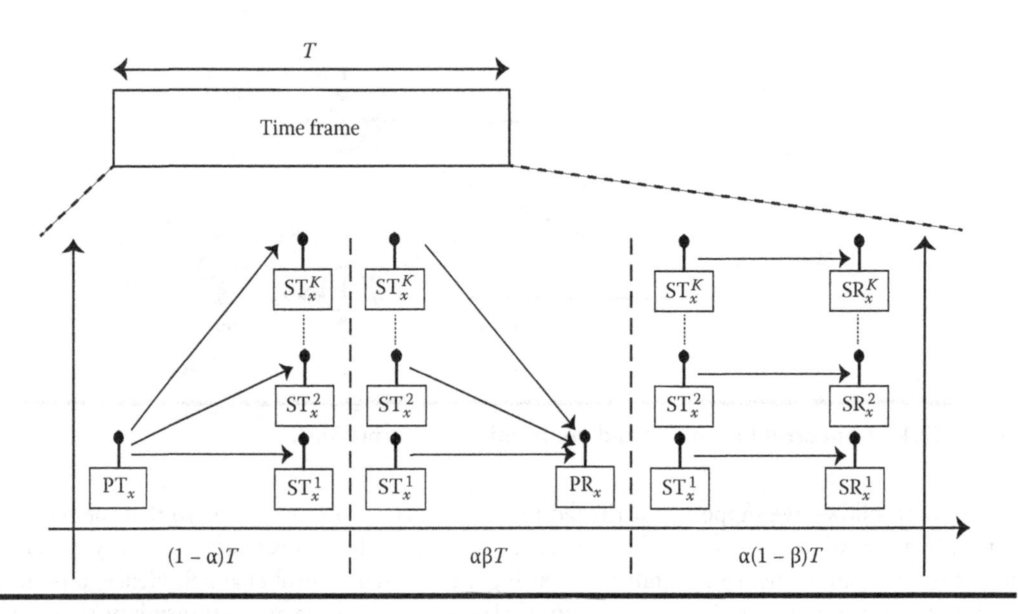

Figure 12.2 Standard three-phase transmission scheme.

stream X_S^i to the destination SR_x^i. This signal encounters a Rayleigh channel fading $h_{SS,i}$ and an additive white Gaussian noise (AWGN) n_{SS} with zero mean and power N_0, along the opportunistic path from the source to the destination. Besides, the secondary pairs sharing the same primary link can harmfully interfere with each other. The output signal Y_S^i subject to fading, noise, and mutual interference can be expressed as

$$Y_S^i = h_{SS,i}X_S^i + \sum_{j=1, j\neq i}^{K} h_{SS,ij}X_S^j + n_{SS}. \quad (12.1)$$

To provide a performance analysis of the described scheme, we characterize the end-to-end throughput at the receiving end for both primary and secondary networks. The throughput metric is an efficient measurement of the quality of a communication link in terms of how much

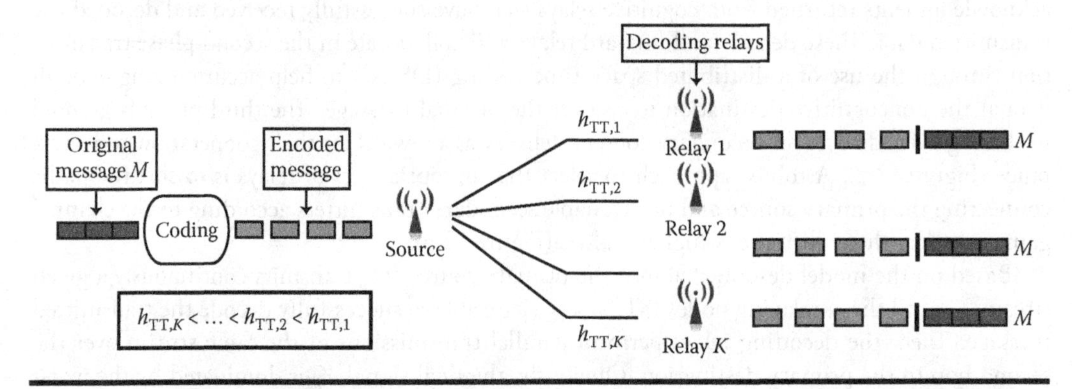

Figure 12.3 Decode-and-forward relays sorted in decreasing order of the channel gain.

data can be correctly transmitted over that link per unit of time. For a simple derivation of the closed-form expression of the throughput, we assume a constant transmit rate which is equal to the ergodic capacity. The set of relaying nodes may be seen and tackled in a holistic manner since all the members behave as a single destination dominated by the worst member, from the source node viewpoint. The same set of relays acts as an unified structure with accumulated energy at the destination node. Accordingly, the throughput of the primary communication is the minimum of the throughput in the two first phases and can be derived as

$$\eta^{\mathrm{P}}_{\mathrm{SISO}} = \min\{(1-\alpha)C^{\mathrm{TT}}_{\mathrm{SISO}},\ \alpha\beta C^{\mathrm{SP}}_{\mathrm{SISO}}\}. \tag{12.2}$$

with capacities $C^{\mathrm{TT}}_{\mathrm{SISO}}$ and $C^{\mathrm{SP}}_{\mathrm{SISO}}$ formulated as

$$C^{\mathrm{TT}}_{\mathrm{SISO}} = \log_2\left(1+\frac{\min_{i\in\{1,\ldots,K\}} |h_{\mathrm{TT},i}|^2 P_{\mathrm{p}}}{N_0}\right). \tag{12.3}$$

and

$$C^{\mathrm{SP}}_{\mathrm{SISO}} = \log_2\left(1+\frac{\sum_{i=1}^{K} |h_{\mathrm{SP},i}|^2 P_{\mathrm{s}}^{i}}{N_0}\right). \tag{12.4}$$

where P_{P} and P_{s}^{i} are the transmit power of the primary transmitter and the i-th secondary transmitter, respectively. The secondary transmitter is subjected to power constraint, $P_{\max}$.

Next, the normalized throughput at the secondary receiver is obtained from Equation 12.1 as

$$\eta^{\mathrm{s}}_{\mathrm{SISO}} = \alpha(1-\beta)\log_2\left(1+\frac{|h_{\mathrm{SS},i}|^2 P_{\mathrm{s}}^{i}}{N_0+\sum_{j=1,j\neq i}^{K} |h_{\mathrm{SS},ij}|^2 P_{\mathrm{s}}^{j}}\right). \tag{12.5}$$

The above expression represents the capacity of the cognitive link weighted by the fraction of time granted for the unlicensed utilization of spectrum.

The PU has full rights on its spectrum and is responsible for deciding on the appropriate parameters in view of time splitting (α, β) and the number of cooperating nodes K that maximize its utility in terms of the achieved throughput. Secondary links observing the decision made by the primary link act subsequently and choose the optimal power levels $(P_{\mathrm{s}}^{i})_{1\le i\le K}$ in return for the leased fraction of time β, while seeking to maximize their individual utilities too, that is, throughputs. In either case, different players attempt to find a good revenue/payment trade-off. Both the selection of the appropriate strategy and the optimization of the expected utility issues become more tractable by transforming the problem into a Stackelberg game with the primary link as a leader and the followers as the secondary links. This approach is widely adopted in the literature. However, a major handicap of such frameworks is the fact that the factor β is imposed by the primary network, which has substantial impacts on the secondary network reliability especially in the case of bandwidth-consuming applications. The primary link has tendency to choose a large β which reduces the time fraction dedicated to secondary communications and as a result secondary devices may become reluctant to cooperate, whereas cognitive users are incapable of providing high transmit powers to gain access to a larger portion of time (smaller β values) because they have tight power constraints.

In this chapter, to circumvent this bottleneck, we proceed differently by assuming a negotiated spectrum sharing (NSS) approach in which licensed users ensure explicit arrangements with interested parties to grant them spectrum access rights in exchange for monetary and nonmonetary compensations. This option allows the unlicensed users to trade their power requirements and their utility target for better spectrum sharing. The licensees may be very interested in allowing high transmit power levels at the unlicensed network side, but only with strong technical assurances that cognitive devices guarantee an interference-free environment and as long as a part of the secondary power will be used to assist the primary traffic. Moreover, the additional monetary revenues will encourage the licensed operators to satisfy and deal with the unlicensed traffic demands; in such case, these monetary incomes can be invested in strengthening the resilience of the licensed infrastructure against potential interference incidents.

In the following, we look to the predefined secondary transmit power dictated by the technical requirements of the unlicensed context and referred to as the nominal secondary power. Each secondary user targets at optimizing an utility function that represents its transmission rate and is equal to the achievable throughput minus the energy cost during the time duration of the allocated subslot. This utility is given by

$$U_{\mathrm{s}}^{i} = \alpha(\eta_{\mathrm{SISO}}^{\mathrm{s}} - cP_{\mathrm{s}}^{i}). \tag{12.6}$$

with c being the cost per unit transmission energy.

According to Reference 14, this function is concave and thus the maximum achieved utility is attained at the point where the first derivative is null. By computing the first-order derivative of U_{s}^{i} with respect to P_{s}^{i}, the above utility function reaches its maximum at

$$P_{\mathrm{s}}^{i} = \begin{cases} P_{\max}, & \text{if } P_{\mathrm{s}}^{i} \geq P_{\max} \\ \dfrac{1-\beta}{c} - \dfrac{N_0}{|h_{\mathrm{SS},i}|^2} - \sum_{j=1,j\neq i}^{K} \dfrac{|h_{\mathrm{SS},ij}|^2}{|h_{\mathrm{SS},i}|^2} P_{\mathrm{s}}^{j}, & \text{if } 0 < P_{\mathrm{s}}^{i} < P_{\max} \\ 0, & \text{if } P_{\mathrm{s}}^{i} \leq 0 \end{cases} \tag{12.7}$$

This indicates that the maximum achievable rate depends closely and only on the value of β as α appears as a multiplier in both terms in the left-hand side of Equation 12.6. Therefore, given the strict power values $(\widehat{P_{\mathrm{s}}^{j}}, j \in \{1, \ldots, K\})$ at the cognitive nodes enforced by the secondary use case, the corresponding $\hat{\beta}$ can be deduced as

$$\hat{\beta} = 1 - \frac{cN_0}{|h_{\mathrm{SS},i}|^2} - \widehat{P_{\mathrm{s}}^{i}} - \sum_{j=1,j\neq i}^{K} \frac{|h_{\mathrm{SS},ij}|^2}{|h_{\mathrm{SS},i}|^2} \widehat{P_{\mathrm{s}}^{j}}. \tag{12.8}$$

Once the optimal $\hat{\beta}$ is calculated, we need to determine the value of α in accordance with the primary throughput defined in Equation 12.2. The throughput from source to relays is a decreasing function of α and the relays-to-destination throughput is monotonically increasing as the value of α increases, thereby the optimal $\hat{\alpha}$ is obtained when both uplink and downlink achieve a same throughput value.

$$(1-\hat{\alpha})C_{\text{SISO}}^{\text{TT}} = \hat{\alpha}\hat{\beta}C_{\text{SISO}}^{\text{SP}}. \tag{12.9}$$

Hence, $\hat{\alpha}$ can be calculated as

$$\hat{\alpha} = \frac{C_{\text{SISO}}^{\text{TT}}}{C_{\text{SISO}}^{\text{TT}} + \hat{\beta}C_{\text{SISO}}^{\text{SP}}}. \tag{12.10}$$

By substituting Equation 12.10 in Equation 12.2, the optimal primary throughput is expressed as

$$\widehat{\eta_{\text{SISO}}^{\text{P}}} = \frac{\hat{\beta}C_{\text{SISO}}^{\text{TT}}C_{\text{SISO}}^{\text{SP}}}{C_{\text{SISO}}^{\text{TT}} + \hat{\beta}C_{\text{SISO}}^{\text{SP}}}. \tag{12.11}$$

To incentivize the existing licensed networks to tolerate overlay cognitive communications, the throughput achieved through the use of overlay schemes need to be larger than the throughput of the standard scheme defined as the throughput of a direct transmission between the primary transmitter and the primary receiver during the duration of one time frame:

$$\eta_{\text{DIR}}^{\text{P}} = \log_2\left(1 + \frac{|h_{\text{PP}}|^2 P_{\text{P}}}{N_0}\right). \tag{12.12}$$

In practice, the cooperative single-antenna strategy needs a radical change in spectrum regulations in terms of when spectrum bands and at which locations are enabled for secondary usage; however, it is very hard to reach such an agreement between different incumbents. Moreover, secondary users that have cooperated to help communicating the primary traffic over the first two stages have gained the spectrum right to temporarily use the licensed radio channel but the earned revenue (throughput) remains limited due to the concurrent access. An\other defect of such schemes is that only the temporal dimension has been considered for spectrum assignment and traffic scheduling which has imposed a sequential execution of the activities two and three, thereby penalizing the effective transmission time of both primary and secondary users. The temporal dimension can be combined with other forms of diversity namely, the spatial DoF, to help overcoming these drawbacks.

12.3.3 Cooperative Multiantenna Schemes

Equipped with multiple antennas, secondary transmitters can carry out their own traffics in parallel to their relaying operations, owing to the integration of the spatial dimension. Accordingly, phase two and phase three may be merged into a single phase during which the incumbent user can deliver its traffic and every secondary user may be responsible of conveying its own stream as well as acting as a relay for the PU. One of the beauties of this solution is that it allows the coexistence of the primary and secondary networks without any hardware requirements at the primary side. Besides, the primary source exploits the second phase to improve the signal strength at the receiving end. A simplified time frame structure is presented in Figure 12.4.

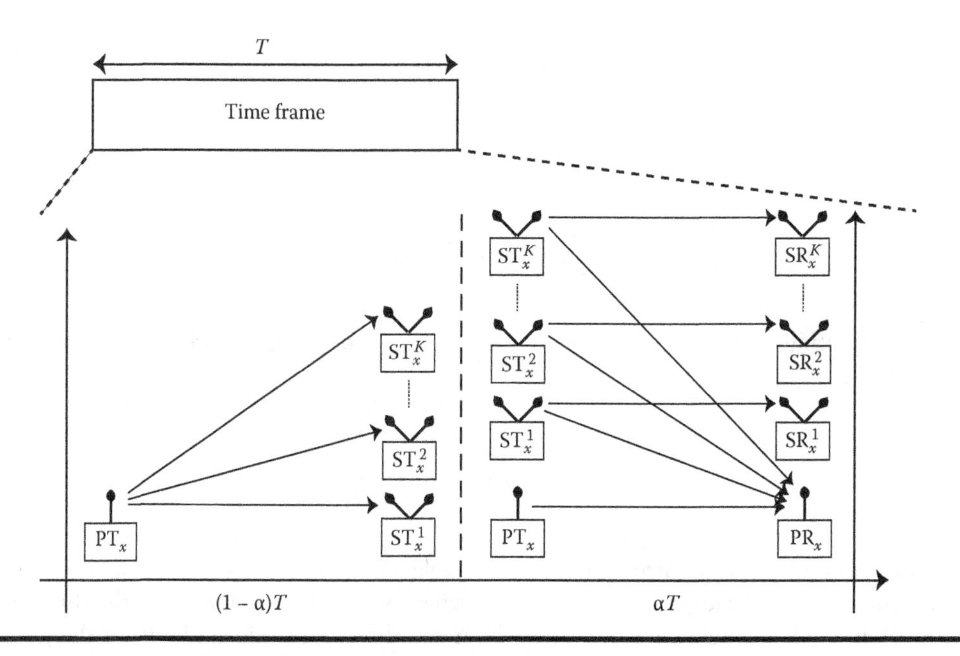

Figure 12.4 Proposed two-phase transmission scheme.

Despite the efficient use of the temporal and spatial resources in the proposed overlay scheme, the primary and secondary communications may interfere with each other causing severe attenuation of both signals. This observation shed lights on the need for robust interference mitigation techniques to combat the negative impact of interuser competition and boost the end-to-end throughput.

After successfully conducting the first phase, the primary signal is completely known to the secondary relays. The same signal is forwarded to the primary receiver during the second phase by both the primary transmitter and the secondary relays, causing harmful interference to the neighboring cognitive communications subject to power constraint, P_{max}. Each secondary transmitter can exploit the prior knowledge of the interfering signal and employ some appropriate coding schemes to precode its private message so that the interinterference with the primary link is totally canceled at the secondary receiver side. One of the most commonly adopted solutions is Dirty Paper Coding (DPC), this name arises due to the resemblance between the transmission of a stream on a frequency band with cochannel interference and the process of writing a message on a piece of paper covered with normally distributed dirt spots. In this case, however, the dirt is interference, the paper is the channel, the writer on the paper is the transmitter, and the reader is the receiver. These codes have the surprising capability of adapting the transmitted signal in accordance with the expected estimate of the interference level in order to maintain the same capacity as an interference-free environment. It is worth recalling that one drawback of using DPC is the complexity of its implementation at both the transmitter and receiver sides. However, the complexity aspect of DPC is an active research field and recent advances in this topic are expected to provide the basis for DPC frameworks with moderate computational complexity [19]. A low-complexity solution is beamforming which is an alternative technique for interference mitigation, where the signal energy is concentrated in a given direction [20]. Technically, the user's stream is multiplied by an appropriate weight vector to adjust its phase and amplitude. However, the tricky part in this technique is computing the optimal weight vectors especially in multiuser scenarios.

As the interfering primary signal is completely canceled at the secondary receiver side using appropriate DPC, the expression of the received signal at the secondary destination takes into account only the mutual interference with the concurrent secondary pairs and can be described as

$$Y_S^i = \sqrt{P_s^i}\, h_{SS,i} X_S^i + \sum_{j=1, j\neq i}^{K} \sqrt{\delta P_s^j}\, h_{SS,ij} X_P + \sum_{j=1, j\neq i}^{K} \sqrt{(1-\delta) P_s^j}\, h_{SS,ij} X_S^j + n_{SS} \tag{12.13}$$

where δ is the fraction of the power P_s^j devoted to relaying the primary message.

Assuming the receiving PU accumulates the energy from its associated primary transmitter as well as the relaying nodes by exploiting the maximum ratio combining (MRC) method, the resulting SNR is the sum of all the SNRs of each transmitter. Accordingly, the capacity of the second hop is reformulated as

$$C_{MIMO}^{SP} = \log_2\left(1 + \frac{|h_{PP}|^2 P_p + \sum_{i=1}^{K} |h_{SP,i}|^2 \delta P_s^i}{N_0 + \sum_{i=1}^{K} |h_{SP,i}|^2 (1-\delta) P_s^i}\right). \tag{12.14}$$

whereas the first-hop capacity remains invariant, that is, $C_{MIMO}^{TT} = C_{SISO}^{TT}$. The formula of the achievable primary throughput is always the same:

$$\eta_{MIMO}^{P} = \min\{(1-\alpha) C_{MIMO}^{TT}, \alpha C_{MIMO}^{SP}\}. \tag{12.15}$$

Using the expression of the output cognitive signal in Equation 12.13, the secondary throughput takes the form

$$\eta_{MIMO}^{s} = \alpha \log_2\left(1 + \frac{|h_{SS,i}|^2 (1-\delta) P_s^i}{N_0 + \sum_{j=1, j\neq i}^{K} |h_{SS,ij}|^2 \delta P_s^j + \sum_{j=1, j\neq i}^{K} |h_{SS,ij}|^2 (1-\delta) P_s^j}\right). \tag{12.16}$$

The optimal power allocation is a fundamental issue in multiple transmit-antennas communication systems. The power can be unequally or equally allocated to each antenna depending on whether or not the transmitter is provided with complete CSI. The first case relies on the concept of "water-filling," which assigns more power to the channel with good fading conditions and less to the worst channels. On the contrary, the latter case corresponds to the situation when all transmitting antennas have the same input power irrespective of the CSI, unknown to the transmitter device. Power control mechanisms are beyond the scope of this chapter, as all the transmitting relays have the same power level, P_s. Further, we assume equal power allocation at the transmitting antennas ($\delta = 1/n_T$, with n_T the number of the transmitting antennas at each secondary source). The optimal fraction α is computed in a similar manner as the previous scheme and is given by the point for which both hops achieve equal throughputs.

To corroborate the above theoretical design, some experimental simulations have been undertaken in the following to investigate the throughput benefits of the examined schemes. These simulations have been performed using a large number of random fading samples generated through Monte Carlo experiments.

For these experiments and unless otherwise stated, a primary transmitter–receiver pair selects a set of four competing secondary transmitter-receiver pairs whose transmitter is willing to cooperate with the primary network. In order to achieve optimal cooperation, we assume that all the chosen relays are located at the same position on the straight line between the primary transmitter and the primary receiver, at a normalized distance $d = 0.3$ from the primary transmitter. For the relaying operation, we assume a large-scale path loss model with a coefficient $\zeta = 2$, and as a result the average power gains on the source-to-relay and the relay-to-destination channels are given as $E[|h_{TT,i}|^2] = 1/d^{\zeta}$ and $E[|h_{SP,i}|^2] = 1/(1-d)^{\zeta}$ ($\forall i \neq j \in \{1, \ldots, K\}$), respectively. For the secondary competing links, the small scale Rayleigh fading gives rise to the following average channel gains: $E[|h_{SS,i}|^2] = 20$ dB and $E[|h_{SS,ij}|^2] = 1$ ($\forall i \neq j \in \{1, \ldots, K\}$). Likewise, the average channel gain on the primary link is $E[|h_{PP}|^2] = 1$. The transmit power for the primary source as well as the maximum tolerated power at the secondary transmitters is fixed at $P_{p,dB} = P_{max,dB} = 1$. The variance of noise is $N_0 = 1$ and the cost per unit energy is $c = 0.1$.

Figure 12.5 examines the impact of the secondary transmit power on the average primary throughput for the case where only the best cognitive node, the one with the highest channel gain from the primary source, is aiding the primary communication ($K = 1$). The throughput of the direct transmission on the primary link, η^{P}_{DIR}, is also provided as a reference measure. For different antenna configurations, we state that the obtained throughput remains smaller than the throughput of a direct transmission between the primary source and the intended destination. Single-relay aided overlay schemes do not leverage the benefits of cooperation. The multirelay case is worth inspecting since a set of cooperating nodes can assist the destination to promptly gather enough packets for original content recovery.

In what follows, we assume multirelay diversity scenario ($K = 4$).

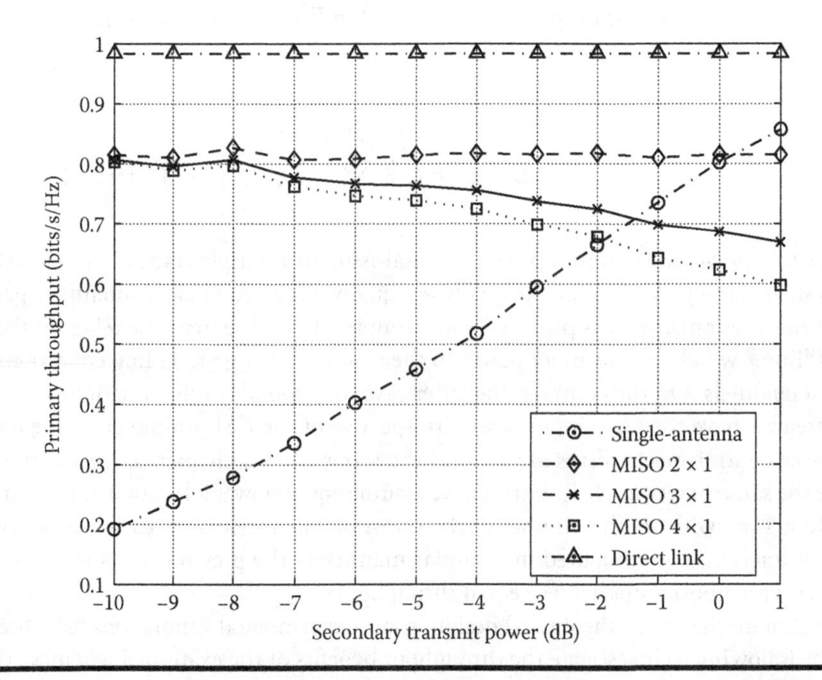

Figure 12.5 Primary throughput versus transmit power for various overlay designs without cooperation.

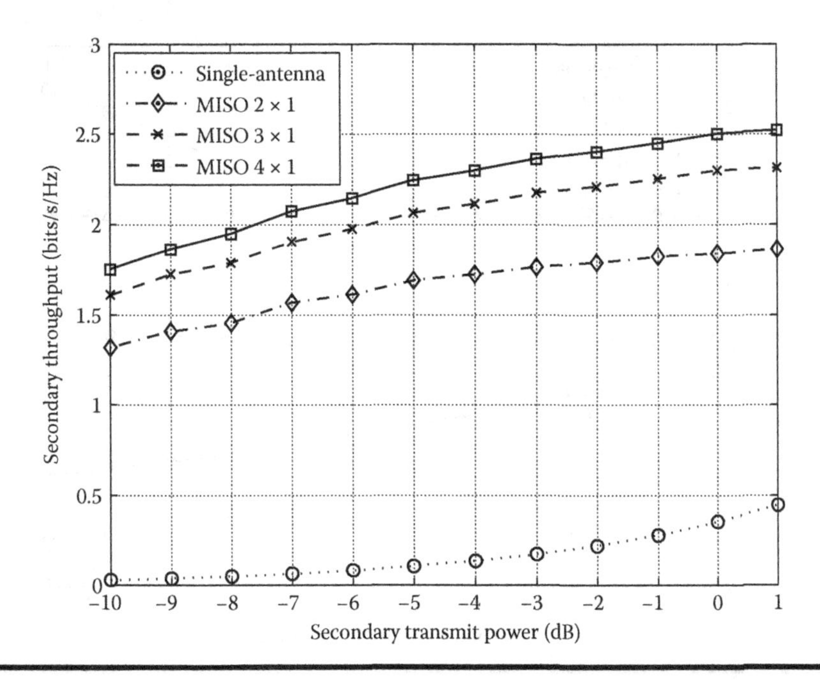

Figure 12.6 Secondary throughput versus transmit power for various overlay designs without interference-avoidance techniques.

Figure 12.6 illustrates the secondary throughput of the conventional overlay scheme compared to that achieved by the proposed MIMO-based overlay framework as a function of the secondary transmit power. It is clearly observed that the proposed MIMO-assisted scheme yields better efficacy compared to the conventional architecture. As a matter of fact, secondary devices exploit the extra diversity offered by the spatial dimension to acquire a larger fraction of time rather than being constrained to a restricted interval. Moreover, the following conclusions are drawn concerning the cooperative single-antenna model. In fact, the lower the transmit power of the secondary device, the shorter the time leased for the cognitive transmission which degrades the throughput of the secondary link. Conversely, higher transmit powers are compensated by a larger time interval dedicated for the cognitive communication, resulting in an improved secondary throughput. On the other hand, for higher values of P_s ($P_{s,dB} \geq 0$ dB), we noticed that the secondary transmit power has little impact on the throughput of the cooperative MIMO scheme because there is no specific portion of time devoted to cognitive transmissions, and the secondary transmit power is spent for simultaneous relaying and transmitting. Meanwhile, the secondary throughput of the cooperative multiantenna scheme gets higher as long as more antennas are implemented regarding the fact that the secondary signal is sent over all the transmitting antennas leading to enhanced SNR at the receiving end.

The primary throughput performance of the two schemes under different values of the secondary transmit power are plotted in Figure 12.7. We state that for higher values of the secondary transmit power, P_s ($P_{s,dB} \geq -6.3$ dB), the conventional cooperative system outperforms the proposed multiantenna design, despite increasing the number of the transmitting antennas (up to 4). The cooperative mode becomes gainful only for transmit power values greater than −3.3 dB in the single-antenna scenario, as the achieved throughput surpasses the throughput of a direct transmission between the primary source and the primary destination. This is due to the fact that

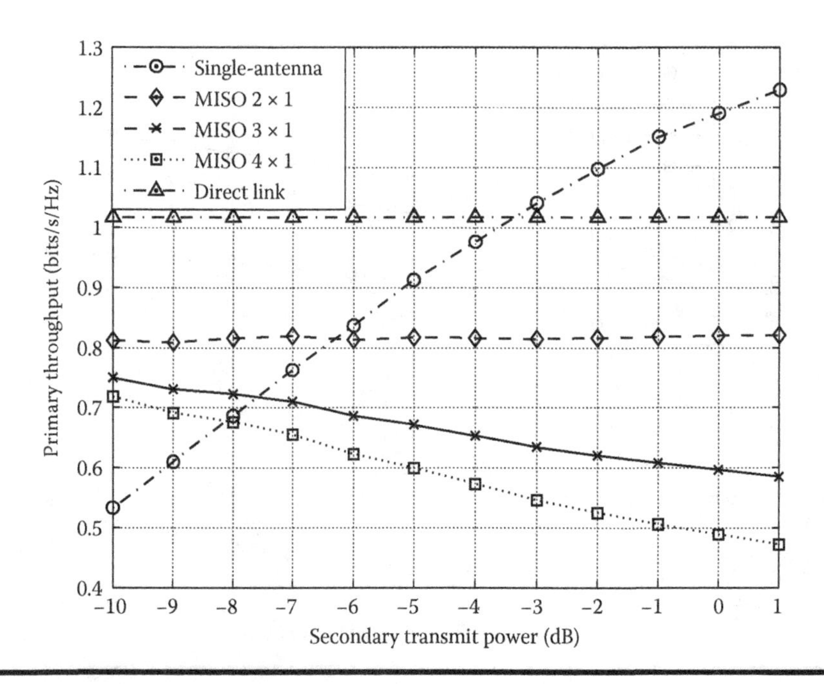

Figure 12.7 Primary throughput versus transmit power for various overlay designs without interference-avoidance techniques.

with sufficient number of single-antenna relays collaborating during the second hop transmission, sufficient data packets can be promptly accumulated at the receiver side and thus the throughput target $\eta_{\mathrm{DIR}}^{\mathrm{P}}$ becomes more likely to be achieved or even exceeded. The MIMO-based cooperation, on the contrary, delivers better results for lower levels of transmit power ($P_{\mathrm{s,dB}} \leq -6.3$ dB) but never reaches the throughput of a direct primary link transmission. The primary signal is severely affected by interference whose source is the neighboring secondary links, which prevents the licensed user from communicating at larger throughput values. Under its current form, a MIMO-based cooperative CRN is not a viable solution.

12.3.4 Cooperative Multiantenna Schemes with TDMA Scheduling

TDMA schemes have found great popularity as being immune to interference conditions. A transmission schedule based on TDMA sharing is adopted for the concurrent unlicensed access. In particular, the second phase will be partitioned into K time slots with the same duration T/K, separately allocated to competing cognitive communications. Each SU will continuously relay the primary traffic during all time slots, despite being assigned one time slot for its private communication (Figure 12.8). This approach helps avoiding the interference originated by the remaining secondary links, albeit less time is left for each cognitive communication. After eliminating the term of the secondary interinterference, the capacity of the relays-to-destination hop simplifies as

$$C_{\text{T-MIMO}}^{\text{SP}} = 1/K \sum_{i=1}^{K} \log_2 \left(1 + \frac{h_{\mathrm{PP}}^2 P_{\mathrm{p}} + \sum_{i=1}^{K} |h_{\mathrm{SP},i}|^2 \delta P_{\mathrm{s}}^i}{N_0 + |h_{\mathrm{SP},i}|^2 (1-\delta) P_{\mathrm{s}}^i} \right). \tag{12.17}$$

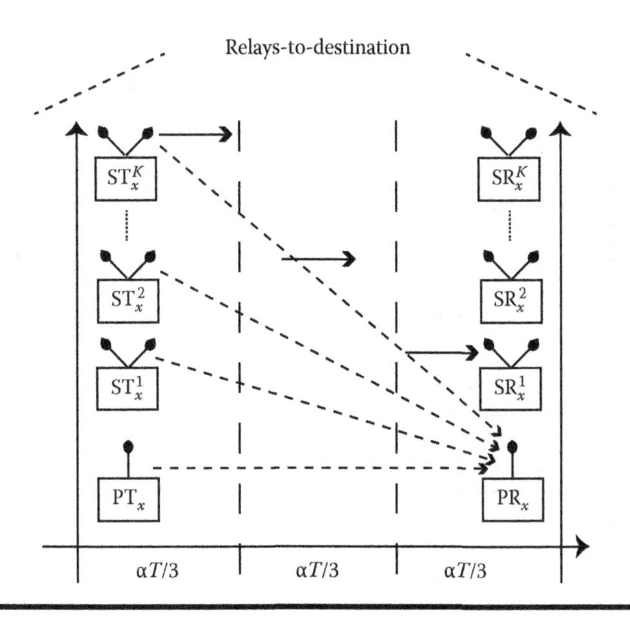

Figure 12.8 Proposed TDMA-based two-phase transmission scheme.

The expression of the source-to-relays link capacity has not changed, that is, $C_{\text{T-MIMO}}^{\text{TT}} = C_{\text{MIMO}}^{\text{TT}}$, and the primary throughput is that of the hop with the lowest ergodic capacity:

$$\eta_{\text{T-MIMO}}^{\text{P}} = \min\{(1-\alpha)C_{\text{T-MIMO}}^{\text{TT}}, \alpha C_{\text{T-MIMO}}^{\text{SP}}\}. \tag{12.18}$$

Afterwards, the secondary throughput will become

$$\eta_{\text{T-MIMO}}^{\text{s}} = \frac{\alpha}{K}\log_2\left(1+\frac{|h_{\text{SS},i}|^2(1-\delta)P_{\text{s}}^{i}}{N_0+\sum_{j=1,j\neq i}^{K}|h_{\text{SS},ij}|^2\,\delta P_{\text{s}}^{j}}\right). \tag{12.19}$$

To confirm whether the proposed interference mitigation mechanism fits well in the cooperative MIMO setup we introduced, Figure 12.9 assesses the performance of the proposed TDMA-based cooperative MIMO scheme in terms of the average secondary throughput versus the cooperative SISO scheme, plotted against the secondary transmit power. It is worth noting that the incorporation of the TDMA sharing in the cooperative MIMO scheme has led to a sharp deterioration of the secondary throughput, albeit being always better than the secondary throughput achieved by the cooperative SISO scheme. Separating the secondary users in time throughout the second phase of the cooperative MIMO scheme reduces the amount of time available for each cognitive communication which negatively impacts the resulting throughput. Nevertheless, the TDMA-based cooperative MIMO design still performs better than the cooperative SISO model and the throughput gain is more predominant for lower transmit power values. Notice also that the more the number of antennas, the larger the secondary throughput because all these antennas (except for the one dedicated to relaying) are involved in the secondary service provision.

Figure 12.10 shows the evolution of the primary throughput as a function of the secondary transmit power for different antenna configurations. Cooperation and MIMO coupled

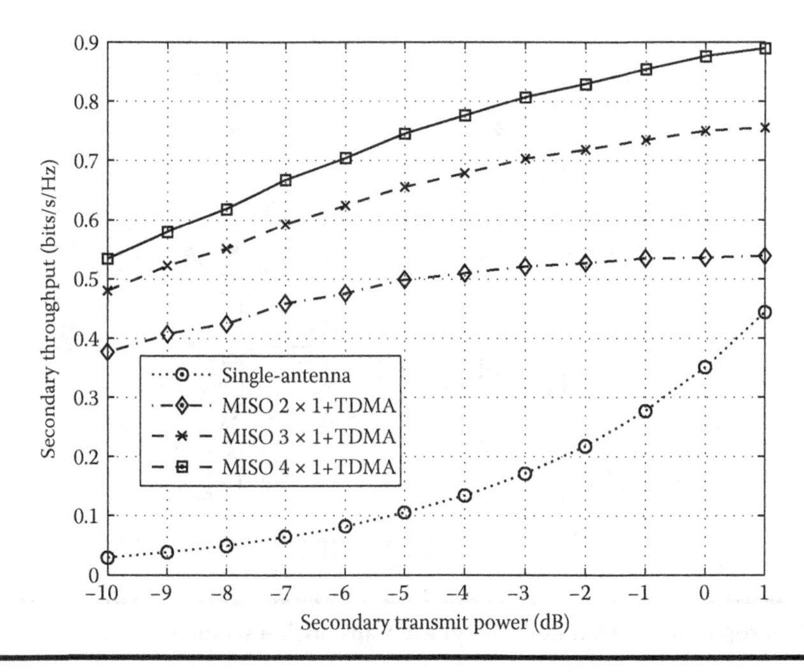

Figure 12.9 Secondary throughput versus transmit power for various overlay designs with TDMA sharing.

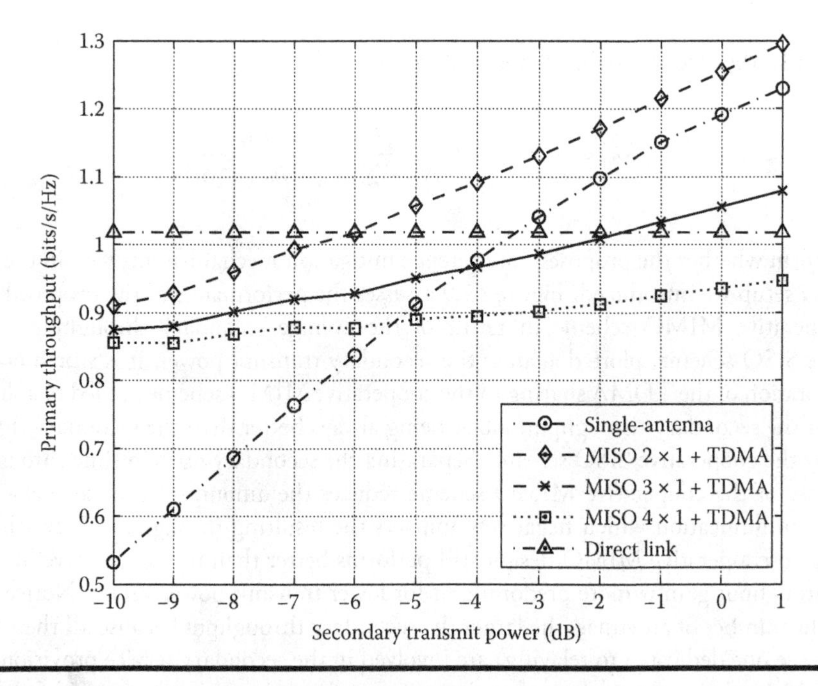

Figure 12.10 Primary throughput versus transmit power for various overlay designs with TDMA sharing.

with TDMA multiplexing have entailed the primary throughput amelioration and the obtained throughput for the two-antenna configuration exceeded those achieved by the direct primary link transmission and the classical single-antenna setup for $P_{s,dB} \geq -6$ dB. Another interesting observation we make is the fact that the achieved throughput drops as the number of transmitting antenna increases. In this way, the common belief that deploying more antenna at the transmitter side provides added capacity to communication systems not necessarily apply in CR scenarios. The number of transmitting antennas should be adjusted carefully in accordance with various environmental parameters to avoid the problem of multipath propagation.

Figure 12.11 depicts the impact of the secondary transmit power on the optimal parameters $\hat{\alpha}$ and $\hat{\beta}$ for different antenna designs with TDMA scheduling. It is clearly noticed that the parameter $\hat{\beta}$ decreases as the secondary transmit power gets increased for the single-antenna configuration. The parameter $\hat{\alpha}$ presents a similar behavior with a varying decrease tendency for each scheme. In fact, the parameter $\hat{\alpha}$ decreases slowly in the proposed cooperative MIMO framework as opposed to the cooperative SISO scheme where the decrease is fast and more significant. Clearly, by increasing the secondary transmit power in the cooperative SISO setup a larger time need to be leased to the cognitive transmission and thus the duration of the relaying phase becomes shorter. However, higher secondary transmit powers cut down the time needed by the destination to gather necessary packets because the energy is summed over all the intermediate nodes, and as a direct consequence more time may be granted to all relays to accomplish receiving enough data packets from the primary source. Extending the duration of the third phase constitutes an incentive for secondary devices to cooperate, but shortens the duration of the primary communication. This observation translates into a trade-off between the throughput gained by secondary users and the overhead added to the licensed communication.

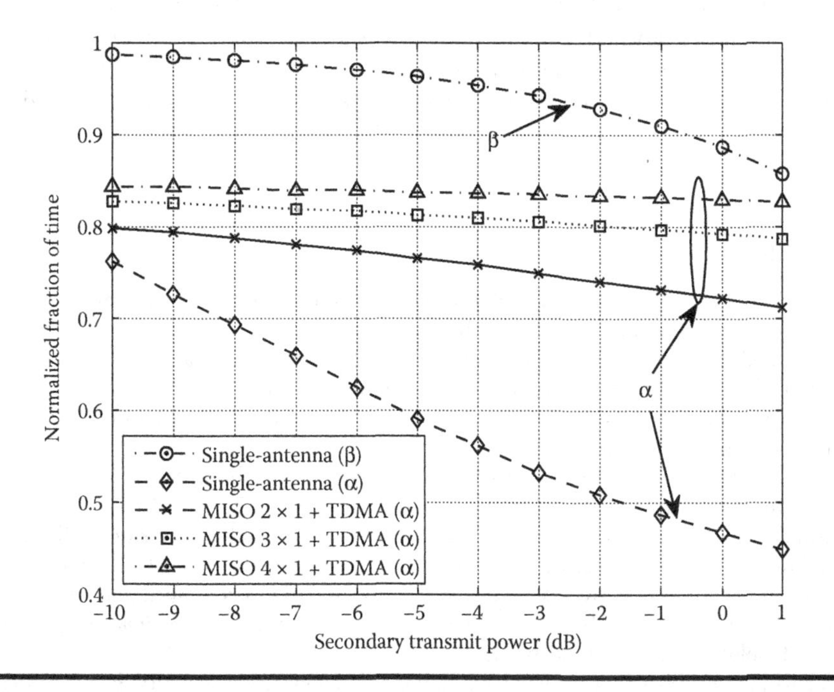

Figure 12.11 Time fractions, α and β, for various overlay designs with TDMA sharing.

12.4 Conclusions

In this chapter, the embrace of MIMO features within CR networks has been investigated in the context of cooperative dual-hop multirelay systems. A transparent cohabitation of primary and secondary systems over the same wireless channel has been made possible using the principle of dynamic spectrum leasing. In particular, the licensed transmitter forwards its traffic to a chosen set of multiantenna cognitive nodes during the first hop, these intermediate nodes act as relays and transmit the licensed traffic to the target destination during the second hop while simultaneously each relay uses the same channel for its own transmission purposes as a compensation for its assistance. For further insight into the proposed framework performance, secondary communications are scheduled according to a TDMA sharing to reduce the effects of interference originated by the competing secondary pairs, in which case only one SU can access the wireless channel for its private transmission at any given time while all SUs keep relaying the primary traffic during the whole time frame. A theoretical model has been developed with the objective of finding an adequate time splitting mechanism, which converges into a maximization problem of a given utility function, namely the overall throughput, for both primary and secondary networks. Besides, representative numerical simulations are performed to analyze the throughput of the cohabiting systems separately. As a first result, we assert that for cooperation to be beneficial, the primary network needs to recruit more than one secondary relay. In addition, it has been shown that the proposed MIMO framework without TDMA access suffers harmful interference due to the dynamic access of multiple secondary users, which entails a sharp degradation of the primary throughput up to 50% for the two-antenna case in comparison with the single-antenna configuration. The use of TDMA has lowered the generated interference and consequently the throughput gain has reached 5% and 23% for the primary network and the secondary network, respectively, at high secondary transmit powers using a two-antenna setup compared to a single-antenna setup. Moreover, the common observation that the throughput increases linearly with the number of antennas not necessarily applies to overlay CR frameworks; however, the two-antenna design has shown to be superior over the other configurations with respect to the optimal primary–secondary throughput trade-off. These deep discussions highlight the need for accurately setting different system parameters to find out a good compromise between the achieved throughput and the resulting interference.

References

1. Mitola, J., III and G. Q. Maguire. 1999. Cognitive radio: Making software radios more personal, *IEEE Personal Communications*, 6(4), 13–18.
2. Weiss, T. and F. Jondral. 2004. Spectrum pooling: An innovative strategy for the enhancement of spectrum efficiency, *IEEE Communications Magazine*, 42, 8–14.
3. Summit conclusions. Available at: <http://www.3gpp.org/news-events/conferences/1515-etsi-summit-on-future-mobile-and>.
4. Hong, X., J. Wang, C. X. Wang, and J. Shi. 2014. Cognitive radio in 5G: a perspective on energy-spectral efficiency trade-off, *IEEE Communications Magazine*, 52(7), 46–53.
5. Goldsmith, A., S. A. Jafar, I. Maric, and S. Srinivasa. 2009. Breaking spectrum gridlock with cognitive radios: An information theoretic perspective, *Proceedings of the IEEE*, 97(5), 894–914.
6. Hesammohseni, S. A., K. Moshksar, and A. K. Khandani. 2014. A combined underlay and interweave strategy for cognitive radios. In *Proceedings of the IEEE International Symposium on Information Theory (ISIT)*, Honolulu, HI, USA, June, pp. 1396–1400.
7. Sendonaris, A., E. Erkip, and B. Aazhang. 2003. User cooperation diversity. Part I. System description, *IEEE Transactions on Communications*, 51(11), 1927–1938.

8. Ahmed, M. H. and S. S. Ikki. 2011. To Cooperate or not to Cooperate? That is the Question!. In *Cooperative Networking*, eds. M. S. Obaidat and S. Misra, pp. 21–33. John Wiley & Sons, Ltd, Chichester, UK.
9. Biglieri, E., R. Calderbank, A. Constantinides, A. Goldsmith, A. Paulraj, and H. V. Poor. 2007. *MIMO Wireless Communications*. Cambridge, UK, Cambridge University Press.
10. Jindal, N., J. G. Andrews, and S. Weber. 2009. Rethinking MIMO for wireless networks: Linear throughput increases with multiple receive antennas. In *Proceedings of IEEE International Conference on Communications (ICC)*, Dresden, Germany, June, pp. 1–6.
11. Wang, B., Y. Wu, and K. J. Liu. 2010. Game theory for cognitive radio networks: An overview, *Computer Networks*, 54(14), 2537–2561.
12. Wang, X., W. Chen, and Z. Cao. 2009. A rateless coding based multi-relay cooperative transmission scheme for cognitive radio networks. In *Proceedings of IEEE Global Telecommunications Conference (GLOBECOM)*, Honolulu, HI, Nov. 30 2009–Dec. 4 2009, pp. 1–6.
13. Chaoub, A. and E. Ibn-Elhaj. 2014. Multimedia transmission over cognitive radio networks using decode-and-forward multi-relays and rateless coding. In *Proceedings of International Conference on Communications and Networking (ComNet)*, Hammamet, Tunisia, March, pp. 1–5.
14. Simeone, O., I. Stanojev, S. Savazzi, Y. Bar-Ness, U. Spagnolini, and R. Pickholtz. 2008. Spectrum leasing to cooperating secondary ad hoc networks, *IEEE Journal on Selected Areas in Communications*, 26(1), 203–213.
15. Zhang, J. and Q. Zhang. 2009. Stackelberg game for utility-based cooperative cognitive radio networks. In *Proceedings of the Tenth ACM International Symposium on Mobile ad hoc Networking and Computing*, New Orleans, LA, USA, May, pp. 23–32.
16. Hua, S., H. Liu, M. Wu, and S. S. Panwar. 2011. Exploiting MIMO antennas in cooperative cognitive radio networks. In *Proceedings of IEEE INFOCOM*, Shanghai, China, April, pp. 2714–2722.
17. Scutari, G., D. P. Palomar, and S. Barbarossa. 2008. Cognitive MIMO radio, *IEEE Signal Processing Magazine*, 25(6), 46–59.
18. Ghamari Adian, M. and M. Ghamari Adyan. 2014. Optimal and suboptimal resource allocation in MIMO cooperative cognitive radio networks, *Journal of Optimization*, 2014, 13p.
19. Shilpa, G., A. Thangaraj, and S. Bhashyam. 2010. Dirty paper coding using sign-bit shaping and LDPC codes. In *Proceedings of IEEE International Symposium on Information Theory (ISIT)*, Austin, Texas, USA, June, pp. 923–927.
20. Zheng, G., S. Ma, K. K. Wong, and T. S. Ng. 2010. Robust beamforming in cognitive radio, *Wireless Communications*, 9(2), 570–576.

ADVANCED TOPICS

V

Chapter 13

Directional Antennas and Beamforming for Cognitive Radio-Based Wireless Networks

Abdelaali Chaoub and Elhassane Ibn-Elhaj

Contents

13.1 Introduction

Directional antenna is a revolutionary paradigm that has gained high user popularity during the last decades. Research community started to realize the potential of directional antennas to spur further achievements in emerging radio frequency (RF) technologies such as cognitive radio (CR). This will permit cognitive terminals to intelligently direct and concentrate the transmitted signal toward a particular destination in order to maximize the energy benefit and reduce the unintended interference to users in vicinity.

This chapter aims primarily at highlighting the improvements brought by innovative directional antenna designs to spectrum sharing-based contexts. This implies tackling some pertinent questions regarding the inefficiencies and limitations related to secondary traffic scheduling over spectrum holes. Recent research advances suggest the use of mixed spectrum sharing strategies to leverage the benefits of interweave, underlay, and overlay paradigms all together. On this basis, we employ a hybrid solution referred to as "Always Transmit," wherein the cognitive node both senses its surrounding environment and underlays its signal over the shared portion of spectrum at the beginning of the time frame. For the rest of the time, the cognitive node continues to transmit using either the underlay or the interweave mode contingent on the outcome of the sensing operation. Two separate antennas are used to accomplish both sensing and transmit functions. For further performance enhancement, an improved agile strategy is introduced in which each cognitive node will incorporate multiple directional antennas with frequency-agile capabilities to support higher-traffic quantities. Moreover, we advocate for the concept of beamforming due to its low complexity over other similar solutions for energy optimization and interference alleviation. We reformulate the expressions of the overall ergodic capacity of the primary and secondary networks separately according to these suggested changes. In light of the above theoretical analysis, a simulation campaign has been performed and insightful conclusions are reported regarding the ability of directional antennas to boost the secondary capacity while necessarily keeping the primary capacity intact.

The rest of this chapter is organized as follows. Section 13.2 enumerates the motivations behind this study. A brief recapitulation of antenna designs, their merits and limitations, is provided. In particular, we invoked the concept of wideband and narrowband antennas in CR contexts. Besides, this chapter describes contributions and publications related to the current research topic. We present in Section 11.3 the time frame model that is adopted and give a detailed description of the new proposed framework. In particular, the properties of directional antenna are investigated to extend a state of the art hybrid strategy for spectrum sharing called AT. Later, analytical formulations of the primary and secondary ergodic capacities are computed. Numerical simulations are conducted in Section 13.4 to assess the performance of the proposed framework in terms of the ergodic capacity compared to the conventional AT scheme. Insightful comments are made concerning the potential of directional antennas to evolve unlicensed networks to support higher-traffic volumes without harming the licensed infrastructure. At the end, Section 13.5 draws the conclusions of this chapter.

13.2 Directional Antenna-Based Cognitive Radio Systems

13.2.1 Cognitive Radio: Preliminaries

The radio spectrum is the most valuable resource in telecommunication industry. This resource, naturally finite, is today rare and expensive due to many factors such as the enormous proliferation of "quality hungry" services and the inefficiency of the actual spectrum regulatory policies. As a matter of fact, various measurement campaigns for spectrum occupancy statistics conducted at many locations and during many time intervals have revealed that spectrum bands are underutilized in several instances due to the inefficient use of frequencies, mostly in rural and sparsely populated areas. On the other hand, the traditional policy for spectrum governance based on exclusive and static spectrum assignments prohibits the access to the allocated frequencies even those who are poorly and scarcely used. Since the radio spectrum is the main drive for all communication technologies, it becomes increasingly imperative to provide a new frequency exploitation process. The concept of CR [1] has proven to be the answer for most of these demands concerning radio and wireless technologies.

Wireless technologies worldwide require smart and agile radios which can connect, associate, link, and transact opportunistically. Such capabilities require an advanced level of intelligent sensing and decision making, and this is only possible with CRs or the so-called software defined radios (SDRs). The concept of CR was first proposed by Joseph Mitola in his publications and his PhD thesis [2] during 1999. This term refers to a radio device that is aware of its surrounding environment and use different mechanisms to adapt its internal parameters to the variation of the spectrum availability. Cognitive devices can be reconfigured independently to choose the best network and the best frequency bands to ensure optimal service to unlicensed users. It can be reasonably argued that the low and discontinuous use of the licensed (primary) spectrum leaves white spaces that can be exploited by other, unlicensed (secondary) traffics, provided that each secondary user (SU) promptly evacuates the wireless channel as soon as the corresponding primary user (PU) is detected [3].

In the recent past, there has been an intense research in the area of CR technical standardization [4] to bridge the gap between theory and practical achievements. For instance, numerous attempts are made to establish universal standards for CR namely, IEEE 802.22, IEEE 802.11af, and IEEE 802.19.1. IEEE 802.22 aims at deploying wireless regional area network (WRAN) infrastructures in rural and remote areas and is able to attain an aggregate data rate of up to 23 Mbps. It is the most mature CR standard at this time since it has started the process of standardization way back in October 2004. The IEEE 802.11af task group is the IEEE standard working on the issue of providing WIFI services over television (TV) bands to achieve higher throughput data rates and faster connections, and this task group was formed during December 2009. While using a certain band of spectrum, the secondary system must not only avoid disturbing the neighboring licensed network but also obtain awareness about other coexisting secondary systems. Therefore, standards like IEEE 802.19.1 have been proposed to address the coexistence issue between several secondary technologies operating over TV bands, and this project was approved way back in December 2009. Further, the ECMA-392 standard deals with the Media Access Control (MAC) and physical (PHY) parts to enable the operation of portable devices over TV white spaces, and it was published by the end of 2009. Moreover, the IEEE has recently (September 2011) launched the task group IEEE 802.15.4m (802.15.4 over TVWS) with the goal of enabling wireless personal area network (WPAN) infrastructures over TV bands.

13.2.2 Fundamental Tasks of the Cognition Cycle

CR standards must endow their compliant devices with the capability to periodically sense the spectrum and to communicate subsequently. However, to ensure safe delivery of the secondary content over the available spectrum holes, two crucial aspects must be handled. On the one hand, to initiate the secondary transmission under the most favorable conditions, the surrounding environment must be observed to predict the state of each frequency and seek for the best spectrum opportunities. On the other hand, the opportunistic access of cognitive devices to shared spectral resources is a source of conflicts between primary and secondary systems and between the competing secondary systems themselves, thus each cognitive transmission must alleviate the interference caused to its surrounding environment. To meet the first challenge, an omnidirectional antenna is placed on each cognitive device to be able to obtain awareness about the spectrum occupancy over multiple dimensions such as time, space, frequency, code, and azimuth angle. Spectrum sensing is one of the most critical functions of a CR since it imposes tight requirements in terms of accuracy in detecting the primary activity to be able to make real-time decisions about the free spectrum. With respect to the second challenge, the secondary source must steer the energy toward the direction of the intended destination by virtue of using directional antenna to easily orient the transmitter beam and as a result avoiding the unwanted radiation.

Based on the above analysis, spectrum scanning and opportunistic transmission are two fundamental tasks performed on the physical air interface of wireless CR networks. This interface is necessarily served by at least one antenna with its indispensable circuitry. To date, antenna designs are gaining further momentum worldwide. An antenna is a special metallic component used for transmitting or receiving radio waves and most modern telecommunication technologies are concerned with the integration of one or multiple antenna elements to avoid the need for wiring. A recent technological trend in telecommunication industry is to manufacture small and less cumbersome antennas owing to the explosive growth of smartphones and the need for miniaturization to stack many antennas in a small volume.

13.2.3 Radio Elements of CR Networks

Nowadays, the antenna technology and its directional properties [5] have great potential to handle the problem of spectrum crunch by enabling the reuse of licensed frequencies and thus conserving spectral resources, bandwidth, and transmit power. For compact antenna systems, it is preferred to have a single antenna performing both sensing and transceiver tasks; however, the development of such antennas poses technical difficulties and significantly impedes the throughput efficiency. Parallel execution of both operations can be ensured by means of separate sets of antennas, namely, wideband (omnidirectional) antennas for RF environment sensing and narrowband (directional) antennas for communication purposes (with at least 0 dBi gain). The second set of antenna must be reconfigurable [6] to support operating and switching between multiple frequencies of interest. The reconfiguration property includes many advanced features as the choice of the most suitable frequency for an upcoming transmission, the maximum transmission power, the modulation rate, the angle of arrival for directional transmissions, the number of the antennas, and so on. Recent progress in hardware designs, on the other hand, has allowed wideband antenna setups using a combination of a sufficient number of directional antennas each with a different radiation pattern that is, different direction of the major lobe. Hence, we believe that directional antennas constitute the baseline component for emerging CR networks.

In research literature, CR networks are categorized according to three transmission strategies: interweave, underlay, and overlay. This classification depends on the way the spectrum is shared between cognitive and noncognitive users while higher priority is given to the latter type of users. The interweave CR interprets the collected sensing measurements to infer a global view of the surrounding environment and the most accurate in terms of spectrum occupancy in order to fill the spectrum gaps. Secondary devices must be evicted from these free channels whenever the corresponding primary traffic returns. The underlay paradigm allows the unlicensed users to communicate concurrently with the licensed users provided that the secondary transmit power stays below the interference temperature constraint of the primary system. Exceeding the predefined tolerable interference threshold may degrade dramatically the primary signal. Using the overlay approach, the secondary network should maintain a close interaction with the primary system as a significant part of the secondary transmit power will be exploited to strengthen the primary signal. All the primary codebooks are first transferred to the secondary system, which acts as a relay and will subsequently disseminate data to the intended destination. Afterward, the secondary devices are granted channel access for a given portion of time for their private communications as a compensation for their assistance. Interweave, underlay, and overlay paradigms can be mixed to provide efficient holistic approaches, considering a CR as a combination of the aforementioned transmission modes. Mixed schemes leverage the advantages of the conventional strategies for better service provision and adaptability to fast-varying spectrum conditions.

13.2.4 Related Works and Background

In recent studies, the topic of directional antenna designs for CR networks has raised huge interest among researchers.

In an invited article published online on the website of the IEEE Antennas and Propagation Society [7], Christodoulou asked a pertinent question and wondered "what role can antenna designs play in the context of cognitive radio networks?" The answer was positive and, typically, a fundamental requirement is to be able to manufacture innovative antennas with various resonant frequencies and supporting multiple standards. The ideal CR antenna should allow both to uniformly scan the environment over a wide frequency range and also to communicate over a set of desired frequencies, probably in a specific direction while eventually excluding some frequency channels in order to limit the resulting interference. Moreover, the antenna system should be able to learn and adaptively reconfigure itself.

An optimal detector with a prior knowledge about channel gains, noise variance, and licensed signal variance is introduced in Reference 8 exploiting multiple sensing antennas. Besides, a general likelihood ratio (GLR) detector is derived for the practical case where some or all of the above parameters are unknown. The performance of the proposed detectors has been evaluated in terms of false alarm and detection probabilities.

In Reference 9, two novel antenna designs are proposed, wherein wideband and narrowband antennas are integrated into a limited volume in order to fit well the specific needs of modern smartphones. In the first design, the wideband antenna is a coplanar waveguide (CPW) fed wine-glass shape monopole and the narrowband one is a shorted microstrip patch antenna printed on the reverse side of the substrate. This type of integrated antennas enables handling parallel sensing and communication. The second antenna model is a monopole antenna combined with a microstrip patch antenna along with switches. This antenna has the ability of transparently switching between wideband and narrowband modes. Negligible discrepancies are observed between simulations (software tool) and real measurements (prototype antenna).

Frequency and direction have been considered jointly in Reference 10 as a potential degree-of-freedom to be monitored. Compact multiband multiport antenna arrays together with a well designed feed network are used to exploit these unconventional degrees-of-freedom, which can be simultaneously as well as separately accessible. For instance, a generic antenna system design with three frequency bands and three directions, that is, nine degrees-of-freedom is proposed and briefly examined. Besides, a baseline multipath scenario has been considered along with conducting a suite of radio measurements using laboratory versions of the proposed antennas to demonstrate that the directional dimension open up new horizons for more spectrum access opportunities.

Authors in Reference 11 provide a brief review of frequency-agile antenna designs with reconfigurability features and investigate three new antenna implementations with integrated research and operation capabilities. Some numerical simulations are conducted to outline the efficiency of the novel antenna configurations in providing a large range of frequency-tuning possibilities.

Directional antennas, besides mitigating the interference generated in the direction of the primary links, offer many advantages over omnidirectional antennas regarding their potential to face the negative impacts of mobility [12]. Numerical results have confirmed the theoretical expectations and shown that directional antenna provided significant improvements in terms of throughput and network connectivity aspects. Despite increasing the signaling overhead needed for updating the location information and keeping direction tracking, a careful tuning of the antenna beamwidth helps bypassing this thorny issue.

As evidenced by the rich literature on that topic, the keen interest in directional antennas for CR can be substantiated by multiple motivations essentially related to the way the spectrum can be efficiently deregulated and how to comply with dynamic spectrum sharing constraints. However, research literature usually neglects to consider the impact of directional antennas on the capacity of combined spectrum sharing schemes particularly the effects exerted on the licensed traffic. As a result, this chapter aims at assessing the performance of using multiple directional antennas to enable mixed transmission strategies in license-exempt networks, in particular a state-of-the-art scheme called AT [13].

13.3 Proposed Network Model

13.3.1 Principle and Problem Setup

As mentioned above, the power of CR lies in cognition that allows devices to automatically make intelligent decisions about when and where to utilize a particular segment of spectrum. The cognition cycle comprises mainly two tasks: spectrum sensing and traffic transmission. Conventional CR networks conduct these two phases consecutively using two separate antennas. However, spectrum scanning incurs a transmission overhead that reduces the amount of time left for the effective data delivery. Furthermore, the above system topology prevents cognitive devices from efficiently exploiting the spatial opportunity eventually offered by the use of two separate antennas. It can be clearly seen that this antenna system is not operating at its full capacity since it switches in time between sensing and communication.

These inefficiencies emanate the need for truly opportunistic schemes using a combination of interweave, underlay, and overlay paradigms. A state-of-the-art hybrid strategy was introduced in Reference 13, wherein each time frame is divided into two intervals. During the first time interval, each cognitive device performs spectrum sensing and simultaneously underlay its traffic with that of the primary network. The transmission strategy throughout the second time interval depends closely on the sensing outcome, the cognitive device either continues to underlay its signal if a false alarm occurred or a licensed user appeared, or operates in the interweave mode otherwise. This scheme uses two separate antenna for parallel research and operation.

It is worth noting that recent advances in hardware industry has allowed the fabrication of directional antennas with sophisticated beamforming characteristics. As a result, the scheme previously discussed can be further improved and additional transmit antennas with well-designed beamformers can be deployed to accommodate higher secondary traffic volumes.

13.3.2 System Description and Analysis

Spectrum occupancy measurements undertaken on the band [30 MHz, 3 GHz] in New York city [14] have revealed that the overall usage of spectrum is only about 13%. For small-size antenna and compact terminals, we assume a baseline scenario of a CR infrastructure collocated with a licensed technology operating at the higher frequencies of the aforementioned band.

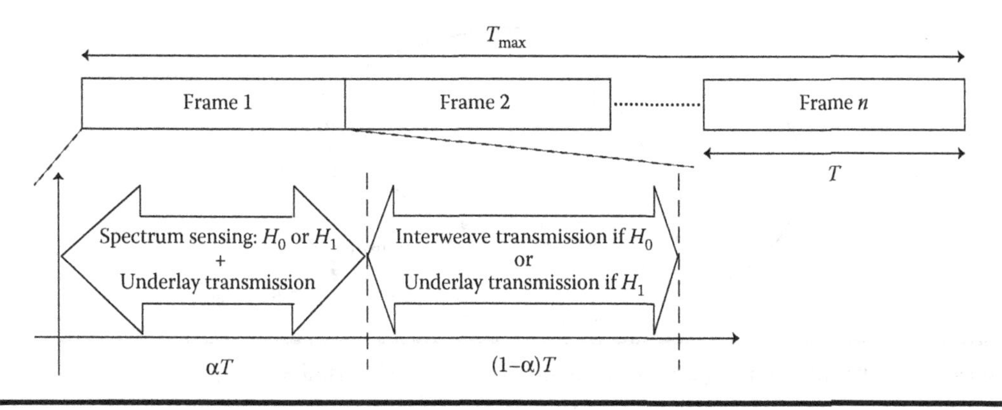

Figure 13.1 Time frame structure.

More specifically, a secondary communication between a cognitive source ST_x equipped with $M+1$ antennas (with one antenna dedicated to spectrum sensing) and a single-antenna cognitive destination SR_x is taking place within the range of a nearby primary transmitter–receiver pair (PT_x–PR_x) according to the AT strategy. The cognitive transmission divides time into fixed size frames of duration T. Each frame is in turn partitioned into two parts: a sensing phase and a transmission phase, with duration α and $1 - \alpha$, respectively (Figure 13.1). The on–off evolution of the channel state is supposed to follow a Markov distribution with a transition probability of $P_t = 1 - \exp[-(1-\alpha)\lambda]$, λ is the characteristic parameter (Figure 13.2). All the wireless channels in this environment can be considered as Gaussian and the channel gains between different nodes are supposed to be Rayleigh distributed. Assuming a frequency flat block fading process over the wireless links, the channel gain remains static during each frame but changes from frame to frame. We assume also an additive white Gaussian noise (AWGN) with zero mean and unit power ($N_0 = 1$) over all the wireless channels. We also define P_p and P_s as the transmit power of the primary transmitter and the secondary transmitter, respectively. The secondary transmitter is subject to the power constraint, P_{max}.

For notational convenience, the channel gain from the primary transmitter to the sensing antenna of the secondary transmitter is denoted as h_{TT} (1×1). Further, h_{SP} is the vector of channel gains between the secondary transmitter and the primary receiver $(M \times 1)$ and h_{PS} is the channel gain between the primary transmitter and the secondary receiver (1×1). The vector of channel gains of the secondary link $ST_x - SR_x$ is denoted as h_{ss} $(M \times 1)$. $|.|$ is the absolute value in the case of a scalar; and it is the Frobenius norm in the case of a vector.

During the first stage, the cognitive transmitter thoroughly senses and analyses the surrounding environment to identify all nearby radio activities. Spectrum sensing is a major concern in

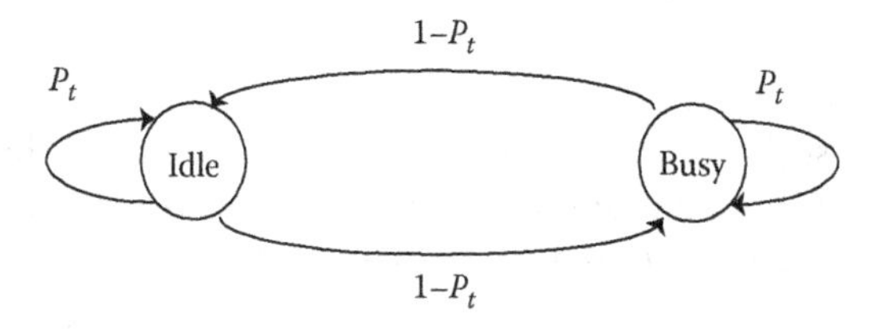

Figure 13.2 Channel state using Markov model.

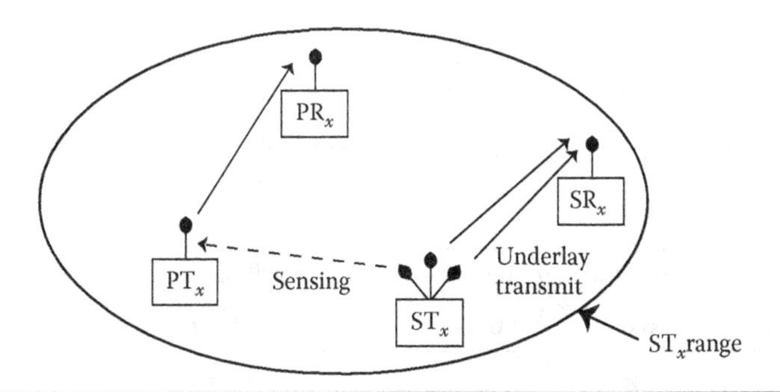

Figure 13.3 Parallel sensing and transmission using the proposed mixed policy.

wireless CR networks. This task consists of actively monitoring and searching for spectrum rooms and imposes severe requirements in terms of accuracy and swiftness in noticing the presence of primary activity.

A flurry of works on sensing techniques have been developed, including intensive research efforts on the concept of energy detection. This technique is among the most common ways of spectrum sensing due to its simplicity and ease of implementation. For instance, at the sensing node there is a given energy threshold above which the primary signal is considered as present. If the sensed signal energy is below the threshold, the channel is assumed to be vacant and available for secondary use. The predefined threshold is noise floor dependent and PU pattern independent.

In this study, the energy detector is implemented as the local sensing detector and we assume that only one antenna (among $M+1$) is responsible for carrying out the spectrum scanning task (Figure 13.3). In fact, while exploring the strengths and weaknesses of each sensing method, the energy detector has shown to be very efficient in the sense that it has a low-time complexity and does not require any prior knowledge about the primary signal pattern, which makes it the ideal choice for the present generic framework designed to work for heterogeneous licensed traffics. Implementing more sophisticated spectrum sensing detectors may lengthen the sensing time and need acquiring more information about the primary signal characteristics. Moreover, digital signal processors need to be faster and more robust in order to fulfill the challenge of low-power consumption to improve the battery life of cognitive mobile devices. On the contrary, the energy detection is inappropriate for low signal-to-noise (SNR) ratio values; thus, we assume a shared context under moderate or high-SNR conditions.

Two critical parameters are involved in characterizing the accuracy of spectrum sensing over a given wireless channel, the false alarm probability P_f and the miss detection probability P_m. A false alarm occurs when a channel, which is free, is sensed as being occupied due to background noise. In contrast, the fact that there exists a failure in detecting the presence of a primary activity constitutes a missed detection because the channel will be sensed as being idle as a result of the weak energy of the sensed primary signal.

We define H_0 and H_1 such as H_0 refers to the state of the frequency channel when it is vacant and eligible for unlicensed communications, and H_1 represents the channel state when it is busy. It follows that the two probabilities P_f and P_m are defined as

$$P_f = \text{Probability(sensing outcome is } H_1 / H_0 \text{ is true).} \tag{13.1}$$

and

$$P_{\mathrm{m}} = \text{Probability}(\text{sensing outcome is } H_0 / H_1 \text{ is true}). \tag{13.2}$$

The cognitive node collects the energy of the licensed signal during a given time interval T over the range of frequencies where the secondary use of spectrum is tolerated, denoted as W. The collected energy measurements follow either a central or a noncentral Chi-square distribution depending on the channel state whether it is empty or not [15]. Afterward, the measured value of energy is compared to a prefixed threshold θ to make a decision about the presence of the primary signal over the bandwidth of interest, W. We assume that the primary signal has unit variance.

Accordingly, the false alarm and the miss detection probabilities on a Rayleigh fading channel can be expressed as Reference 15

$$P_{\mathrm{f}} = \frac{\Gamma(N/2,\theta)}{\Gamma(N/2)}. \tag{13.3}$$

and

$$\begin{aligned} P_{\mathrm{m}} &= \exp(-\theta/2)\sum_{i=1}^{N-2}\frac{1}{p!}\left(\frac{\theta}{2}\right)^{p} \\ &+\left(\frac{1+\bar{\gamma}}{\bar{\gamma}}\right)^{N-1}\times\left[\exp(-\theta/2(1+\bar{\gamma}))-\exp(-\theta/2)\sum_{i=1}^{N-2}\frac{1}{p!}\left(\frac{\theta\bar{\gamma}}{2(1+\bar{\gamma})}\right)^{p}\right]. \end{aligned} \tag{13.4}$$

with N the number of samples of the discrete primary signal given by $N = 2WT$ and $\bar{\gamma}$ is the SNR of the primary signal activity perceived at the cognitive device acting as an energy detector. $\Gamma(a, x)$ and $\Gamma(a)$ are the incomplete gamma function and the gamma function, respectively.

The SNR of the primary signal measured at the secondary receiver side is given by

$$\bar{\gamma} = \frac{P_{\mathrm{p}} \cdot |h_{\mathrm{TT}}|^2}{N_0}. \tag{13.5}$$

During the sensing part, M transmit antennas are envisioned whereby the cognitive source conveys its stream in parallel using the underlay approach. This ensures that the primary traffic remains unaffected as the channel state is still unknown provided that the secondary transmitter properly complies with the interference constraint of the primary receiver.

To guard well the licensed traffic against interference risks, one interesting feature of directional antenna is beamforming, which is a kind of spatial shaping used to mitigate or even cancel the interference generated to other traffics in vicinity. We perform a precoding process by multiplying the secondary signal with an appropriate vector before transmission. The encoding vector, denoted as ω, must be carefully chosen to minimize the resulting interference at the unintended users. A noteworthy implication that follows this assumption is the fact that the capacity of the used wireless channel depends closely on the varying channel state and the independent beamformer as well. More precisely, let $C^1_{\mathrm{I,U}}(\mathrm{SU})$ ($C^1_{\mathrm{B,U}}(\mathrm{SU})$, respectively) be the capacity of the underlay transmission when the channel is idle (busy, respectively). These two ergodic capacities can be formulated as

$$C^1_{\mathrm{I,U}}(\mathrm{SU}) = \log_2\left(1 + \frac{(P_s/M)\cdot|h^*_{\mathrm{SS}}\cdot\omega|^2}{N_0}\right). \tag{13.6}$$

and

$$C^1_{\mathrm{B,U}}(\mathrm{SU}) = \log_2\left(1 + \frac{(P_s/M)\cdot|h^*_{\mathrm{SS}}\cdot\omega|^2}{N_0 + P_p\cdot|h_{\mathrm{PS}}|^2}\right). \tag{13.7}$$

with (.)* the Hermitian transpose.

In the underlay approach, the interference level originated by the secondary source must obey the interference-temperature limit, denoted as I. Thereby, ω should be adequately tuned to satisfy the following inequality

$$\frac{P_s}{M}|h^*_{\mathrm{SP}}\cdot\omega|^2 < I. \tag{13.8}$$

ω is supposed to be unit vector.

On the other hand, the capacity of the primary link after adopting the beamforming principle can be written, independently on the transmission mode, as

$$C^1_{\mathrm{B},x}(\mathrm{PU}) = \log_2\left(1 + \frac{P_p\cdot|h_{\mathrm{PP}}|^2}{N_0 + (P_s/M)|h^*_{\mathrm{SP}}\cdot\omega|^2}\right). \tag{13.9}$$

The inequality (13.8) provides a lower bound on the primary rate as

$$C^1_{\mathrm{B},x}(\mathrm{PU}) \geq R_p. \tag{13.10}$$

with

$$R_p = \log_2\left(1 + \frac{P_p\cdot|h_{\mathrm{PP}}|^2}{N_0 + I}\right).$$

In light of the aforementioned analysis, the fundamental requirement of the cognitive user is to maximize its transmission rate subjected to transmit power constraint while enforcing a minimum rate for the licensed traffic. Thus, the transmission context can be completely characterized as an optimization problem simplified as

$$\max_{\omega, P_s} \frac{P_s}{M}\cdot|h^*_{\mathrm{SS}}\cdot\omega|^2$$

$$\log_2\left(1+\frac{P_p \cdot |h_{PP}|^2}{N_0+(P_s/M)|h_{SP}^* \cdot \omega|^2}\right) \geq R_p$$

$$\|\omega\| = 1, \quad 0 \leq P_s \leq P_{max}. \tag{13.11}$$

We assume that noise has unit variance.

In analogy with a similar issue tackled in References 16 and 17, the problem (13.11) can be solved by

$$P_s = P_{max}. \tag{13.12}$$

and

$$\omega = \sqrt{\beta}\frac{\pi_{h_{SP}} h_{SS}}{\|\pi_{h_{SP}} h_{SS}\|} + \sqrt{1-\beta}\frac{\pi_{h_{SP}}^{\perp} h_{SS}}{\|\pi_{h_{SP}}^{\perp} h_{SS}\|}. \tag{13.13}$$

with

$$\beta = \begin{cases} \beta_{MRT}, & \text{if } \beta_{MRT} \leq \dfrac{z}{\|h_{SP}\|^2 P_{max}} \\ \dfrac{z}{\|h_{SP}\|^2 P_{max}}, & \text{otherwise} \end{cases}. \tag{13.14}$$

where

$$\beta_{MRT} = \frac{\|\pi_{h_{SP}} h_{SS}\|^2}{\|\pi_{h_{SP}} h_{SS}\|^2 + \|\pi_{h_{SP}}^{\perp} h_{SS}\|^2} \quad \text{and} \quad z = MI.$$

$$\pi_{h_{SP}} = h_{SP}(h_{SP}^* h_{SP})^{-1} h_{SP}^* \quad \text{and} \quad \pi_{h_{SP}}^{\perp} = I_d - \pi_{h_{SP}}$$

with I_d the identity matrix.

The capacity of the secondary link during the first phase is inferred from Equations 13.6 and 13.7 as

$$C^1(\text{SU}) = P_a C_{I,U}^1(\text{SU}) + (1-P_a) C_{B,U}^1(\text{SU}). \tag{13.15}$$

with P_a is the probability that the frequency of interest is available during the first phase with respect to a binomial distribution.

Likewise, the overall capacity of the primary link during the first phase is formulated as

$$C^1(\text{PU}) = (1-P_a) C_{B,x}^1(\text{PU}). \tag{13.16}$$

Afterward, the most suitable transmission strategy is chosen during the second phase according to the sensing outcome reported by the former phase. To avoid the collapse of the whole infrastructure, the uncertainty of sensing measurements should be taken into account. The secondary device can freely transmit over the selected frequency using the interweave approach provided that either the associated primary traffic is still far away or a missed detection has happened. Otherwise, the secondary device chooses to operate in an underlay fashion to ensure that the unlicensed communication is not detrimental to the licensed traffic. In either case, the secondary transmitter performs spatial shaping since its decision relies on the sensing result considered as not totally trustworthy; this ensures keeping the interference level at the primary receiver capped regardless of the secondary transmission policy. We define $C_{\mathrm{I,I}}^{2}(\mathrm{SU})$ ($C_{\mathrm{B,I}}^{2}(\mathrm{SU})$, respectively) as the capacity of the interweave cognitive access in the case of idle (busy, respectively) channel state. Likewise, we use the following notations: $C_{\mathrm{I,U}}^{2}(\mathrm{SU})$ and $C_{\mathrm{B,U}}^{2}(\mathrm{SU})$ to denote the channel capacities for the underlay mode. Thereafter, the aforementioned ergodic capacities are given as

$$C_{\mathrm{I,I}}^{2}(\mathrm{SU}) = C_{\mathrm{I,U}}^{2}(\mathrm{SU}) = C_{\mathrm{I,U}}^{1}(\mathrm{SU}) = \log_2\left(1 + \frac{(P_s/M)\cdot|h_{\mathrm{SS}}^{*}\omega|^2}{N_0}\right). \tag{13.17}$$

and

$$C_{\mathrm{B,I}}^{2}(\mathrm{SU}) = C_{\mathrm{B,U}}^{2}(\mathrm{SU}) = C_{\mathrm{B,U}}^{1}(\mathrm{SU}) = \log_2\left(1 + \frac{(P_s/M)\cdot|h_{\mathrm{SS}}^{*}\omega|^2}{N_0 + P_p\cdot|h_{\mathrm{PS}}|^2}\right). \tag{13.18}$$

In both underlay and overlay transmission modes, the ergodic capacity of the primary link with a spatially shaped secondary stream has the same expression as Equation 13.9

$$C_{\mathrm{B},x}^{2}(\mathrm{PU}) = C_{\mathrm{B},x}^{1}(\mathrm{PU}). \tag{13.19}$$

By partitioning the global event space into a set of mutually exclusive subevents according to the probability of having a spectrum whole, the probability of a false alarm or a missed detection and the probability of channel state transition, the capacity of the wireless channel during the second phase is a linear combination of the capacities given in Equations 13.17 and 13.18 and takes the form

$$\begin{aligned} C^2(\mathrm{SU}) = {} & [P_a(1-P_t)(1-P_f) + (1-P_a)P_t(1-P_d)]C_{\mathrm{I,I}}^{2}(\mathrm{SU}) \\ & + [P_aP_t(1-P_f) + (1-P_a)(1-P_t)(1-P_d)]C_{\mathrm{B,I}}^{2}(\mathrm{SU}) \\ & + [P_a(1-P_t)P_f + (1-P_a)P_tP_d]C_{\mathrm{I,U}}^{2}(\mathrm{SU}) \\ & + [P_aP_tP_f + (1-P_a)(1-P_t)P_d]C_{\mathrm{B,U}}^{2}(\mathrm{SU}). \end{aligned} \tag{13.20}$$

where P_d is the probability of detection defined as the complement of P_m.

The capacity of the primary link during the second phase transmission is the ergodic capacity $C_{\mathrm{B},x}^{2}(\mathrm{PU})$ given in Equation 13.19 weighted by the probability that the channel is occupied during the second phase

$$C^2(\mathrm{PU}) = (P_aP_t + (1-P_a)(1-P_t))C_{\mathrm{B},x}^{2}(\mathrm{PU}). \tag{13.21}$$

The overall capacity C_T(SU) of the secondary wireless channel during the whole frame is a weighted sum of capacities C^1(SU) and C^2(SU) obtained in Equations 13.15 and 13.20, respectively. C_T can be easily derived as

$$C_T(\mathrm{SU}) = \alpha C^1(\mathrm{SU}) + (1-\alpha)C^2(\mathrm{SU}). \tag{13.22}$$

Similarly, the overall capacity of the primary link through the entire frame is calculated according to Equations 13.16 and 13.21

$$C_T(\mathrm{PU}) = \alpha C^1(\mathrm{PU}) + (1-\alpha)C^2(\mathrm{PU}). \tag{13.23}$$

13.4 Simulations and Numerical Results

The effectiveness and the performance of the proposed multiple directional-antennas scheme are mainly assessed through simulation results whose purpose is to strengthen the above analytical study.

For these experiments and unless otherwise stated, we assume a two-antenna secondary transmitter conveying its stream to a single antenna secondary receiver in the range of a licensed system with an interference temperature constraint of 1 dB. The average power gains on the ST_x–PT_x and the ST_x–PT_x channels are given as $E[|h_{TT}|^2] = 1$ and $E[|h_{SP}|^2] = 2$. For the secondary link, the average channel gain is $E[|h_{SS}|^2] = 10$ dB. Likewise, the average channel gain on the primary link is $E[|h_{PP}|^2] = 1$. The transmit power for the primary source as well as the maximum tolerated power at the secondary source is fixed to $P_{p,dB} = P_{max,dB} = 1$ dB. The variance of noise is $N_0 = 1$.

Time is separated into frames each of length $T = 20$ ms. We assume a total bandwidth of $W = 5$ KHz. The probability of false alarm, the binomial parameter, and the Markovian characteristic parameter are fixed to $P_f = 0.1$ $P_a = 0.8$ and $\lambda = 1$, respectively.

We assume that the SU has good enough knowledge about the surrounding environment.

Figure 13.4 examines the impact of the sensing time fraction on the secondary ergodic capacity for various antenna configurations. It can be observed that the proposed framework bypasses the single-antenna AT setup regarding the ergodic capacity of the secondary link and the achieved capacity gain is up to 13% in the case of the two-antenna configuration ($M = 2$). This capacity improvement is due to the implementation of the beamforming mechanism, which has enabled concentrating the secondary signal through the desired direction where the energy is really needed. It is also interesting to note that the more the number of antennas mounted on the secondary device, the higher the capacity of the secondary network. Deploying more transmit antennas allows the CR infrastructure to pass larger traffic volumes compared to a single-antenna infrastructure and thus expands the secondary network capacity.

To further analyze the impact of directional antennas on the performance of spectrally agile contexts, Figure 13.5 displays the influence of the sensing phase duration on the ergodic capacity of the primary traffic for a varying number of antennas. The proposed multiple antenna scheme performs better than the conventional AT strategy and the resulting graph reveals roughly 16.4% in capacity increase when precoding the secondary signal prior to a two-antenna transmission. The proposed scheme has the key property that larger capacities are achieved for smaller sensing overheads ($\alpha \leq 0.7$). However, for $\alpha > 0.7$ both schemes achieve seamlessly comparable capacity. As a direct consequence, the proposed framework is less demanding in terms of sensing time.

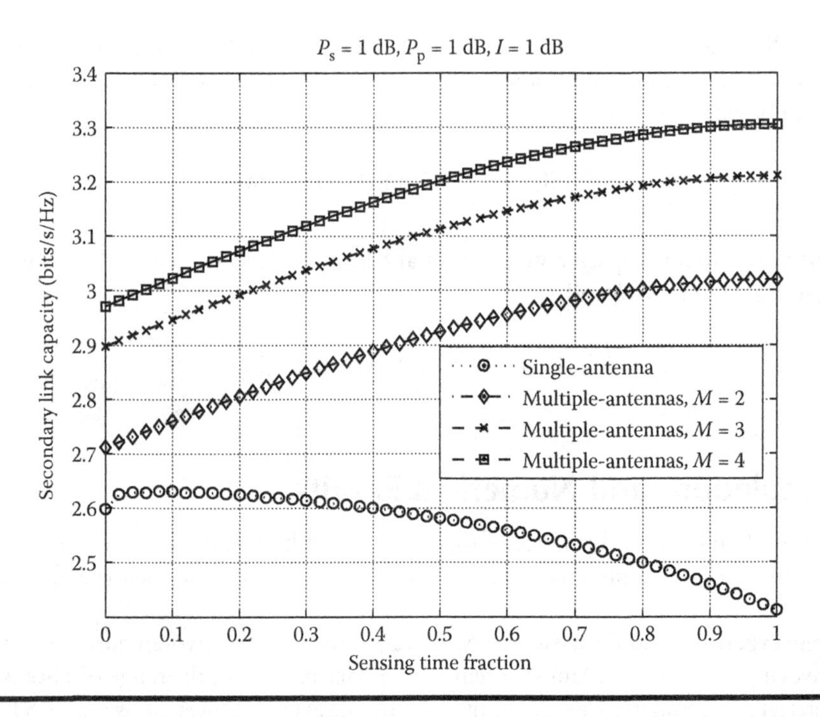

Figure 13.4 Secondary ergodic capacity versus sensing time fraction for various antenna configurations.

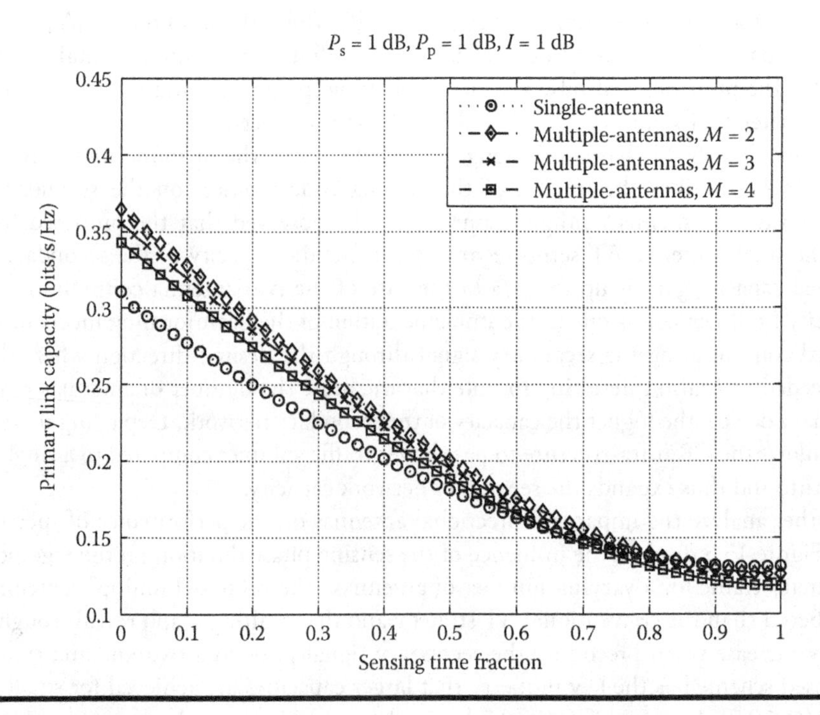

Figure 13.5 Primary ergodic capacity versus sensing time fraction for various antenna configurations.

Spatial shaping provides a safe environment for PUs since it reduces the undesired radiations and keeps the interference level below the acceptable threshold. Another aspect worth mentioning is that as long as the secondary device has more antenna, the capacity of the primary network slightly degrades. Significantly increasing the number of transmit antennas will be inadequate to keep up with the primary traffic requirements in terms of reliability due to multipath propagation issues. From a directional antenna perspective, the secondary transmitters need to choose the more appropriate antenna configuration to fully leverage the multiantenna benefits. The common method of implementing higher order antenna systems to improve system capacity does not hold in spectrum-sharing contexts.

The secondary capacity performance of the two schemes under different values of the secondary transmit power is plotted in Figure 13.6 against the fraction of time devoted to spectrum sensing. As can be seen, the proposed multiple antenna strategy outperforms the conventional AT scheme and the obtained capacity gain is more noticeable for low values of the secondary transmit power (i.e., 16% for $P_s = -1$ dB vs. 13% for $P_s = 1$ dB). Further, the proposed scheme shows a clear tendency toward achieving high-capacity values compared to the conventional AT scheme as long as more time is spent in the sensing phase. In fact, for small α-values (ideal sensing), the secondary device transmits more during the second phase according to either the underlay or the interweave policy. However, the secondary transmitter employs the beamforming technique whatever the transmission mode and so it does not leverage the benefit of the interweave method (relaxed interference constraints) and misses opportunities to boost its capacity. The opposite is true for higher α-values. This beamforming strategy which disregards the transmission mode, conversely, ensures increased protection to licensed users as the interference level never exceeds the prefixed threshold. Besides, we state that the novel scheme performs better as long as the SU increases its

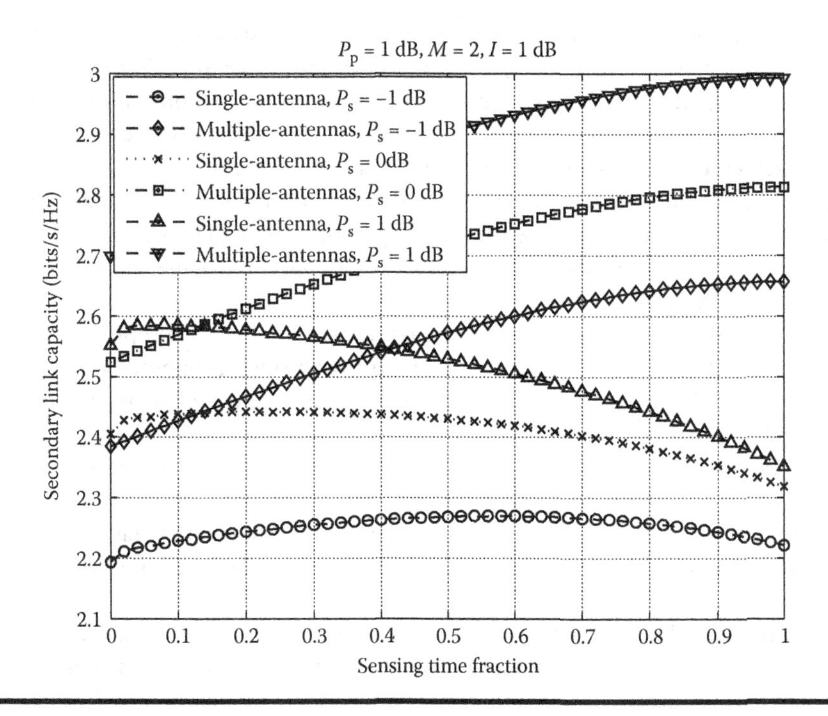

Figure 13.6 Ergodic capacity versus sensing time fraction for various secondary transmit power values.

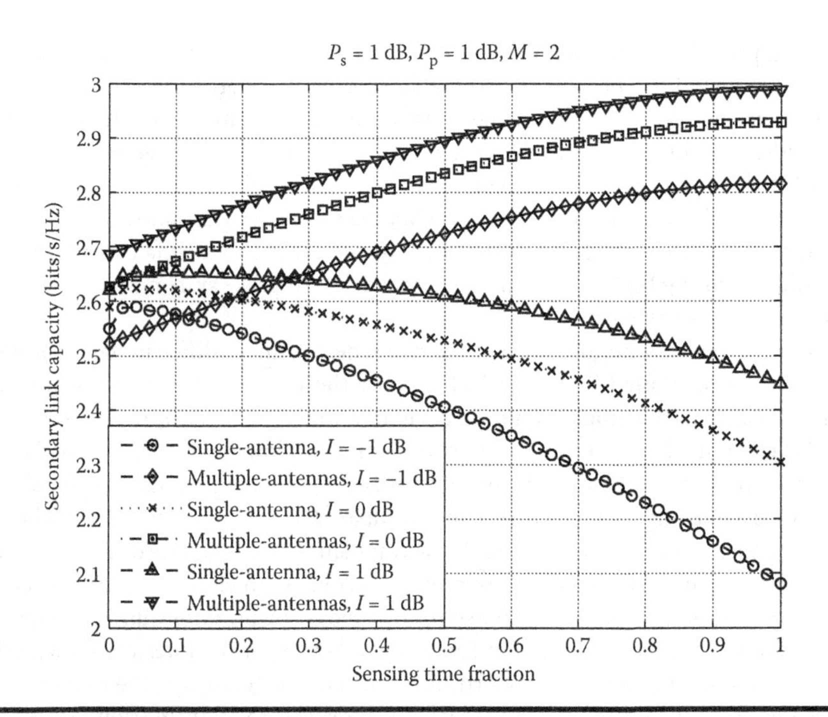

Figure 13.7 Ergodic capacity versus sensing time fraction for various interference temperature thresholds.

transmitted power. The ability of the proposed multiantenna strategy to address the incoming unlicensed traffic depends closely on the emitted secondary power to avoid drastically degrading the licensed service quality. Increasing the secondary transmit power, on the contrary, allows the secondary network to increase its capacity to convey considerable amounts of unlicensed data; in such case the achieved ergodic capacity rapidly attains higher values.

Figure 13.7 depicts the impact of the interference temperature on the secondary ergodic capacity for different sensing overhead values. It is clearly noticed that the proposed multiple antenna configuration is more capacity efficient than the conventional AT scheme and the capacity gain is increasingly sharp as long as the interference temperature limit gets higher (i.e., 8.8% for $I = -1$ dB vs. 12.8% for $I = 1$ dB). For a fixed interference temperature limit, the proposed multiple-antennas approach necessitates a shorter time allocated to the first stage as opposed to the conventional AT scheme, because the received energy is summed over all the transmitting antennas and thus improved speed and capacity are ensured. Moreover, we observe that while increasing the maximum interference limit the proposed setup yields better results. The interference temperature as a measurement of the acceptable interference level at the incumbent user, when sufficiently raised, provides more relaxed constraints and allows the underlay connections to last long enough until satisfying a larger capacity. Sensitivity to interference is a major bottleneck for licensed traffics and needs to be carefully tackled from the CR point of view.

To have a closer look at the influence of the statistical pattern of the licensed traffic on the behavior of both schemes, Figure 13.8 assesses the performance of the proposed multiantenna AT scheme in terms of the secondary ergodic capacity versus the conventional AT scenario, plotted against the sensing time fraction. We noticed that the proposed framework yields better capacity efficacy in comparison to the conventional AT strategy and the capacity benefit attains 16.4% for

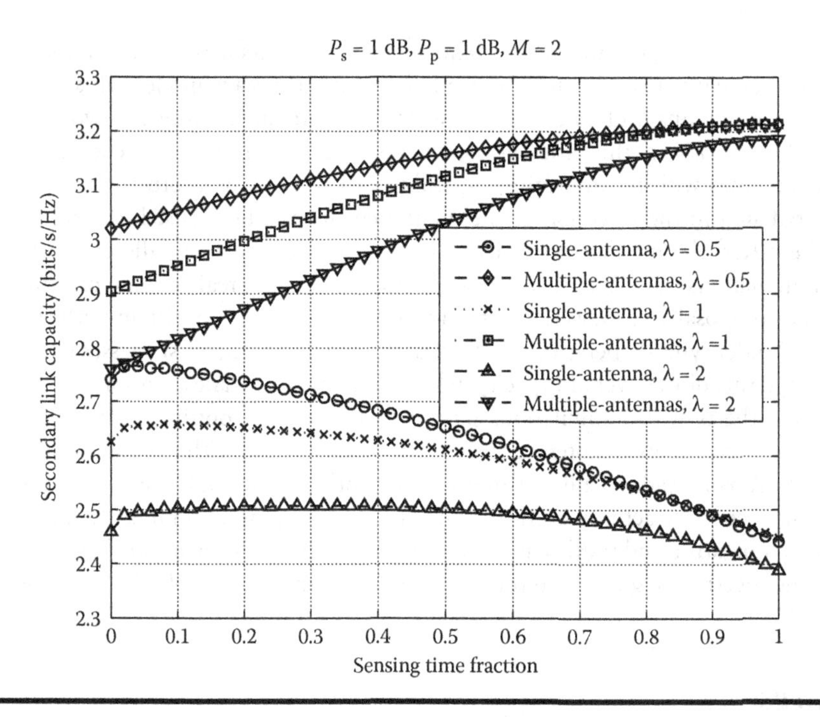

Figure 13.8 Ergodic capacity versus sensing time fraction for various Markov-modeled primary traffic profiles.

the two-antenna setup. The use of multiple directional antennas with beamforming properties helps reducing the needlessly radiated energy and extends markedly the system capacity, since many antennas collaborate to aggregate sufficient energy at the receiver in the aim of obtaining a single stream with superior reliability. Furthermore, the smaller the value of λ the higher the capacity achieved by our beamforming proposal. This is an interesting result because for small λ the primary traffic is slowly varying and thus can be easily tracked to predict the channel state in the near future, and this helps in determining the temporal spectrum availability and foster the spectrum sharing-based coexistence. The proposed framework is devised for wireless environments where channels are tending to remain in a given state for a relatively moderate duration and do not evolve in a fast way. More importantly, fast varying licensed traffics are more likely susceptible to be disrupted by cognitive devices in the vicinity and thus combined transmission schemes may be prohibited in such contexts. Naturally, service providers of such technologies are cautious about giving spectrum access rights to a third party.

13.5 Conclusions

This chapter raises some interesting issues related to the embrace of directional antennas in emerging CR networks. A directional antenna not only enables routing the traffic destined to or from cognitive nodes, but it also allows and fulfills other fundamental functions such as interference mitigation and energy saving, thus enhancing the reliability of end-to-end CR systems. To guarantee a peaceful coexistence between primary and nearby secondary devices, the concept of dynamic spectrum sharing has been adopted to opportunistically reuse the underutilized frequencies.

This is done by granting spectrum access rights to cognitive users with respect to the so-called AT scheme, which employs a combination of underlay and interweave modes. This scheme exploits a single transmit antenna and an additional wideband antenna exclusively devoted to spectrum monitoring. On the basis of this observation, we proposed an improved AT scheme using multiple directional antennas with beamforming capabilities to be able to absorb the unwanted secondary energy radiated in the direction of the incumbent user to be focused toward the secondary receiver. Thereafter, the mathematical derivation of the secondary ergodic capacity is inferred. Numerous simulations using extensive Monte Carlo based fading realizations confirmed the effectiveness of our proposal and showed that the numerical results are consistent with the theoretical expectations. Under typical operating conditions, the proposed scheme surpasses the conventional framework in terms of the overall capacity, while simultaneously allowing for reasonable sensing times. Besides, the provided simulations permit to retrieve the optimal antenna configuration among the tested taxonomy of antenna setups. This is in contrast with the commonly known fact that transmit diversity leads to a linear uplift in network capacity. Likewise, the obtained results stated that the secondary activity need to adjust its transmit power and must definitely adhere to interference temperature and traffic pattern constraints of the primary network, and this may help to reassure the spectrum regulators and license holders of the benefits of CR.

References

1. Mitola, J., III and G. Q. Maguire, 1999. Cognitive radio: Making software radios more personal, *IEEE Personal Communications*, 6(4), 13–18.
2. Mitola, J., III, 2000. *Cognitive radio: An integrated agent architecture for software defined radio.* PhD thesis KTH Royal Institute of Technology.
3. Weiss, T. and F. Jondral, 2004. Spectrum pooling: An innovative strategy for the enhancement of spectrum efficiency, *IEEE Communications Magazine*, 42, 8–14.
4. Sherman, M., A. N. Mody, R. Martinez, C. Rodriguez, and R. Reddy, 2008. IEEE standards supporting cognitive radio and networks, dynamic spectrum access, and coexistence, *IEEE Communications Magazine*, 46(7), 72–79.
5. Winters, J. H., 1998. Smart antennas for wireless systems, *IEEE Personal Communications*, 5(1), 23–27.
6. Alexiou, A. and M. Haardt, 2004. Smart antenna technologies for future wireless systems: Trends and challenges, *IEEE Communications Magazine*, 42(9), 90–97.
7. Christodoulou, C. G., 2009. Cognitive radio: The new frontier for antenna design, *IEEE Antennas and Propagation Society Feature Article*, available at: www.ieeeaps.org.
8. Taherpour, A., M. Nasiri-Kenari, and S. Gazor, 2010. Multiple antenna spectrum sensing in cognitive radios, *IEEE Transactions on Wireless Communications*, 9(2), 814–823.
9. Kelly, J. R., E. Ebrahimi, P. S. Hall, P. Gardner, and F. Ghanem, 2008. Combined wideband and narrowband antennas for cognitive radio applications. In *Proceedings of the IET Seminar on Cognitive Radio and Software Defined Radios: Technologies and Techniques*, September, London, UK, pp. 1–4.
10. Murtaza, N., R. K. Sharma, R. S. Thoma, and M. A. Hein, 2013. Directional antennas for cognitive radio: Analysis and design recommendations, *Progress in Electromagnetics Research*, 140, 1–30.
11. Hall, P. S., P. Gardner, J. Kelly, E. Ebrahimi, M. R. Hamid, F. Ghanem, F. J. Herraiz-Martnez, and D. Segovia-Vargas, 2009. Reconfigurable antenna challenges for future radio systems. In *Proceedings of the 3rd European Conference on Antennas and Propagation (EuCAP)*, March, Berlin, Germany, pp. 949–955.
12. Murawski, R., E. Ekici, V. Chakravarthy, and W. K. McQuay, 2011. Performance of highly mobile cognitive radio networks with directional antennas. In *Proceedings of the IEEE International Conference on Communications (ICC)*, June, Kyoto, Japan, pp. 1–5.

13. Ridouani, M. and A. Hayar, 2013. General scheme for always transmit solution in cognitive radio systems. In *Proceedings of the 20th International Conference on Telecommunications (ICT)*, May, Casablanca, Morocco, pp. 1–5.
14. McHenry, M. A., P. A. Tenhula, D. McCloskey, D. A. Roberson, and C. S. Hood, 2006. Chicago spectrum occupancy measurements & analysis and a long-term studies proposal. In *Proceedings of the First International Workshop on Technology and Policy for Accessing Spectrum (TAPAS)*, August, Boston, MA, USA, Article No.: 1.
15. Digham, F. F., M. S. Alouini, and M. K. Simon, 2007. On the energy detection of unknown signals over fading channels, *IEEE Transactions on Communications*, 55(1), 21–24.
16. Jorswieck, E. A., E. G. Larsson, and D. Danev, 2008. Complete characterization of the Pareto boundary for the MISO interference channel, *IEEE Transactions on Signal Processing*, 56(10), 5292–5296.
17. Lv, J. and E. A. Jorswieck, 2011. Spatial shaping in cognitive system with coded legacy transmission. In *Proceedings of the International ITG Workshop on Smart Antennas (WSA)*, February, Aachen, Germany, pp. 1–6.

Chapter 14

Multicast Algorithm Design for Energy-Constrained Multihop Wireless Networks with Directional Antennas

Yi Shi, Y. Thomas Hou, Hanif D. Sherali, and Wenjing Lou

Contents

Abstract

The beam-forming property associated with directional antennas introduces some unique problems that do not exist for omni-directional antennas and therefore significantly increases the design space for routing algorithms. A problem that can be solved when omni-directional antennas are used may become Non-deterministic Polynomial-time hard (NP-hard) when directional antennas are used. In this chapter, we discuss some algorithmic challenges associated with directional antennas and then present an in-depth case study. In the case study, we consider a wireless ad hoc network where each node employs a single-beam directional antenna and is provisioned with limited energy. We are interested in an online routing algorithm for successive multicast communication requests with the aim of maximizing network lifetime. We provide some important theoretical understanding on various multicast problems and deduce that even an offline version of this problem is NP-hard. Then we develop a highly competitive online routing algorithm that takes network lifetime consideration directly into iterative calculations and show that this algorithm provides consistently better performance than the current state-of-the-art algorithm considering only remaining energy. The theoretical results and routing algorithm in this chapter offer some important insights on algorithm design for energy-constrained wireless ad hoc networks with directional antennas.

14.1 Introduction

The use of directional antennas in wireless communication has enabled new approaches to energy saving for energy-constrained wireless networks. Indeed, the use of directional antennas allows concentration of the beam toward the intended destination without wasting energy in unwanted directions. A node's transmission coverage area can be effectively controlled by adjusting the power level, beamwidth, and beam orientation of the directional antennas. Further, because the beam is generated only toward a certain direction, it creates less interference to other nodes that are outside the beam, which enables greater information capacity in the network. Finally, since nodes outside the beam coverage cannot receive the source's signal, privacy concerns associated with omni-directional broadcast can be somewhat alleviated. As a result, it is expected that the use of directional antennas has a great potential in wireless networks.

Depending on the specific wireless environment and a node's hardware and software implementation, each node's energy consumption behavior may be different. In our theoretical development (Section 14.2) and algorithm design (Section 14.3), we model the transmission energy at a node u as a function of ρ, θ, and ω, where ω is the beam orientation, ρ is the reachable distance along this orientation, and θ is the beamwidth. Denote $p_u^T(\rho,\theta,\omega)$ as the beam transmission cost function, which is node dependent. Without loss of generality, we assume this function is an increasing function of ρ and θ, that is,

$$p_u^T(\rho_1,\theta,\omega) < p_u^T(\rho_2,\theta,\omega) \quad \text{if } \rho_1 < \rho_2, \tag{14.1}$$

$$p_u^T(\rho,\theta_1,\omega) < p_u^T(\rho,\theta_2,\omega) \quad \text{if } \theta_1 < \theta_2. \tag{14.2}$$

Further, to better model the wireless environment in practice, we do not assume uniform path loss in all directions (ω). Instead, we let $p_u^T(\rho,\theta,\omega)$ depend not only on ρ and θ, but also depend on beam orientation ω. Therefore, it is possible that $p_u^T(\rho,\theta,\omega_1) \neq p_u^T(\rho,\theta,\omega_2)$ if $\omega_1 \neq \omega_2$. Due to the non-uniform path loss along different directions, the beam coverage may not be a uniform sector, although for ease of illustration, we use a uniform sector (e.g., in Figure 14.1) to represent the coverage of a directional beam in all figures.

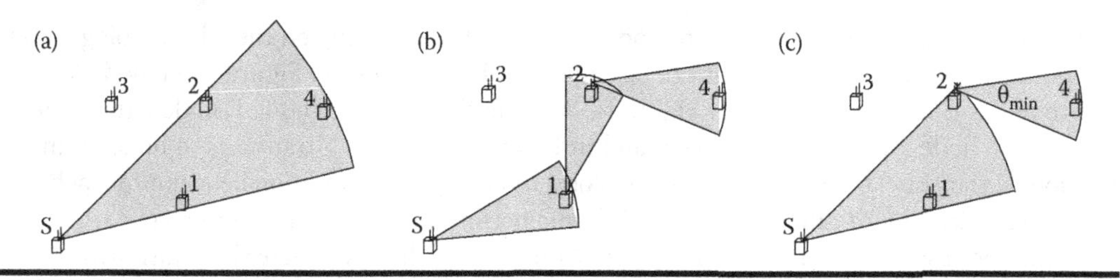

Figure 14.1 Three different multicast routing solutions for the same multicast session. (a) Change the beam radius, (b) change the beam orientation, and (c) change the beamwidth.

Since energy is also consumed for other nodal processing functions, for each node u, we denote p_u^P as the transmission processing energy. Denote p_u^R as the reception energy for each unit data. Then, the energy consumed at a node u is:

$$C_u(\rho,\theta,\omega) = \begin{cases} p_u^T(\rho,\theta,\omega) + p_u^P & \text{if } u \text{ is a transmitter;} \\ p_u^R + p_u^P & \text{if } u \text{ is a receiver.} \end{cases} \quad (14.3)$$

From a theoretical perspective, the use of directional antennas has introduced some unique challenges in algorithm design, particularly when each node is assumed to generate a *single* directional beam.* This is because a single directional beam provides *partial* broadcast to those nodes that are within the beam coverage. Unlike the case of omni-directional antennas, where the design space depends solely on the radius (i.e., communication range), the algorithm design space for directional antennas now encompasses three components: *beam radius* ρ, *beamwidth* θ, and *beam orientation* ω and various solutions can be achieved by adjusting these three parameters (see Figure 14.1a–c). Thus, a directional antenna based routing problem needs to address the assignment of these three parameters on each node in the network.

In this chapter, we consider a wireless ad hoc network consisting of N nodes located over a region. For wireless communication, we assume that each node is equipped with directional antennas for transmission and an omni-directional antenna for reception.† Similar to Reference 26, we assume that each node's transmitter has power control capability. That is, by adjusting the transmission power level, the sender could reach destination nodes located at different distances ρ. Further, we assume that each node could control the beamwidth θ and beam orientation ω of its directional antenna [21].

In its most general form, any node in the multicast may need to transmit to a subset of other nodes in the network. A source node could generate a single beam to reach all nodes in this subset in a single hop (e.g., Figure 14.1a). Although simple, this approach is not energy efficient, particularly for large-sized networks, due to the power consumption behavior in Equations 14.1 and 14.2. As a result, it is important to explore a multihop relaying approach to extend network lifetime (e.g., Figure 14.1b and c).

We use an example to illustrate the multicast communication problem for a particular time instance. In Figure 14.1, suppose that the multicast communication request arrives at node S

* Although multiple beams can be formed by directional antenna arrays, the hardware complexity and energy consumption are much higher than those for single-beam directional antenna [22].

† It is possible to use a directional antenna for reception as well, although its energy saving may not be as significant as that for transmission, particularly for large-sized networks.

(source node) and wishes to transmit to nodes 1, 2, and 4. Depending on the relay topology, there is various transmission behavior that can be employed. For example, in Figure 14.1a, node *S* transmits to nodes 1, 2, and 4 directly; in Figure 14.1b, node *S* transmits to node 1, node 1 transmits to node 2, and node 2 transmits to node 4; and in Figure 14.1c, node *S* transmits to nodes 1 and 2, and node 2 transmits to node 4. Clearly, topology and energy consumption behavior for each case are different, leading to different network lifetime performance. Note that in practice, there is a minimum beamwidth requirement θ_{min} for a beam, even in the case where the transmitting node may have only one downstream neighbor.* Further, we denote θ_{max} as the maximum beamwidth.

For an ad hoc network where each node is provisioned with limited energy (also called an energy-constrained network), it is now well understood that *an energy efficient routing usually cannot provide the best result for network lifetime performance* [26]. This important result has led to the separation of two lines of research: one focuses on minimizing the total energy required to maintain a tree (broadcast/multicast) [2,4,5,13,24,25] and the other focuses on how to perform routing so that the network lifetime can be prolonged as much as possible [14,26]. We follow the second line of research in this chapter.

We consider the important problem of multicast routing with directional antennas and the objective of maximizing network lifetime. The significance of this problem lies in that not only it is a general problem (with unicast and broadcast as special cases), but also it is a generalized problem for omni-directional antennas, which is a special case of directional antenna with a 2π beamwidth.

We assume that the directional antennas at each node can only form a single beam where the beam radius ρ, beamwidth θ, and beam orientation ω are adjustable. Instead of looking for an optimal routing solution for a single multicast session, we are interested in an *online algorithm* for the problem, where multicast requests arrive and depart over time and there is no prior knowledge of the future request arrivals. Our case study includes both theoretical understanding and algorithm design for the multicast routing problem. From the theoretical perspective, we show that an *offline* version of this multicast routing problem is NP-hard. By "offline," we mean that we have complete knowledge of multicast requests over time. This result builds upon several intermediate results, each of which has its own significance and offers important understanding of some closely related problems. In one intermediate result (Theorem 14.1), we show that, when directional antennas are used, the static maximum-lifetime tree problem for a single multicast (or broadcast) session is NP-complete. The proof of this result gives insights on how directional antennas increase the computational complexity in algorithm design.

Since even an offline multicast routing problem is NP-hard, for an online algorithm, only a heuristic approach is feasible. In the second half of this chapter, we aim to develop an online multicast routing algorithm to maximize network lifetime. Wieselthier et al. [26] made a major step in the systematic study of the online multicast routing problem. In particular, they designed an algorithm called D-MIP, which incorporated nodal residual energy into the local cost metric for routing. Although this algorithm offered good performance, there is a very subtle detail in this algorithm that motivates us to pursue further investigation. We find that nodal residual energy is indeed closely related to node lifetime (and, thus, network lifetime). But it is even more important to take lifetime consideration *directly* into algorithm design. Based on this understanding, we make the following conjecture in our investigation: if we incorporate lifetime consideration explicitly into the design of an online multicast routing algorithm, then we should expect to

* Typically, the smaller the minimum beamwidth requirement, the more complex and costly the directional antennas [22].

have an algorithm that outperforms the D-MIP algorithm. To prove this conjecture, we design a new algorithm called MLR-MD (for maximum lifetime routing for multicast with directional antennas). The design experience for MLR-MD is quite interesting and offers understanding on beam-forming behavior under single-beam directional antennas. Through simulation results, we conclusively demonstrate that the proposed MLR-MD algorithm offers consistent performance improvement over the D-MIP algorithm.

14.2 Understanding the Multicast Routing Problem

In this section, we explore some theoretical understanding of the multicast routing problem with energy constraint. Our investigation builds upon several intermediate results, each of which studies the level of computational complexity of some closely related problems.

14.2.1 Minimum Power Routing

Das et al. [6] studied a single-session minimum-power broadcast tree problem with omni-directional antennas. Although the authors proposed three mixed integer linear programming (MILP) formulations, no explicit solutions were given. It is well known that an MILP problem is NP-hard in general [10]. Guo and Yang [11] studied the single-session minimum-power multicast tree problem in the context of directional antennas (with fixed beamwidth). They also formulated the problem into an MILP. Again, there was no explicit analytic solution (due to NP-hardness). Although there is software available to solve MILP problems, one can obtain solutions only for small-sized problems.

It is important to realize that the minimum-power broadcast tree problem cannot be translated into a spanning-tree problem, which can be solved in polynomial time. The spanning-tree problem addresses a connected graph with predefined edges and associated costs. The objective is to select edges with the minimum total cost that connects all vertices in the graph, where the total cost is the sum of costs of selected edges. But in wireless networks, which are broadcast in nature (or partial broadcast in the case of directional antennas), the total cost at a node is not a simple summation of the cost of its outgoing links.

Cagalj et al. [4] proved that, for omni-directional antennas, the minimum-power broadcast tree problem is NP-complete. Since (i) broadcast is a special case of multicast and (ii) omni-directional antenna is a special case of the directional antenna, we conclude that the minimum-power multicast tree problem is NP-hard under either omni-directional or directional antennas. We summarize this conclusion in the following lemma.

Lemma 14.1

For either directional or omni-directional antennas, the problem of finding a static minimum-power multicast tree is NP-hard. ■

14.2.2 Maximum Lifetime Routing

It has been well recognized that minimum-power routing usually cannot provide good network lifetime performance. As such, there have been separate efforts on exploring multicast routing, with an objective of maximizing network lifetime, for energy-constrained ad hoc networks. The problems along this line of research can be classified into three problems:

- *Problem 1*: maximizing the lifetime of a single static multicast tree.
- *Problem 2*: maximizing the lifetime of a single multicast tree with dynamic topology updates.
- *Problem 3*: maximizing the lifetime for a sequence of requests, each of which will generate a multicast tree, with dynamic topology updates for each multicast tree.

The third problem is the focus of this chapter. Note that the first and second problems can be considered special cases of the third problem. We now discuss each problem separately.

Problem 1. The first problem addresses the network lifetime of a single static multicast tree (without dynamic topology updates). There are polynomial time algorithms [7,14] to solve this problem for the broadcast case with omni-directional antennas. Floreen et al. [9] proved that this problem can be solved in polynomial time for omni-directional antennas. All these prior results were obtained under the assumption that omni-directional antennas have uniform path loss behavior in all directions (i.e., a node's coverage is a disc). We now extend the proof in Reference 9 for the general case where path loss may be non-uniform (see discussion in Section 14.1). Since broadcast is a special case of multicast, we only need to show the result for the multicast case in the following lemma.

Lemma 14.2

For omni-directional antennas, the problem of finding a static maximum-lifetime tree for a single multicast (or broadcast) session can be solved in polynomial time.

Proof. Suppose we have N nodes in the network. For each node in a multicast tree, the energy consumed on p_u^P term (source node) or $p_u^P + p_u^R$ term (non-source node) is deterministic. Now, we consider the p_u^T term. For the case of omni-directional antennas, given the value of p_u^T, the set of covered nodes is unique. To be energy efficient, we only need to consider $O(N)$ values of p_u^T term at each node, which correspond to the number of power levels to cover i neighbors $(0 \le i \le N-1)$, where $i = 0$ represents the special case that the node does not transmit data to any node. Thus, there are $O(N)$ different total power consumption levels at each node, which correspond to $O(N)$ different node lifetimes. Since we have a total of N nodes, there exists a maximum of $O(N^2)$ different lifetime values. We only need to check which value among these $O(N^2)$ lifetime values yields a maximum feasible lifetime solution for this multicast.

We now check if a given lifetime value t is feasible. If this t is feasible, then there exists a multicast tree such that each node has a lifetime of at least t. For each node, we first assume that it is on the multicast tree and subtract the energy consumed on p_u^P term (for source node) or $p_u^P + p_u^R$ term (for non-source node) over t. For each node that has negative remaining energy after this subtraction, we check to see if it is the source node or a destination node in the multicast session. If yes, we can declare immediately that this t is infeasible. Otherwise, this node cannot be a node in the multicast tree and is thus removed from further consideration. For the remaining nodes, we first compute the maximum transmission powers based on t and their remaining energy. Then, we can compute their transmission coverage. Based on the coverage of each node ($O(N^2)$ complexity), we can quickly determine if a multicast tree exists (e.g., via depth first search) in $O(N)$ time. The complexity for this feasibility check is therefore $O(N^2)$.

Since we have $O(N^2)$ different lifetime values, we can sort them in $O(N^2 \log N)$ complexity. Then we use binary search ($O(\log N)$ times) to find the maximum-lifetime tree. The overall complexity, that is, $O(N^2 \log N) + O(\log N)\, O(N^2)$, is $O(N^2 \log N)$. ■

For the case of directional antennas, Problem 1 becomes much harder. Its complexity is addressed in the following theorem.

Theorem 14.1

For directional antennas, the problem of finding a static maximum-lifetime tree for a single multicast (or broadcast) session is NP-complete.

Proof. Instead of proving the maximum-lifetime tree problem is NP-complete, it is sufficient to prove that the lifetime feasibility problem is NP-complete. This is because if we can find the maximum lifetime t^* in polynomial time, then for any given t, the lifetime feasibility problem can be solved by comparing t and t^*. On the other hand, if we can determine the feasibility of any given t in polynomial time, with a similar analysis on the possible maximum network lifetime as that in Lemma 14.2, the maximum lifetime t^* can be obtained by binary search in polynomial time.

We begin with the directed Hamilton path problem with a given starting vertex. That is, for a given directed graph $G(V, E)$, with vertex set V and directed edge set E, and a designated vertex S, we want to find if there exists a directed Hamilton path with starting vertex S to each of the other vertices in V. It is well known that the directed Hamilton path problem (with any starting vertex) is NP-complete [19]. Note that by creating a dummy starting node and directed edges from this node to all other nodes, we can show that the directed Hamilton path problem with a given starting vertex is NP-hard. It is easy to show that this problem is also in NP. Thus, the directed Hamilton path problem with a given starting vertex is NP-complete.

Now, given any instance of the directed Hamilton path problem with starting vertex S, we show how to reduce it to an instance of the lifetime feasibility problem in polynomial time. In the lifetime feasibility problem, we consider broadcast, a special case of multicast, where a node S needs to transmit data to all other nodes. We let $\theta_{min} = \theta_{max}$ and denote it as θ_f, that is, the beamwidth is fixed. For the given directed graph G, denote N_u the set of vertices that are "outgoing" neighbors of vertex u such that for any $q \in N_u$, there is a directed edge from u to q, that is, $u \to q$.

Now, we assign the parameter values of an N-node network for the feasibility problem. First, we arrange the N nodes such that there are no more than two nodes on the same line. As a result, we can set a value for θ_f such that a single beam from any node (with any ρ and ω) can cover at most one node. Further, we can arrange an energy consumption function $C_u(\rho, \theta, \omega)$ with such a property that for each node u, the energy cost to cover any node $q \in N_u$ is smaller than the energy cost to cover any node $z \notin N_u$. Now, for an arbitrarily given lifetime $t > 0$, we can always initialize the energy of each node u such that it can transmit to any node $q \in N_u$ over time t while it is unable to transmit to any node $z \notin N_u$ for the entire duration of t. As an example of how to define $C_u(\rho, \theta, \omega)$ and set initial energy for each node u, we can let $C_u(\rho_{uq}, \theta_f, \omega_{uq}) = a_u$ for every node $q \in N_u$, and $C_u(\rho_{uz}, \theta_f, \omega_{uz}) = b_u$ for every node $z \notin N_u$, where a_u and b_u are constants and $b_u > a_u > 0$, ρ_{uq} and ρ_{uz} are distances from node u to q and z, ω_{uq} and ω_{uz} are beam orientations from node u to q and z, respectively. The value of $C_u(\rho, \theta, \omega)$ at other locations are not of our concern and thus can be defined arbitrarily. The only requirement that $C_u(\rho, \theta, \omega)$ should have is that it is an increasing function of ρ for any fixed ω as we discussed in Equation 14.1 (note that θ is already fixed as θ_f earlier). Also, as we discussed in Section 14.1, due to potential non-uniform path loss

along different beam orientations ω in practical wireless environment, $C_u(\rho, \theta, \omega)$ also depends on beam orientation ω. In particular, even if $\rho_1 > \rho_2$, it is possible that $C_u(\rho_1, \theta_f, \omega_1) < C_u(\rho_2, \theta_f, \omega_2)$ if $\omega_1 \neq \omega_2$. Now we can set the initial energy at each node u in the network to be $t \cdot r \cdot a_u$, where r is the transmission data rate. Thus, node u can transmit to any node $q \in N_u$ over time t while is unable to transmit to any node $z \notin N_u$ for the entire duration of t since $a_u < b_u$. This completes the example.
Under the above setting, it follows that the lifetime t is feasible if and only if G has a directed Hamilton path with starting vertex S to all other vertices. Therefore, any instance of the directed Hamilton path problem with starting vertex S can be reduced to an instance of the lifetime feasibility problem. Thus, the lifetime feasibility problem is NP-hard. It is easy to show that the lifetime feasibility problem is in NP. Thus, the lifetime feasibility problem is NP-complete. As a result, for directional antenna case, our static maximum-lifetime tree problem for multicast is also NP-complete. This completes the proof. ■

Problem 2. We now move onto the discussion of the second problem, which addresses how to maximize the lifetime of one multicast tree under dynamic routing (i.e., routing topology may change over time for this single multicast). For omni-directional antennas, Floreen et al. [9] showed that this problem is NP-hard. Since directional antenna is a general case of omni-directional antenna, the problem of maximizing the lifetime for one multicast tree under dynamic routing is also NP-hard.

Problem 3. The third lifetime problem addresses how to perform multicast routing when successive multicast requests arrive to the network. In practice, multicast communication requests arrive in the network over time. For a given source node, the multicast group can change from request to request. For each request, there is an amount of data (also varies from multicast session to multicast session) that needs to be sent to the respective multicast group. Our objective is to pursue an optimal transmission behavior (assignment of beam radius, beamwidth, and beam orientation at each node) so that the network lifetime is maximized.

This problem is substantially more difficult than multicast routing for a single request (e.g., [4,6,9,11]) in that we are not interested in the maximum-lifetime tree for one request, but rather, we are interested in the network lifetime performance when successive multicast session requests (generated at different nodes in the network and with different multicast groups) arrive and depart over time. We need to design an "online" algorithm *without any knowledge of future request arrivals* with the aim of maximizing network lifetime.* This is in contrast to the "offline" optimization for maximum network lifetime problem, which assume the future multicast requests are known *a priori*. Li et al. [17] proposed an online routing algorithm to maximize network lifetime for unicast case. They showed that an online algorithm does not have a constant competitive ratio† over an offline optimum. Since unicast is a special case of multicast, we conclude that online algorithms for multicast routing do not have a constant competitive ratio.

As the online optimization cannot be solved analytically, one might ask whether it is possible to pursue an "offline" optimization algorithm. By "offline," we mean that we first record successive multicast request arrivals over time. Then assuming we can "go back" in time with the knowledge

* Recall that network lifetime is defined as the first time instance when a multicast communication fails, either due to energy depletion at sender or any receiver of the multicast group.

† The competitive ratio of an online algorithm is the ratio between the performance of this online algorithm and an optimal offline algorithm.

of all these future arrivals, we attempt to pursue routing optimally for each successive request such that the network lifetime is maximized. Note that in the extreme case, when all multicast requests have the same source node and destination nodes, this problem reduces to Problem 2 that we discussed earlier, which is NP-hard. Therefore, we conclude that offline multicast routing for successive multicast requests is also NP-hard.

Theorem 14.2

For both omni-directional and directional antennas, an offline problem of optimal routing to maximize the network lifetime for a wireless ad hoc network with successive multicast requests is NP-hard. ■

Since the offline problem for multicast routing is NP-hard, for an online problem, only heuristic approach is feasible. We present such an algorithm in Section 14.3.

14.3 Lifetime-Centric Design for an Online Multicast Routing Algorithm

14.3.1 Background and Motivation

Wieselthier et al. [26] made a major step in their study of online multicast routing for energy-constrained ad hoc networks. In particular, they examined the source-initiated, session-based multicast problem for successive requests and proposed an online heuristic algorithm (D-MIP) that was shown to have good performance in terms of network lifetime and traffic volume. In the design of D-MIP, the authors explicitly incorporated nodal residual energy into local routing cost metric. Then they used a spanning-tree like technique to obtain a broadcast tree, which they called broadcast incremental power (BIP) algorithm. A multicast tree can be obtained by pruning the unnecessary links. The algorithm for directional antenna case was called directional multicast incremental power (D-MIP) algorithm.

Although the D-MIP algorithm may be applied to the multicast routing problem, there is a very subtle detail in its design that calls for further investigation. Specifically, although nodal residual energy indeed is closely related to node lifetime (and thus network lifetime), it may not be as effective as if we take network lifetime metric directly into iterative calculations. We make the following conjecture in our investigation. *If we incorporate lifetime consideration directly into the iterative calculation of the online multicast routing algorithm, we should have an algorithm that outperforms an algorithm based on nodal residual energy (e.g., D-MIP).* In this section, we develop an online multicast routing algorithm based on this approach, and in Section 14.4, we use simulation results to demonstrate performance improvement. We name our algorithm MLR-MD, which is to contrast with traditional minimum power routing (MPR) [24,25] or variants of minimum cost routing [26]. It is worth pointing out that problems for either broadcast or omni-directional antennas are considered as special cases under multicast or directional antennas, respectively.

14.3.2 Basic Idea

For a given multicast request, the basic idea of the MLR-MD algorithm is to start with a multicast routing solution first (e.g., a single beam from the source covering all multicast destination nodes) and then iteratively improve lifetime performance of the current solution by identifying the node with the smallest lifetime* and revising routing topology as well as corresponding beam-forming behavior for an increased network lifetime. For directional antennas with power control capability, a node's lifetime can be increased via two techniques: *narrowing beamwidth* θ and *reducing beam radius* ρ. A direct consequence of such operation is that some nodes in the multicast tree that are covered by the original beam could be exposed (uncovered) under the new beam with reduced beamwidth or beam radius. The MLR-MD algorithm has several approaches to "re-attach" these exposed nodes back in the multicast tree. Since a re-attachment operation would decrease some other node's lifetime, a decision must be taken on whether a re-attachment operation is feasible. Naturally, a re-attachment operation is feasible only if the new network lifetime is increased. For the next iteration, we repeat the same process, that is, identifying the node among all the nodes in the network with the minimum lifetime and attempting to revise routing topology and beam-forming behavior to increase network lifetime.

When nothing can be done to further improve the current lifetime, we move on to consider the node with the second smallest lifetime and attempt to increase its lifetime, under the condition that the lifetime for the first node (with minimum lifetime) will not decrease. In particular, MLR-MD does not increase the lifetimes of downstream nodes of the first node. The motivation for attempting to re-configure the node with the second smallest lifetime is the following. Although the increase of this second smallest node lifetime may not increase the minimum node lifetime, it will enable the multicast routing topology to evolve to a better structure, thereby creating new optimization space for the first node (with minimum node lifetime) in the next iteration.

If nothing can be done for the node with the second smallest node lifetime, the MLR-MD algorithm will continue to try the node with the third smallest node lifetime and so forth. The algorithm terminates after it has tried all the nodes (in the order of nondecreasing node lifetime) and cannot increase lifetime for any of the nodes. The pseudocode of this basic idea is shown in Figures 14.2 and 14.3.

```
1. Source node generates one beam to cover all destination nodes;
2. Sort all nodes in non-decreasing lifetime order and arrange the sorted list
3. with a stack L (with the top node having the smallest lifetime);
4. set Removed[i][j] = 0 for 1 ≤ i,j ≤ N;
5. while (L != ∅){
6.      i = pop(L);
7.      Identify the logical downstream one-hop neighbors of node i that are on
8.      the border on node i's beam;
9.      Sort such nodes in non-increasing improvement order and arrange the
10.     sorted list with a stack L_i (with the top node contributing to the largest
11.     lifetime increase on node i if removed);
12.     Improved = 0;
13.     while (L_i != ∅){
14.          j = pop(L_i);
15.          if (RemoveLink(i, j)==1){
16.               Removed[i][j] = 1;
17.               Improved = 1;
18.               break;}}
19.     if network lifetime increases, set Removed[i][j] = 0 for 1 ≤ i,j ≤ N;
20.     if (Improved==1)
21.          Sort all nodes in non-decreasing lifetime order and arrange the
22.          sorted list with a stack L (with the top node having the smallest
23.          lifetime);}
```

Figure 14.2 Main program.

* Node lifetime is calculated based on the current routing topology.

```
1. int RemoveLink(int i, int j){
2.      Remove link i → j;
3.      if (Case1B(i, j)==1) return 1;
4.      if (Case1A(i, j)==1) return 1;
5.      if (Case2A(i, j)==1) return 1;
6.      if (Case2B(i, j)==1) return 1;
7.      Add link i → j back, recover node i's beam;
8.      return 0;}
9. int Case1A(int i, int j){
10.     Identify a node v (v ≠ i) so that
11.     1) Removed[v][j] = 0;
12.     2) v is in the multicast tree or ∃k, Removed[k][v] = 0, v is covered
13.     by node k's beam, and k's lifetime>node i' old lifetime (in this case,
14.     add link k → v); and
15.     3) after adding link v → j, v's new lifetime> i's old lifetime;
16.     if such node exists, then choose the best node (having the largest new
17.     lifetime) and return 1; else return 0;}
18.int Case1B(int i, int j){
19.     Identify a node v in the multicast tree (v ≠ i) so that Removed[v][j]
20.     = 0, j is covered by node v's beam, and v's lifetime>node i's
21.     old lifetime;
22.     if such node exists, then choose any node and return 1; else return 0;}
23.int Case2A(int i, int j){
24.     Identify a pair of nodes (u, v) (u ≠ i) so that
25.     1) Removed[u][v] = 0 and Removed[v][j] = 0;
26.     2) u is in the multicast tree or ∃k, Removed[k][u] = 0, u is covered
27.     by node k's beam, and k's lifetime>node i's old lifetime (in this case,
28.     add link k → u); and
29.     3) v is not covered by any beam of the multicast tree; and
30.     4) after adding links u → v → j, the pair lifetime (the smaller new
31.     lifetime of nodes u and v)>node i's old lifetime;
32.     if such pair exists, then choose the best pair (having the largest pair
33.     lifetime) and return 1; else return 0;}
34.int Case2B(int i, int j){
35.     Identify a node v so that
36.     1) Removed[i][v] = 0 and Removed[v][j] = 0;
37.     2) v is not covered by any beam of the multicast tree; and
38.     3) after adding links i → v → j, the pair lifetime (the smaller new
39.     lifetime of nodes i and v)>node i's old lifetime;
40.     if such node exists, then choose the best node (having the largest pair
41.     lifetime) and return 1; else return 0;}
```

Figure 14.3 Auxiliary functions.

14.3.3 Algorithm Details

We now consider some details in the MLR-MD algorithm. An important concept in designing routing algorithms for wireless networks is the distinction between *physical one-hop neighbor* and *logical one-hop neighbor* [14]. To illustrate these two concepts in the context of multicast routing with directional antennas, we use the example in Figure 14.4. In Figure 14.4a, we have an ad hoc network and a multicast request is initiated by node S, with multicast destination nodes being 4, 5, and 6. Figure 14.4b shows a particular logical multicast routing topology that can support this multicast communication session, where nodes 1 and 3 are used as relay nodes in the multicast tree. Figure 14.4c shows the corresponding physical beam-forming behavior at each node of the multicast tree. Note that in this example, a single beam from the source node S can cover not only nodes 1 and 3, but also node 2. Although node 2 is not a logical one-hop neighbor to node S on the multicast tree, it is a *physical one-hop neighbor* to node S. An important application of physical one-hop neighbors is that should it become necessary to re-configure a new multicast tree, these physical one-hop neighbor nodes can be added (attached) to the logical multicast tree without changing the current beam forming at any node.

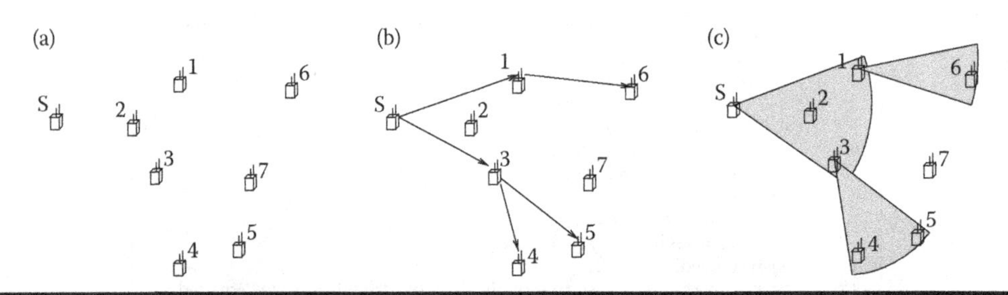

Figure 14.4 A logical multicast routing tree and a physical beam forming behavior for a multicast session. (a) Network topology, (b) a logical routing tree, and (c) a beam forming pattern.

As described earlier, upon identifying a minimum-lifetime node at an iteration, we attempt to reduce either its beamwidth or beam radius in order to increase its lifetime. The immediate consequence of this operation is that some nodes along the border of the original beam are being pushed out of the new beam's coverage. We use an example to illustrate this point. Figure 14.5a shows the logical one-hop links on node 1, while Figure 14.5b shows the beam-forming behavior on node 1. Suppose that we wish to extend node 1's lifetime by reducing either its beamwidth or beam radius. Under either technique, it is only necessary to consider the three border nodes 2, 4, and 5. In the case of beam radius reduction, we can consider to expose node 5 and let the beam cover only nodes 2, 3, and 4 (see Figure 14.5d). This will result in a new downstream logical topology for node 1 in Figure 14.5c, where the previous logical link between nodes 1 and 5 is removed. Since node 5 (or one node in node 3's subtree) may belong to the multicast group, it has to be reattach back to the multicast tree through another link, by means of a procedure that we describe shortly. In the case of beamwidth reduction, we can consider to remove node 2 or 4 following the same approach. As either beam radius reduction (i.e., remove coverage for node 5) or beamwidth reduction (i.e., remove coverage for node 2 or 4) will increase node 1's lifetime, a decision must be

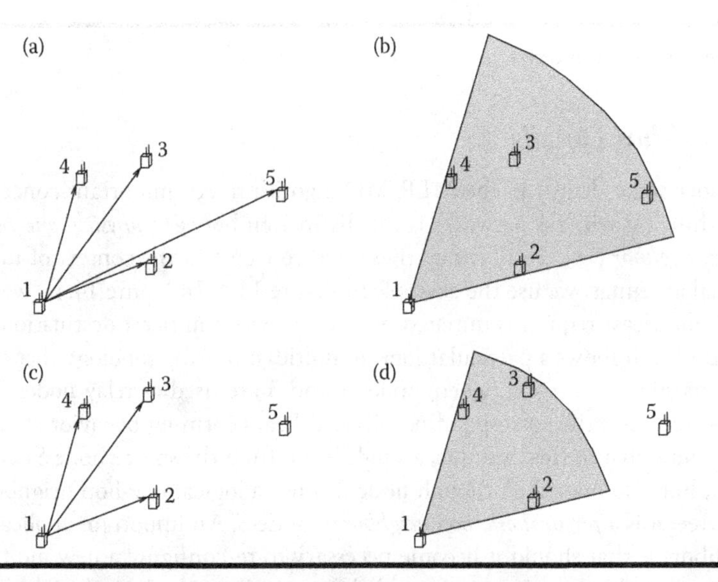

Figure 14.5 An example of reducing node 1's beam coverage. (a) Logical one-hop links, (b) beam from node 1, (c) new logical one-hop links, and (d) new beam from node 1.

made as which node we should remove (2, 4, or 5). In our implementation (Figure 14.2), we rank the order of these three possibilities (nodes 2, 4, and 5), in terms of how much improvement each will bring to node 1's lifetime. We first try to remove the node that yields the largest increase in nodes 1's lifetime. If the re-attachment of this node (node 5 in example) is feasible, then we are done. Otherwise, we declare removing this node as a failure and we consider to remove the node that will yield the second largest increase in node 1's lifetime and so forth. From the perspective of logical one-hop neighbor, any of these node removal operations, either due to beamwidth reduction or due to beam radius reduction, is equivalent to breaking a logical link to one of the logical one-hop neighbors. This observation is important in coding and implementation in the sense that a link removal subroutine (RemoveLink() in Figure 14.3) can be used by either the beam radius reduction operation or the beamwidth reduction operation.

We now discuss another important property associated with nodes that are not on the logical multicast tree but are within the coverage of one of the directional beams associated with the multicast tree (i.e., physical one-hop neighbor). Referring to Figure 14.6, suppose node S is the source node and nodes 3, 4, 5, 6, and 8 are the multicast destination nodes. Figure 14.6a shows a multicast tree topology for a particular routing solution and Figure 14.6b shows the areas that is being covered by the beams of the multicast routing solution. For those nodes that are not on this multicast tree but are within the coverage of these beams (e.g., nodes 7 and 9), we claim that there exists a path from the source of the multicast to each of them. For example, a path for node 7 is $S \rightarrow 2 \rightarrow 5 \rightarrow 7$, where logical link $5 \rightarrow 7$ can be added under node 5's current beam since node 7 is the physical one-hop neighbor of node 5. That is, if there is a need to add one of these nodes onto the multicast tree, all we need to do is to add one logical link in the multicast tree without any change to existing physical beams. We formally state this important property associated with directional antenna-based multicast routing as follows.

Property 1 (multicast beam coverage)

Consider a node that is not in the logical multicast tree but is a physical one-hop neighbor of a node within the multicast routing tree. This node can be attached to the logical multicast tree by adding one logical link, without any change to the existing beam-forming structure in the network. ■

We now discuss how the MLR-MD algorithm handles the "re-attachment" operation, that is, the re-connection of an exposed node back to the multicast routing tree. This operation requires the re-configuration of existing beam-forming structures in the network and can be classified into two cases: (1) without the use of intermediate relay nodes (Case I) and (2) with the use of

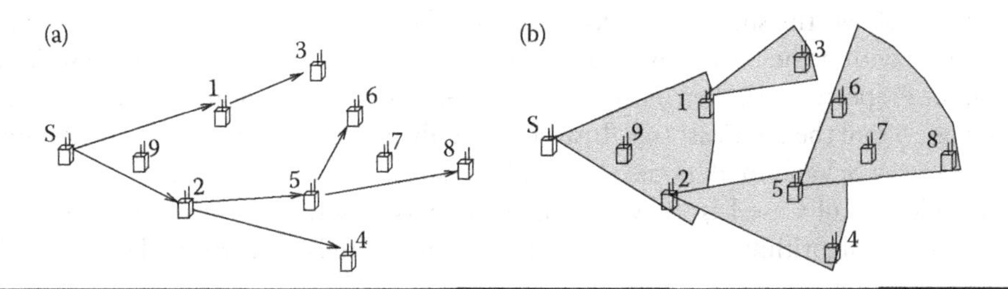

Figure 14.6 Concept of logical one-hop neighbor and physical one-hop neighbor. (a) A logical multicast tree and (b) a physical beam-forming behavior at each node.

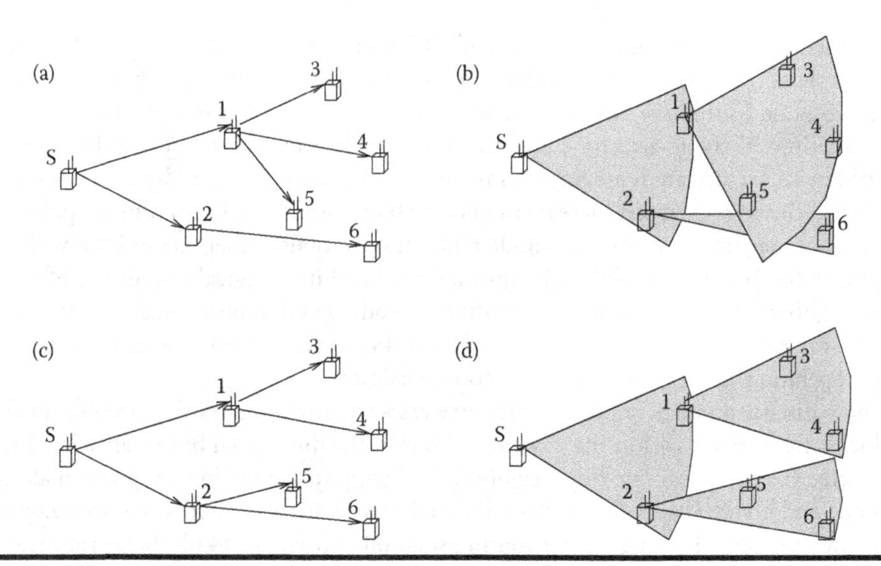

Figure 14.7 Case I—Re-attachment without intermediate relay nodes. (a) Multicast tree, (b) beam on each node, (c) new multicast tree, and (d) new beam on each node.

intermediate relay nodes (Case II). An intermediate relay node is a node that is currently not within the coverage of any beam and is chosen as a relay for re-attachment.

Case I: Re-attachment without intermediate relay nodes. This case is best explained with an example. Suppose that we have a logical multicast tree in Figure 14.7a with a physical beam-forming solution in Figure 14.7b, where node S is the source and nodes 3 to 6 are multicast destination nodes. Now we want to increase node 1's lifetime by pushing out node 5 from its beam. Consequently, a new beam can be formed to cover nodes 3 and 4 only and node 5 is exposed. It is necessary to re-attach node 5 back to the multicast tree. Under Case I, no intermediate relay nodes are used. We only consider to adjust the beam at one of nodes S, 2, 3, 4, or 6 to cover 5, that is, nodes in the multicast tree or nodes covered by a node in the multicast tree, excluding node 1, and the new lifetime of the corresponding node (with a modified beam) will decrease. The re-attachment operation is considered a success only if this node's new lifetime (with modified beam) is larger than node 1's lifetime before pushing out node 5. In the case when there are multiple successful re-attachments, we choose the re-attachment that yields the longest node lifetime. For example, in Figure 14.7c, suppose that node 2's new lifetime is the largest among others, then MLR-MD will choose node 2 to connect node 5, with a new beam-forming shown in Figure 14.7d.

Figure 14.8 shows the special case that one can take advantage of when one node (i.e., node 5) already falls within the beam coverage of another node (i.e., node 2). In this case (recall our discussion for Property 14.1), there is no need to generate new beams or update beams in order to re-attach node 5 into the multicast tree. Instead, it is only necessary to update the logical multicast tree (see Figure 14.8c) and mark node 5 to be a downstream node of node 2.

The pseudocode of Case I is shown in Figure 14.3 as CaseIA() and the special case of Case I is CaseIB(). The algorithm tries CaseIB() first because there is no lifetime decrease under this special case.

Case II: Re-attachment with intermediate relay nodes. Again, this case is best explained with an example. Suppose that we have a logical multicast tree in Figure 14.9a where node S is the source

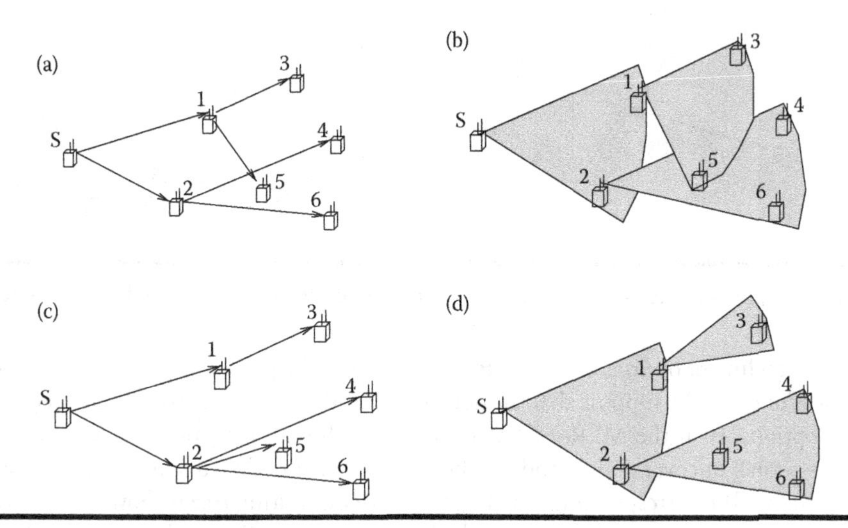

Figure 14.8 A special case of Case I. (a) Multicast tree, (b) beam on each node, (c) New multicast tree, and (d) new beam on each node.

and nodes 1, 2, 3, and 6 are multicast destination nodes. Figure 14.9b shows a beam-forming solution of this multicast tree. Now we want to increase node 1's lifetime by pushing out node 6 from its beam. Consequently, we re-generate a new beam from node 1 (with beamwidth θ_{min}) to just cover node 3. Since node 6 is now exposed, we need to re-attach it back to the multicast tree. Under Case II, we consider to employ one intermediate relay node (node 4 or 5) during the re-attachment process. In particular, we adjust the beam on one of nodes *S*, 2, and 3 to cover the intermediate relay node, that is, nodes in the multicast tree or nodes covered by a node in the multicast tree, excluding node 1. For the pair of adjusted node and intermediate relay node, define the pair-lifetime as the smaller lifetime of their node lifetimes. The r-eattachment is successful only if the pair-lifetime is larger than node 1's lifetime before pushing out node 6. If there are multiple

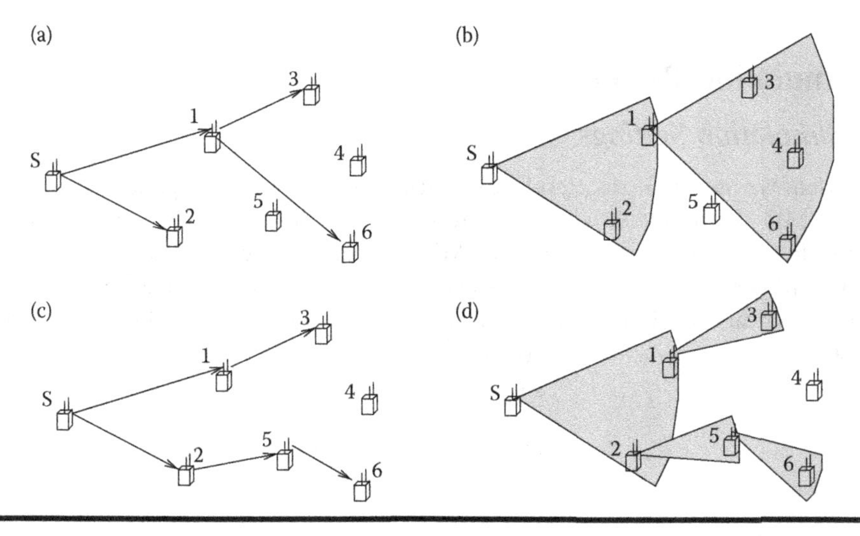

Figure 14.9 Case II—Re-attachment with intermediate relay nodes. (a) Multicast tree, (b) beam on each node, (c) new multicast tree, and (d) new beam on each node.

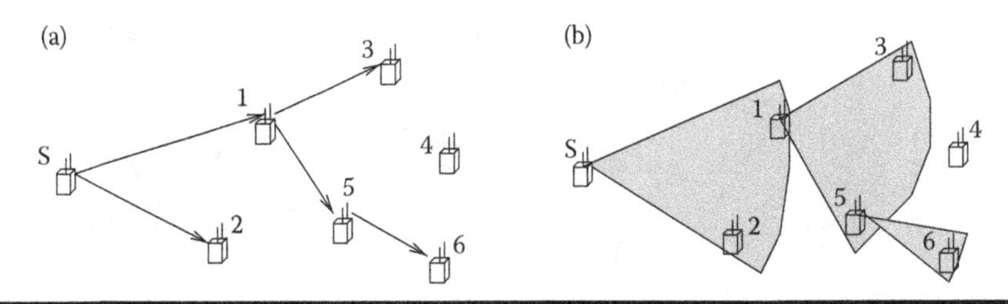

Figure 14.10 A special case of Case II. (a) New multicast tree and (b) new beam on each node.

successful re-attachment options to choose from, we choose the pair of nodes that yields the largest pair-lifetime. For example, suppose that the node pair 2 and 5 yields the largest pair-lifetime among all possible options, then the MLR-MD algorithm will choose this pair of nodes and generate a new beam at node 2 to cover node 5 and another new beam at node 5 to cover node 6, respectively (see Figure 14.9d). The corresponding new logical multicast routing tree is shown in Figure 14.9c.

In the special case, suppose that we find the best option is to choose the node pair 1 and 5. In this case, we need to re-adjust the beam on node 1 to cover the intermediate relay node (i.e., node 5). Since each node is allowed to generate one beam for the directional antenna under our investigation, we have to modify the existing beam from node 1 to cover both node 5 and node 3. In this case, node 1's beamwidth is increased (after first decreasing its beam radius to push out node 6). This solution is shown in Figure 14.10b, with corresponding logical multicast routing tree shown in Figure 14.10a.

The pseudocode of Case II is shown in Figure 14.3 as Case2A() and the special case of Case II is shown as Case2B(). In the special case, node *i*'s lifetime is smaller than that in Case II, so MLR-MD tries Case2B() last.

Between two network lifetime increases, it is possible that MLR-MD removes one logical link from a logical multicast tree and after some iterations, MLR-MD tries to add this link back to an advanced logical multicast tree. We further limit that once MLR-MD removes a logical link, it cannot use this link until the network lifetime is increased. It is easy to verify that the computational complexity of the MLR-MD algorithm is strictly polynomial.

14.4 Simulation Results

14.4.1 Simulation Settings

In this section, we use simulation results to illustrate the behavior and performance of the proposed MLR-MD algorithm and compare it to the D-MIP algorithm. For comparison, we also show results for multicast routing under the MPR paradigm, where a broadcast tree is obtained first by a spanning-tree like technique and then is pruned to a multicast tree [25].

In our numerical investigation, we assume that energy consumption in Equation 14.3 is independent of ω. Further, we define $p_u^T(\rho,\theta)$ as follows [26]:

$$p_u^T(\rho,\theta) = \max\left\{\frac{\theta}{360}\rho^{\alpha}, p_{\min}\right\}, \tag{14.4}$$

where α is the path loss index and is typically within $2 \le \alpha \le 4$ [20], and $p_{\min}$ is the minimum power that is needed to generate a beam.

We consider networks of various sizes consisting of either 10, 20, 50, or 100 nodes. For 10-, 20-, and 50-node networks, we assume that the nodes are randomly deployed in a 5 unit by 5 unit square region, where the distance unit is consistent to that for ρ in Equation 14.4. For 100-node networks, we assume the nodes are randomly deployed over a 15 unit by 15 unit square region. In all cases, we assume that each node starts with 200 units of energy, with the energy unit consistent to that in Equation 14.4.

We are interested in an online operation where multicast requests arrive sequentially over time. The source of the multicast request is chosen at random and the multicast group is also a random group of nodes in the network (excluding the source node). For each multicast request, the amount of data generated by the source node is uniformly chosen between [10,100] units and transmission rate at the source node is 10 units of data per time unit. In our simulation, we assume $\alpha = 4$ in Equation 14.4, $p_u^T \gg p_u^R$, and $p_u^T \gg p_u^P$. That is, radio frequency transmission energy is the dominant source of energy consumption. We also assume $p_{\min} = 0$ in Equation 14.4. For the bounds of beamwidth for directional antennas, we set $\theta_{\min} = 30$ and $\theta_{\max} = 360$ (both in degrees).

For both MLR-MD and D-MIP algorithms, routing topology is dynamically changed every time unit as discussed in Section 14.4 (if there is remaining data to send at the source node), where time unit can be defined to reflect practical settings. For MPR, routing is only performed for each multicast request and remains static (fixed routing).

There are various definitions for network lifetime [3]. For simplicity, we define network lifetime as *the time instance when the network can no longer support a multicast communication session.* This happens when either the source node or any multicast receiving node runs out of energy during a multicast communication session. Clearly, the idle periods where there are no multicast sessions in the network should not be considered as part of the network lifetime since there is no energy expenditure during these periods. That is, network lifetime should only consist of the time intervals where there are active multicast sessions in the network.

In the simple case where there is no time overlap between consecutive multicast communication sessions in the network, the accounting for network lifetime is the sum of successive time intervals for multicast communication sessions. In the case that a new multicast session request arrives (at a different source node) before the previous multicast session terminates. That is, we may have multiple multicast sessions in the network at the same time. Our online MLR-MD algorithm (so does the D-MIP algorithm) still works since it will consider multicast routing for each session independent from other ongoing sessions in the network. Although multiple concurrent sessions do not pose any difficulty to our algorithm, it does introduce a subtle issue in accounting of total network lifetime. A logical approach to address this issue is to consider data volume that is being transmitted. As data transmission rate is common for all nodes (10 units of data/time unit), the total amount of data that has been transmitted successfully by each multicast source node should be directly related to network lifetime calculation. Following this reasoning, any time overlap of multiple multicast communication sessions should be counted multiple times corresponding to the time overlap of multiple concurrent multicast sessions.

14.4.2 Results

For each network size (10, 20, 50, 100), we run simulations 100 times by generating 100 network topologies randomly and run the three algorithms on each topology. Before we show complete numerical results, we illustrate how the multicast routing looks like for each algorithm at some time instances under a particular topology. Figure 14.11a and b shows a 20-node network and the multicast tree under MPR, D-MIP, and MLR-MD for the first multicast session request

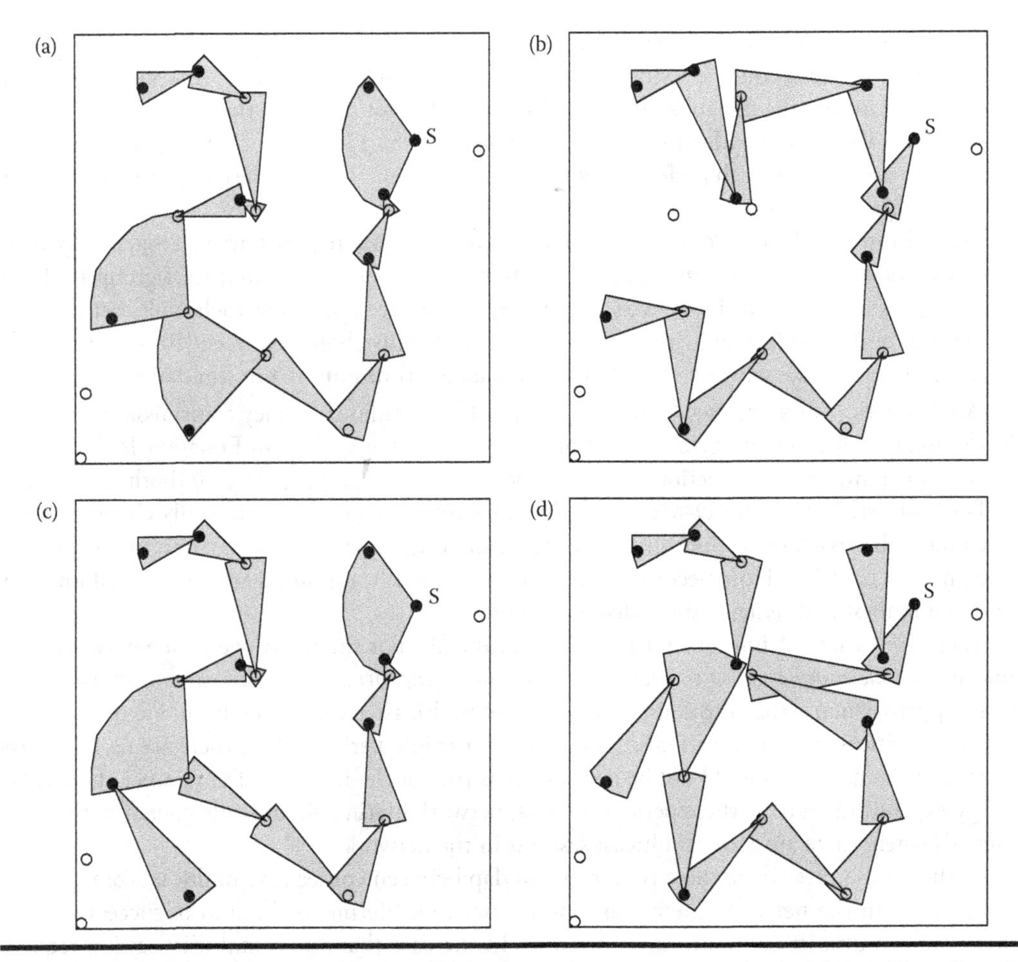

Figure 14.11 Multicast tree under MPR, D-MIP, and MLR-MD. (a) MPR and D-MIP at time 0, (b) MLR-MD at time 0, (c) D-MIP at time 8, and (d) MLR-MD at time 2.

(at time 0). In this multicast, node S is the source node, and there are eight destination nodes (filled in black color). The total data volume that needs to be sent at the source is 90 units for the first multicast session. At time 0, the remaining energy at each node is its initial energy (200 units). Also at time 0, it happens that D-MIP has the same multicast tree as MPR (Figure 14.11a), while MLR-MD has a different multicast tree topology (Figure 14.11b).

Recall that both D-MIP and MLR-MD update routing topology dynamically and the routing algorithm is executed every time unit. At time 2, MLR-MD updates its multicast tree as shown in Figure 14.11d. Under the D-MIP algorithm, it turns out that multicast tree remains the same as that constructed at time 0 until time advances to 8, with a new topology shown in Figure 14.11c. Recall that for MPR, its multicast routing tree is *fixed* throughout this multicast, that is, the multicast tree in Figure 14.11a is used for MPR until the end of this multicast. Upon successive multicast requests over time (with source and destination nodes randomly chosen for each request), we calculate the network lifetime for MPR, D-MIP, and MLR-MD, respectively.

We now summarize our numerical results (by performing 100 simulations for each network size). Instead of showing the absolute network lifetime values, we find that it is more meaningful

Table 14.1 Statistical Data of Normalized Network Lifetime Performance of MLR-MD and D-MIP Algorithms with Respect to MPR

N	*Algorithm*	*Average*	*Best*	*Worst*	*Standard Deviation*
10	D-MIP	245.68	1466.67	100.00	63.69
	MLR-MD	336.54	1861.76	98.08	93.02
20	D-MIP	276.91	1210.11	100.00	77.79
	MLR-MD	359.72	1821.35	100.00	117.42
50	D-MIP	383.15	1203.24	154.57	138.07
	MLR-MD	504.38	1894.44	161.32	213.27
100	D-MIP	307.62	973.91	98.98	167.39
	MLR-MD	424.67	1132.61	105.26	242.34

to show normalized network lifetime for easy comparison. Define normalized network lifetime as the network lifetime obtained by MLR-MD or D-MIP divided by the network lifetime obtained by MPR. The average, best case, worst case, and standard deviation (all in percentage) are shown in Table 14.1. For 100-node networks, D-MIP obtains 207% improvement while MLR-MD is able to achieve 324% improvement on average compared to MPR. For the best case, MLR-MD obtains 1032% improvement while D-MIP obtains 873% improvement. Recall that D-MIP takes explicit consideration of a node's remaining energy in routing and is a *cost-based* algorithm. On the other hand, the proposed MLR-MD algorithm directly addresses the lifetime issue in algorithm design and thus is able to achieve better network lifetime performance over D-MIP on average. Simulations on 10, 20, and 50-node network show similar results. To get a sense of what actual network lifetimes look like under different multicast routing algorithm in real time unit, Table 14.2 shows the first 20 sets of results for the 50-node network under the MPR, D-MIP, and MLR-MD algorithms.

Table 14.2 Network Lifetime of MPR, D-MIP, and MLR-MD for a 50-Node Network

Index	*MPR*	*D-MIP*	*MLR-MD*	*Index*	*MPR*	*D-MIP*	*MLR-MD*
1	246.5	699.8	872.7	11	207.1	576.9	642.4
2	74.4	191.9	249.3	12	71.5	707.8	883.3
3	167.2	497.9	730.2	13	145.0	425.9	579.4
4	165.2	552.9	684.4	14	92.8	231.9	293.3
5	147.3	328.9	505.0	15	311.5	774.2	775.4
6	93.1	268.9	699.1	16	237.8	738.9	950.5
7	207.1	610.9	820.2	17	55.3	344.9	488.1
8	161.6	412.8	424.3	18	21.6	259.9	409.2
9	69.8	490.6	549.0	19	72.5	491.8	676.3
10	158.2	337.9	401.9	20	114.7	496.7	540.5

References

1. Adamou, M. and S. Sarkar, 2002. A framework for optimal battery management for wireless nodes. In *Proceedings IEEE INFOCOM*, June 23–27, New York, NY, pp. 1783–1792.
2. Agarwal, M., J. H. Cho, L. X. Gao, and J. Wu, 2004. Energy efficient broadcast in wireless ad hoc networks with hitch-hiking. In *Proceedings IEEE INFOCOM*, March 7–11, Hong Kong, China.
3. Blough, D. and S. Paolo, 2002. Investigating upper bounds on network lifetime extension for cell-based energy conservation techniques in stationary ad hoc networks. In *Proceedings ACM MobiCom*, September 23–28, Atlanta, GA, pp. 183–192.
4. Cagalj, M., J. P. Hubaux, and C. Enz, 2002. Minimum-energy broadcast in all-wireless networks: NP-completeness and distribution issues. In *Proceedings ACM MobiCom*, September 23–26, Atlanta, GA, pp. 172–182.
5. Cartigny, J. D. Simplot, and I. Stojmenovic, 2003. Localized minimum-energy broadcasting in ad-hoc networks. In *Proceedings IEEE INFOCOM*, April 1–3, San Francisco, CA, pp. 1001–1010.
6. Das, A. K., R. J. Marks, M. El-Sharkawi, P. Arabshahi, and A. Gray, 2003. Minimum power broadcast trees for wireless networks: Integer programming formulations. In *Proceedings IEEE INFOCOM*, April 1–3, San Francisco, CA, pp. 2210–2217.
7. Das, A. K., R. J. Marks, M. El-Sharkawi, P. Arabshahi, and A. Gray, 2003. MDLT: A polynomial time optimal algorithm for maximization of time-to-first-failure in energy constrained broadcast wireless networks. In *Proceedings IEEE GLOBECOM*, December 1–5, San Francisco, CA, pp. 362–366.
8. Das, A. K., R. J. Marks, M. El-Sharkawi, P. Arabshahi, and A. Gray, 2003. R-shrink: A heuristic for improving minimum power broadcast trees in wireless networks. In *Proceedings IEEE GLOBECOM*, December 1–5, San Francisco, CA, pp. 523–527.
9. Floreen, P., P. Kaski, J. Kohonen, and P. Orponen, 2003. Multicast time maximization in energy constrained wireless networks. In *Proceedings 2003 Joint Workshop on Foundations of Mobile Computing*, September, San Diego, CA, pp. 50–58.
10. Garey, M. R. and D. S. Johnson, 1979. *Computers and Intractability: A Guide to the Theory of NP-Completeness*. New York: W. H. Freeman and Company.
11. Guo, S. and O. Yang, 2004. Antenna orientation optimization for minimum-energy multicast tree construction in wireless ad hoc networks with directional antennas. In *Proceedings ACM MobiHoc*, May 24–26, Tokyo, Japan, pp. 234–243.
12. Irani, S. and A. R. Karlin, 1997. On online computation. In *Approximation Algorithms for NP-Hard Problems, ACM SIGACT News Archive*, ed. Dorit Hochbaum, vol. 28(2), New York, NY: ACM Press, June, pp. 521–564.
13. Kang, I. and R. Poovendran, 2003. A novel power-efficient broadcast routing algorithm exploiting broadcast efficiency. In *Proceedings IEEE Vehicular Technology Conference*, October 6–9, Orlando, FL, pp. 2926–2930.
14. Kang, I. and R. Poovendran, 2005. Maximizing network lifetime of broadcast over wireless stationary ad hoc networks, *Springer Mobile Networks and Applications (MONET)*, 10(6), 879–896.
15. Kar, K., M. Kodialam, T. V. Lakshman, and L. Tassiulas, 2003. Routing for network capacity maximization in energy-constrained ad-hoc networks. In *Proceedings IEEE INFOCOM*, April 1–3, San Francisco, CA, pp. 673–681.
16. Leonardi, S. 1998. On-line network routing. In *Online Algorithms: The State of the Art*, eds. A. Fiat and G. Woeginger, Springer LNCS 1442, pp. 242–267.
17. Li, Q., J. Aslam, and D. Rus, 2001. Online power-aware routing in wireless ad-hoc networks. In *Proceedings ACM MobiCom*, July 16–21, Rome, Italy, pp. 97–107.
18. Li, Q., J. Aslam, and D. Rus, 2003. Distributed energy-conserving routing protocols. *36th Annual Hawaii International Conference on System Sciences*, January 6–9, Hawaii: Big Island.
19. Papadimitriou, C. H. 1994. *Computational Complexity*. Boston, MA: Addison-Wesley Publishing Company.
20. Rappaport, T. S. 1996. *Wireless Communications: Principles and Practice*. Upper Saddle River, NJ: Prentice Hall.

21. Rappaport, T. S. 1998. *Smart Antennas: Adaptive Arrays, Algorithms, and Wireless Position Location.* Piscataway, NJ: Institute of Electrical and Electronics Engineers.
22. Sarkar, T. K., M. C. Wicks, M. Salazar-Palma, and R. J. Bonneau, 1996. *Smart Antennas.* Hoboken, NJ: John Wiley & Sons.
23. Wan, P-J., G. Calinescu, X-Y. Li, and O. Frieder, 2001. Minimum-energy broadcast routing in static ad hoc wireless networks. In *Proceedings IEEE INFOCOM,* April 22–26, Anchorage, AK, pp. 1162–1171.
24. Wang, B. and S. K. S. Gupta, 2003. G-REMiT: An algorithm for building energy efficient multicast trees in wireless ad hoc networks. In *IEEE International Symposium on Network Computing and Applications,* April 16–18, Cambridge, MA, pp. 265–272.
25. Wieselthier, J. E., G. D. Nguyen, and A. Ephremides, 2002. Energy-efficient broadcast and multicast trees in wireless networks, *Springer Mobile Networks and Applications (MONET),* 7(6), 481–492.
26. Wieselthier, J. E., G. D. Nguyen, and A. Ephremides, 2002. Energy-aware wireless networking with directional antennas: The case of session-based broadcasting and multicasting, *IEEE Transactions on Mobile Computing,* 1(3), 176–191. June–September.

Chapter 15

Connectivity of Large-Scale Wireless Networks with Directional Antennas

Chi-Kin Chau, Richard J. Gibbens, and Don Towsley

Contents

Abstract

In multi-hop wireless networks, per-hop forwarding strategies that optimize local transmissions can have a subtle impact on network performance. Motivated by a number of scenarios for improving signal strength or mitigating interference, we study a fundamental problem that arises in a wireless ad hoc network with directional antennas, where nodes are randomly placed with their transmission footprints (each as a sector) aligned toward the destinations. Only the nodes located in the transmission footprint of a transmitter act as forwarders. Our study addresses the connectivity of this setting. We observe that there is a critical spread angle, above which there is little impact on these properties. Analytically, we derive upper and lower bounds for the critical spread angle.

15.1 Introduction

In wireless networks, transmissions are broadcast in nature. Thus, per-hop forwarding operations that minimize interfering transmitters or receivers will improve the per-hop transmission performance. Nonetheless, local forwarding operations often have a subtle impact on the overall network-wide performance. A useful approach for improving per-hop transmission performance is directional transmission, which invokes the next-hop forwarder only in a certain direction, for example, by directional antennas, or signal processing techniques using beamforming [13]. In directional antennas, transceivers can focus their signal strength in a particular direction, for enhancing signal quality at the intended receivers and reducing interference with other transmitters. In this chapter, we study a fundamental problem that arises in a wireless ad hoc network with directional antennas, where nodes are randomly placed with their transmission footprints (each as a sector) aligned toward the destinations. Only the nodes located in the transmission footprint of a transmitter act as its forwarders.

Besides directional antennas, our problem is also related to several applications that require directional transmission decisions:

1. *Opportunistic Routing*: Instead of using pre-determined paths in traditional routing protocols, opportunistic routing allows intermediate nodes that overhear a transmission to form ad hoc forwarders. However, opportunistic routing faces the problem of duplicate simultaneous transmissions, when there are multiple overhearing forwarders. A recent solution [12] was proposed that only the nodes near the direction toward the destination are strategically selected as forwarders, while other nodes ignore the transmission.
2. *In-Network Information Processing*: Sensor networks for information gathering are more efficient, if they can carry out in-network information processing to aggregate the information. In-network information processing often assumes the formation of acyclic graph that propagates information in a direction toward the sink.

In these settings, directional transmission decisions are required for forwarding the packets in a certain direction toward one or more targets.

In light of the usefulness of directional antennas, we address a fundamental question—*how do directional antennas affect the overall connectivity of large-scale wireless ad hoc networks, when the locations of nodes are unknown?* Because of the directed transmission links, connectivity is inevitably diminished. But we can adjust the transmission footprints to control the overall connectivity. Thus, we investigate the problem of controlling directional antennas to effectively optimize local transmissions, while maintaining connectivity in a wireless network. To be specific, we model each transmission footprint by a sector with a spread angle ϕ and communication range r. We study how ϕ and r affect connectivity to the destinations from an arbitrarily far-away region via multi-hop forwarding in a large network. The results of this study can shed light on the feasibility of establishing information flows to the destinations. By these results, we provide a foundation for optimizing the performance of large-scale wireless networks with directional antennas.

For clarity, we study two representative network settings: (1) A *strip* network (Figure 15.1a), where there are multiple destinations placed on a boundary at one end of a rectangular region. (2) A *radial* network (Figure 15.1b), where a single destination is placed at the center. We assume that the locations of destinations are known to the nodes and that the directions of the transmission footprints can be appropriately adjusted by the nodes to orient toward the destinations. Although the transmission is directional, we assume that the reception of a node is omnidirectional.

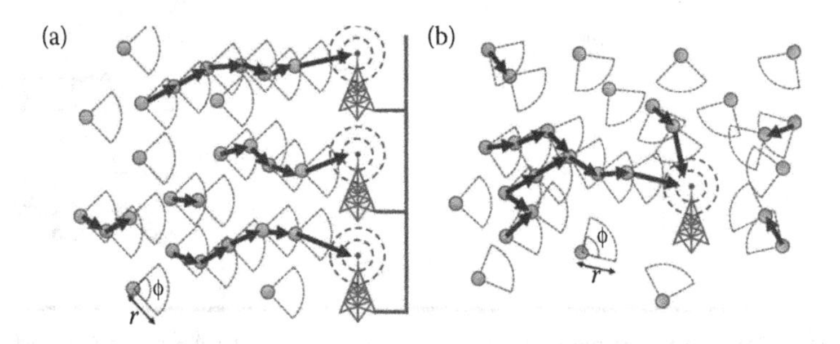

Figure 15.1 Random nodes align their transmission footprints (each as a sector of spread angle ϕ and radius r) toward the destinations. (a) A strip network, where destinations are located at one end of a region. (b) A radial network, where a destination is located at the center.

Equivalently, one can consider directional reception in the direction as the information flow to the destinations.

From the best of our knowledge, this problem has not been studied in the literature. Traditional studies of connectivity in large-scale wireless ad hoc networks are based on percolation theory [2–4,10,16]. Our consideration of transmission footprints, however, introduces a new challenge of handling directed transmission links in contrast to the traditional omnidirectional antennas links. Hence, our results draw on more sophisticated but less well-known results from *oriented percolation theory* [5,9,11].

Briefly, we outline the sections of this chapter as follows:

- In Section 15.2, we present the model and briefly survey oriented percolation theory.
- In Section 15.3, we observe that there is a critical spread angle, above which a further increase has little impact on connectivity.
- In Section 15.4, we derive upper and lower bounds for the critical spread angle threshold, which characterize the feasibility region of information flows in the presence of directional antennas.

15.2 Model and Preliminaries

First, we present the model and background of (oriented) percolation theory, and outline its relevance to our work.

15.2.1 Percolation Theory

In the existing literature, a random wireless network with omnidirectional antennas is often modeled as a 2D Poisson point process of normalized unit density in an $x \times y(x)$ region (see Figure 15.2a). Under a fixed power constraint, two nodes communicate with each other if and only if they are within a constant range r.

Connectivity of the above random model has been extensively studied by continuum percolation theory [4], which characterizes the critical threshold r_c, such that there exists infinite connected components of x and $y(x) \to \infty$ when connecting pairs of random points that locate within a range $r \geq r_c$ of each other. The critical threshold is important to the performance of large-scale wireless networks, such as wireless network capacity [2,3,8], secure communications [15], and energy-efficient

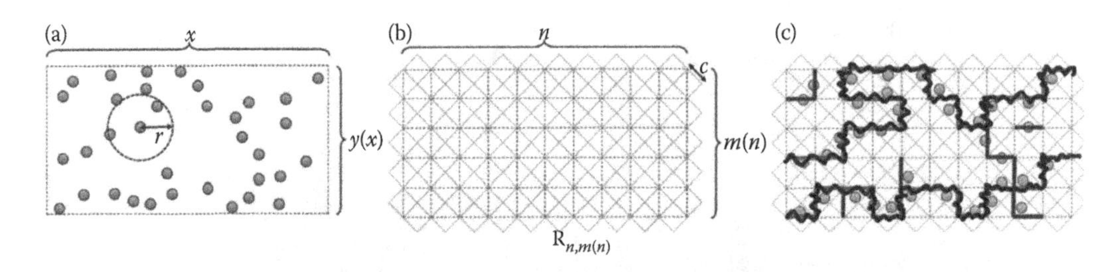

Figure 15.2 (a) A Poisson point process in an $x \times y(x)$ region. (b) An $n \times m(n)$ undirected lattice $R_{n,m}(n)$ (where links are represented by dashed lines) are laid out by an array of tilted square cells (in solid lines). (c) Poisson point process induces random links in the lattice, such that a link is enabled (as thick edge) if a node lies in the cell. The random left-to-right paths are highlighted in fluctuating lines.

duty cycling [6,7]. On the other hand, we are also interested in the presence of (undirected) left-to-right paths that traverse the region, where each hop takes a distance of at most r.

A natural approach to the above study is to map the system to the bond percolation model of a 2D lattice [4]. Such a mapping is based on an array of square cells of length c, laid out on the same area, where each cell represents a possible link in either the vertical or horizontal orientation (represented as dashed links in Figure 15.2b). These links form an $n \times m(n)$ lattice, where $n = \left\lceil \frac{x}{c} \right\rceil$. For convenience, we also denote the dimension of a Poisson point process by $n \times m(n)$.

In the lattice, we enable a link in a cell if and only if at least one node lies in the cell. Suppose $r \geq 2\sqrt{2}c$. If there exists a left-to-right path in the lattice, then there exists a left-to-right path formed by neighboring nodes within distance r of each other (see Figure 15.2c). The presence of a link is characterized by a link probability p equal to the probability of a node being present in a cell in the Poisson point process:

$$p = 1 - e^{-A} \tag{15.1}$$

where $A = c^2$ is the area of a cell.

Denote by $\{v_0 \leftrightarrow \infty\}$ the event that there exists an infinitely long undirected path starting at the center v_0. We sometime say the system percolates when v_0 is connected to an infinite connected component. Let the *percolation probability* for bond percolation model at link probability p be

$$\theta^{bp}(p) @ P\{v_0 \leftrightarrow \infty\} \tag{15.2}$$

From bond percolation theory [4], there exists a critical threshold, $p^{*bp} = 1/2$ such that whenever $p > p^{*bp}$, we have $\theta^{bp}(p) > 0$. The critical probability p^{*bp} characterizes a feasibility region of information flows in a large network.

Next, denote by $R_{n,m(n)}$ an $n \times m(n)$ undirected lattice, and $\{R^{\leftrightarrow}_{n,m(n)}\}$ the event that there exists an undirected left-to-right path in $R_{n,m(n)}$. Based on [4, Section 11.5], it can be shown that

$$P\{R^{\leftrightarrow}_{n,a\log n}\} \to 1, \quad \text{as } n \to \infty \tag{15.3}$$

provided that $p > p^{*bp}$ where $a > a^*(p)$ is a positive function decreasing in p. The implication is that it suffices for the width to be proportional to $\log n$ to ensure connectivity between the vertical ends of the strip.

15.2.2 Oriented Percolation Theory

Unlike previous literature, this chapter studies a random wireless network with directional antennas, where nodes are distributed according to a 2D Poisson point process of normalized unit density. Node v can transmit to node u if u lies in the transmission footprint of v, which is a sector of spread angle ϕ and radius r. We consider two types of networks:

1. A *strip network* where all transmission footprints are oriented in the same direction toward one boundary of a rectangular region (as in Figure 15.1a).
2. A *radial network* where the directional antennas are oriented toward the center of a square region (as in Figure 15.1b).

But there is a new challenge of handling directed links (because of directional antennas). Thus, we draw on oriented percolation theory [5,9], which studies a conventional type of directed lattice, denoted by $L_{n,m(n)}$, with directed links in uniform orientations as either rightwards or upwards (see Figure 15.3).

Directed lattices are considerably more difficult to analyze than undirected lattices. In particular, many results of (unoriented) percolation theory rely on the notion of a dual graph,* which does not apply to directed lattices. We next present some results of oriented percolation theory, which will be useful later in our study.

Denote by v_0 the site located at (0,0) in the directed lattice $L_{n,n}$ labeled $[0, n] \times [0, n]$, and $\{v_0 \to \infty\}$ the event that v_0 lies on an infinite long directed path as $n \to \infty$ (see the green path in Figure 15.3 for an illustration). The percolation probability for the bond percolation model in $L_{n,n}$, as $n \to \infty$ is denoted by

$$\theta_{\text{dir}}^{\text{bp}}(p) @ P\{v_0 \to \infty\} \tag{15.4}$$

We define the critical threshold of p by

$$p_{\text{dir}}^{*\text{bp}} @ \inf\{p \geq 0 : \theta_{\text{dir}}^{\text{bp}}(p) > 0\} \tag{15.5}$$

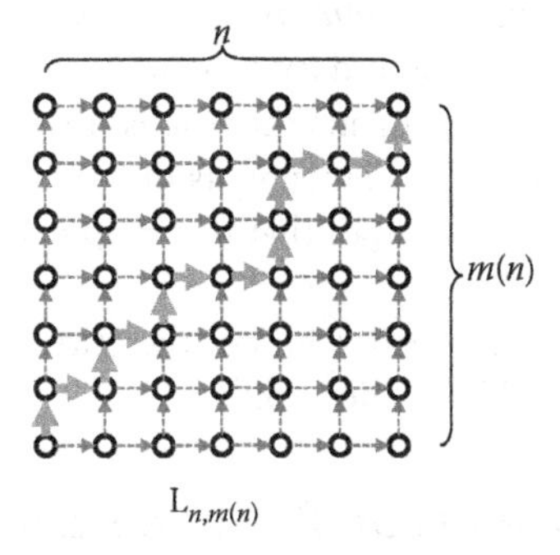

Figure 15.3 $L_{n,m(n)}$ is the conventional lattice studied in oriented percolation theory.

* A dual graph results by interchanging the roles of edges and nodes of a planar graph. Because of the existence of a dual graph in the 2D lattice, the existence of an infinite path is equivalent to the absence of a finite closed circuit [4].

In addition to the bond percolation model, where each link is independently enabled with probability p, there is the site percolation model, where each site (intersection of links) is independently enabled with probability p. Similarly, we define $\theta_{\text{dir}}^{\text{sp}}(p)$ and $p_{\text{dir}}^{*\text{sp}}$ for the site percolation model.

Theorem 15.1

(See [9,5,11]) The critical thresholds are bounded by

$$\frac{1}{3} \le p_{\text{dir}}^{*\text{bp}} \le \frac{2}{3}, \quad \frac{2}{3} \le p_{\text{dir}}^{*\text{sp}} \le \frac{3}{4} \tag{15.6}$$

and $\theta_{\text{dir}}^{\text{bp}}(p)$ *is a continuous increasing function of* p.

Note that the gap between the upper and lower bounds in the site percolation model is smaller than that of the bond percolation model. Hence, we obtain more accurate results on the critical thresholds, if we map a Poisson point process with directional antennas to a site percolation model.

15.3 Empirical Studies and Observations

Before presenting the analytical results, let us obtain a number of insights from simulation studies.

1. *Probability of Left-to-Right Path in Strip Networks*: First, we simulated a strip network of length $n = 100$ and width $m = 4, 8, 12$. We evaluated over 1000 runs the probability that there exists a left-to-right path via footprints of directional antennas for some specific values of ϕ and r, as plotted in Figure 15.4. This probability converges to the percolation probability, when the width and length of the region become very large. We observe that there is a critical threshold for ϕ that depends on r, at which there appears a rapid increase in the probability of a left-to-right path from zero. Note that such a probability is close to 1, when the strip width m is sufficiently large, but is much smaller than the length n. Furthermore, we plot the empirical values of the critical threshold of ϕ in Figure 15.9, and observe that they lie within the bounds that we derive in Section 15.4.
2. *Probability of Path to the Center in Radial Networks*: We also evaluated the probability of a path to reach the center from the boundary of a 100×100 area in a radial network, as plotted in Figure 15.5. Such a path represents the reachability of the sink at the center from the boundary. We observe a similar phenomenon as in a strip network. However, the values of the critical threshold of ϕ are lower than that of a strip network.

Furthermore, we also plot the empirical values of the critical thresholds of ϕ's in Figure 15.9, which lie within the bounds derived in Section 15.4.

15.3.1 Case Study of Transmission Power for Directional Antennas

The results of the above empirical studies are useful to certain optimization problems subject to connectivity constraints. In this section, we present a case study of optimizing the directional antennas spread angle of directional antennas with respect to the transmission power.

Suppose that a receiver lies in the transmission footprint of a transmitter with spread angle ϕ, and the transmitter–receiver distance is r. Denote the power consumption at the transmitter

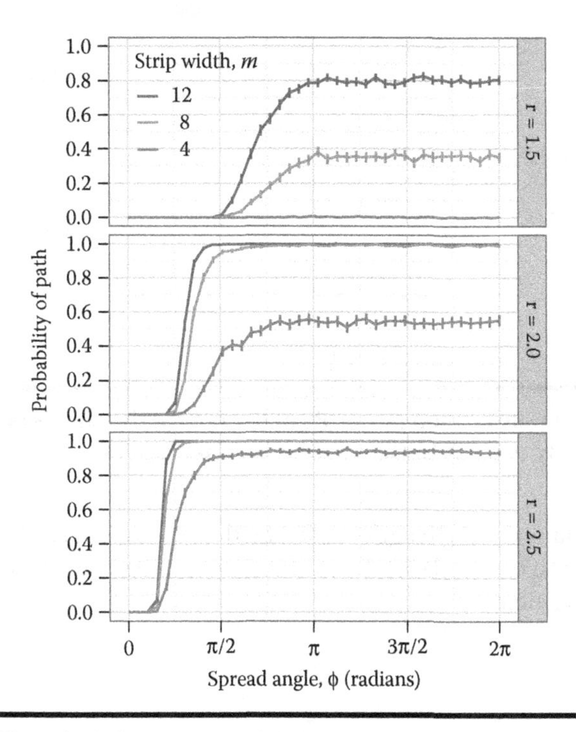

Figure 15.4 Probability of a left-to-right path in a strip network.

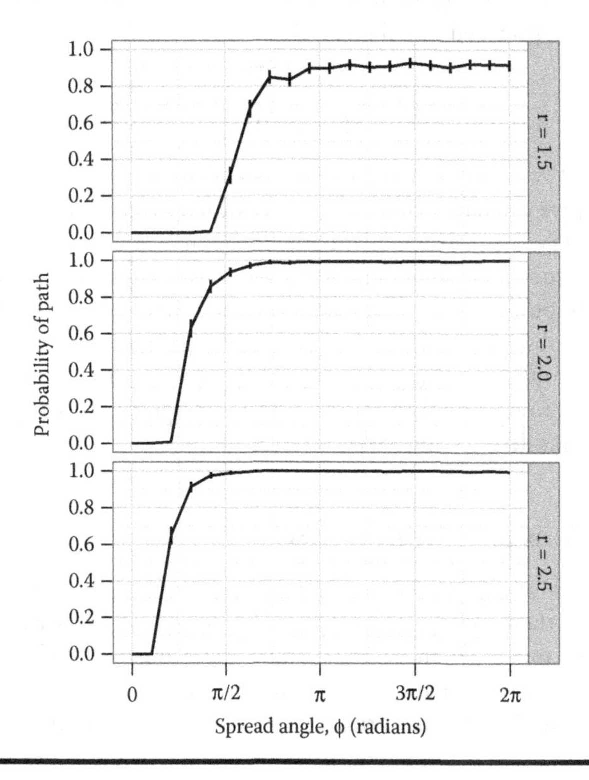

Figure 15.5 Probability of a path to reach the center from the boundary in a radial network.

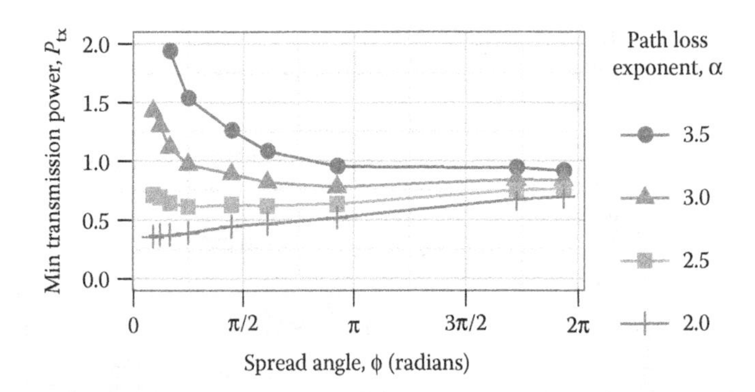

Figure 15.6 Minimum transmission power P_{tx} at the critical thresholds of ϕ and r for maintaining a positive probability of a left-to-right path in a strip network.

by P_{tx}, and the minimum signal strength needed by the receiver to decode successfully by γ. Let the path-loss exponent be α. For clarity, we assume a simple directional antenna model [14], and obtain the following relation:

$$\gamma = \frac{2\pi}{\phi} P_{tx} \cdot r^{-\alpha} \tag{15.7}$$

Hence, the minimum power to maintain a transmission footprint with spread angle ϕ and radius r that satisfies a normalized signal strength $\gamma = 1$ is given by

$$P_{tx} = \frac{1}{2\pi} \phi r^{\alpha} \tag{15.8}$$

From Figure 15.4, we evaluated the empirical critical spread angles ϕ's for radii r's, for having a positive probability of a left-to-right path in a strip network (which are also plotted in Figure 15.9). We then substitute these angles ϕ's and radii r's into Equation 15.8 to obtain the minimum transmission power P_{tx} that is subject to connectivity, and plot them in Figure 15.6. For values of $\alpha = 2$–3.5, we observe that when $\alpha = 2$, a smaller spread angle ϕ can reduce the transmission power, whereas a larger spread angle is needed for $\alpha = 3$, at the critical thresholds of ϕ and r. Therefore, our study provides useful decision criteria for optimizing the power consumption of directional antennas considering connectivity.

15.3.2 Observations and Motivations

Briefly, the above observations provide several motivations for later analytical studies. From Figures 15.4 and 15.5, there exists critical spread angles ($\phi < \pi$) in strip networks and radial networks respectively, above which there is a positive probability of a cross-area path. The critical spread angle is estimated for solving optimization problems subject to connectivity constraints (e.g., Section 15.3.1). We provide upper and lower bounds of the critical spread angle in Section 15.4.

15.4 Critical Percolation Threshold

15.4.1 Strip Networks

In this section, we derive bounds for the critical spread angle threshold ϕ in a strip network for a given radius r. Similarly, we also derive those for r for a given ϕ. Denote the percolation probability for a Poisson point process with directional antennas in a strip network by $\theta_{\text{Po}}^{\text{stp}}(\phi, r)$, such that there exists an infinite directed path via a footprint of directional antennas at each hop from left to right path in a strip network (as n, $m(n) \to \infty$). We define the critical thresholds of ϕ and r that induce percolation as follows:

$$\phi_c^{\text{stp}}(r) @ \inf\{0 \le \phi \le 2\pi : \theta_{\text{Po}}^{\text{stp}}(\phi, r) > 0\}, \tag{15.9}$$

$$r_c^{\text{stp}}(\phi) @ \inf\{r \ge 0 : \theta_{\text{Po}}^{\text{stp}}(\phi, r) > 0\} \tag{15.10}$$

The basic idea is to relate a Poisson point process with directional antennas to a directed site percolation model, and then apply Theorem 15.1. So we introduce two mappings, MU and ML, that construct two types of directed site percolation models as in Figures 15.7 and 15.8, which allow us to use the upper and lower bounds respectively from Theorem 15.1.

We describe the two mappings as follows:

MU: We define a cell as a sector of angle ϕ and radius $r/2$, containing a site (depicted as a black circle in Figure 15.7a). The cells are placed in the region of Poisson point process, and the sites are connected by directed links as in Figure 15.7b. Nodes lying outside any cells are ignored.

ML: We define a cell as a rectangle of width r (see Figure 15.8a). The cells are laid out as in Figure 15.8b.

We enable a site in a cell, if and only if a node generated from Poisson process lies in the cell. Note that both directed lattices induced by MU and ML are isomorphic to the conventional directed lattice $L_{n,n}$ in Figure 15.3, being rotated by an angle $\pi/4$ and deformed with an inner angle ϕ.

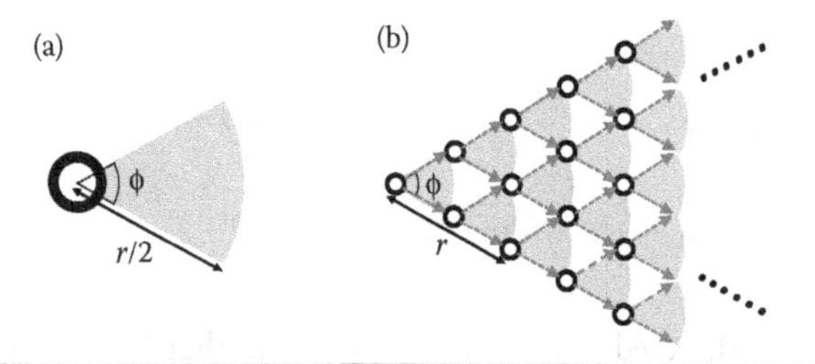

Figure 15.7 Mapping MU constructs a directed lattice by sector cells of radius *r*/2. A cell is depicted in (a). Connectivity of the directed lattice implies connectivity of nodes using directed transmission. (a) One cell; (b) connections of multiple cells.

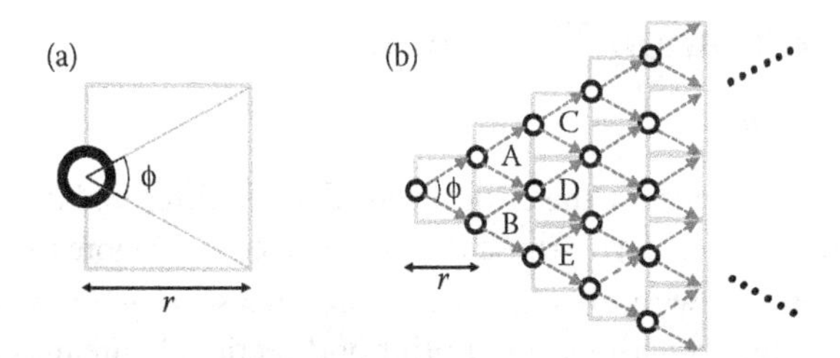

Figure 15.8 Mapping ML constructs a directed lattice by rectangular cells of width *r*. A cell is depicted in (a). Disconnectivity of the directed lattice implies disconnectivity of nodes using directed transmission. (a) One cell; (b) disconnections of multiple cells.

Theorem 15.2

The critical thresholds for a strip network are bounded by

$$2\tan^{-1}(\log 3/2r^2) \le \phi_c^{stp}(r) \le 8\log 4/r^2 \tag{15.11}$$

$$\sqrt{\log 3/\left(2\tan\frac{\phi}{2}\right)} \le r_c^{stp}(\phi) \le \sqrt{8\log 4/\phi} \tag{15.12}$$

Proof: (Sketch) The upper bounds are established by mapping MU, while the lower bounds are by mapping ML.

Upper Bounds: In the directed lattice induced by MU, we note that the distance between a pair of nodes individually lying in a pair of neighboring cells respectively is at most r. Hence, if a pair of neighboring sites are enabled, then there exists a pair of nodes individually lying in the respective cells that are connected by directional antennas with spread angle ϕ and range r. Namely, connectivity of the directed lattice implies connectivity of nodes using directed transmission. Thus, we obtain

$$\theta_{dir}^{sp}(p) \le \theta_{Po}^{stp}(\phi, r) \tag{15.13}$$

where $p = 1 - e^{-A}$ and $A = \phi(r/2)^2/2$_ is the cell area.

By Theorem 15.1, there exists $p = p_c^{sp} \le (3/4)$, such that $\theta_{dir}^{sp}(p) > 0$. Therefore, we obtain

$$1 - e^{-A} \le \frac{3}{4} \Rightarrow \phi_c^{stp}(r) \le 8\log 4/r^2 \quad \text{and}$$
$$r_c^{stp}(\phi) \le \sqrt{8\log 4/\phi} \tag{15.14}$$

Lower Bounds: In the directed lattice by ML, we note that it is not possible to connect via directional antennas a pair of nodes lying in different cells separated by more than one level. For instance, if node x can transmit to node y and y lies in cell D in Figure 15.8b, then x can only lie in cells A, B, C, D, or E. Hence, if there does not exist a connected sequence of enabled sites via

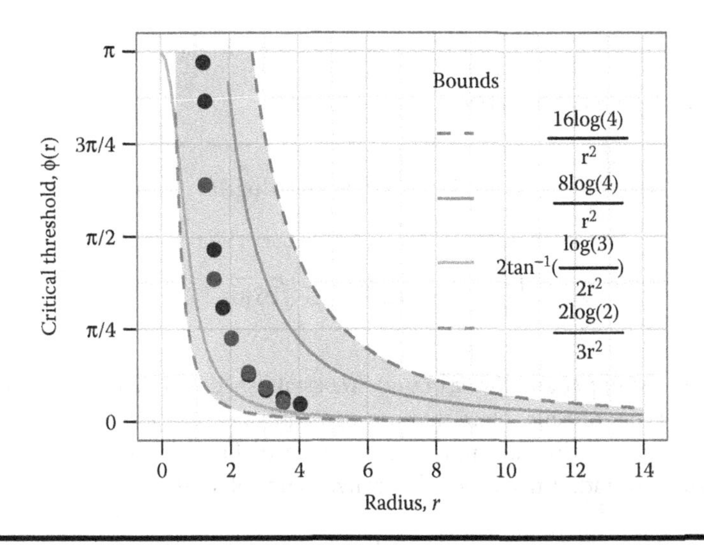

Figure 15.9 Solid lines are the upper and lower bounds for strip networks from Theorem 15.2, while the dashed lines are the ones for radial networks from Theorem 15.3. The black dots are empirical values of the critical spread angles from Figure 15.4 from the simulation in Section 15.3, while the gray dots are the ones for radial networks from Figure 15.5.

directed links in the lattice, then there does not exist a connected sequence of nodes lying in the respective cells using directional antennas with spread angle ϕ and range r. Namely, disconnectivity of directed lattice implies disconnectivity of nodes using directed transmission. Thus,

$$\theta_{\text{dir}}^{\text{sp}}(p) \geq \theta_{\text{Po}}^{\text{stp}}(\phi, r) \tag{15.15}$$

where $p = 1 - e^{-A}$ and $A = 2r^2 \tan \phi/2$ is the cell area.

By Theorem 15.1, there exists $p = (2/3) < p_c^{\text{sp}}$, such that $\theta_{\text{dir}}^{\text{sp}}(p) = 0$. Therefore, we obtain

$$\begin{aligned} &1 - e^{-A} \geq (2/3) \Rightarrow \phi_c^{\text{stp}}(r) \geq 2\tan^{-1}(\log 3/2r^2) \\ &r_c^{\text{stp}}(\phi) \geq \sqrt{\log 3/(2\tan(\phi/2))} \end{aligned} \tag{15.16}$$

To corroborate with simulation, we plot the upper and lower bounds of Theorem 15.2 in Figure 15.9 (solid lines), together with the empirical values (black dots) from Figure 15.4, which lie within the bounds.

15.4.2 Radial Networks

We next derive bounds for the critical threshold of spread angle ϕ in a radial network. Denote the percolation probability for a Poisson point process with directional antennas to the center v_0 in a radial network by $\theta_{\text{Po}}^{\text{rad}}(\phi, r)$ (as $n \to \infty$), and the critical thresholds by $\phi_c^{\text{rad}}(r)$ and $r_c^{\text{rad}}(\phi)$.

Theorem 15.3

The critical thresholds for a radial network are bounded by

$$\frac{2}{3}\log 2/r^2 \leq \phi_c^{\text{rad}}(r) \leq 16\log 4/r^2 \tag{15.17}$$

$$\sqrt{\frac{2}{3}\log 2/\phi} \leq r_c^{\text{rad}}(\phi) \leq \sqrt{16\log 4/\phi} \tag{15.18}$$

Proof: (Sketch) The upper and lower bounds are established in a manner similar to Theorem 2, but using different types of cells. Note that one can equivalently consider reception footprints in the direction as the information flow to the center, instead of transmission footprints. We consider reception footprints for radial networks. For convenience of analysis, we assume that $2\pi/\phi$ is an integer.

Upper Bounds: We consider a sector of angle $\phi/2$ at the center. It suffices to show that percolation occurs in such a sector for sufficiently large ϕ. We use a similar mapping as MU. But the inner angle at the center is $\phi/2$ and the cells are deformed to fit in the layers of rings around the center (see Figure 15.10). Note that the inner angle of each cell is at most $\phi/2$. Hence, each cell can be contained in the reception footprint with spread angle ϕ at each site, oriented toward to the center.

Thus, if a pair of neighboring sites are enabled, then there exists a pair of nodes individually lying in the respective cells connected by directional reception footprints aligned toward the center with a spread angle no more than ϕ and range r. Although the cells have different areas, the minimum cell area $A \geq ((\phi/2)(r/2)^2)/2$. Next, we establish the upper bounds in a manner similar to Theorem 15.2.

Lower Bounds: We consider a directed radial lattice of layers of rings around the center, as depicted in Figure 15.11, in which every sector of angle ϕ (colored in dark gray in Figure 15.11) resembles the directed lattice in Figure 15.8b in mapping ML.

Thus, that if there does not exist a connected sequence of enabled sites, then there does not exist a connected sequence of nodes lying in the respective cells using reception footprints with spread angle at most ϕ and range r, where each reception footprint is aligned toward the center. Namely, the disconnectivity of directed lattice implies the disconnectivity of nodes based on directed reception. The maximum area A of a cell (excluding the cell of center) is bounded by

$$A \leq \frac{(4\pi r^2 - \pi r^2)\phi}{2\pi} = \frac{3r^2\phi}{2} \tag{15.19}$$

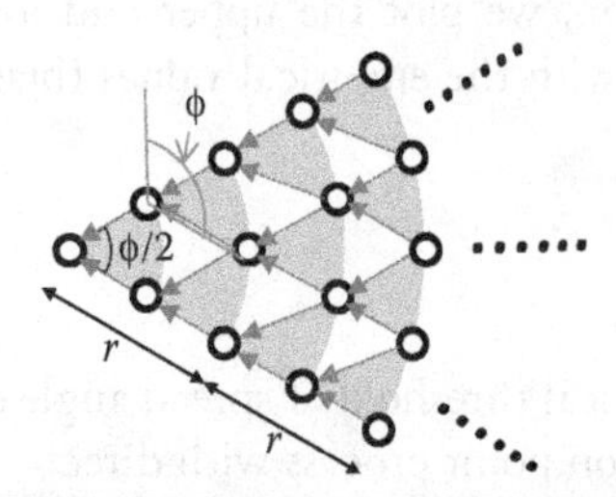

Figure 15.10 We consider cells to be laid out in a similar fashion as MU, but the reception footprints are aligned toward the center with an inner angle $\phi/2$.

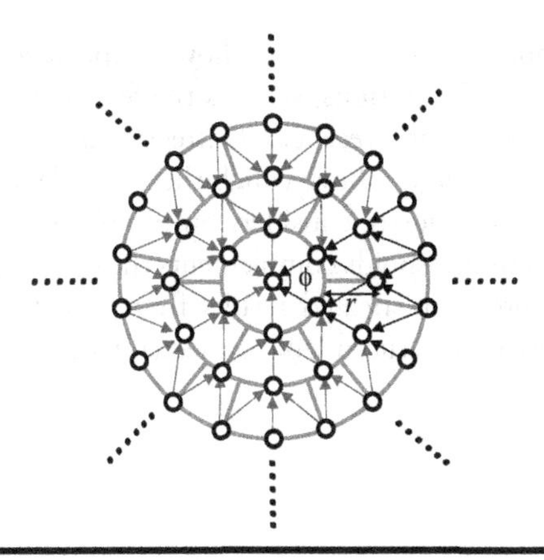

Figure 15.11 A directed radial lattice induced by a radial network. A sector of angle ϕ, colored in dark gray, resembles the directed lattice in Figure 15.8b for ML.

Let $N(n)$ be the expected number of paths in the directed radial lattice with n layers. Note that a site in the i-th layer cannot reach another node in the j-th layer (where $j < i$) in another sector. For instance, the sites in the sector colored in dark gray in Figure 15.11 can only reach sites that also lie within the same sector. Therefore, $N(n)$ is upper bounded by

$$N(n) \le \frac{2\pi}{\phi} 2^n p^n \tag{15.20}$$

Note that if $N(n) \to 0$ as $n \to \infty$, then $P\{v_0 \to \infty\} \to 0$ as $n \to \infty$. A sufficient condition for $N(n) \to 0$ as $n \to \infty$ is $p < 1/2$. Together, we obtain

$$1 - e^{-A} < \frac{1}{2} \Rightarrow \frac{3r^2\phi}{2} < \log 2 \tag{15.21}$$

Hence,

$$\phi_c^{\mathrm{rad}}(r) \ge \frac{2}{3}\frac{\log 2}{r^2} \text{ or } r_c^{\mathrm{rad}}(\phi) \ge \sqrt{\frac{2}{3}\frac{\log 2}{\phi}} \tag{15.22}$$

To corroborate with simulation, we plot the upper and lower bounds of Theorem 15.3 in Figure 15.9 (dashed lines), together with the empirical values (gray dots) from Figure 15.5, which lie within the bounds.

15.5 Conclusion

In this chapter, we have studied the impact of directional antennas on multi-hop wireless ad hoc network. Directional antennas can reduce the interference from transmitters, improve signal quality at intended receivers, and alleviate energy consumption. Our analysis of connectivity properties in such a setting not only sheds light on how to optimize transmission footprints, but also addresses several fundamental questions, such as the feasibility and capacity of information flows in large ad hoc networks in the presence of directional antennas. Our empirical studies show that there is a critical spread angle, above which there is little impact on connectivity; a new contribution to the literature that mostly considers omnidirectional antennas. Our analysis for bounding the critical percolation thresholds complements the empirical studies, and demonstrates a novel application of the more sophisticated results from oriented percolation theory. In our future work, we will characterize the structure of connected paths in large-scale wireless networks with directional antennas [1].

References

1. Chau, C-K., R. Gibbens, and D. Towsley, 2012. Impact of directional transmission in large-scale multi-hop wireless ad hoc networks. In *Proceedings IEEE INFOCOM.*
2. Franceschetti, M., O. Dousse, D. Tse, and P. Thiran, 2007. Closing the gap in the capacity of wireless networks via percolation theory, *IEEE Trans. Information Theory*, 53(3), 1009–1018.
3. Franceschetti, M. and R. Meester, 2008. *Random Networks for Communication—From Statistical Physics to Information Systems*, Cambridge University Press.
4. Grimmett, G. 1999. *Percolation*, Cambridge University Press.
5. Grimmett, G. and P. Hiemer, 2001. Directed percolation and random walk. In *In and Out of Equilibrium*, Birkhäuser, pp. 273–297.
6. Guha, S., C-K. Chau, and P. Basu, 2010. Green wave: Latency and capacity-efficient sleep scheduling for wireless networks. In *Proceedings IEEE INFOCOM.*
7. Guha, S., C-K. Chau, P. Basu, and R. Gibbens, 2011. Green wave sleep scheduling: Optimizing latency and throughput in duty cycling wireless networks, *IEEE J. Selected Areas Communications*, 29(8), 1595–1604.
8. Li, S. Y. H. Liu, and X-Y. Li, 2008. Capacity of large scale wireless networks under Gaussian channel model. In *Proceedings ACM MobiCom.*
9. Liggett, T. M. 1995. Survival of discrete time growth models, with applications to oriented percolation, *Annals Applied Probability*, 6(3), 613–636.
10. Meester, R. and R. Roy, 2008. *Continuum Percolation*, Cambridge University Press.
11. Pearce, C. E. M. and F. K. Fletcher, 2005. Oriented site percolation, phase transitions and probability bounds, *J. Inequalities Pure Appl. Math.*, 6(5), 1–15.
12. Rozner, E., J. Sheshadri, Y. A. Mehta, and L. Qiu, 2009. SOAR: Simple opportunistic adaptive routing protocol for wireless mesh networks, *IEEE Trans. Mobile Computing*, 8(12), 1622–1635.
13. Sun, C., J. Cheng, and T. Ohira, 2008. *Handbook on Advancements in Smart Antenna Technologies for Wireless Networks*, IGI.
14. Tse, D. and P. Viswanath, 2005. *Fundamentals of Wireless Communication*, Cambridge University Press.
15. Vasudevan, S., D. Goeckel, and D. Towsley, 2010. Security versus capacity trade-offs in large wireless networks using keyless secrecy. In *Proceedings ACM MobiHoc.*
16. Xue, F. and P. R. Kumar, 2006. On the theta-coverage and connectivity of large random networks, *IEEE Trans. Information Theory*, 52(6), 2289–2399.

Chapter 16

Bounds on the Lifetime of Wireless Sensor Networks with Lossy Links and Directional Antennas

Juan M. Alonso, Amanda Nordhamn,
Simon Olofsson, and Thiemo Voigt

Contents

16.1 Introduction

Energy-efficiency and longevity are of major importance in wireless sensor networks (WSN) as they have a large impact on the usefulness and cost of the target application. While there has been a tremendous amount of research on energy-efficient WSNs, relatively few studies have been performed that have aimed at quantifiying the gains of using directional antennas [1]. Toward this end, we describe a method to bound the lifetime of WSN developed earlier [2,3], that can be extended to cover directional antennas. The method's aim is to find bounds on the energy consumption of nodes, regardless of the routing used. The main problem is to find a *lower* bound, which is then used to obtain an *upper* bound on the lifetime.

We begin with an informal description of the method in its simplest form, where networks contain no directional antennas and links are not lossy but perfect. We then proceed to explain the method in full generality, where networks can contain directional antennas and links are lossy. In both cases, definitions and results are stated rigorously and explained carefully, but no proofs are offered. The interested reader is referred to References 2 and 3 for some of the proofs. The proof of the most general result, as well as other examples of application, will be published elsewhere.

In the final sections the relevance of the explicit bounds obtained in the theory is demonstrated by simulations with a practical Medium Access Control (MAC) layer.

16.1.1 General Assumptions on the Network

The nodes in the network are of two types: *sensor nodes* and *base nodes*. Sensor nodes (or, simply, nodes) are of low-energy type and have very limited memory and processing capabilities, whereas base nodes (often called sinks) are of high-energy type and have significantly more processing power and memory capacity than sensor nodes. We make the assumption that there is an underlying hierarchic architecture whereby the base nodes control the sensor nodes deciding, in particular, which *routing* to use. We use the term routing to denote a specific set of paths (or multipaths) that packets take through a network. A routing is the result of the particular routing algorithm used.

The sensor nodes take readings and send them to the bases using other sensor nodes as relays, that is, communication occurs in a multihop fashion. This process is *iterated* until nodes eventually die, breaking connectivity and making the network nonoperational. While time is not an explicit parameter in our model, we do implicitly assume a weak form of synchronization: Each iteration takes a certain amount of time that may be different for different iterations, and within each time interval every node makes its reading and transmission/reception. Observe that the time each iteration takes may vary from seconds to days or even months. There is no assumption that any of these events occur simultaneously. Also, routings may be different in each iteration: packets can take different paths in different iterations.

These assumptions are suitable for sensor networks with a *continuous data delivery model*. In their taxonomy, Tilak et al. [4] classify sensor networks in accordance to their data delivery model into continuous, event-driven, observer-initiated, and hybrid. Continuous networks are those in which "the sensors communicate their data continuously at a prespecified rate."

No data aggregation is done in the network: all data gathered is sent unchanged to the base nodes. Data aggregation is certainly an important way to save energy and to extend the lifetime of sensor networks. We leave it out of our analysis because our main interest is in the effect that routing has on energy consumption. We also feel that data aggregation is too intimately related to specific applications.

16.1.2 Spheres and Discs

Spheres were introduced in Reference 2, and play a crucial role in our approach. They are defined inductively. The sphere of radius zero, S_0, is by definition the set of base nodes. Notice that networks are allowed to have many base nodes. The sphere of radius one, S_1, is defined to be the set of nodes *exactly* one hop away from a base node. In general, the sphere S_i of radius $i \geq 1$ consists of all nodes that can reach some base node in *exactly i* hops. The word "exactly" means that there is a way to reach some base node in i hops, *and* that no base node can be reached in *less* than i hops. It follows that spheres of different radius are disjoint sets. By definition D_i, the *disc of radius i*, is the union of all spheres of radius $\leq i$, that is,

$$D_i = S_0 \amalg S_1 \amalg \ldots \amalg S_i, \tag{16.1}$$

In particular, $D_0 = S_0$, is the set of base nodes; to simplify the notation, we set $B := D_0$. The square union signs in Equation 16.1 denote *disjoint* union and are used to make explicit, in this case, that a disc is a union of sets that are disjoint. Nodes not in D_i will be said to lie *outside* D_i.

The key idea is to consider S_i as a bottleneck for traffic originating in nodes outside D_i, trying to reach B. More formally, fix $i \geq 1$, a routing, and a packet that originates outside of D_i. Consider the path toward some base point followed by the packet, as prescribed by the routing. Eventually, the packet will have to *go through* S_i, that is, visit a node of S_i, and continue its journey *inside* D_{i-1}, never again touching S_i. Figure 16.1 illustrates different situations that might happen in a trajectory followed by a packet toward a base node. In the picture, the packet first touches S_i only tangentially (i.e., it continues its journey outside D_i), then goes temporarily through the sphere (i.e., afterward it goes again outside D_i) and then, eventually, goes through S_i, at node w.

It follows that, in each iteration, all packets that originate outside of D_i must go through S_i, regardless of the routing used.

We use this fact to do some accounting. Assume there are N nodes in the network, and let $s_i = |S_i|$ denote the number of nodes in the sphere of radius i. If $d_i = |D_i|$, then $d_i = s_0 + \cdots + s_i$. There are $N - d_i$ nodes *outside* of D_i. Thus, in each iteration, there are $N - d_i$ packets that must go through S_i,(i = 1, 2, …). Hence, the nodes of S_i receive and transmit *at least*

$$N - d_i \tag{16.2}$$

packets.

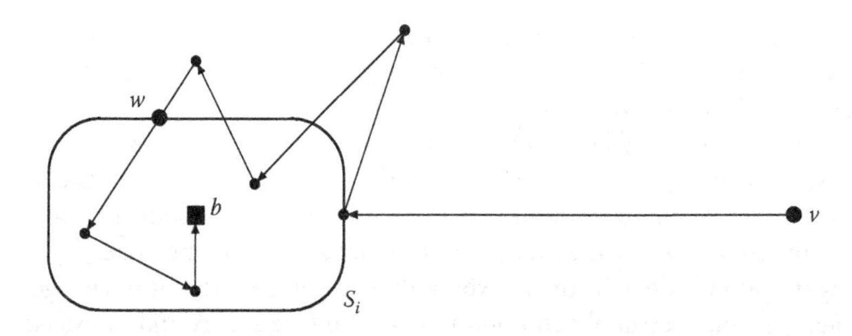

Figure 16.1 Trajectory toward a base node *b* of a packet originating at a node *v* that lies outside the sphere S_i. The packet *goes through* S_i at node *w*.

16.1.3 Routings and Their Energy Consumption

We now take a more formal view. The network is modelled by a *directed graph* $G = (V, E)$, where V is the set of vertices (or nodes) and E is the set of oriented edges (we call them simply edges, or links). We assume that there is at most one *directed* edge connecting two different vertices. Thus edges are ordered pairs of vertices $e = (v, w)$, for $v, w \in V$, and $v \neq w$. Observe that if $e = (v, w) \in E$, it is possible that also $\overline{e} = (w, v) \in E$.

The network operates with the following traffic pattern. For each iteration t, $1 \leq t \leq T$, every node sends a packet of a certain length to some base node. Informally, each way to do this is a routing. More formally, a *routing* is a vector

$$y = (y_e^t)_{1 \leq t \leq T, e \in E}$$

where y_e^t represents the total number of packets (destined to some base node) that are transmitted through e during the t:th iteration. Observe that we can think of the routing y as being a sequence $y = (y^1, \ldots, y^T)$, where y^t is the routing used during the t:th iteration. The only restriction we place on routings is that they should be *effective*, in the sense of not having loops. A routing *has no loops* if for all $1 \leq t \leq T$, the following holds: for every node, the directed path used to send its packet to a base node, visits a node *at most* once. Observe that the routing suggested in Figure 16.1, which one might consider to be very inefficient, is effective in the sense of this definition because it has no loops. Let

$$\mathcal{R}^T := \{y = (y_e^t)_{1 \leq t \leq T, e \in E} \mid y \text{ has no loops}\}. \quad (16.3)$$

The condition that routings have no loops gives a bound similar to Equation 16.2, but in the opposite direction: we claim that any nonbase node v will receive [resp., transmit] *at most*:

$$N - s_0 - 1 \quad [\text{resp.}, N - s_0]$$

packets in each iteration. Indeed, a packet originating at $w \neq v$ can visit v at most once, and there exactly $N - s_0 - 1$ nodes like w.

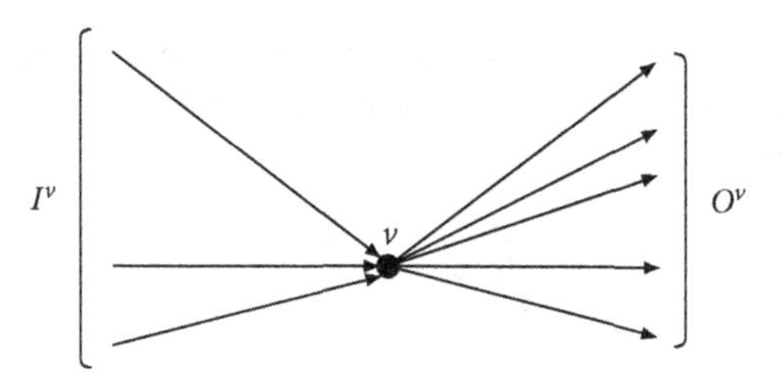

Figure 16.2 Ingoing (I^v) and outgoing (O^v) links at a node v.

The energy consumption of a routing y will be measured by the following *cost function* $f^T : \mathcal{R}^T \to \mathrm{R}_+$:

$$f^T(y) := \max_{v \in V \setminus B} \left\{ \sum_{t=1}^{T} \left(\sum_{e \in I^v} \rho y_e^t + \sum_{e \in O^v} \tau y_e^t \right) \right\} \tag{16.4}$$

where ρ [resp. τ] is the cost for the reception [resp. transmission] of one packet, I^v is the set of *incoming links* of v, $I^v := \{(i, j) \in E | j = v\}$, and O^v is the set of *outgoing links* of v, $O^v := \{(i, j) \in E | i = v\}$ (see Figure 16.2).

Thus, $f^T(y)$ measures the maximum energy used by nodes when transmitting and receiving according to routing y. In using f^T we are disregarding the energy consumption of base nodes. This is in line with the assumption that base nodes have no shortage of energy. Observe that we disregard the routing protocol overhead and concentrate instead on the energy consumption of the application-related communication. We think this is a reasonable simplification to make in networks with a continuous data delivery model. In practice, this situation is achieved when the network, after an initial setup phase, reaches a steady state.

Set

$$\mathrm{MIN}^T := \min_{y \in \mathcal{R}^T} f^T(y), \quad \mathrm{MAX}^T := \max_{y \in \mathcal{R}^T} f^T(y).$$

Then $\mathrm{MIN}^T \le f^T(y) \le \mathrm{MAX}^T$, for all routings $y \in \mathcal{R}^T$. Observe that there are routings $y_0, y_1 \in \mathcal{R}^T$, such that $f^T(y_0) = \mathrm{MIN}^T$, and $f^T(y_1) = \mathrm{MAX}^T$. In other words, the maximum and minimum values are always achieved. Our aim is to find bounds on the size of the interval [MIN^T, MAX^T], the most important being lower bounds, denoted (*l.b.*) for short, see Figure 16.3.

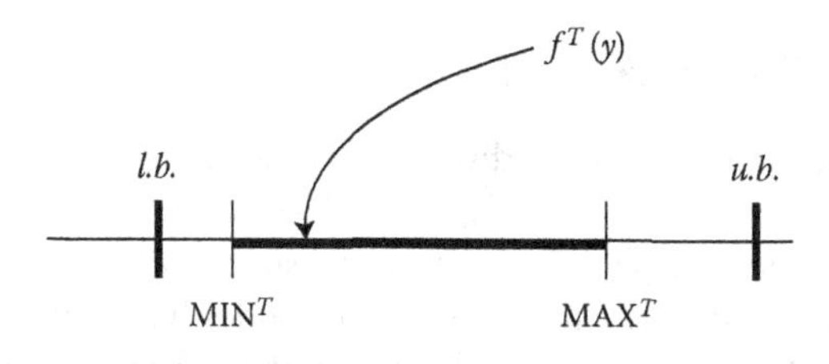

Figure 16.3 Values of f^T lie in the interval [MIN^T, MAX^T], and have lower bounds (*l.b.*) and upper bounds (*u.b.*). Theorem 16.1 computes a specific (*l.b.*) and (*u.b.*).

The condition that there is no data aggregation in the network implies that all routings considered conserve flow. This can be expressed rigorously as follows. For all t and every v that is not a base node, the equations

$$\sum_{e \in O^v} y_e^t = 1 + \sum_{e \in I^v} y_e^t \tag{16.5}$$

hold. Indeed, the left-hand side computes the number of packets transmitted by node v during iteration t, while the right-hand side equals one (the packet produced and transmitted by v at each iteration) plus the number of packets received by v during iteration t. Note that for leaf nodes, the right-hand-side of Equation 16.5 equals 1, since routings are effective.

16.2 The Bounds in the Simplest Case

Besides the general assumptions we have described so far, which will hold throughout, we assume in this section that the following extra *restrictions* hold: all antennas are omnidirectional (i.e., there are *no directional antennas*), all nodes transmit at the *same constant cost*, and all links are *perfect*, with no loss. Later we lift these restrictions to obtain more general results.

Suppose the network has $n + 1$ spheres, so that $N = |V| = |S_0| + \cdots + |S_n| = s_0 + \cdots + s_n$, and all $s_i > 0$. The lower bound we seek is expressed in terms of the following *sphere numbers* introduced in Reference 2:

$$m(S_i) := \frac{N - d_i}{s_i}\rho + \frac{N - d_i + s_i}{s_i}\tau \tag{16.6}$$

Observe that $m(S_n) = \tau$. The simplest case of our main result is as follows:

Theorem 16.1

For networks satisfying, besides the general requirements of Section 16.1.1, the extra restrictions of this section, the following inequalities hold:

i. $\mathrm{MIN}^T \geq T\max\{m(S_1), \ldots, m(S_n)\}$
ii. $\mathrm{MAX}^T \leq T(\rho(N - s_0 + 1) + \tau(N - s_0))$
iii. $\mathrm{MAX}^T \leq Ts_1\,\mathrm{MIN}^T + T\rho(s_1 - 1)$ ■

16.2.1 Remarks on Theorem 16.1

a. Theorem 16.1(i) shows that the bound we seek is the largest of the $m(S_i)$ (more precisely, it is this number multiplied by T). By Equation 16.6, $s_i\, m(S_i)$ is the cost, for each iteration, of receiving $N - d_i$ packets, plus the cost of transmitting $N - d_i + s_i$ packets. Recall from Section 16.1.2 that $N - d_i$ is the total number of nodes lying outside D_i. Thus $m(S_i)$ is the average cost incurred, in each iteration, by nodes of S_i for reception and transmission of an amount of packets equal to those created outside D_i plus, in the case of transmission, of the extra packets they produce themselves.

It follows that $s_i\ m(S_i)$ is the minimum consumption of the i-sphere, and this minimum is achieved when traffic through S_i can be balanced, so that each of its nodes consumes the exact same amount, namely $m(S_i)$.

b. Observe that to define the $m(S_i)$ we need only know the parameters:

$$\rho, \tau, s_0, s_1, \ldots, s_n, \tag{16.7}$$

that is, the reception and transmission costs per packet (determined by the hardware and MAC layer), and the number of nodes in each sphere (determined by the geometry of the network).

c. It is instructive to compare our approach with a traditional linear program. Recall that traffic in networks has been modelled as a linear program long ago (see e.g., [5,6]), more specifically, as a *(multi)commodity flow problem*. We have used bits of this model in Section 16.1.3 (for the particular form we have in mind, see [7,8]).
In a linear program one uses rather detailed information on the network in order to formulate a program, which is then solved. The result is a number and a routing: the number is the optimal value of f^T, that is, MIN^T, and the routing is optimal, in the sense that its energy consumption is the computed optimal one. Most changes made to the network will require a reformulation of the program, which will have to be solved again, to produce a new optimal value and routing.
By contrast, in our approach we do not solve a linear program, but through a simple computation obtain instead *bounds* on the energy consumption of all possible routings. The only computation needed is that of the numbers $m(S_1), \ldots, m(S_n)$ and their maximum. In its turn, these numbers require rather general information about the network, given by the parameters (16.7). In fact, a given set of parameters is information shared by many networks, some of them quite different (see Figure 16.4). On the other hand, our approach says nothing about the quality of the bounds, or about any routings that might approach those bounds. These matters have to be discussed separatedly for each network, see for example Section 16.4.1.1.

d. As mentioned in (c) above, the quality of the lower bound can vary: it is sometimes sharp, and sometimes not. Let (*l.b.*) denote the lower bound of the theorem. Recall (*l.b.*) is called *sharp* if (*l.b.*) = MIN^T, that is, if there is no gap between the bound and the minimum possible consumption. Observe that when (*l.b.*) is sharp, there is a routing y_0 with $f^T(y_0) = (l.b.)$. Conversely, if there is a routing y_0 such that $f^T(y_0) = (l.b.)$, then (*l.b.*) is sharp, since:

$$f^T(y_0) = (l.b.) \leq \text{MIN}^T \leq f(y_0)^T$$

where the first inequality is given by Theorem 16.1(i). See Section 16.2.4 for further remarks and examples relating to Theorem 16.1.

e. The strength of the theorem derives from its generality, as its results apply, regardless of routing or radio energy model, to any two networks that have the same parameters (16.7) and, of course, satisfy the conditions and requirements of Sections 16.1.1 and 16.2. Inequalities (i–iii) of Theorem 16.1 can be seen as providing fundamental limits to the possible amount of improvement in energy consumption that can be derived from changes in the routing algorithm, as well as benchmarks to compare your favorite routing(s) against.

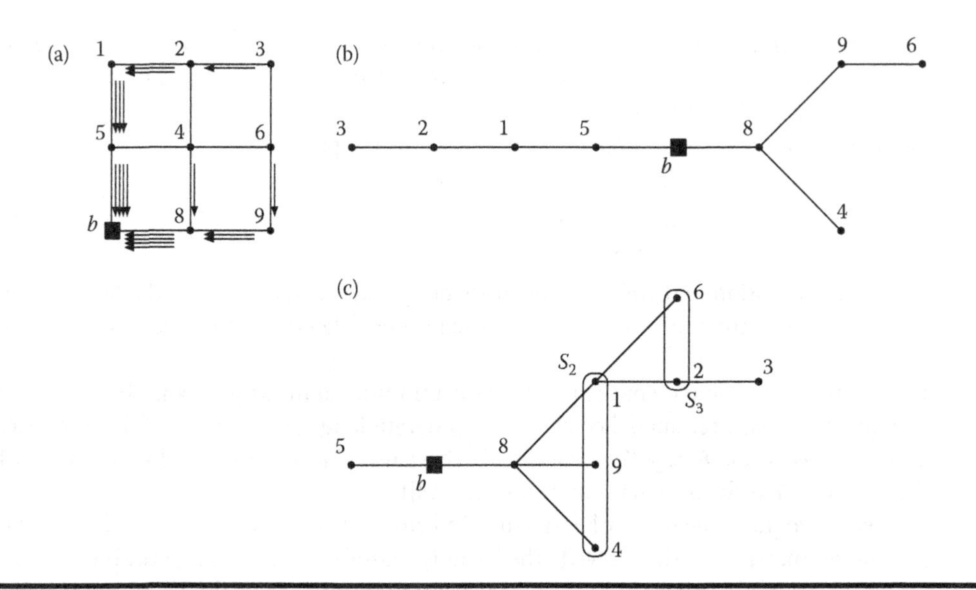

Figure 16.4 Three quite different networks with $s_0 = 1$, $s_1 = 2$, $s_2 = 3$, $s_3 = 2$, and $s_4 = 1$. Spheres S_2 and S_3 are marked in (c). (a) Grid-like topology; (b) tree-like: less data fusion nodes; and (c) tree-like: more data fusion nodes.

16.2.2 Bounds on the Lifetime of Networks

An immediate application of Theorem 16.1 is to bound the lifetime of networks. Suppose each node has the same amount $\mathbb{E}$ of energy, and we use a routing y in a traffic pattern consisting of T iterations. The network will be operational as long as $f^T(y) \leq \mathbb{E}$. Define $T_{\max}$ to be the largest integer T for which $\text{MIN}^T \leq \mathbb{E}$. Thus, when Theorem 16.1(i) is sharp, say $\text{MIN}^T = Tm(S_i)$, for some i, then $T_{\max} = \lfloor \mathbb{E}/m(S_i) \rfloor$, where $\lfloor x \rfloor$ is, by definition, the largest integer $\leq x$, for instance $\lfloor 3.21 \rfloor = 3$. However, Theorem 16.2 bounds lifetime.

Theorem 16.2

The maximum number $T_{\max}$ of readings a sensor network can take under the given assumptions, is bounded as follows:

$$\frac{\mathbb{E}}{\rho(N - s_0 + 1) + \tau(N - s_0)} - 1 \leq T_{\max} \leq \frac{\mathbb{E}}{\max\{m(S_1), \ldots, m(S_n)\}}$$

The validity of this theorem was checked experimentally in Reference 9 with very good results. ■

16.2.3 Other Corollaries

The initial motivation for this work was a question posed by the authors of one of the first sensor networks deployed: that of Great Duck Island [10]. Discussing different routing algorithms, the authors write (in Section 6.2 of Chapter 6): "Although these methods provide factors of *2 to 3 times* [our italics] longer network operation, our application requires a factor of 100 times longer network operation ...." We thought this was intriguing: *What factor is reasonable to expect of a routing algorithm?* The following corollaries of Theorem 16.1 give answers to this question. More specifically, the factor $u(k)$ computed in Equation 16.9 is not far from the "factors of *2 to 3 times*" mentioned above.

One can distinguish two cases in Theorem 16.1, according to whether or not $n = 1$. Consider first the rather trivial case when $n = 1$, that is, when all nodes are one hop away from a base node. The network deployed in Great Duck Island in the summer of 2002 was of this type; the one deployed in 2003 was multihop. Inequality (i) in Theorem 16.1 reduces to $\text{MIN}^T \geq T\tau$, that is, the minimal energy use after T iterations is the transmission cost times T. It is easy to find an optimal routing, that is, a routing achieving this minimum: for each node, select a base node one hop away, and transmit the node's unique packet to the chosen base node; repeat T times. In this case, the upper bound for MAX^T, $T[(\rho + \tau)s_1 - \rho]$, can be achieved if, for instance, the nonbase nodes can use each other to transmit their packets to a specified nonbase node that receives all the packets minus its own, and transmits all s_1 packets to a base node; notice that this routing would be effective. Summarizing:

Corollary 16.1

In the special case when every node is only one hop away from a base node, we have:

i. $\text{MAX}^T \leq T[(\rho + \tau)s_1 - \rho]$
ii. $\text{MIN}^T \geq T\tau$.

Moreover, (ii) is sharp, that is, there is a routing $y \in \mathcal{R}^T$ *with* $f^T(y) = \text{MIN}^T$.

We now consider cases that are somewhat opposite to the previous one. Suppose $n > 1$, and let $r > 0$. Consider the inequality:

$$r(N - d_1) = r(s_2 + \ldots + s_n) \geq s_1$$

This number r can be seen a *relative* measure of the "thickness" of S_1, when compared to the rest of the network. Thus S_1 is "thin" when r is small, and gets "thicker" as r increases. In the next result, the interesting case is when r is the smallest value satisfying $r(N - d_1) \geq s_1$, that is, $r = s_1/(N - d_1)$. ■

Corollary 16.2

If $r(N - d_1) \geq s_1$, *for some* $r > 0$, *then:*

$$\frac{\text{MAX}^T}{\text{MIN}^T} \leq (1 + r)(s_1 - 1) + 1 \tag{16.8}$$

As an example, consider a square network with $N = (2k + 1)^2$ nodes, and with $d_0 = 1$, $s_1 = 4$. Figure 16.5 (d) shows an example with $k = 2$. Then, Equation 16.8 gives:

$$\frac{\text{MAX}^T}{\text{MIN}^T} \leq 3\left(1 + \frac{1}{k^2 + k - 1}\right) + 1 := u(k) \searrow 4 \tag{16.9}$$

This factor $u(k)$ is the "theoretical" bound to the gain one can expect from routing alone. It starts at $u(1) = 7$, and decreases fast to the limit 4, as $k \rightarrow \infty$. ■

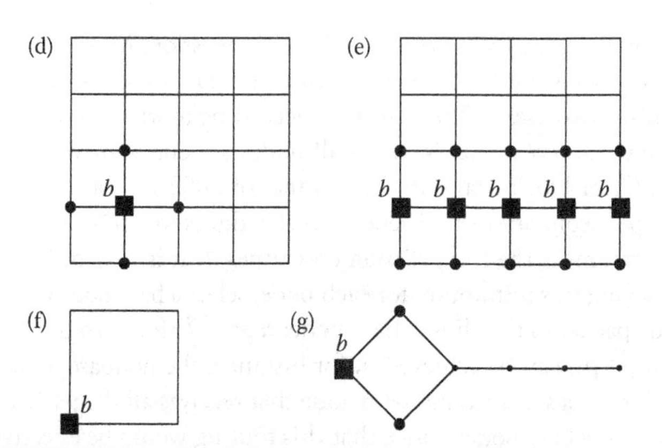

Figure 16.5 More examples. (d) One sphere with grid shape; (e) multiple spheres with grid shape; (f) just one center node; and (g) one center node plus multiple side nodes.

16.2.4 Example

In this section we discuss inequality (*i*) of Theorem 16.1. We show via examples (mostly for $T = 1$, in which case we write MIN, instead of MIN^1) that (*i*) can be sharp and not sharp. We also show that, even in those cases where (*i*) is sharp, it may happen that $\max\{m(S_1), \ldots, m(S_n)\} = m(S_i)$, for some $i \geq 2$, that is, it is not necessarily true that the heaviest communication burden is carried by S_1. Not so long ago this fact went against popular belief and rather convincing, but wrong, arguments (see e.g., [11]).

If (*i*) were always sharp it would mean that the explicit formula

$$\max\{m(S_1), \ldots, m(S_n)\} \tag{16.1.1}$$

computes the optimum for all possible networks that satisfy the conditions of Section 16.1.1. This is plainly too much to expect, so it should not be difficult to find examples with $MIN > \max\{m(S_1), \ldots, m(S_n)\}$. On the other hand, there are many cases where Equation 16.1.1 does indeed compute the optimal value, that is, cases where (*i*) is sharp. So one can think of Equation 16.1.1 as the "theoretical" optimum: it is the value that "should" be optimal. It is achieved in networks where traffic can be balanced evenly between the elements of the sphere of appropriate radius. Example (*f*) of Figure 16.5 is a simple instance where traffic cannot be evenly balanced in a strict sense, but only "on average." We consider this case as one in which the optimum *is* achieved.

We start with examples (a), (b), and (c) of Figure 16.4, which shows different networks that share the same parameters (16.7) (we assume that, besides the geometry of the spheres, they share the same reception and transmission costs, i.e., have the same ρ and τ). It follows that all three networks have exactly the same sphere numbers:

$$m(S_1) = 3\rho + 4\tau, \quad m(S_2) = \frac{3}{2}\rho + \frac{5}{2}\tau, \quad m(S_3) = \frac{1}{2}\rho + \frac{3}{2}\tau, \quad m(S_4) = \tau$$

Clearly, $m(S_1) > m(S_2) > m(S_3) > m(S_4)$, so that $m(S_1)$ is the maximum of these values. By Theorem 16.1, $\text{MIN} \geq m(S_1)$. We now look closer to the similarities and differences between (a), (b), and (c).

The numbers that identify the nodes in the three networks are meant to preserve the spheres: for instance, the two elements of the 1-spheres in (a), (b), and (c) denote 5 and 8. We start with example (a). The arrows next to some of the edges are meant to indicate a specific routing. For example, the three lines near edge (1,5) mean that node 1 transmits three packets, its own plus the two packets it receives from node 2, in every iteration. Packets are routed along a tree. Indeed, this tree is precisely network (b), and the numbering of the nodes help visualise this fact. Let k_i denote the cost incurred by node i in each iteration (cf. Equation 16.15). We have $k_5 = 3\rho + 4\tau = k_8$, so that this routing is optimal. On the other hand, $k_1 = 2\rho + 3\tau = k_6 = \tau$, and $k_9 = \rho + 2\tau$. The fact that $m(S_2) < k_1$, shows that traffic through S_2 is not balanced.

Consider now (b). It has only one effective routing, which is also optimal. The reader can easily check that the k_i just computed for (a) is equal to the corresponding ones for network (b) and this routing. Hence, the numbers $m(S_i)$ and k_i are equal in both networks. So, *is there any difference between (a) and (b)?* Indeed there is! To the obvious "visual" ones, we add an important communication difference. As already pointed out, (b) has only one optimal routing. In contrast, (a) has several optimal, tree-like, routings. So if node 1, say, were to fail and die, one could easily reroute traffic in (a) to another (near) optimal routing, but in (b) traffic from nodes 2 and 3 would be irremediably lost. Hence, from the point of view of *reliability*, (a) is much to be preferred to (b): here is a concrete, objective difference. Note also that the different optimal routings of (a) give rise to several different tree-like networks similar to (b).

In case (c) we have $k_5 = \tau$, and $k_8 = 6\rho + 7\tau$. Hence node 5 consumes less, and node 8 more, than the optimum $m(S_1)$. In contrast to (a) and (b), where the lower bound of Theorem 16.1 is sharp, (c) is an example where the bound is far from sharp. The reason is that in (c) traffic through S_1 cannot be balanced.

Consider Figure 16.5. Example (d) is characteristic for square networks with "judicious" choice of base nodes in that the bound is sharp. Example (e) shows, however, that size and placement of the base are important parameters to consider if this is to hold. In both (d) and (e) only the discs of radius 1 are marked, but there is a node at every intersection. Despite the off-center choice of base node in (d), this network can balance traffic and bound (*i*) of Theorem 16.1 is sharp. We leave it to the reader to check that for this network, $\text{MIN} = m(S_1) = \max\{m(S_1), \ldots, m(S_n)\}$, and that $m(S_1) = 5\rho + 6\tau$, $m(S_2) = (7/3)\rho + (10/3)\tau$, $m(S_3) = (4/3)\rho + (7/3)\ \tau$, $m(S_4) = (3/5)\rho + (8/5)\tau$, $m(S_5) = (1/2)\rho + (3/2)\tau$, and $m(S_6) = \tau$. The opposite holds for (e): the network cannot take advantage of the 10 nodes of S_1 in order to balance traffic, and the bound is far from sharp.

Example (f) is a simple case to illustrate that using the same routing at each iteration can be far from optimal. Here $m(S_1) = (1/2)\rho + (3/2)\tau$, $m(S_2) = \tau$, and m(S1) = $\max\{m(S_1), m(S_2)\}$, but $\text{MINT}^T \neq Tm(S_1)$. However, if we let $y = (y_1, y_2, y_1, y_2, \ldots)$, then $f^T(y) = Tm(S_1)$ "on average," in the sense that, when $T \to \infty$, $f^T(y)/T \to m(S_1)$. Indeed,

$$f^T(y) = \begin{cases} Tm(S_1) & \text{if } T \text{ is even} \\ \dfrac{\rho(T+1)}{2} + \dfrac{\tau(3T+1)}{2} & \text{if } T \text{ is odd} \end{cases}$$

Network (g) is an example where $\max\{m(S_1), \ldots, m(S_n)\} = m(S_2) > m(S_1)\}$. In this example it is not S_1 but S_2 that supports the heaviest burden of communication. It is easy to generalise the example to push the most heavily burdened sphere further away from the base nodes, thus obtaining examples where $\max\{m(S_1), \ldots, m(S_n)\} = m(S_i)$, for $i > 2$ (of course, i will be much smaller than n). Indeed, assuming that traffic is evenly split at the bifurcation, we have:

$$m(S_1) = 2\rho + 3\tau = m(S_3), \quad m(S_2) = 3\rho + 4\tau, \quad m(S_4) = \rho + 2\tau, \quad m(S_5) = \tau$$

It follows that $\max\{m(S_1), \ldots, m(S_n)\} = m(S_2) > m(S_1)\}$, regardless of the actual values of ρ, $\tau > 0$. The explanation is simple: the node at the bifurcation is the one that transmits and receives most packets in the network. It is easy to see that the bound is sharp, so MIN $= m(S_2)$.

16.3 Networks with Directional Antennas

We now remove the restrictions imposed at the beginning of Section 16.2. For the rest of the chapter, we assume that networks can have nodes with *directional* antennas and nodes with *omnidirectional* antennas, that transmission costs are *not necessarily equal* for all nodes, and that links are *lossy* instead of perfect.

That sensor nodes can be equipped either with a directional or with an omnidirectional antenna is modeled by having two *types* of vertices: *directional* (d-nodes, for short) or *omnidirectional* (o-nodes). The type of vertex v is denoted as $\ddot{v}$. Set $V_o := \{v \in V \mid \ddot{v} = o\}$, and $V_d := \{v \in V \mid \ddot{v} = d\}$. Thus

$$V = V_o \coprod V_d$$

Nodes in $V_d \backslash B$ transmit *and* receive directionally. Similarly, nodes in $V_o \backslash B$ both transmit *and* receive omnidirectionally. Base nodes, on the other hand, can be d-nodes or o-nodes only in the sense that their reception is likewise.

Recall that for every nonbase node v, I^v denotes the set of incoming edges of v, $I^v = \{(i, j) \in E \mid j = v\}$, and O^v denotes the set of outgoing edges of v, $O^v = \{(i, j) \in E \mid i = v\}$. We further classify edges as follows. Outgoing edges according to the type of their receiving node:

$$O_d^v := \{(i,j) \in O^v \mid j \in V_d\}, \quad \text{and} \quad O_o^v := \{(i,j) \in O^v \mid j \in V_o\},$$

and incoming edges according to the type of their originating node:

$$I_d^v := \{(i,j) \in I^v \mid i \in V_d\}, \quad \text{and} \quad I_o^v := \{(i,j) \in I^v \mid i \in V_o\}.$$

Clearly, $O^v = O_d^v \coprod O_o^v$, and $I^v = I_d^v \coprod I_o^v$.

16.3.1 A Simplifying Assumption on Predecessors

We make the following simplifying assumption:

$$I^v = I_d^v, \quad \text{or} \quad I^v = I_o^v, \quad \text{for all } v \in V \backslash B. \tag{16.10}$$

The condition prevents nodes from having both d-nodes *and* o-nodes as predecessors, see Figure 16.6. It is a technical condition that simplifies the mathematics.

Observe that condition $I^v = I_d^v$ is equivalent to $I_o^v = \emptyset$. Similarly, $I^v = I_o^v$ is equivalent to $I_d^v = \emptyset$. With this condition, the reception cost depends, again, only on v, since it is the same for all incoming edges at v. These reception costs, denoted μ_v ($v \in V \backslash B$), are given by:

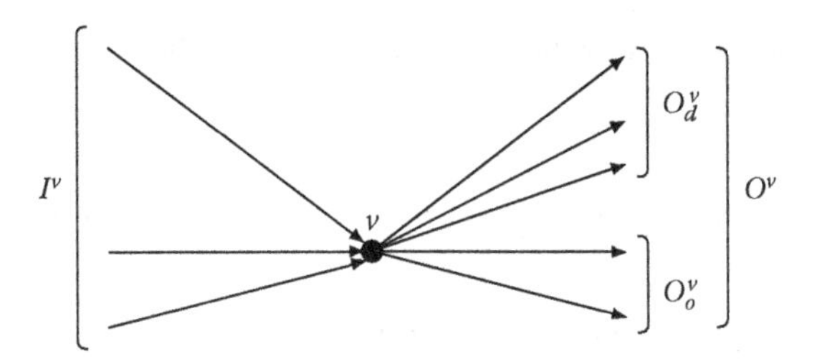

Figure 16.6 Ingoing links have only one type, but outgoing links can be both directional and omnidirectional.

$$\mu_v := \begin{cases} \rho_v^{d\ddot{v}}, & \text{if } I^v = I_d^v \\ \rho_v^{o\ddot{v}}, & \text{if } I^v = I_o^v \end{cases}, \tag{16.11}$$

where $\ddot{v} \in \{d, o\}$, is the type of v.

16.3.2 Lossy Links

We model lossy edges by counting the number of packets that must be transmitted at v through $e = (v, w)$, in order to actually receive the packet at w. Let $p_e \in (0, 1]$ denote the *probability of success* for e, meaning that when a large number L of packets is transmitted through e, then $p_e \cdot L$ will be received at w, and $(1 - p_e) \cdot L$ will be lost. Thus, p_e is the same as *PDR*, the *packet delivery rate* of e.

For $e = (v, w)$, we make the assumption that p_e depends *only* on the type of its end nodes v, w. Thus, p_e will take one of the following four values, for all $e \in E$:

$$p(oo) \leq p(od) \quad \text{and} \quad p(do) \leq p(dd). \tag{16.12}$$

So, for instance, if $v \in V_o$ and $w \in V_d$, then $p_e = p(\ddot{v}\ddot{w}) = p(od)$. It should be clear that the indicated inequalities hold. Regarding $p(od)$ and $p(do)$, theory says that they are equal because antennas are reciprocal [12]. In practice, however, they tend to be close with $p(od) \leq p(do)$, and this is the way we model these probabilities in the simulations.

Let τ_v denote the transmission cost of v (i.e., the energy spent by v to transmit a packet). We assume that τ_v depends only on v, that is, it is the same for all links $e \in O^v$. But some of the links from v are more lossy than others. The transmission cost so that *all* packets transmitted at v through $e = (v, w)$, are received at w, is

$$\tau_v \frac{y_e^t}{p_e}, \quad \text{or} \quad \frac{\tau_v}{p_e} y_e^t$$

The first identity computes the cost by multiplying the per-packet cost, τ_v, by the larger number of packets, y_e^t/p_e, that must be transmitted so that y_e^t packets are received at w. Of course, the last identity gives the same value, but can be interpreted as saying that we transmit exactly the amount of packets required by the routing y (i.e., y_e^t) but at a higher transmission cost, namely

$$\tau_v^e := \frac{\tau_v}{p_e}$$

Thus, the transmission cost is now not only a function of v, but also of $e \in O^v$. We also use the notation:

$$\tau_v^{dd} := \frac{\tau_v}{p(dd)} \leq \tau_v^{do} := \frac{\tau_v}{p(do)}, \quad \text{and} \quad \tau_v^{od} := \frac{\tau_v}{p(od)} \leq \tau_v^{oo} := \frac{\tau_v}{p(oo)} \tag{16.13}$$

For reception, there is the extra cost of listening to the retransmissions caused by lossy links, that we also take into account. Let ρ_v denote the energy spent by v to receive a packet. We assume that ρ_v depends only on v, that is, it is the same for all edges $e \in I^v$. But, again, some of the edges into v are more lossy than others. The reception cost so that *all* packets transmitted at w through $e = (w, v)$, are received at v, is

$$\rho_v \frac{y_e^t}{p_e}, \quad \text{or} \quad \frac{\rho_v}{p_e} y_e^t$$

The last identity can be interpreted as saying that we receive exactly the amount of packets required by the routing y (i.e., y_e^t) but the reception cost is higher, namely

$$\rho_v^e := \frac{\rho_v}{p_e}$$

As before, the reception cost is now not only a function of v, but also of $e \in I^v$. We use the notation

$$\rho_v^{dd} := \frac{\rho_v}{p(dd)} \leq \rho_v^{do} := \frac{\rho_v}{p(do)}, \quad \text{and} \quad \rho_v^{od} := \frac{\rho_v}{p(od)} \leq \rho_v^{oo} := \frac{\rho_v}{p(oo)} \tag{16.14}$$

In summary, our model for lossy links takes into account the need for retransmission *and* the need to spend more time listening to retransmitted packets, both caused by packet loss. On the side of simplification, we asume a link's lossyness depends only on the type of its end nodes.

16.3.3 The Cost Function

We introduce the following notation. The total costs $k_v^t = k_v^t(y)$, and $K_v = K_v(y)$, for a node v when transmitting and receiving according to a given a routing y, are given by:

$$k_v^t := \sum_{e \in I^v} \rho_v^e \cdot y_e^t + \sum_{e \in O_d^v} \tau_v^e \cdot y_e^t + \sum_{e \in O_o^v} \tau_v^e \cdot y_e^t, \quad \text{and} \quad K_v := \sum_{t=1}^{T} k_v^t \tag{16.15}$$

where ρ_v^e, and τ_v^e were defined in Section 16.3.2. Observe that we have used condition (16.10) in the first summation term: without the condition, the sum over I^v would have to be split into a sum over I_d^v and a sum over I_o^v. The first summand in the definition of k_v^t calculates the total cost for reception, and the last two take care of the transmission costs, given that we now have two different transmission costs, depending on the two types of outgoing edges. After all iterations are done, the total cost for v is K_v.

The energy consumption of a routing y will be measured by the following cost function $f^T : \mathcal{R}^T \to \mathbf{R}_+$:

$$f^T(y) := \max_{v \in V \setminus B} \left\{ \sum_{t=1}^{T} k_v^t(y) \right\} = \max_{v \in V \setminus B} \{K_v(y)\}$$

Clearly, $f^T(y)$ is the same function (16.4), only adapted via (16.10) to the present, more general, situation. With this new f^T, we define as before:

$$\text{MIN}^T := \min_{y \in \mathcal{R}^T} f^T(y), \quad \text{MAX}^T := \max_{y \in \mathcal{R}^T} f^T(y)$$

16.3.4 Spheres Revisited

Recall that the set of nodes of the network is partitioned into $n + 1$ disjoint spheres $S_0, \ldots, S_n$, with $B = S_0$ the set of base nodes. In this more general context, the sphere numbers $m(S_i)$ are more difficult to define, as they involve a lot more information that in the previous case.

Given a sphere S_i, let

$$\mu(S_i) = (\mu_v)$$

denote the s_i-tuple indexed by the vertices of S_i, consisting of the numbers μ_v of Equation 16.11.

We consider now transmission costs, and introduce the notion of *forced nodes*. Each node has only one reception cost, but two transmission costs. Forced nodes help in assigning transmission costs attuned with each node's actual transmission capabilities. The end result is that the lower bound of Theorem 16.3 will be sharper. For a subset of base nodes $W \subseteq B$, we let $S_1(W)$ denote the set of nodes at exactly one hop from some $w \in W$. Observe that $S_1(W) \subseteq S_1(B) = S_1$. When $W = \{w\}$, we simply write $S_1(w)$, instead of $S_1(\{w\})$.

Definition 16.1

Suppose $B \cap V_o = \{w_1, \ldots, w_k\}$, *for some* $0 \le k \le s_0$, *and let* $\hat{S}_1(w_j) := S_1(w_j) \setminus S_1(B \cap V_d)$. *We define* V^f, *the set of* forced nodes *as follows:*

$$V^f := \{v \in V \setminus B \mid O_d^v = \emptyset\} \cup \hat{S}_1(w_1) \cup \cdots \cup \hat{S}_1(w_k)$$

■

Essentially, forced nodes are those that cannot "choose" to transmit cheaply, but are forced to transmit to o-nodes, at the most expensive rate. To these, we have added the nodes of the sets $\hat{S}_1(w_j)$ which consist, by definition, of nodes of S_1 that are exactly one hop away from w_j, but cannot reach any d-node of B in one hop, see Figure 16.7. We want to consider nodes in $\hat{S}(w_j)$ also as forced because, although they might be able to transmit to a neighboring d-node (in S_1, or maybe backwards, in S_2), we think of them as being forced to transmit to the o-node w_j. The case $k = 0$ means that the sets $\hat{S}_i$ are empty.

The definition of $\hat{S}_1(w)$ is illustrated by the portion of a network shown in Figure 16.7. It has a base node b, and nodes 1,3,4 lie in S_1. Nodes b,1,3,4 are o-nodes, and 2,5,6 are among the d-nodes.

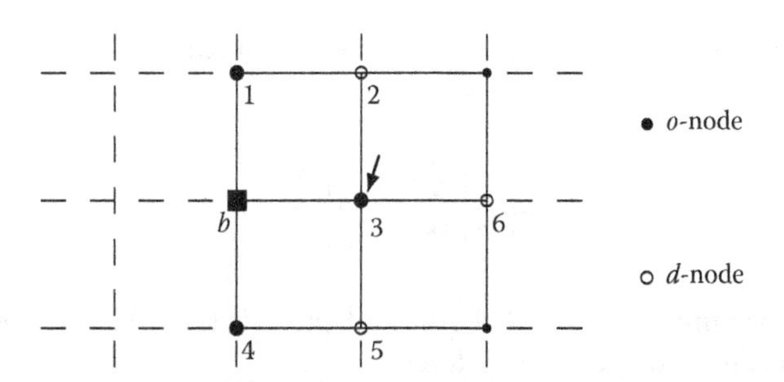

Figure 16.7 Node 3 ∈ $\hat{S}_1(b)$ is *forced*: Even though it might transmit cheaply to nodes 2, 5, or 6, we force it to take the highest cost, as if it were to transmit to the more expensive base node.

Node 3, marked in the picture with an arrow, has a choice: it can transmit its packets at a cheap rate to nodes 2,5,6, or at an expensive rate, to the base node b. By declaring it a forced node, we take its contribution to $m(S_1)$ at the higher cost, as if it had sent its traffic to b.

Let $S_i(f) := S_i \cap V^f$ denote the forced nodes of S_i, and $S_i(nf) := S_i \backslash S_i(f)$ denote the *nonforced* nodes of the sphere. This decomposes the sphere as a disjoint union

$$S_i = S_i(f) \coprod S_i(n\,f)$$

that reflects the classification of the nodes of S_i according to their outgoing edges: forced or not forced. Let $\ell_i := |S_i(f)|$, for $i = 1, \ldots, n$, denote the number of forced nodes of the i-sphere.

Next, we introduce auxiliary numbers $z(S_i, \mu(S_i))$:

$$z(S_i, \mu(S_i)) := \begin{cases} \displaystyle\sum_{v \in S_i(f)} \frac{\mu_v}{\beta_v} + \sum_{v \in S_i(nf)} \frac{\mu_v}{\alpha_v} & \text{when } \ell_i \geq 1 \\ \displaystyle\sum_{v \in S_i} \frac{\mu_v}{\alpha_v} & \text{when } \ell_i = 0 \end{cases} \tag{16.16}$$

where $\beta_v := \mu_v + \tau_v^{\ddot{v}o}$ (for $v \in S_i(f)$), $\alpha_v := \mu_v + \tau_v^{\ddot{v}d}$ (for $v \in S_i(n\,f)$), and $\ddot{v}$ is the type of v. Replacing μ by $\mathbf{1} = (1, \ldots, 1)$ in $z(S_i, \mu(S_i))$, we obtain $z(S_i, \mathbf{1})$; in practical terms, this means that we replace each μ_v by 1, in the definition of z. We are now ready to generalize the sphere numbers:

Definition 16.2

For each sphere S_i, define sphere numbers:

$$m(S_i) := \frac{N - d_{i-1} - z(S_i, \mu(S_i))}{z(S_i, \mathbf{1})} \tag{16.17}$$

■

Remark 16.1

The sphere numbers of Equation 16.17 are a far-reaching generalization of the sphere numbers introduced in Equation 16.6, as we now show. When the reception costs are equal for all v, say $\mu = \mu_v = \rho$ for all $v \in V \backslash B$, we have $z(S_i, \mu(S_i)) = \rho z(S_i, \mathbf{1})$. Hence,

$$m(S_i) = \frac{N - d_{i-1}}{z(S_i, 1)} - \rho \quad (16.18)$$

To simplify even further, assume, moreover, that no link is lossy. Then the transmission costs depend only on v (i.e., not on the particular link used to transmit from v). In this case, $\beta_v = \alpha_v = \rho + \tau_v$ and, regardless of ℓ_i,

$$z(S_i, \mu(S_i)) = \rho z(S_i, 1) = \rho \sum_{v \in S_i} \frac{1}{\rho + \tau_v}$$

so that

$$m(S_i) = \frac{N - d_{i-1}}{\sum_{v \in S_i} (1/(\rho + \tau_v))} - \rho$$

which is exactly the form of $m(S_i)$ given in [3]. If, moreover, $\tau_v = \tau$ for all v, then

$$m(S_i) = \frac{N - d_{i-1}}{s_i/(\rho + \tau)} - \rho = \frac{N - d_i}{s_i}\rho + \frac{N - d_{i-1}}{s_i}\tau$$

and we recover definition (16.6) of $m(S_i)$. ■

Remark 16.2

To define the sphere numbers (16.17) we need the following parameters:

$$S_0, \{S_i, S_i(f), \mu(S_i) \mid i = 1, \ldots, n\}, \{\beta_v \mid v \in S_i(f)\}, \{\alpha_v \mid v \in S_i(nf)\} \quad (16.19)$$

Having computed the spheres and their subsets of forced nodes, one obtains the rest of the needed parameters: s_o, N, s_i, d_i, and ℓ_i, for $i = n, \ldots, n$. Comparing (16.19) with (16.7) is a good gauge of the greater complexity of the general sphere numbers. ■

16.3.5 The Main Theorem for Networks with Directional Antennas and Lossy Links

Theorem 16.3 below is our main result. The proof of the theorem and further examples of its application will be published elsewhere.

Theorem 16.3

Suppose we have a network that satisfies the general requirements of Section 16.1.1, and those of Section 16.3. If, moreover, the following condition

$$z(S_i, \mu(S_i)) - \mu_v z(S_i, 1) \le N - d_{i-1} \quad (16.20)$$

holds for all $v \in S_i$, *and* $1 \le i \le k$, *for some* $k \le n$. *Then*

$$\mathrm{MIN}^T \ge T \cdot \max\{m(S_1), \ldots, m(S_k)\}$$

■

Remark 16.3

a. As explained in Remark 16.1, the general sphere numbers generalize the simple ones defined by Equation 16.6. Similarly, Theorem 16.3 above generalizes Theorem 16.1, except possibly for the extra condition (16.20) required for the former theorem but not for the latter. However, the condition is trivially satisfied under the assumptions of Theorem 16.1, as shown by (b) below. Thus, Theorem 16.1 is a consequence of Theorem 16.3.
b. Observe that condition (16.20) is trivially satisfied when μ is constant in each sphere. For suppose $\mu(S_i) = (\mu, \ldots, \mu) = \mu\mathbf{1}$. Then

$$z(S_i, \mu(S_i)) = z(S_i, \mu\mathbf{1}) = \mu z(S_i, \mathbf{1}).$$

c. The right-hand-side of Equation 16.20, $N - d_i - 1$, is a decreasing function of i. The parameter k introduced in Equation 16.20 is meant to make the theorem more applicable, by restricting the inequality's validity to the first k radii, instead of necessarily checking all the way to n. Indeed, it is usually not relevant if (16.20) does not hold for large i, since the nodes on these far-out spheres usually consume only a small portion of their total energy, before the network stops operating due to energy depletion in the "inner" spheres (i.e., the spheres with smaller radius).
d. Remark (a) of Section 16.2.1 is also valid for Theorem 16.3 because f^T has the same form in both theorems, and the spheres play a similar role. Note, however, that the sentence "... *and this minimum is achieved when traffic through* S_i *can be balanced, so that each of its nodes consumes the exact same amount of energy, namely* $m(S_i)$." while correct in both cases, has a different concrete meaning. In the case of Theorem 16.1, all nodes of S_i consume the same *amount of energy* if and only if there is a routing sending the same *amount of packets* through each node of the sphere, that is, "energy consumption" reduces to "number of packets." In the case of Theorem 16.3 it must be taken literally: traffic must be sent in such a way that each node of S_i consumes the same amount of energy (cf. Sections 16.4.1.1 and 16.4.2.1). ■

16.3.6 Bounds on the Lifetime of Networks with Directional Antennas and Lossy Links

Just as we did before, Theorem 16.3 can be used to bound the lifetime of networks. The only formal difference with Section 16.2.2 is that we now have only a lower bound, and that we need to add the extra hypothesis of this theorem. Note, however, that $m(S_i)$ is now given by Equation 16.17 instead of Equation 16.6.

Theorem 16.4

The maximum number $T_{\max}$ *of readings a sensor network can take, assuming the hypothesis of Theorem 16.3, is bounded as follows:*

$$T_{\max} \le \frac{\mathbb{E}}{\max\{m(S_1), \ldots, m(S_k)\}}$$

■

16.4 An Application to Networks with Directional Antennas and Lossy Links

The main theorem can be used to compare the lifetime of networks when (a) all nodes are omnidirectional, and (b) some nodes are equipped with directional antennas. We use the network of Figure 16.8 to illustrate this. A similar analysis, to be published elsewhere, can be applied to other networks, such as the infinite family of square networks of size $N = |V| = (2q + 1)^2$, for $q = 1, 2, \ldots$.

The network of Figure 16.8 has seven nodes, and five spheres: $S_0 = \{1\}$, $S_1 = \{2, 3, 4\}$, $S_2 = \{5\}$, $S_3 = \{6\}$, and $S_4 = \{7\}$. The distance between pairs of connected nodes is one, except for the pairs b,2 and b,4, which is $\sqrt{3}$. We assume further that the transmission cost of nodes is proportional to the length of the edges. Thus, the transmission cost of nodes 3,5,6,7 is τ, and that of 2,4 is $\sqrt{3}\tau$.

This network was studied in Reference 3, Example VII.8, under the assumption that all links were perfect. We consider now two instances: one where all nodes are o-nodes, and one where only node 5 has a directional antenna. In each case, we compute the theoretical optimum, then investigate the question of sharpness and, finally, study in some detail a specific routing. We use this information to compare the lifetimes of the network with, and without, the directional antenna.

From now on, we use the following abbreviations (cf. Equation 16.12):

$$p_1 := p(oo), \quad p_2 := p(od), \quad p_3 := p(do), \quad p_4 := p(dd)$$

16.4.1 The Case When All Nodes Are o-Nodes

We start with the theoretical optimum. To distinguish between cases, we decorate the sphere numbers with o or d, to indicate that we are dealing with the omnidirectional or the directional case: $m^o(S_i)$, $m^d(S_i)$.

We collect the information (16.19) necessary to compute the constants $m^o(S_i)$. All nodes are forced, $V^f = \{2, 3, \ldots, 7\}$, since there are only o-nodes. Hence, $S_i = S_i(f)$ and $\ell_i > 0$, for all i. By

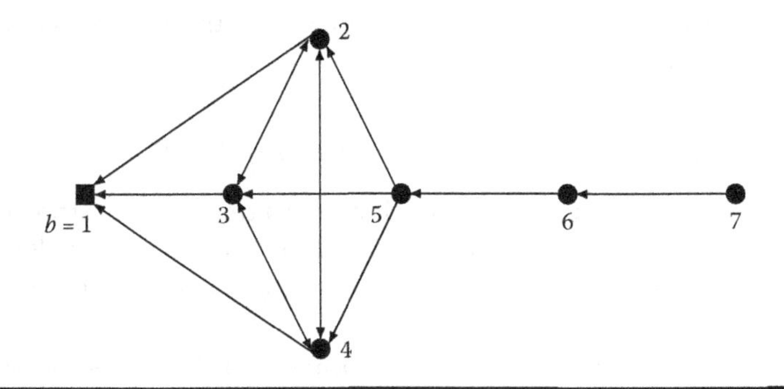

Figure 16.8 Nodes 2, 3, and 4 lie in a circle of radius 1 centered at node 5.

Equation 16.14, we have: $\rho_v = \rho/p_1$, for all nodes. The simplifying assumption of 3.1 is of course satisfied (there is no mixture of nodes!), and $\mu_v = \rho/p_1$ for all nodes v, according to Equation 16.11, hence $\mu(S_1) = (\rho/p_1, \rho/p_1, \rho/p_1)$, $\mu(S_2) = (\rho/p_1) = \mu(S_3) = \mu(S_4)$.

The transmission cost for nodes 3,5,6,7 is τ, and for nodes 2,4 is $\sqrt{3}\tau$ (this is based on the assumption that the range of a node varies linearly with the transmission cost τ). Using formula (16.13), we have:

$$\tau_3 = \tau_5 = \tau_6 = \tau_7 = \frac{\tau}{p_1}, \quad \text{and} \quad \tau_2 = \tau_4 = \frac{\sqrt{3}\tau}{p_1}$$

Finally, $\beta_v = (1/p_1)(\rho + \tau_v)$, for all nodes v.

The $m(S_i)$ can be computed using formula (16.18), Thus,

i. $$z(S_1,1) = \frac{2p_1}{\rho+\sqrt{3}\tau} + \frac{p_1}{\rho+\tau}, \quad \text{and} \quad m^o(S_1) = \frac{6/p_1}{(2/(\rho+\sqrt{3}\tau)) + (1/(\rho+\tau))} - \frac{\rho}{p_1} = \frac{m_1}{p_1}.$$

ii. $$z(S_2,1) = \frac{1}{\beta_4} = \frac{p_1}{\rho+\tau}, \quad \text{and} \quad m^o(S_2) = \frac{3(\rho+\tau)}{p_1} - \frac{\rho}{p_1} = \frac{1}{p_1}(2\rho+3\tau) = \frac{m_2}{p_1}.$$

iii. $$z(S_3,1) = \frac{1}{\beta_5} = \frac{p_1}{\rho+\tau}, \quad \text{and} \quad m^o(S_3) = \frac{2(\rho+\tau)}{p_1} - \frac{\rho}{p_1} = \frac{1}{p_1}(\rho+2\tau) = \frac{m_3}{p_1}.$$

iv. $$z(S_4,1) = \frac{1}{\beta_6} = \frac{p_1}{\rho+\tau}, \quad \text{and} \quad m^o(S_4) = \frac{1}{p_1}(\rho+\tau-\rho) = \frac{\tau}{p_1} = \frac{m_4}{p_1}.$$

Note that we have expressed the sphere numbers in terms of numbers m_1, ..., m_4. The m_i are the sphere numbers computed in Reference 3 under the additional hypothesis that all links are perfect. Thus, when all links are perfect, that is, when $p_1 = 1$, we recover the m_i. Also, the formulas above imply, as in Reference 3, that there are for all values of ρ and τ,

$$m^o(S_2) = \max\{m^o(S_1), \ldots, m^o(S_4)\} \tag{16.21}$$

To be able to invoque Theorem 16.3 and claim that $\text{MINT}^T \geq m^o(S_2)$, that is, that (16.21) is the theoretical optimum, we need to check condition (16.20). But this follows from Remark 16.2(b), since μ is constant on spheres.

16.4.1.1 How Sharp Is the Theoretical Optimum (16.21)?

Since the "tail" of the network (i.e., nodes 5,6,7) is part of a tree, in that portion there is only one effective routing, namely: node 7 sends its packet to node 6 which, in turn, sends its two packets to node 5. For this (partial) routing, node 5 consumes* $k_5 = m^o(S_2)$, in the notation of Equation 16.15,

* This is so because transmission costs at 5 are equal, that is, nodes 2, 3, 4, the only possible receptors of the three packets of 5 (routings are effective!), are all o-nodes. Hence, all possible continuations of the routing will have the same value for k_5.

and one is tempted to argue that the bound is sharp. However, the routing, which is unique so far, can be continued toward b in different ways. It could happen that, for some of these continuations, $\kappa(S_1) := \max\{k_2, k_3, k_4\} > m^o(S_2)$ (cf. Section 16.4.1.2 for an example). If this were to happen for *all* possible continuations, then MIN $> m^o(S_2)$, and the bound would not be sharp.

We prove now that $m^o(S_2)$ is indeed sharp by constructing a routing whose maximum cost occurs at S_2. The routing is defined by continuing the partial routing constructed above, and it is "mathematical" in the sense that it prescribes transmitting portions of packets (possibly even irrational portions!) at each iteration, instead of whole ones, in a way resembling that of example (*f*) of Section 16.2.4. It turns out that the optimal routing thus constructed (see Equation 16.23) will depend on the factor λ that we now introduce: $\lambda := \rho/\tau$. A small [resp. large] value of λ means that transmission [resp. reception] is much more expensive than reception [resp. transmission], while $\lambda = 1$ means transmission and receptions costs are equal. Observe that λ is determined by the MAC layer and the hardware.

To obtain practical routings we discretize, to obtain approximations to the optimal, "mathematical" routings. The discretized routings very nearly balance traffic in S_1, so that their consumption is close to $m^o(S_1)$. We show how to do this in the following.

Consider the following continuation of the partial routing above: after T iterations, it has transmitted X packets to node 2, Y packets to node 3, and X packets to node 4. This defines a unique routing, once we have determined the unknowns X, Y, since from these nodes there is only one way to transmit traffic to b. The following set of linear expressions determines X, Y uniquely:

$$\begin{cases} \tau(\lambda X + \sqrt{3}(X+T)) & \text{this equals } p_1 k_2 \\ \tau(\lambda Y + Y + T) & \text{this equals } p_1 k_3 \\ 2X + Y = 3T & \end{cases} \tag{16.22}$$

The first two expressions compute, up to multiplication by p_1, the consumption of nodes 2, 4 (first expression) and node 3 (second expression), while the third equation is conservation of flow. We require the first two expressions to be equal (traffic balance). In this way we obtain a system of two linear equations, that has the unique solution (note the denominator is always >0):

$$\frac{X}{T} = \frac{3\lambda + 4 - \sqrt{3}}{3\lambda + 2 + \sqrt{3}}, \quad \text{and} \quad \frac{Y}{T} = \frac{3\lambda - 2 + 5\sqrt{3}}{3\lambda + 2 + \sqrt{3}} \tag{16.23}$$

This gives one solution for every $0 < \lambda < \infty$. When λ runs through the interval $(0, \infty)$, X/T say, will run through $(0.607\ldots, 1)$, where $0.607\ldots \simeq (4-\sqrt{3})/(2+\sqrt{3})$.

In a concrete situation we would know the value of λ and would use it to discretize the obtained "mathematical" routing. To illustrate, suppose $\lambda = 2$. Then $X/T \simeq 0.849558$, and $Y/T \simeq 1.300882$. If we take, say, $T = 10$, we can approximate X, Y to $X = 8$, and $Y = 14$. A larger value, say $T = 100$, gives the finer approximation $X = 85$, $Y = 130$. And $T = 100{,}000$ will give the even finer approximation $X = 84{,}956$ and $Y = 130{,}088$.

We now concentrate in case $T = 100$, $X = 85$, and $Y = 130$, and define an optimal routing y, as promised. Let y^1 denote the following routing: node 5 sends one packet to each of nodes 2,3,4, which, in their turn, transmit two packets each to the base node (the beginning of the routing is as explained at the beginning of this section). Routing y^2 transmits all three packets from node 5 to node 3, and node 3 sends four packets to b. Nodes 2 and 4 send one packet each to b.

There are many ways to define y. For the first 100 iterations we could take, for example:

$$y^{100} = (\overbrace{y^1, \ldots, y^1}^{85 \text{ times}}, \overbrace{y^2 \ldots, y^2}^{15 \text{ times}})$$

and then repeat this block: $y = (y^{100}, y^{100}, \ldots)$. There are many variations for y^{100}, by changing the order in which y^1, y^2 appear; the crucial thing is that there are 85 appearences of y^1, and 15 of y^2. For any such y^{100}, we have $p_1 k_2 \simeq \tau 490.42$, $p_1 k_3 = 490$, and $Tp_1 m^o(S_1) \simeq 490.26$. On the other hand, $Tp_1 m^o(S_2) \simeq \tau 700$, when $\lambda = 2$. Thus $\kappa(S_1) \simeq m^o(S_1) < m^o(S_2) \simeq \kappa(S_2)$ and $m^o(S_2)$ is sharp.

16.4.1.2 A Certain Routing

Consider the (suboptimal) routing that sends one packet from node 5 to each node 2, 3, 4, at each iteration. We compute how much nodes 2, 3 and 5 consume (recall we denote the consumption of node i by k_i). We have, $k_5 = 2(\rho/p_1) + 3(\tau/p_1) = m(S_2)$, $k_3 = \rho/p_1 + 2(\tau/p_1)$, and $k_2 = k_4 = \rho/p_1 + 2\sqrt{3}(\tau/p_1)$. It follows that*

$$\kappa(S_1) := \max\{k_1, k_2, k_3\} = \frac{1}{p_1}(\rho + 2\sqrt{3}\tau)$$

Hence,

$$\kappa(S_1) \le m(S_2) \quad \text{if and only if} \quad (2\sqrt{3} - 3)\tau \le \rho$$

or, in terms of $\lambda := \rho/\tau$, if and only if $0.46 \simeq 2\sqrt{3} - 3 \le \lambda$. In conclusion†: (a) if $0.46 \le \lambda$, then node 5 dies first, and (b) if $\lambda < 0.46$, then nodes 2 and 4 die first. So even when Equation 16.21 shows that node 5 "should" die first, the routing we have considered forces other nodes to die first, depending on λ.

Recall that a small λ means that reception is cheap (in relation to transmission). So a small λ means that transmission costs determine the behavior of the network. It is not clear, however, which node has the heaviest transmission burden: node 5 transmits the most packages, a total of 3, but nodes 2, 4 have the most expensive transmission cost, and still transmit 2 packets each. Which one will dominate the other and, under which conditions? The above result answers this question precisely.

As a final observation, notice that it follows from Equation 16.23 that when λ is extremely large, X and Y are very close to 1, that is, the routing constructed in this section, which is "obviously" suboptimal on S_1, becomes nearly optimal on S_1. This is because in this case reception dominates the behaviour of the network, hence 5 dies first (since it has the heaviest reception burden), and all nodes of S_1 consume almost exactly the same amount (since they all receive one packet), making the routing nearly optimal on S_1. Of course, the routing is optimal for all $\lambda > 0.46$ because its maximum consumption occurs on S_2, and equals the predicted theoretical one.

---

* Recall that the specific routing is implicit in the notation $\kappa(S_1)$.

† This is precisely Equation 19 of Reference 3, and we can derive the same conclusions now as we did in that paper. Although the present situation is more general in that links are now lossy, not perfect, we derive the same equation because loosyness is assumed to be uniform.

16.4.2 The Case with Only One Directional Node (Node 5)

We consider now the case when node 5 in Figure 16.8 is a *d*-node, while all others are *o*-nodes. As we did in Section 16.4.1, we label sphere numbers thus: $m^d(S_1)$, etc. The transmission cost for nodes 3, 6, 7 is τ, for node 5 is $2\tau/3$ (this is based on the hypothesis that the range of node 5 when transmitting at level τ, equals 3/2 the range of the *o*-nodes), and for nodes 2, 4 is $\sqrt{3}\tau$. Using formula (16.13), we have:

$$\tau_3 = \tau_6 = \tau_7 = \frac{\tau}{p_1}, \quad \tau_5 = \frac{2\tau}{3p_3}, \quad \text{and} \quad \tau_2 = \tau_4 = \frac{\sqrt{3}\tau}{p_1}$$

Observe that the simplifying assumption of Section 16.3.1 is satisfied. Using Equation 16.11, we compute the following reception costs:

$$\mu_2 = \mu_3 = \mu_4 = \frac{\rho}{p_3}, \quad \mu_5 = \frac{\rho}{p_2}, \quad \text{and} \quad \mu_6 = \mu_7 = \frac{\rho}{p_1}$$

Hence $\mu(S_1) = (\rho/p_3, \rho/p_3, \rho/p_3)$, $\mu(S_2) = (\rho/p_2)$, and $\mu(S_3) = \mu(S_4) = (\rho/p_1)$; so by Remark 16.3, condition Equation 16.20 of Theorem 16.3 is satisfied. We have $V^f = \{2, 3, 4, 5, 7\}$, that is, the only nonforced node is 6. Hence, $S_i = S_i(f)$ and $\ell_i > 0$, for $i = 1, 2, 4$, while $\ell_3 = 0$. Hence, we need α_6 and β_v for $v = 2, 3, 4, 5, 7$. These are given by:

$$\beta_2 = \beta_4 = \frac{\rho}{p_3} + \sqrt{3}\frac{\tau}{p_1}, \beta_3 = \frac{\rho}{p_3} + \frac{\tau}{p_1}, \beta_5 = \frac{\rho}{p_2} + \frac{2}{3}\frac{\tau}{p_3}, \alpha_6 = \frac{\rho}{p_1} + \frac{\tau}{p_2}, \beta_7 = \frac{1}{p_1}(\rho + \tau)$$

We are now ready to compute the $m^d(S_i)$ using formula (16.8). In terms of λ (recall that $\lambda\rho = \lambda$), we have,

$$m^d(S_1) = \frac{6}{z(S_1,1)} - \frac{\rho}{p_3} = \frac{\tau}{p_1 p_3}\left[\frac{3p_1^2\lambda^2 + p_1 p_3 \lambda(4 + 5\sqrt{3}) + 6\sqrt{3}p_3^2}{3p_1\lambda + (2 + \sqrt{3})p_3}\right]$$

where

$$z(S_1,1) = 2\frac{1}{\beta_2} + \frac{1}{\beta_3} = \frac{2}{(\rho/p_3) + (\sqrt{3}\tau/p_1)} + \frac{1}{(\rho/p_3) + (\tau/p_1)}$$

For $m^d(S_2)$,

$$m^d(S_2) = \frac{3}{z(S_2,1)} - \frac{\rho}{p_2} = \frac{2\tau}{p_2 p_3}(p_3\lambda + p_2)$$

where

$$z(S_2,1) = \frac{1}{\beta_5}$$

For $m^d(S_3)$,

$$m^d(S_3) = \frac{2}{z(S_3,1)} - \frac{\rho}{p_2} = \frac{\tau}{p_1 p_2}[\lambda(p_2 - p_1) + p_1]$$

where

$$z(S_3,1) = \frac{1}{\alpha_6}$$

Finally, for $m^d(S_4)$,

$$m^d(S_4) = \frac{1}{z(S_4,1)} - \frac{\rho}{p_1} = \frac{\tau}{p_1}$$

where

$$z(S_4,1) = \frac{1}{\beta_7}$$

Setting $p_1 = \ldots = p_4 = 1$, we get $m^d(S_i)$, where m_i $(i = 1, \ldots, 4)$, are the constants computed in Section 16.4.1.

The question now is to compute $\max\{m^d(S_1), \ldots, m^d(S_4)\}$. This involves rather complicated expressions involving λ, p_1, p_2, and p_3. To make the calculations more manageable, we consider *scenarios* defined in terms of the probabilities, thus obtaining formulas involving only λ.

We consider two basic scenarios. Scenario I is the case where $p_1 = 0.8$, $p_2 = 0.85$, and $p_3 = 0.9$, whereas in Scenario II, $p_1 = 0.3$, $p_2 = 0.55$, and $p_3 = 0.6$. These scenarios model opposite situations: in Scenario I links are moderately lossy, while in Scenario II they are extremely lossy.

Consider first the inequality:

$$m^d(S_1) \le m^d(S_2)$$

In terms of λ, the difference $m^d(S_1) - m^d(S_2) = 0$ gives the second degree equation:

$$3p_1^2(p_2 - 2p_3)\lambda^2 + \left(p_1 p_2 p_3\left(4 + 5\sqrt{3}\right) - 6p_1^2 p_2 - p_1 p_3^2\left(4 + 2\sqrt{3}\right)\right)\lambda$$
$$+ 6\sqrt{3}p_2 p_3^2 - \left(4 + 2\sqrt{3}\right)p_1 p_2 p_3 = 0$$

Solving this equation in both scenarios, we obtain the following results.

In Scenario I, $m^d(S_1) \le m^d(S_2)$, if and only if $\lambda \ge 1.09819 \cong 1.1$, whereas in Scenario II, $m^d(S_1) \le m^d(S_2)$, if and only if $\lambda \ge 3.2024 \cong 3.2$.

Similarly,

$$m^d(S_2) \ge m^d(S_3) \tag{16.24}$$

is equivalent to:

$$p_3\lambda(3p_1 - p_2) \geq p_1(p_3 - 2p_2) \tag{16.25}$$

Assuming $3p_1 - p_2 > 0$, a condition satisfied both in Scenarios I and II, Equation 16.25 is equivalent to:

$$\lambda \geq \frac{p_1}{p_3}\frac{p_3 - 2p_2}{3p_1 - p_2} \tag{16.26}$$

In both scenarios, I and II, the right-hand side of Equation 16.26 is negative, so that (16.24) holds for all values of λ.

Consider $m^d(S_3) \geq m^d(S_4)$. This is equivalent to $(\tau/p_1p_2)[\lambda(p_2 - p_1) + p_1] \geq \tau/p$, which, in its turn, is equivalent to $\lambda(p_2 - p_1) \geq p_2 - p_1$, an inequality valid for $\lambda \geq 1$. For $\lambda < 1$, we have $m^d(S_4) \leq m^d(S_2)$ in both scenarios.

Putting it all together, we see that in Scenario I,

$$\max\{m^d(S_1), \ldots, m^d(S_4)\} = \begin{cases} m^d(S_2) & \text{when } \lambda \geq 1.1 \\ m^d(S_1) & \text{when } 0 \leq \lambda \leq 1.1 \end{cases} \tag{16.27}$$

In Scenario II,

$$\max\{m^d(S_1), \ldots, m^d(S_4)\} = \begin{cases} m^d(S_2) & \text{when } \lambda \geq 3.2 \\ m^d(S_1) & \text{when } 0 \leq \lambda \leq 3.2 \end{cases} \tag{16.28}$$

Observe that this only means that the first node(s) to die *should* lie in the sphere corresponding to $\max\{m^d(S_1, \ldots, m^d(S_4)\}$, and that the lifetime of the network *should* be $\leq \mathbb{E}/\max\{m^d(S_1), \ldots, m^d(S_4)\}$. In order to remove the word "should," we have to show that there actually exists a routing that makes (16.27) [resp., Equation 16.28], a valid equality. This is exactly what we show in Section 16.4.2.1.

16.4.2.1 How Sharp Are the above Bounds?

We proceed as in Section 16.4.1.1, and use the same notation. In this case, Equation 16.22 becomes:

$$\begin{cases} \dfrac{\lambda}{p_3}X + \dfrac{\sqrt{3}}{p_1}(X+T) & \text{this equals } \dfrac{k_2}{\tau} \\ \dfrac{\lambda}{p_3}Y + \dfrac{1}{p_1}(Y+T) & \text{this equals } \dfrac{k_3}{\tau} \\ 2X + Y = 3T & \end{cases}$$

As before, we set the first two expressions to be equal and, using the third, obtain a system of two linear equations, that has the unique solution (note the denominator is always >0):

$$\frac{X}{T}=\frac{3p_1\lambda+(4-\sqrt{3})p_3}{3p_1\lambda+(2+\sqrt{3})p_3}, \quad \text{and} \quad \frac{Y}{T}=\frac{3p_1\lambda+(5\sqrt{3}-2)p_3}{3p_1\lambda+(2+\sqrt{3})p_3}$$

With these solutions we can proceed as in Section 16.4.1.1 to construct routings. As an illustration, in Scenario I we consider the case when $T=100$, $X=78$, $Y=144$, and take $\lambda=1.1$, slightly above the break value of Equation 16.27. For the corresponding routing, $k_2/\tau \simeq 480.71$, $k_3/\tau \simeq 481.00$, and $(100/\tau)m^d(S_1) \simeq 480.83$. On the other hand, $(100/\tau)m^d(S_2) \simeq 481.04$. Thus, $\kappa(S_1) \simeq m^d(S_1) < m^d(S_2) = \kappa(S_2)$, and the routing is sharp. In Scenario II we take for instance $T=100$ and $\lambda=1$, and let $X=75$, $Y=156$. Then $k_2/\tau \simeq 1113.04$, and $k_3/\tau \simeq 1113.33$. On the other hand, $(100/\tau)m^d(S_1) \simeq 1113.17$, and $(100/\tau)m^d(S_2) \simeq 696.97$. Thus, $\kappa(S_1) \simeq m^d(S_1) > m^d(S_2) = \kappa(S_2)$, and the routing is sharp.

16.4.2.2 A Certain Routing

Consider the same suboptimal routing of Section 16.4.1.2. Its per node consumption is: $k_5 = 2(\rho/p_2 + \tau/p_3) = 2\tau(\lambda/p_2 + 1/p_3)$, $k_2 = k_4 = \tau(\lambda/p_3 + 2\sqrt{3}/p_1)$, and $k_3 = \tau(\lambda/p_3 + 1/p_1)$. Hence, $\kappa(S_1) = k_1 = \tau(\lambda/p_3 + 2\sqrt{3}/p_1)$. Then, $\kappa(S_1) \leq m^d(S_2)$ if and only if

$$\lambda \geq \frac{2p_2(\sqrt{3}p_3 - p_1)}{p_1(2p_3 - p_2)} \tag{16.29}$$

Equation 16.29 holds, in Scenario I, when $\lambda \geq 1.69 \ldots \simeq 1.7$, and in Scenario II, when $\lambda \geq 4.17 \ldots \simeq 4.17$. We have a situation similar to that of Section 16.4.1.2, except that now the dividing line (16.29) is not a constant but a function of λ. We conclude that (a) if $1.7 \leq \lambda$, in Scenario I, or if $4.17 \leq \lambda$, in Scenario II, then node 5 dies first, and (b) if $\lambda < 1.7$, in Scenario I, or if $\lambda < 4.17$, in Scenario II, then nodes 2 and 4 die first. So this routing forces nodes other than those predicted by Equations 16.27 and 16.28 to die first, depending on the value of λ.

The last two paragraphs of Section 16.4.1.2 apply essentially unchanged in the present situation.

The theoretical computation of "mathematical" routings in Sections 16.4.1.1 and 16.4.2.1 give, in a rough first approximation, the near optimal routings we use in the simulations of Section 16.5.4.

16.4.3 Comparison with, and without, a Directional Antenna

We now compare the lifetimes of the network with, and without, a directional antenna in node 5. For this purpose, we consider the quotient

$$Q = Q(\lambda) = \frac{\max\{m^o(S_1), \ldots, m^o(S_4)\}}{\max\{m^d(S_1), \ldots, m^d(S_4)\}} \tag{16.30}$$

where $m^o(S_i)$ [resp., $m^d(S_i)$] denotes $m(S_i)$ in case all nodes are o-nodes [resp., when node 5 is the only d-node of the network]. Note that Equation 16.30 measures the percentual increase in the lifetime of the network when one compares its performance in case 5 is a d-node, with the case when all nodes are o-nodes.

The explicit form of Equation 16.30 is, in Scenario I,

$$Q(\lambda) = \begin{cases} \dfrac{m^o(S_2)}{m^d(S_2)} & \text{when } \lambda \geq 1.1 \\ \dfrac{m^o(S_2)}{m^d(S_1)} & \text{when } 0 \leq \lambda \leq 1.1 \end{cases}$$

and, in Scenario II,

$$Q(\lambda) = \begin{cases} \dfrac{m^o(S_2)}{m^d(S_2)} & \text{when } \lambda \geq 3.2 \\ \dfrac{m^o(S_2)}{m^d(S_1)} & \text{when } 0 \leq \lambda \leq 3.2 \end{cases}$$

The general form of these quotients, given hereunder, can be specialized to the two scenarios by choosing the appropriate values for p_1, p_2, p_3. We have:

$$Q_{2,2}(\lambda, p_1, p_2, p_3) := \frac{m^o(S_2)}{m^d(S_2)} = \frac{p_2 p_3(2\lambda + 3)}{2p_1(p_3\lambda + p_2)}$$

and

$$Q_{2,1}(\lambda, p_1, p_2, p_3) := \frac{m^o(S_2)}{m^d(S_1)} = \frac{p_3(2\lambda + 3)(3p_1\lambda + p_3(2 + \sqrt{3}))}{3p_1^2\lambda^2 + p_1 p_3(4 + 5\sqrt{3})\lambda + 6\sqrt{3}p_3^2}$$

The graph of $Q(\lambda)$ can be seen in Figure 16.9. Tables 16.1 and 16.2 show some values of $Q(\lambda)$, in both scenarios. The first row, labelled λ, shows a few values that the quotient $\lambda := \rho/\tau$ can take. The second row gives the corresponding values of Q. For example, in Table 16.1, $1.219 \simeq Q(0.5) = Q_{2,1}(0.5, 0.8, 0.85, 0.9)$.

Table 16.1 Scenario I, Node 5 is a *d*-Node *vs.* All Nodes are *o*-Nodes

λ	0	0.5	1	2	3	5	10	∞
Q	1.077	1.219	1.332	1.262	1.212	1.161	1.116	1.0625

Table 16.2 Scenario II, Node 5 is a *d*-Node *vs.* All Nodes are *o*-Nodes

λ	0	0.5	1	2	3	3.5	5	10	∞
Q	1.077	1.304	1.497	1.808	2.050	2.075	2.014	1.931	1.833

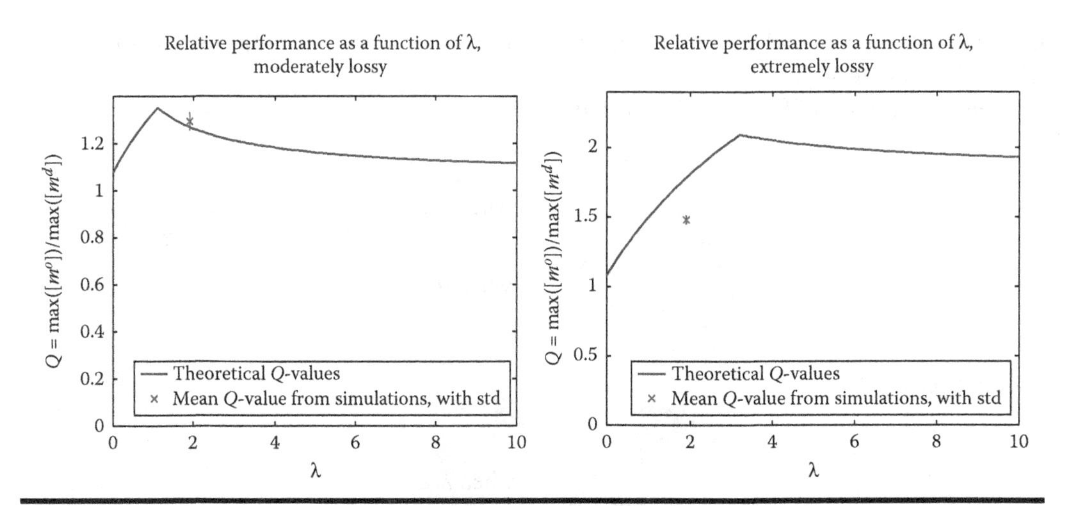

Figure 16.9 Our simulation results show that the performance improvements Q with a directional antenna match the theory.

It follows from Equations 16.27 and 16.28 that, in both scenarios,

$$Q(0) = \left.\frac{m^o(S_2)}{m^d(S_1)}\right|_{\lambda=0} = \frac{2+\sqrt{3}}{2\sqrt{3}} \simeq 1.077$$

Also, lim $Q(\lambda) = p_2/p_1$, when $\lambda \to \infty$, in both scenarios. This gives the values corresponding to $\lambda = \infty$ in the tables. In Scenario I, $Q(1.91) \simeq 1.269$, and in Scenario II, $Q(1.91) \simeq 1.784$; these theoretical values correspond to the simulations in Figure 16.9.

The maximum of $Q(\lambda)$ is attained at the cut points, $\lambda \simeq 1.1$ with a gain of approx. 35% in Scenario I, and $\lambda \simeq 3.2$ with a gain of approx. 100% in Scenario II. When λ is small, transmission costs dominate behavior and the maximum consumption occurs at S_1 (since S_1 is heavily burdened by transmission); hence, there should be little gain in having a directional antenna at S_2. On the contrary, when λ is large we know the maximum consumption occurs at S_2. In this network, node 5 is heavily burdened both with transmission and reception, so for large λ the savings in transmission should be insignificant while those in reception should be important, specially when link (6, 5) is extremely lossy. We know the theoretical behavior of the gain for large λ is determined by p_2/p_1 = lim $Q(\lambda)$. This result corroborates and quantifies the above intuitions.

16.5 Application of the Theory

In this section, we demonstrate the practical relevance of the explicit bounds we have derived in Section 16.4.3. In WSNs, the main energy consumer is typically the radio. Since the MAC layer is responsible for turning on and off the radio, it plays a critical role for the energy consumption. Therefore, a huge number of MAC layers has been developed for WSNs. These include asynchronous MAC layers such as X-MAC [13], LPP [14], Ri-MAC [15], protocols with scheduled contention [16] such as T-MAC [17], and S-MAC [18] as well as Time Division Multiple Access (TDMA)-based protocols such as DRAND [19], Dozer [20], and GinMAC [21]. The latter two also are vertically integrated stacks, that is, they integrate both routing and MAC layer functionality.

16.5.1 GinLITE

Recent standards for industrial sensor networks such as WirelessHART [22] and 802.15.4e [23] are TDMA-based protocols which motivates us to do our experiments with a TDMA protocol as well. Another advantage is that the use of directional antennas is straightforward in that there is no need to change a TDMA MAC layer when using directional antennas once the network is setup. In particular, we exploit the property of directional antennas that they transmit more reliably over the same distance than omnidirectional antennas since they can focus their radiated energy and hence the signal-to-noise ratio at the receiver increases. A higher probability of success of a transmission means that less retransmissions are needed which saves energy and increases the network lifetime.

Since there are no open implementations available for WirelessHART, we use GinLITE [24]. GinLITE is a derivative of GinMAC [21] that provides only the basic functionality of the latter. GinMAC has been successfully used in a deployment at an oil refinery in Portugal [25]. GinMAC uses an exclusive TDMA schedule. Nodes within a Ginseng network are usually within interference range of each other which prevents efficient slot sharing. The GinMAC TDMA schedule is determined before network deployment. The schedule has an epoch length of E slots. Within an epoch three types of slots are allocated: *basic* slots for Tx; *additional* slots for retransmissions (RTx) to increase reliability; and *unused* slots to decrease the required duty cycle or to implement supplementary services such as performance debugging. Slots are large enough to contain the data transmission and an acknowledgment from the receiver. Depending on a node's position in the network a node must become active for a subset of Tx and RTx slots. Thus, the layout of the TDMA schedule determines the nodes' energy consumption as well as the achievable overall data delivery delay and latency. Some of the slots are required for administrative tasks such as time synchronization which are not modelled in our analysis.

16.5.2 Setup

We focus on the GinLITE-based data collection network in Figure 16.8, since we have used this network for earlier evaluations [3,9]. As the original GinLITE works only for trees, we had to slightly modify it so that it would also work with the network in Figure 16.8 where node 5 has not only one but three parents. We evaluate our solution in the COOJA simulator [26] but our results are easily transferable to real deployments since GinLITE also works on real nodes and COOJA simulates deployable code. In COOJA, we emulate *T-Mote Sky* [27] using the MSPSim emulator [28]. We evaluate both a moderately lossy and an extremely lossy network. The probability of success of data packet transmissions are defined in Section 16.4. They are $p_1 = p(oo) = 0.8$, $p_2 = p(od) = 0.85$ and $p_3 = p(do) = 0.9$ for moderately lossy networks (Scenario I). For the extremely lossy network in Scenario II these probabilities are $p_1 = 0.3$, $p_2 = 0.55$ and $p_3 = 0.6$. We compare results with and without using a directional antenna. If we use a directional antenna, we equip node 5 with such an antenna. The routing schemes we use are near-optimal in the sense that node 5 sends one packet to each of its successors for $r - 1$ iterations and then in the r-th iteration, sends all packets to node 3 since this node can transmit to the sink at a lower cost than node 2 and node 4.

16.5.3 Lifetime Bounds with One Directional Node

In Section 16.4.3 we have compared the achievable lifetime gain Q with directional antennas compared to omnidirectional antennas.

In the first set of experiments we compare the theoretical gain with the gain we achieve in our simulations. The quotient Q was defined in Equation 16.30. It is a function of $\lambda := \rho/\tau$, where ρ is the cost of a packet reception and τ the cost of a packet transmission. Observe that λ depends on the hardware and the MAC layer used. In the case of the GinLITE on *T-Mote Sky*, $\lambda = 1.91$ as we have shown earlier [3].

Figure 16.9 compares the theoretical Q values with the value we achieve with moderately (left part of the figure) and extremely lossy networks. As the figure shows, in our simulations we achieve an improvement of about 30% in the moderately lossy and around 50% in the extremely lossy scenarios when using a directional antenna. Note that for the reasons discussed above, we have only one value for the simulations. The figure shows a good match between the theoretical and practical results, in particular for the moderately lossy scenario.

16.5.4 Distribution of Energy Consumption

Assuming that all nodes are equipped with the same initial amount of energy, the best one can do to achieve a high lifetime is to distribute the energy consumption equally among the spheres as discussed in Section 16.2.1 (a) and Remark 16.3 (d).

Figures 16.10 and 16.11 compare how well we are able to distribute the energy consumption over the different nodes in the network with and without using a directional antenna at node 5. The theoretical results and the results achieved in simulation are shown both for the moderately and the extremely lossy scenario.

The figures show that the practical results match the analytic results very well for most of the nodes. The exception is node 7 that has a much higher part of the load in simulation. The reason for this discrepancy is that in theory, this node only needs to transmit one packet in every

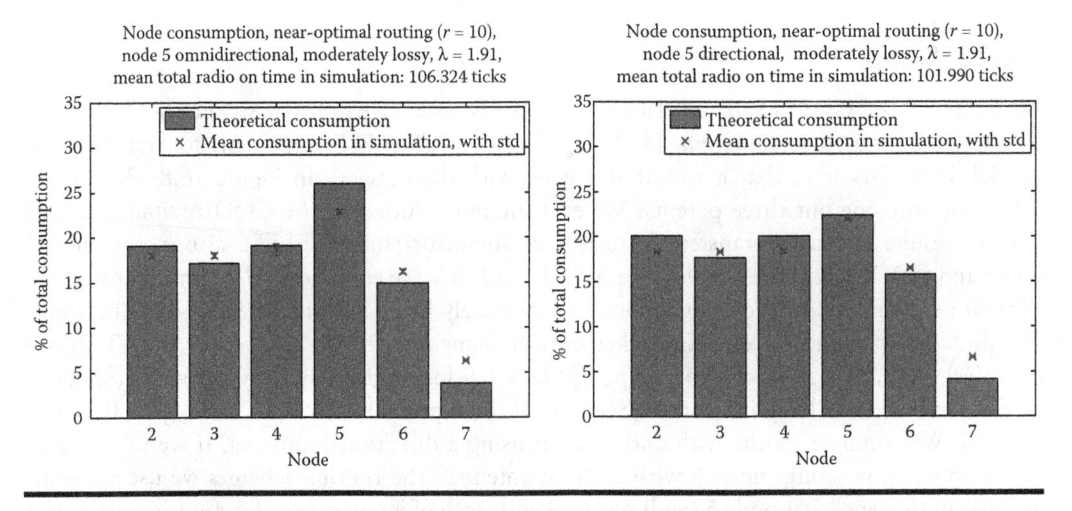

Figure 16.10 Distribution of energy consumption over the individual nodes when the network is moderately lossy (Scenario I). With a directional node, the energy consumption can be better balanced between the nodes and the spheres.

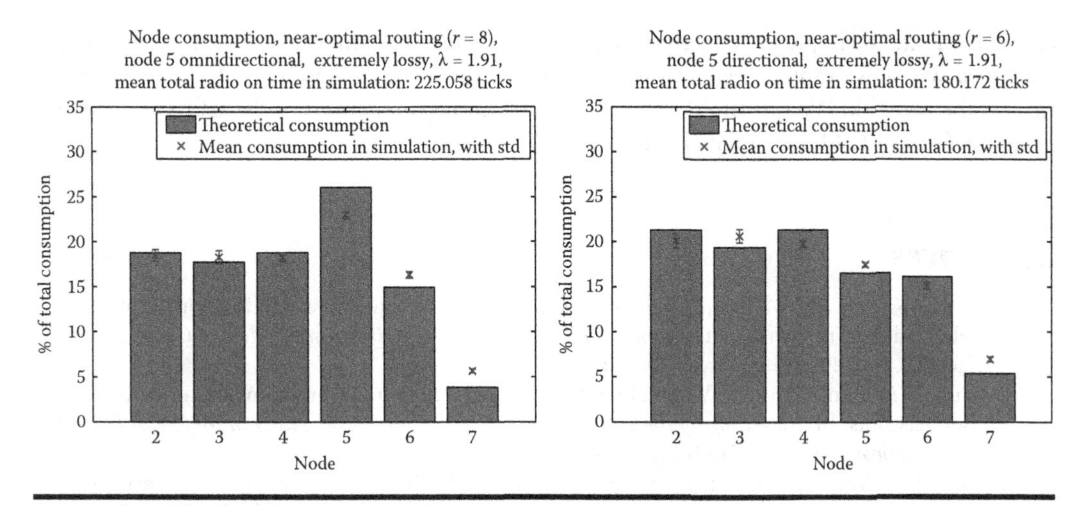

Figure 16.11 Distribution of energy consumption over the individual nodes when the network is extremely lossy (Scenario II). Also here with a directional node, the energy consumption can be balanced better between the nodes and the spheres. Also, the bottleneck sphere becomes S_1 when a directional antenna is used (right part of the figure).

iteration, while in our simulations it also needs to receive packets sent out for time synchronization which increases its energy consumption both in absolute and relative terms. Furthermore, the energy consumption in simulation is slightly more evenly distributed among the nodes than in theory. We also attribute this effect to the additional energy consumption for the maintenance packets which need to be received by all nodes.

The figures also show that when using the directional antenna, we are able to distribute the required energy more evenly among the nodes and hence among the spheres. In Figure 16.10 we see that the energy consumption of node 5 is very high compared to the other nodes which makes this node's sphere the bottleneck for the omnidirectional case (left part of the figure). Furthermore, the overall energy consumption (here measured as radio-on time in ticks) decreases from 106.324 ticks in the omnidirectional case to 101.990 ticks when node 5 is equipped with a directional antenna. As Figure 16.11 shows, the same is true for the extremely lossy network but here it is even more accentuated. In this scenario, the difference in the overall energy consumption is almost 20%.

Moreover, the figures also support the findings about the bottleneck spheres in Section 16.4.3. For $\lambda = 1.91$, the theory predicts that in Scenario I, S_2 is the bottleneck sphere both with and without directional antennas. The same is true for Scenario II without directional antennas. In all these scenarios, node 5 has the highest energy consumption. In Scenario II with directional antennas, however, in both simulation and theory, node 5 no longer has the highest energy consumption and S_1 becomes the bottleneck sphere.

In summary, our simulation results back up the mathematical analysis and hence demonstrate the relevance and applicability of our explicit bounds.

16.6 Conclusions

In this chapter, we have presented a method to compute bounds on the lifetime of WSNs. In contrast to a linear program, our method computes bounds rather than a routing that maximizes the lifetime of the network. It includes lossy links and directional antennas. Using our method, we are able to compute the improvements that are possible by replacing omnidirectional with directional antennas. We show example scenarios where one directional antenna more than doubles the lifetime of the network. We validate the theory by performing simulations in the COOJA simulator using the GinLITE TDMA MAC layer.

References

1. Dai, H., K. Ng, and M. Wu, 2006. An overview of mac protocols with directional antennas in wireless ad hoc networks. In *International Conference on Wireless and Mobile Communications (ICWMC)*, July 2006, Bucharest, Romania.
2. Alonso, J., A. Dunkels, and T. Voigt, 2004. Bounds on the energy consumption of routings in wireless sensor networks. In *Proceedings of the 2nd WiOpt, Modeling and Optimization in Mobile, Ad Hoc and Wireless Networks*, March, Cambridge, UK.
3. Alonso, J., T. Voigt, and A. Varshney, 2013. Bounds on the lifetime of wsns. In *Proceedings of the 9th IEEE International Conference on Distributed Computing in Sensor Systems (IEEE DCOSS 2013)*, May, Cambridge, USA, pp. 367–373..
4. Tilak, S., N. Abu-Ghazaleh, and W. Heinzelman, 2002. A taxonomy of wireless micro-sensor network models, *ACM Mobile Computing and Communications Review (MC2R)*, 6, 28–36.
5. Ford, L. R. and D. R. Fulkerson, 1956. Maximal flow through a network, *Canadian Journal of Mathematics*, 8(3), 399–404.
6. Hu, T. C., 1963. Multi-commodity network flows, *Operations Research*, 11(3), 344–360.
7. Abrahamsson, H., B. Algren, J. Alonso, A. Andersson, and P. Kreuger, 2002. A multi-path routing algorithm for ip networks based on flow optimisation. In *From QoS Provisioning to QoS Charging*, LNCS, Springer, vol. 2511, 2002, pp. 135–144.
8. Alonso, J., H. Abrahamsson, B. Algren, A. Andersson, and P. Kreuger, 2002. Objective functions for balance in traffic engineering, SICS-Swedish Institute of Computer Science, Tech. Rep. T2002:05.
9. Ritter, H., J. Schiller, T. Voigt, A. Dunkels, and J. Alonso, Experimental evaluation of lifetime bounds for wireless sensor networks. In *Proceedings of the Second European Workshop on Wireless Sensor Networks (EWSN2005)*, January, Istanbul, Turkey, pp. 25–32.
10. Mainwaring, A., J. Polastre, R. Szewczyk, D. Culler, and J. Anderson, 2002. Wireless sensor networks for habitat monitoring. In *First ACM Workshop on Wireless Sensor Networks and Applications (WSNA 2002)*, September, Atlanta, GA, pp. 88–97.
11. Mhatre, V. P., C. Rosenberg, D. Kofman, R. Mazumdar, and N. Shroff, 2005. A minimum cost heterogeneous sensor network with a lifetime constraint, *IEEE Transactions on Mobile Computing*, 4(1), 4–15.
12. Stutzman W. L. and W. A. Davis, 1998. *Antenna Theory*. Wiley Online Library.
13. Buettner, M., G. V. Yee, E. Anderson, and R. Han, 2006. X-MAC: A short preamble MAC protocol for duty-cycled wireless sensor networks. In *SenSys '06: Proceedings of the 4th International Conference on Embedded Networked Sensor Systems*, November 2006, Boulder, Colorado, pp. 307–320.
14. Musaloiu-E. R., C.-J. M. Liang, and A. Terzis, 2008. Koala: Ultra-low power data retrieval in wireless sensor networks. In *IPSN '08*, April 2008, St. Louis , MO, USA.
15. Sun, Y., O. Gurewitz, and D. B. Johnson, 2008. Ri-mac: A receiver-initiated asynchronous duty cycle MAC protocol for dynamic traffic loads in wireless sensor networks. In *Proceedings of the 6th ACM Conference on Embedded Network Sensor Systems*, ser. SenSys '08, November 2008, Raleigh, NC, USA.
16. Klues, K., G. Hackmann, O. Chipara, and C. Lu, 2007. A component-based architecture for power-efficient media access control in wireless sensor networks. In *Proceedings of the 5th International Conference on Embedded Networked Sensor Systems*, November 2007, Sydney, Australia.
17. van Dam, T. and K. Langendoen, 2003. An adaptive energy-efficient MAC protocol for wireless sensor networks. In *Proceedings of the First International Conference on Embedded Networked Sensor Systems*, November, Los Angeles, California, pp. 171–180.

18. Ye, W., J. Heidemann, and D. Estrin, 2002. An energy-efficient MAC protocol for wireless sensor networks. In *Proceedings of the 21st International Annual Joint Conference of the IEEE Computer and Communications Societies (INFOCOM 2002)*, June, New York, NY, pp. 1567–1576.
19. Rhee, I., A. Warrier, J. Min, and L. Xu, Drand: Distributed randomized TDMA scheduling for wireless ad hoc networks, *IEEE Trans. Mob. Comput.*, 8(10), 1384–1396.
20. Burri, N., P. von Rickenbach, and R. Wattenhofer, 2007. Dozer: Ultra-low power data gathering in sensor networks. In *IPSN '07*, April 2007, Cambridge , MA, USA.
21. Brown, J., B. McCarthy, U. Roedig, T. Voigt, and C. Sreenan, 2011. Burstprobe: Debugging time-critical data delivery in wireless sensor networks. In *European Conference on Wireless Sensor Networks (EWSN)*, February, Bonn, Germany.
22. Gungor, V. C. and G. P. Hancke, 2009. Industrial wireless sensor networks: Challenges, design principles, and technical approaches, *IEEE Transactions on Industrial Electronics*, 56(10), 4258–4265.
23. IEEE Standard for Local and Metropolitan Area Networks—Part 15.4: Low-Rate Wireless Personal Area Networks (LR-WPANs), IEEE Stdandard 802.15.4e-2012, 2012.
24. Brown, J., and U. Roedig, 2012. Demo abstract: Ginlite—A MAC protocol for real-time sensor networks. In *Proceedings of 9th European Conference on Wireless Sensor Networks (EWSN'12)*, February, Trento, Italy, pp. 34–35.
25. Pöttner, B. et al., 2011. WSN evaluation in industrial environments—First results and lessons learned. In *3rd International Workshop on Performance Control in Wireless Sensor Networks*, June, Barcelona, Spain, pp 1–8.
26. Österlind, F., A. Dunkels, J. Eriksson, N. Finne, and T. Voigt, 2006. Cross-level sensor network simulation with COOJA. In *Proceedings of the First IEEE International Workshop on Practical Issues in Building Sensor Network Applications (SenseApp 2006)*, November, Tampa, Florida, pp. 641–648.
27. Polastre, J., R. Szewczyk, and D. Culler, 2005. Telos: Enabling ultra-low power wireless research. In *Proceeding of the 5th International Conference on Information Processing in Sensor Networks (IPSN)*, April 2005, Los Angeles, CA, USA.
28. Eriksson, J., A. Dunkels, N. Finne, F. Österlind, and T. Voigt, 2007. MSPSim—An extensible simulator for MSP430-equipped sensor boards. In *Proceedings of the European Conference on Wireless Sensor Networks (EWSN), Poster/Demo Session*, January, Delft, the Netherlands.

APPLICATIONS

Chapter 17

Utilization of Directional Antennas in Flying Ad Hoc Networks: Challenges and Design Guidelines

Şamil Temel and İlker Bekmezci

Contents

Abstract

Unmanned air vehicles (UAVs) are designed and used to free human beings from dirty, dull, and dangerous missions. A UAV is defined as a system that can fly autonomously or is remotely piloted. For the last two decades, UAVs have been widely used for both military and civilian missions. Although single-UAV systems have been successfully implemented and used in aviation so far, multi-UAV missions are gaining interest in the research community. Hence, there is a tendency to utilize a group of smaller and cheaper UAVs instead. Most of the arising communication problems can be mitigated with the deployment of directional antennas. However, to the best of our knowledge, the substitution of a single UAV-based mission is not handled thoroughly in the literature yet.

Today, the novel ad hoc network formed by UAVs up in the sky is called as flying ad hoc networks (FANETs) which bear many open research issues and particular challenges. Although FANETs come with many advantages, they still bear many unique challenges. One of the most prominent challenges is to establish a robust communication infrastructure between highly mobile UAVs. Moreover, the awareness of the exact geographical locations of the neighboring UAVs is vital for a safe flight. In addition, the dissemination of the location information among the FANET nodes is of higher importance in the sense of packet transmission, reliability, and collision avoidance (CA).

In this chapter, we investigate the opportunities and challenges of utilization of directional antennas for FANETs, and we propose a FANET model with directional antennas and present some design guidelines along with the cast of our FANET design experience. We also present solutions to some of the controversial issues in FANETs.

17.1 Introduction

Today, unmanned air vehicles (UAVs) are designed and used to free human beings from dirty, dull, and dangerous missions. There has been an enormous technological development for the last two decades with the advance in electromechanics and microprocessors. The versatile and economic advantages over manned flights enabled the usage of UAVs for various missions such as search and destroy [1], search and rescue [2], border security [3], wildfire monitoring [4], relay for wireless networks [5,6], metrological estimations [7], disaster management [8], remote sensing [9], and traffic monitoring [10]. Instead of deploying a high cost UAV for such missions, it is now more resilient to deploy a group of smaller and cheaper UAVs for search and rescue or military surveillance posts [11,12]. Along with the superiority of reduced costs, multi-UAV systems may provide a more fault tolerant mission accomplishment infrastructure. However, it is a non-trivial and a challenging task to accomplish a robust communication infrastructure between highly mobile terminals.

Today, the novel ad hoc network formed by UAVs is called as flying ad hoc networks (FANETs) which bear many open research issues and come with many particular challenges [12]. One such challenge is the difficulty of providing a successful channel access control and achievement of an increased network capacity. When equipped with traditional omnidirectional antennas, the spatial reuse and overall network capacity will remain limited, because the traditional communication protocols, such as collision avoidance (CA), medium access control (MAC) and routing methods, fail to response to the characteristic demands of FANETs. Yet, directional antennas have several advantages over omnidirectional ones which enable them to be a promising alternative for FANETs. One of the main advantages is that the transmission range of a directional antenna is generally longer than the transmission range of an omnidirectional antenna and longer transmission range decreases the hop count which enhances the latency performance. However, utilization of directional antennas necessitates the usage of directional communication protocols.

One of the most vital issues when deploying directional communication protocols for FANETs is the awareness of the exact geographical locations of the neighboring nodes. Also the dissemination of the location information among the FANET nodes, in the sense of packet transmission, reliability and CA is of higher importance. However, most of the studies in the literature assume that the location information is maintained by the upper layers and do not propose a robust solution in the MAC layer. In this study, we investigate the opportunities and challenges of utilization of directional antennas and directional communication protocols in FANETs. Specifically, we present some design guidelines and analytical results with good practices. We also present solutions to some of the controversial issues in FANETs and we believe that this study will contribute to the upcoming knowledge on FANET designs.

The rest of the study is organized as follows: in Section 17.2, we discuss the FANET characteristics and the need for the deployment of directional antennas on FANETs. In Section 17.3, we propose a FANET model and in Section 17.4, we present some design guidelines. The chapter is concluded in Section 17.5.

17.2 FANETs with Directional Antennas

Traditional wireless ad hoc networks such as MANETs (mobile ad hoc networks) and VANETs (vehicular ad hoc networks) are formed without a network infrastructure. The nodes in such networks may constitute peer-to-peer links either with a base station or with another node in the network. The movement of the nodes is often random and there are no central control stations which regulate the movement and communication of the nodes. For MANETs, the IEEE 802.11 a/b/g/n protocols and for VANETs IEEE 802.11 p protocol is used for the *physical* and the *MAC* layer facilities. Such protocols have been successfully implemented in industrial and home-based applications. However, the FANET concept is in its initial steps, and for the time being, there are only a few number of studies for FANETs.

Although FANETs have some common attributes with the aforementioned traditional wireless networking techniques, the concepts of these networking techniques does not fit well for FANETs. FANETs have some distinct and unique characteristics different from MANETs or VANETs which necessitate re-inventing and developing novel techniques [12–15].

Some of the distinct characteristics of FANETs are summarized as follows:

1. In FANETs, nodes move faster than of MANETs or VANETs. The speeds of the nodes in MANETs mimic the speed of humans. In VANETs, the speed of nodes is equal to the speed of automobiles. In both network types, nodes can pause, reverse or make sharp turns. However, in FANETs, the speed of UAVs is between 20 and 60 m/sec which is faster in comparison. Also, the fixed wing UAVs cannot pause, reverse or make sharp turns. Hence, traditional mobility models which are designed mainly for MANETs or VANETs cannot be used for FANETs and novel mobility models have to be developed for FANETs.
2. Unlike MANETs, the network topology formed by UAV nodes in FANETs changes more spontaneously. In VANETs, nodes move over predefined routes, namely on traffic roads, and the movement of the nodes are highly predictable. On the other hand, the number of nodes in FANETs is between 3 and 20 which is generally less than of MANETs and VANETs [12].

However, the spontaneous variation characteristic of the UAVs imposes an extra burden on the overall channel quality. Thus, FANET designers have to focus on link quality assessment issues. For the traditional wireless networks, this problem is addressed with the *static channel allocation* or *time division multiplexing* methods. However, such methods contract with the ad hoc architecture of the FANETs.

3. In FANETs, the distance between the nodes is greater than of MANETs or VANETs. To avoid collisions, UAV nodes repel each other by flying on abstention routes. However, when the nodes in the network are closer, the packet delivery ratios become more effective. Thus, there is a trade-off between keeping distance among nodes and achieving better throughput results.
4. In multi-UAV network scenarios, nodes often carry different and heterogeneous sensor types. Hence, in FANETs, different data merging and transmission techniques have to be developed.

Today, the transceivers in the traditional wireless network nodes are commonly equipped with the COTS (commercial off-the-shelf) omnidirectional antennas. Although omnidirectional antennas are small, cheap and manageable, they cannot provide adequately effective packet delivery success ratios in wireless networks where their topology frequently changes. In addition, the maximum distance that a packet can be delivered successfully with an omnidirectional antenna is very limited [16,17]. Moreover, in multi-hop wireless ad hoc networks, the following problems are frequently encountered [18,19]:

1. Limited data transmission capacity
2. The need of closer distance of communicating nodes
3. Frequent collision of data and control packets
4. Degradation in signal power efficiency

To overcome the aforementioned problems, nowadays, deployment of directional antennas in ad hoc networks has gained a wide interest in vast research areas [16–29]. With deployment of directional antennas, we observe an increase in the spatial reuse and overall network capacity results. In addition, the bandwidth is handled more effectively which results in more effective network performance results [20–25].

On the other hand, directional antenna utilization necessitates directional MAC and routing protocols which come with the well-known directional *deafness*, *asymmetry in gain* and *location estimation* problems [20,24,25]. Although there are some efforts to minimize such problems with broadcasting and flooding of angle-of-arrival estimation methods, to the best of our knowledge, there is no robust solution which addresses both of these problems yet [20]. Thus in this study, we propose a model and present some design guidelines for successful utilization of directional antennas in FANETs.

17.2.1 Challenges of Directional Antenna Deployment in FANETs

Deployment of directional antennas leads to greater bandwidth utilization, lesser interference, effective energy consumption and extended transmission range. Although directional antennas seem more advantageous, they come with a prize. Since the research efforts to minimize the problems are related to directional antenna deployment in FANETs, enlightenment of the ways to minimize such problems has a significant importance. In this section, we summarize some of the prominent challenges and problems that still need to be addressed. In subsections 17.2.1.1 through 17.2.1.3, common well-known problems of utilization of directional antennas are discussed.

17.2.1.1 Directional Hidden Terminal Problem

Traditional hidden terminal problem occurs when two nodes in a wireless ad hoc network try to initiate a communication with a receiver simultaneously [16]. In such cases, packets collide in the receiver, as illustrated in Figure 17.1. According to the figure, both nodes **A** and **C** are in the range of node **B**. However, nodes **A** and **C** cannot "see" each other and they are called ***hidden*** nodes. The problem occurs when nodes **A** and **C** start to send packets simultaneously to node **B**. This is because nodes **A** and **C** are out of range of each other and so they cannot detect a collision. This problem is mitigated with the IEEE 802.11 DCF (Distributed Coordination Function) protocol by an RTS/CTS (Request to Send/Clear to Send) handshaking method. The RTS packet of the transmitter and the CTS packet of the receiver are sent omnidirectionally and because the communication duration is embedded in the RTS and CTS packets, the neighboring nodes are informed about the upcoming transmission. During this communication period, the neighboring nodes defer their transmission to avoid interfering with the ongoing transmission [16].

In the directional case, it is seen that the RTS and CTS packets may not reach to all neighboring nodes making them unaware of the ongoing transmission. When the node **A** is unaware of the transmission between nodes **B** and **C**, and node **B** beamformed its antenna towards **C**, it fails to get and respond to **A**. Hence node **A** is hidden from nodes **B** and **C** [17].

17.2.1.2 Head of Line (HoL) Blocking Problem

In traditional directional MAC protocols, *a priori*tized FIFO (first-in first-out) queue is used which ultimately gives rise to the HoL blocking problem [26]. The HoL blocking occurs when a node is free to establish a communication only in some antenna beam directions, but not in other directions. If a packet is at the top of the MAC queue and if the receiver is busy, the packet is blocked, and the blocked packet will stand in the way of other packets from being transmitted even if their direction is free. This problem is illustrated in Figure 17.2. According to the figure, node **A** is communicating with nodes **B**, **C**, and **D**. There will be packets in node **A**'s queue to be transmitted to nodes **B**, **C** and **D** sequentially. If nodes **B** and **E** are in communication, **A** has to wait until the communication between **B** and **E** is complete. In this example, node **A** must be able to schedule the packet for **C** instead of waiting for **B** [12,26].

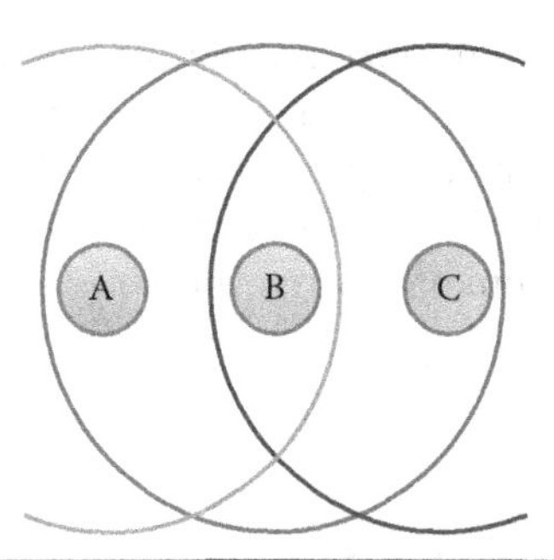

Figure 17.1 Hidden terminal problem.

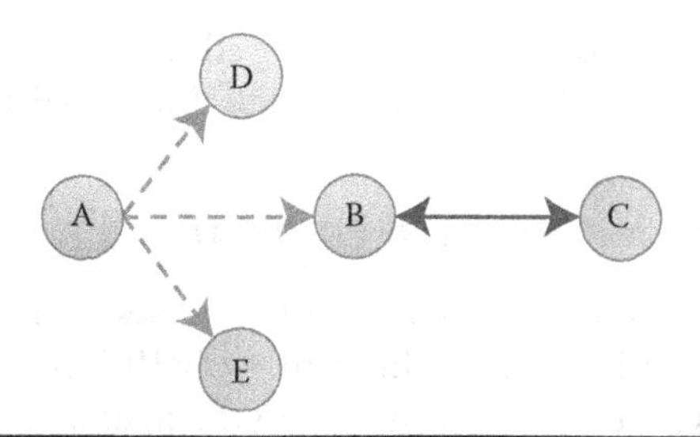

Figure 17.2 Head of line problem.

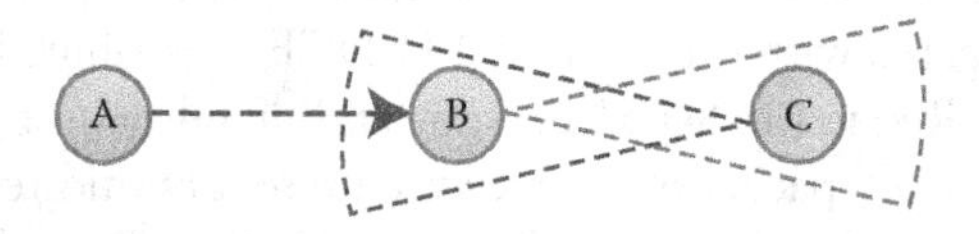

Figure 17.3 Deafness problem.

17.2.1.3 Deafness Problem

In general, a node that uses a directional antenna is considered as "deaf" for all the other directions except the one which is beamformed for Reference 27. Deafness is caused when a transmitter node repeatedly attempts to communicate with a receiver which presently focuses on another antenna beam. At each unsuccessful attempt, the backoff interval is doubled hence degrading network performance. This problem is illustrated in Figure 17.3. According to the figure, node **A** wants to communicate with **B,** but it has beamformed its antenna towards **C.** Hence it is said that **B** is "deaf" to **A.** Deafness problem has a direct impact on the network performance. It leads to higher delay time, unfair assignment of network resources, increase of backoff window and even specific situations of deadlock [24–27].

While traditional wireless networks with omnidirectional antennas bear many problems such as the aforementioned ones, novel and particular design solutions for the directional antenna deployment has to be revisited. Hence, based on our previous studies and simulation results, we propose a model FANET structure and present some design guidelines in the following section.

17.3 Proposed FANET Model

Although there are several studies in the literature for the deployment of directional antennas on FANETS, to the best of our knowledge, there aren't any instructive models to be benchmarks for the research community. In this sense, we propose a FANET model in the cast of our previous studies and analysis results [12,24–25,29]. In this section, we summarize the proposed antenna model, mobility model and communication protocol models which we believe to guide enthusiastic researchers in the field.

17.3.1 Antenna Model

In recent years, the wireless communication research community has turned its attention to the utilization of switched beam antenna arrays in wireless ad hoc networks [24,25]. These types of antennas can focus more energy on a desired direction. They are half-duplex in the sense that at a given time, only one transmission is conducted over one of these directions. It is a well-known fact that, most of the power is radiated towards the main beam direction of a directional antenna. However, some amount of power is also dissipated over the side lobe which is usually called as the *suppression ratio*. This power is regarded as *interference* among the other nodes in the network. If the signal-to-interference ratio (SIR) is high enough, then the receiver can capture the signal but there is a limit on it. The limit can be defined as the sum of all interference from all communication pairs which could easily garble each other's communication [17,20].

When the main lobe and side lobe gains are given as G_m and G_s, respectively, the free-space propagation model can be defined as

$$P_t = P_r \frac{G_m G_s}{d^{\propto} L} (\lambda / 4\pi)^{\propto} \tag{17.1}$$

where, P_t and P_r are the transmit and receive power values in watts, respectively, λ is the wavelength in meters, α is the path loss exponent, d is the distance between transmitter and receiver in meters and L is the loss factor ($L = 1$).

As stated earlier, to successfully receive a signal, the SIR must be greater than or equal to a given threshold value, SIR_{th}. While SIR can be stated as in Equation 17.2, the maximum number of active nodes, N_{act}, (which represents the number of maximum simultaneous communicating nodes) can be given as in Equation 17.3:

$$SIR = P_t c G_t G_r / I_{tot} d^{\alpha} \tag{17.2}$$

$$N_{act} = SIR / SIR_{th} \tag{17.3}$$

where G_t and G_r are the transmitter and receiver antenna gains, respectively, and I_{tot} is the total interference seen on the receiver antenna.

The simulation results which illustrate the impact of the suppression ratio to the maximum number of achievable active nodes with respect to different main beam angles are shown in Figure 17.4. As it can be inferred from the figure, when the main beam angle of the directional antenna increases, the maximum number of possible communicating nodes increases simultaneously. In addition, when the suppression ratio is increased, the number of active nodes decreases. It shows that the side lobe gain must not be underestimated in the capacity analysis of FANETs with directional antennas.

17.3.2 Effect of Speed

Most of the existing studies for MANETs assume that the nodes move with pedestrian speeds at most. When FANETs are the case, mobility is an indispensable effect to analyze. *Doppler Effect* is one of the pioneering phenomena about highly mobile networks. This effect is expressed as follows: when a node is moving with a constant velocity, there will be a frequency shift in the received signal. The amount of the frequency shift is expressed as f_d which can be calculated with Equation 17.4 as follows.

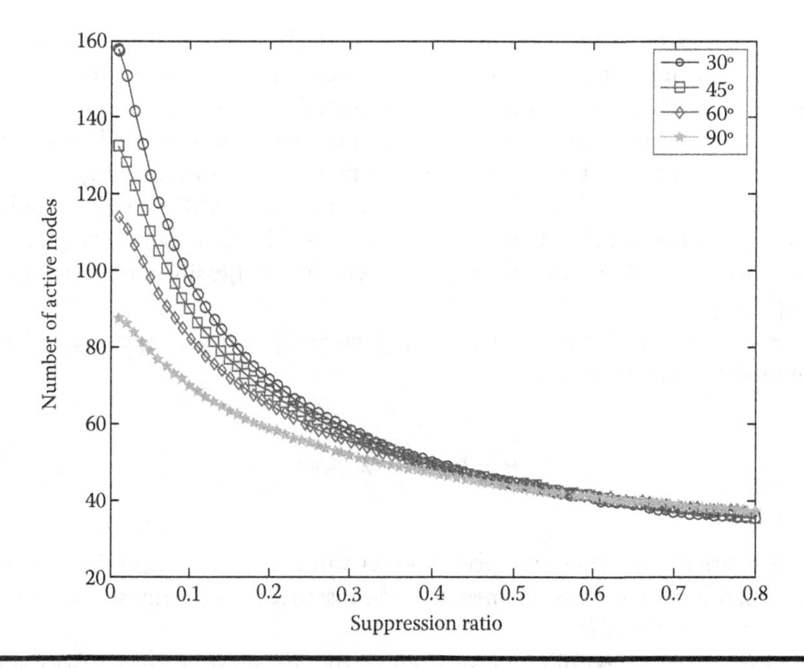

Figure 17.4 Maximum number of active nodes with various main beam angles.

$$f_d = f_0 v / v_c \tag{17.4}$$

where f_0 is the base frequency, v is the relative velocity, and v_c is the speed of light. In high frequencies, this effect becomes negligibly minor and can be minimized with various DSP (digital signal processing) techniques. Hence, we can assume that the Doppler shift has no effect or it is solved with DSP techniques. On the other hand, the relative velocity between two communicating nodes highly affects the network throughput because a receiver is biased to *escape* from the transmitter antenna beam. Relative velocity among transmitter and receiver nodes can be expressed as in Equation 17.5 as follows.

$$V_R = \left| V_{R_i} + V_{\tau_i} \right| \tag{17.5}$$

where V_{R_i} is the speed of the receiver and V_{τ_i} is the speed of the transmitter.

As can be seen from Figure 17.5, at worst case, the receiver may escape from the transmitter's beam with an orthogonal direction to the beam edge.

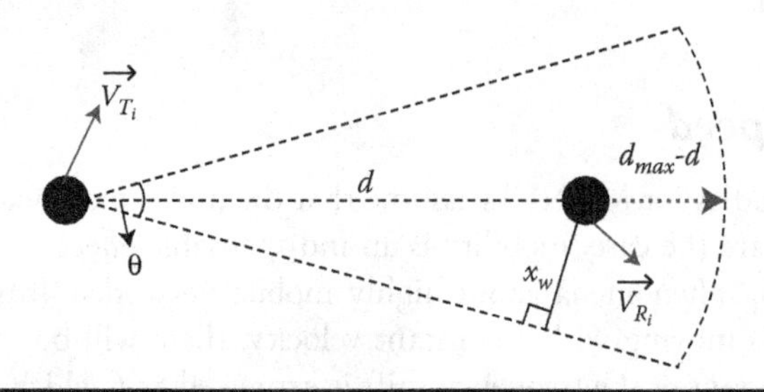

Figure 17.5 Relative velocity between two UAV nodes.

If we assume that the receiver is located on the axis of the beam at distance d, the displacement that the receiver will take is x_w, and the duration to take x_w is t_w, the speed equations can be derived with Equations 17.6 and 17.7.

$$x_w = \sin\frac{\theta}{2}d \quad (17.6)$$

$$t_w = x_w V_{R_i} \quad (17.7)$$

If C is the maximum data rate (which may be chosen to be 11 Mbps indicating a IEEE 802.11b channel), the maximum amount of data, C_{max} to be transferred during t_w can be calculated with Equation 17.8 as follows:

$$C_{max} = t_w C \quad (17.8)$$

The maximum achievable data transfer rate according to relative velocity and main beam angle is shown in Figure 17.6. One important result that can be derived from the simulation results is that there is a need to employ directional antennas with greater directionality and minor suppression effects. Another important effect on maximum data transfer rate is the relative velocity of the transmitter and receiver nodes. Although the Doppler shift can be assumed to be minimized with some DSP techniques, the mobility is still an important fact because nodes are biased to repel each other not to get much closer. As explained in Section 17.2, wider beam angles and smaller relative velocities present better data transfer ranges. Hence, there is a trade-off between choosing the main beam angle and the antenna type.

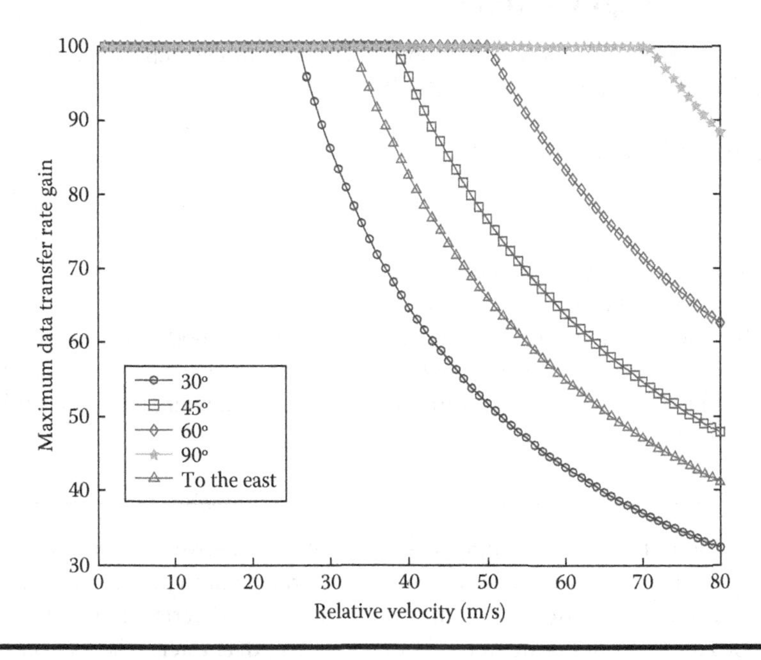

Figure 17.6 Maximum data transfer rate with respect to relative velocity.

For flight simulations, we employed the *distributed pheromone repel* model which assumes that each UAV maintains a local pheromone map to timestamp the last time the pixel was scanned [28]. As a UAV flies over the region, it marks the pixels that it scanned by assigning a numeric pheromone number that is decremented by one on each second in its local pheromone map. Also in this mobility model, each UAV broadcasts its local pheromone map once in every 10 s. All the UAVs within the FANET merge this information to update the global pheromone map. At each second, each UAV conducts a decision sequence to turn left, right or go straight ahead. Instead of making this decision with discrete probabilities, the probabilities are based on the average pheromone smell in three zones, namely left zone, center zone or right zone. Since a UAV should select to go to least-recently visited pixels, it should prefer an area bearing a low pheromone smell amount.

Based on our simulation results, we present the pheromone traces in Figure 17.7. In Figure 17.7a, the overall pheromone traces of 20 UAVs over a 400 × 400 pixel area is shown. The pixel resolution is 30 m which yields a 12 km^2 area. At the very beginning of the simulation, 20 UAVs are placed close to the middle of the field with a random heading direction. The velocity of the UAVs is chosen to be 30 m/s with a maximum turning radius of 45 m. As an example, one of the UAV's (UAV7) unique route trace after 1800 s is illustrated in Figure 17.7b.

In Figure 17.8a, total interference that UAV7 collects along its journey is illustrated. Throughout the simulations, we assigned the total interference power threshold, P_{th}, to −40 db W which is a moderate threshold value for generic access point devices and corresponds to 100 nW. According to the figure, through the first 400 s, we see that the total interference is high respectively. Also, between simulation times of 220 and 230, the total interference represents a peak. In Figure 17.8b, we illustrate the average interference gathered for all of the 20 UAVs in 1800 s. We also observe that the average interference decrease in time. This is because, at the beginning of the flight post, UAVs are close to each other and haven't dissipated enough over the field yet.

17.3 FANET Design Guidelines

In this section we present and summarize some of the prominent challenges and some design guidelines for directional antenna-based FANETs.

17.3.1 Location Estimation

The possession of the exact locations of the neighboring nodes is vital for FANET applications especially where the human life is in concern [12]. In the directional antenna case, a node has to be regularly provided with the location information of the neighboring UAVs, because the antenna of the transmitter has to be beamformed towards the receiver. Most of the studies in the literature assume that the instantaneous location information of the nodes in the FANET is known "somehow" or they assume that this information is provided from an uplink such as a satellite or a high altitude platform. However, these assumptions are far from being reliable on mission critical FANET posts [24,25].

> The dissemination of the location information in FANET networks should be achieved in the MAC layer. The FANET MAC protocol should provide both an effective way to utilize data transmission over directional antennas and a method to determine and disseminate the exact locations of the neighboring nodes in every GPS update interval sequence.

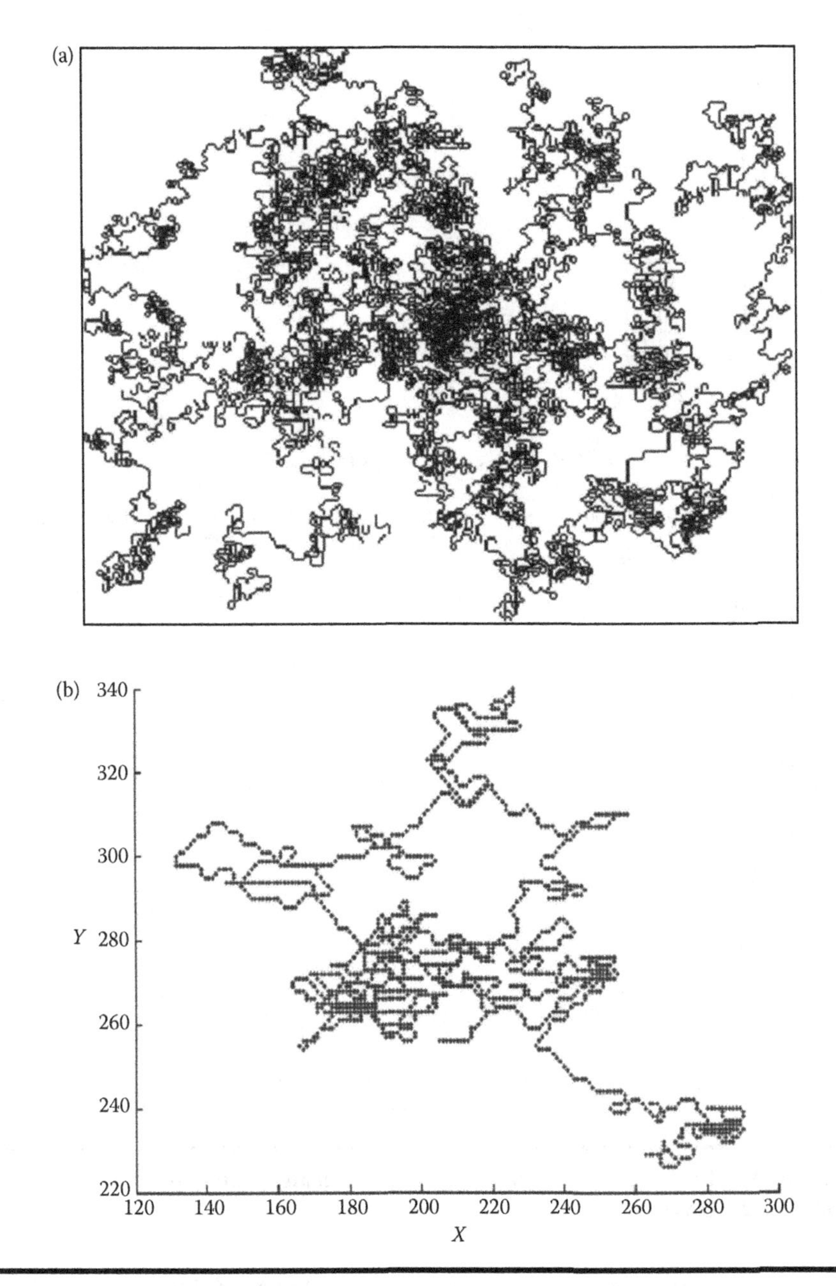

Figure 17.7 (a) Route trace of the pheromone repel mobility model with 20 UAVs on a 12 × 12 km field after 1800 s. (b) Route trace of UAV7 after 1800 s.

17.3.2 Antenna Type

It is a well-known fact that the nodes in the traditional wireless ad hoc networks are generally equipped with omnidirectional antennas. The protocols which have been developed for such networks lack to address the needs of FANET applications, let only the directional antenna-based communication infrastructure needs [12,24,25,29].

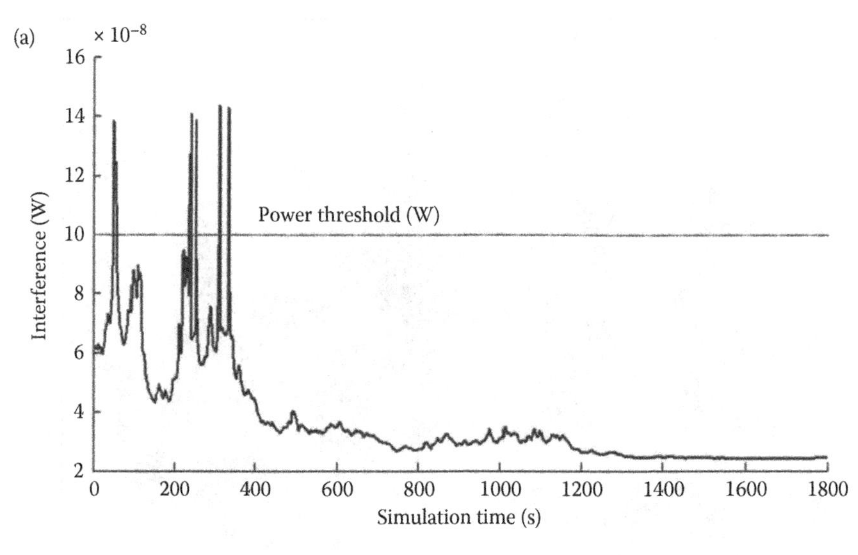

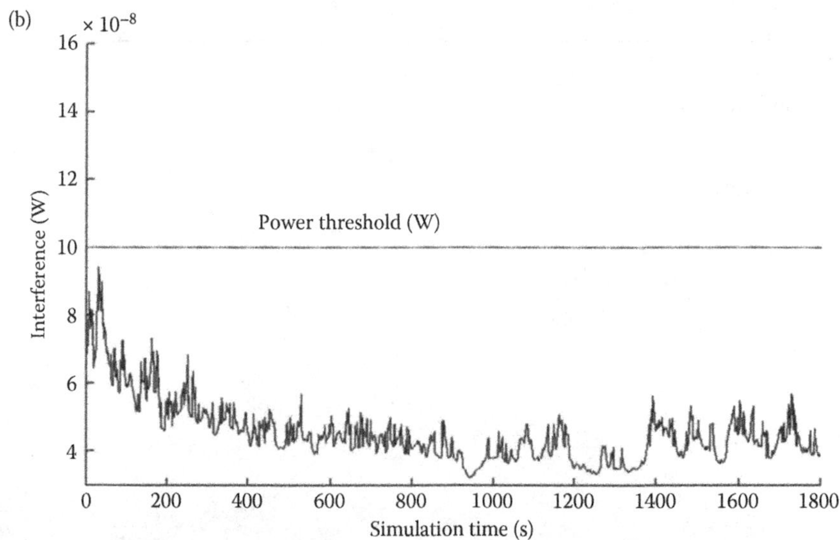

Figure 17.8 (a) Total interference gathered at UAV7 during 1800 s. (b) Average interference during 1800 s for 20 UAVs.

One justification for this judgment is that, in most of the MANET directional MAC protocols, DATA packets are sent directionally whereas RTS/CTS packets are sent omnidirectionally [20]. However, implementing an antenna which bears both omni and directional properties is a very challenging task to be realized in the real world. In addition, sending some type of the packets omnidirectionally and the remaining ones directionally results in *asymmetry* problems in transmission range. As explained earlier, the transmission ranges for directional antennas are further when compared to omnidirectional type of antennas. If a packet is sent omnidirectionally, just the neighboring nodes which are in the transmission range of the omni antenna can catch it. Hence, the communication would be restricted in the omni antenna coverage zone. In this case, the packets sent directionally may garble the communication protocol packets.

On the other hand, if all the packets in the network are sent directionally, the aforementioned asymmetry problem would be minimized. The nodes, which are further than the omni range, can successfully get the packets which results in minimized hop-counts in transmission. As a result, by deploying "only" directional antennas in a FANET gives rise to achieving better overall delay results.

We have analyzed that the number of achievable active nodes in the FANET highly depends on the type of the antenna [24,25]. We observe that when narrower beam antennas are deployed, more UAV nodes could involve in communication. In addition, the side lobe gain of the antenna should never be underestimated or ignored in antenna design facilities.

In addition, it is a well-known fact that with separating control and data channels in wireless networks gives rise to performance benefits [30]. Also, as stated in Section 17.3.1, the location estimation problem can only be mitigated by integrating additional software and hardware on the communicating nodes. Hence, we propose the utilization of a pair of transmitters one of which is only responsible just for location estimation and control packet transmission [24].

> In a FANET, all the packets should be sent directionally and FANET communication protocols should handle two transmitters which utilize switched beam directional antennas on UAV nodes. The first transceiver should be responsible for the control packet exchange and the other one should be dedicated only for the transmission of data packets. By separating the control and the data channels, better throughput can be achieved.

17.3.3 Mobility of Nodes

To evaluate the performance of a MANET, mobility models have an important role. Most of the studies in the literature assume two-dimensional (2D) fight models; however, the altitudes of the UAVs have to be taken into account for protocol designs because communication among UAVs which fly in three-dimensional (3D) space will present different characteristics than in 2D fields [12,24,25,28,29]. In addition, we have observed that the geographical characteristics highly affect the performance of a FANET when the UAV nodes fly over lower altitudes [29]. It is also determined that over harsher terrains, more link failures occur and it degrades the overall network throughput that highly depends on the relative speed of the communicating nodes [24]. The best results are achieved when the relative speed gets closer to zero.

> The performance of the communication protocols which are proposed for FANET usage should be evaluated over 3D assumed mobility modes which necessitate novel 3D flight models.

17.3.4 Mitigation of Deafness

As explained in Section 17.2, when directional antennas are in use, deafness has a direct impact on the network performance leading to higher delay times. To eliminate the deafness problem, there are some methods proposed so far. For example, Gossain et al. [27] proposed a solution for wireless ad hoc networks using directional antennas. They state that that IEEE 802.11 short retry limit (SRL) needs a special handling in the directional environment because of the presence of deafness. They also underline the fact that the deafness problem can only be eliminated by injecting new packets or by providing each of the nodes with their exact location.

Traditional MANETs employ the standard IEEE 802.11 DCF protocol for access control and MAC layer packet exchange. The control packets in DCF protocols are RTS, CTS and ACK (Acknowledged) packets. In DCF, a transmitter will send a request packet (RTS) and the receiver will reply back with a clearance packet (CTS). However, there is no way for a receiver to reply back saying "I'm busy now, so please try again!" Thus, as we have stated earlier, this slavery of the receiver will cause a deafness situation for the transmission.

> The mitigation of the well-known deafness problem can be achieved by defining a novel control packet which enables a busy receiver to respond to a sender that it conducts a transmission and will not be available for a specific amount of time.

17.3.5 Adaptability

In multi-UAV missions, spontaneous parameter changes can be observed quite regularly. These changes may either stem from the atmospheric conditions or from operational failures. Also there could be a need to add new UAVs in the flock of the FANET in any time of a mission flight. Hence, FANETs should be adaptable to the spontaneous changes in the network infrastructure. In addition, the mission itself can be changed while a predefined mission is ongoing. For example, in surveillance and reconnaissance posts, it is very possible that when some hostile activity is detected, most of the UAVs will be directed or focused over the point of the hostile zone. This would easily garble the wireless communication infrastructure resulting in loss of UAVs.

> FANETs should be adaptable to the spontaneous changes in the network infrastructure, mission updates or operational failures.

17.3.6 Network Latency

Latency is a basic and an important performance metric of any wireless ad hoc network. When FANETs are the case and human life is in concern, network latency cannot be tolerated. For example, after a catastrophic event such as an earthquake, hurricane, tsunami, wildfire etc., UAVs must instantaneously augment to inspect survivors and inform the authorities. Also CA from obstructions or other air vehicles necessitate the deployment of delay-sensitive communication protocols.

Single UAV-based missions are generally augmented with satellite communication. However, the inevitable propagation delay, which is caused by the distance from the satellite and atmospheric conditions, results in a non-real-time command and control. As a result, a UAV operator may "see" the effects of his command with a delay. This delay can reach up to a few seconds in harsh conditions which can be regarded as a razor's edge in military missions.

> Single UAV-based missions and traditional MANET routing protocols often fail to respond to the delay-intolerant nature of FANET applications. Deployment of a group of mini-UAVs which are equipped with directional antennas has a positive impact on minimizing the network latency which puts the directional antenna deployment strategy to be a robust alternative.

17.4 Conclusion

In wireless networks, the common communication channel must be shared among several terminals which necessitate the usage of a channel access control and a packet relaying mechanism. Such mechanisms enable terminals to access the physical layer interfaces fairly and without packet collisions. If the nodes in the network tend to content, channel access can be reserved sequentially to terminals. But things get worse when the number of nodes increase within high mobility. The same phenomenon is also valid for FANETs where the communicating nodes are UAVs.

FANETs present many brand new ways for many applications ranging from search and rescue to target monitoring. Today, in most of the FANET applications, UAV nodes are equipped with traditional omnidirectional antennas which result in limited network capacity, transmission distance, and data security. Alternatively, the deployment of directional antennas significantly decreases the aforementioned problems. In addition, nodes of FANET move much more faster than of MANETs. Thus providing the UAV nodes with the exact locations of the neighboring nodes become vital, especially for directional antenna that can be employed for ad hoc scenarios. In this study, the effects of directional antenna deployment on FANETs are discussed thoroughly. Mainly, we summarize and harvest the performance evaluations of FANETs when directional antennas are deployed on UAV nodes and propose some guidelines with best practices. Based on our previous studies on spatial reuse analysis, effects of mobility of UAVs, channel utilization of directional antennas, and effects of side lobe gain, we confidently state that many of the controversial issues over FANET design researches could be minimized by the deployment of a pair of transceivers equipped with smart beam directional antennas on UAV nodes.

Acknowledgement

This work is partly supported by the Scientific and Technological Research Council of Turkey (TUBITAK) with scholarship ID: 2211-C.

References

1. J. George, S. P. B. and J. Sousa, 2011. Search strategies for multiple UAV search and destroy missions, *Journal of Intelligent & Robotic Systems*, 61, 355–367.
2. S. Waharte and N. Trigoni, 2010. Supporting search and rescue operations with UAVs. In *Emerging Security Technologies (EST), 2010 International Conference on*, pp. 142–147.
3. Z. Sun, P. Wang, M. C. Vuran, M. Al-Rodhaan, A. Al-Dhelaan, and I. F. Akyildiz, 2011. BorderSense: Border patrol through advanced wireless sensor networks, Ad Hoc Networks, 9(3), 468–477.
4. C. Barrado, R. Messeguer, J. L'opez, E. Pastor, E. Santamaria, and P. Royo, 2010. Wildfire monitoring using a mixed air-ground mobile network, *IEEE Pervasive Computing*, 9(4), 24–32.
5. E. P. de Freitas, T. Heimfarth, I. F. Netto, C. E. Lino, C. E. Pereira, A. M. Ferreira, F. R. Wagner, and T. Larsson, 2010. UAV relay network to support WSN connectivity, in ICUMT, pp. 309–314, IEEE.
6. F. Jiang and A. L. Swindlehurst, Dynamic UAV relay positioning for the ground-to-air uplink, In *IEEE Globecom Workshops*, 2010.
7. A. Cho, J. Kim, S. Lee, and C. Kee, Wind estimation and airspeed calibration using a UAV with a single-antenna GPS receiver and pitot tube, *IEEE Transactions on Aerospace and Electronic Systems*, 47, 109–117, 2011.
8. I. Maza, F. Caballero, J. Capitán, J. R. Martínez-De-Dios, and A. Ollero, 2011. Experimental results in multi-UAV coordination for disaster management and civil security applications, *J. Intell. Robotics Syst.*, 61, 563–585.

9. H. Xiang and L. Tian, 2011. Development of a low-cost agricultural remote sensing system based on an autonomous unmanned aerial vehicle, *Biosystems Engineering*, 108(2), 174–190.
10. E. Semsch, M. Jakob, D. Pavlícek, and M. Pechoucek, 2009. Autonomous UAV surveillance in complex urban environments, In *Web Intelligence*, 82–85.
11. Cook, K.L.B., 2007. The silent force multiplier: The history and role of UAVs in warfare, In *Aerospace Conference*, 2007 IEEE, pp. 1–7.
12. I. Bekmezci, O.K. Sahingoz and S. Temel, 2013. Flying Ad hoc networks (FANETs): A survey, *Ad Hoc Networks*, 11(3), 1254–1270.
13. E. W. Frew and T. X. Brown, 2009. Networking issues for small unmanned aircraft systems, *Journal of Intelligent and Robotic Systems*, 54(1–3), 21–37.
14. M. Rieke, T. Foerster, and A. Broering, 2011. Unmanned aerial vehicles as mobile multi-platforms, In *The 14th AGILE International Conference on Geographic Information Science. 18-21* April 2011. Utrecht, Netherlands.
15. J. Clapper, J. Young, J. Cartwright, and J. Grimes, 2007. Unmanned systems roadmap 2007–2032, tech. rep., Dept. of Defense.
16. C. E. Perkins, Ad Hoc Networking. *Addison-Wesley*, 2001.
17. R. Ramanathan, On the performance of ad hoc networks with beamforming antennas, In ACM International Symposium on Mobile Ad Hoc Networking and Computing (MobiHoc), Long Beach, California, October 2001, pp. 95–105.
18. J. H. Winters, 2006. Smart antenna techniques and their application to wireless ad hoc networks, *IEEE Wireless Commun.*, 13(4), 77–83.
19. R. Ramanathan, 2004. Antenna beamforming and power control for ad hoc networks, In *Mobile Ad Hoc Networking*, S. Basagni, M. Conti, S. Giordano, and I. Stojmenovic (eds.), Wiley-IEEE Press, pp. 139–174. ISBN: 978-0-471-65688-3.
20. O. Bazan, M. Jaseemuddin, A survey on MAC protocols for wireless ad hoc networks with beamforming antennas, *IEEE Communications Surveys & Tutorials*, 14(2), Second Quarter 2012.
21. J. C. Liberti and T. S. Rappaport, 1999. *Smart Antennas for Wireless Communications.* Prentice Hall, NJ.
22. G. Li, L. L. Yang, W. S. Conner, and B. Sadeghi, 2005. Opportunities and challenges for mesh networks using directional antennas, In *IEEE Workshop on Wireless Mesh Networks (WiMesh)*, Santa Clara, California.
23. S. Yi, Y. Pei, and S. Kalyanaraman, 2003. On the capacity improvement of ad hoc wireless networks using directional antennas, In *ACM International Symposium on Mobile Ad Hoc Networking and Computing (MobiHoc)*, Annapolis, Maryland, pp. 108–116.
24. S. Temel and I. Bekmezci, 2014. Scalability analysis of flying ad hoc networks (FANETs): A directional antenna approach, *IEEE BLACKSEACOM*, 16–31.
25. S. Temel and I. Bekmezci, On the performance of Flying Ad Hoc Networks (FANETs) utilizing near space high altitude platforms (HAPs), *Recent Advances in Space Technologies (RAST)*, 2013 6th International.
26. V. Kolar, S. Tilak, and N. B. Abu-Ghazaleh, 2004. Avoiding head of line blocking in directional antenna, In *IEEE International Conference on Local Computer Networks (LCN)*, Zurich, Switzerland, pp. 385–392.
27. H. Gossain, C. Cordeiro, D. Cavalcanti, and D. P. Agrawal, 2004. The deafness problems and solutions in wireless ad hoc networks using directional antennas, In *IEEE Global Telecommunications Conference (GLOBECOM) Workshops*, pp. 108–114.
28. E. Kuiper, S. Nadjm-Tehrani, 2011. Geographical routing with location service in intermittently connected MANETs, *Vehicular Technology, IEEE Transactions on*, 60(2), 592–604.
29. S. Temel and I. Bekmezci, LODMAC: Location Oriented Directional MAC Protocol for FANETs, to be published in Ad Hoc Networks.
30. Kyasanur, P., Padhye, J., and Bahl, P., 2005. On the efficacy of separating control and data into different frequency bands, *Broadband Networks, 2005. BroadNets 2005. 2nd International Conference on*, 602,611, 1, pp. 3–7.

Chapter 18

Military Networks Enabled by Directional Antennas

Jerry Sonnenberg, Keith Olds, Emil Svatik, and Dave Chester

Contents

Abstract

Designed to meet high capacity large area networking requirements in a mobile tactical environment, the tactical networking waveform allows for a broad band, large-scale network capable of servicing command centers using both static and mobile configurations over the entire battle space. The tactical networking waveform relies on spatially switched directional beam antenna technology to create a robust self-forming and self-healing mesh network, even when command centers are highly dispersed. Utilizing directional instead of omni antennas allows the formation of large networks in a division size operational area using a single 20 MHz frequency assignment to meet their operational needs.

This system can be deployed using a combination of fixed and mobile configurations, allowing the waveform to provide a single robust networking backbone for the transport of C4ISR and IP data from upper echelons down to the tactical edge. To support the dynamic operational needs of mobile commanders, the waveform is designed to interoperate with commercial IP routers, thereby ensuring network interoperability, sustainability, and ease of use through the implementation of open interface and networking IP standards.

This chapter discusses the unique aspects of the media access control (MAC) layer that allows the robust operation of the waveform, along with network protocol adaptations that allow for the routing environment to maximize bandwidth usage.

18.1 Introduction

This chapter discusses the reasons for the development of the tactical networking waveform, some specific operational needs that dictated the functions of the waveform, design of the media access layer for directional antenna nodes, and the networking layer adaptations built into the waveform to take advantage of the enhanced bandwidth and topology options. Tactical radio networks have a number of characteristics that are not found in traditional radio networks. They must not rely on infrastructure—meaning that cell towers and other high-visibility single points of failure cannot be part of their design. The nodes of a tactical radio network must automatically find each other and form a network. The network must automatically adjust to loss of capacity, links, and nodes and continue to provide prioritized service to the critical traffic of a deployed environment. The nodes must operate at the halt and on the move and adjust to varying physical conditions with no manual intervention.

18.1.1 Waveform Overview

Traditional networks using omni-directional antennas radiate radio frequency (RF) energy in all directions, making it straightforward to get from any given source node to an arbitrary destination. However, the price for this is a steady decrease in available network capacity as more and more nodes enter the network. This is because, regardless of the media access method used, the more radio neighbors that exist in a mesh network, the fewer opportunities for transmission exist for a given node. In a single-radio channel mesh, each mesh node acts as an access point that supports local client access and forwards traffic wirelessly to other mesh nodes. Wireless mesh networks are almost never fully connected, meaning that each node in a network can directly communicate with any other node. The nodes must forward packets and act as an access point for its radio neighbors who are more RF disadvantaged. So the same radio channel is used for access and wireless backhaul.

This option represents the lowest cost entry point in the deployment of a wireless mesh network infrastructure. However, because each mesh access point uses an omni-directional antenna to allow it to communicate with any of its neighbors, almost every packet generated by local clients must be repeated on the same channel to send it to at least one neighboring mesh access point.

This packet forwarding generates a lot of traffic. As more mesh access points and clients are added, a higher percentage of the wireless traffic in any mesh neighbor group is dedicated to forwarding. Less and less of the channel capacity is available to support single hop users.

There is an impact of mesh forwarding on actual throughput. The capacity analysis is somewhere between $1/N$ times the channel capacity and $(1/2)^N$ times the channel capacity, where N is the number of wireless hops in the longest path between a client and the ultimate destination. Figure 18.1 illustrates the $1/N$ phenomenon for an early version of Wi-Fi, where the ultimate destination has access to the wired infrastructure. This network paradigm will not scale up as more nodes are added, the number of hops increases per end-to-end packet delivery.

The tactical networking waveform is a modular, directional TDMA/DAMA (time division multiple access/demand-assigned multiple access) waveform optimized for use as the physical and data link layers of a mobile ad hoc battlefield network. The waveform differs from many ad hoc networking waveforms in that it is designed to operate with narrow-beam directional antennas. To support this form of ad hoc networking, the waveform includes a directional TDMA (DTDMA) protocol that schedules access to the communications channel. The DTDMA not only controls the time at which a link between nodes has access to the communications channel, but also controls the direction over which that link operates.

Since the narrow-beam antennas provide isolation in directions other than their beam direction, this permits multiple links to operate on the same channel frequency at the same time. This frequency reuse greatly increases the spectral efficiency of the waveform.

The waveform provides aggregate data capacity to the network layer based on the modulation order of the waveform, which is determined at the physical layer by measured signal-to-noise ratio (SNR). The aggregate data capacity has grown with the development evolution of the waveform. Early versions have a capacity of 6–54 Mbps and a recent version supports an aggregate data capacity of 27–135 Mbps.

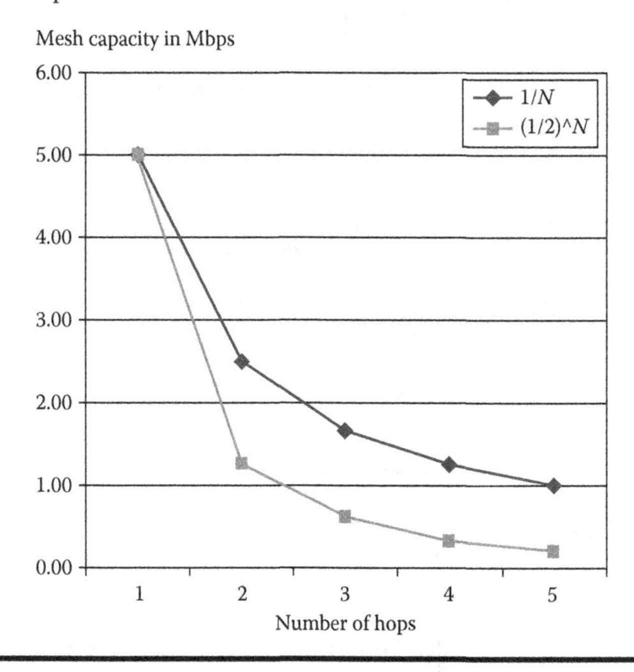

Figure 18.1 Single-channel wireless mesh capacity. (Adapted from BelAir Networks, *Capacity of Wireless Mesh Networks*, White Paper, 2006, http://www.belairnetworks.com.)

Section 18.2 covers some of the detail of the media access control (MAC) layer that utilizes these unique physical layer capabilities.

18.2 Media Access Control

18.2.1 TDMA Frame and Slot Structure

The waveform software is hosted in the tactical networking radio and uses TDMA technology to transfer IP data within the radio core network. As shown in Figure 18.2, the TDMA timing hierarchy consists of epochs, epoch groups, hail timeslots, rendezvous timeslots, and data timeslots.

All tactical networking waveform nodes are synchronized via a global positioning system (GPS) time standard. Each second consists of an integral number of time epochs. An epoch contains an integral number of timeslots where each timeslot in an epoch is assigned a unique number.

Within each timeslot, there exist guard times for hardware jitter. Early versions of the waveform also add guard time allowance for transmit latency based on distance. Recent versions of the waveform provide a range compensation technique that avoids the need for this transmit latency guard time. The IP-based data is loaded from radio memory into a timeslot before it is transmitted over the air. Then at the assigned time, as defined by the MAC schedule, the data within that timeslot is transmitted. The latency of the data received is a result of the transmission propagation delay based on the distance between two nodes. Once the data is received, it is unloaded from the radio and placed on one of the Ethernet ports as an IP packet.

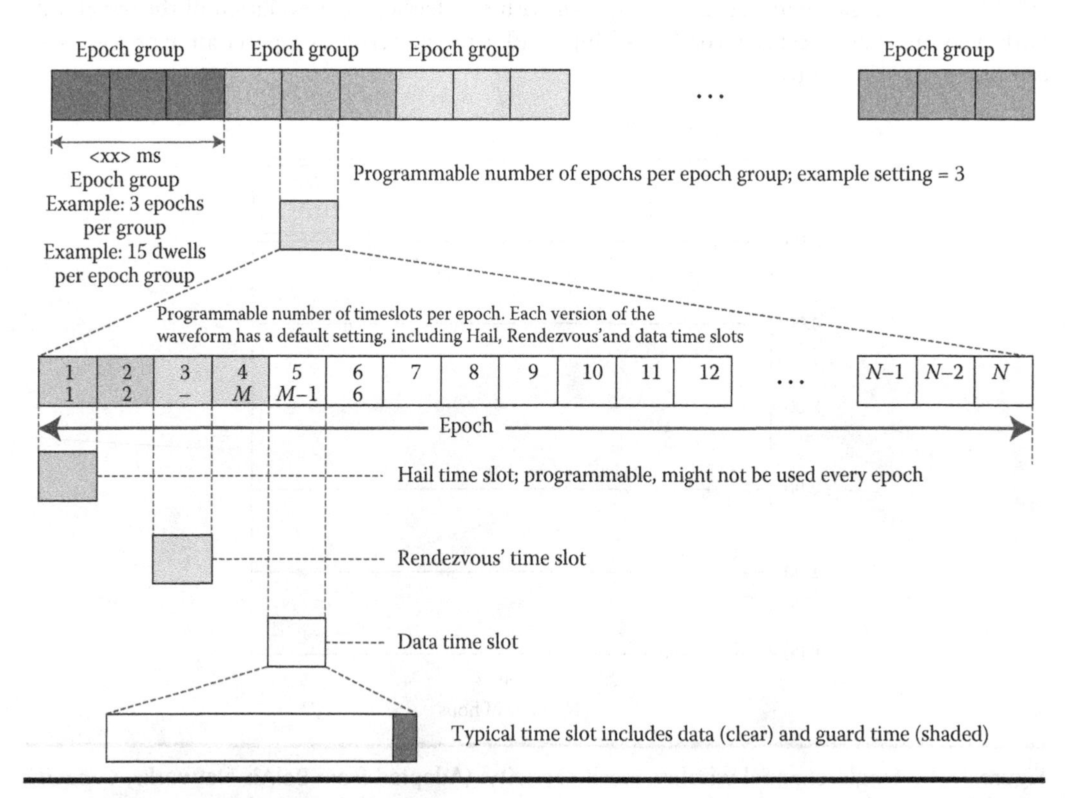

Figure 18.2 TDMA timeslot structure.

An epoch consists of data, hail, and rendezvous timeslots. Data timeslots are dynamically assigned by the link scheduler based on traffic demand, interference, timeslot quality, and other RF parameters. Early versions of the waveform support up to 40 data timeslots in an epoch. The most recent version provides up to 700 timeslots per epoch. Data timeslots consist of IP-based packets, such as TCP and UDP data, and data-slot control information. Hail timeslots transmit hail messages, which contain position information and are used for neighbor discovery and antenna pointing coordination. Rendezvous timeslots contain information to synchronize and associate new nodes entering a core network.

18.2.2 MANET Operation

The tactical networking waveform operates in a mobile ad hoc network (MANET) mode as shown in Figure 18.3. In this mode, link control and resource management is performed using distributed control techniques. This mode is generally employed for line-of-sight interconnection of the terrestrial nodes.

As a distributed, ad hoc waveform, there is no central control entity. Bandwidth allocations are accomplished in the data link layer through peer negotiations between nodes. This waveform mode anticipates a constantly changing network topology due to node mobility. The MANET functions in the data link layer also include ad hoc neighbor discovery and link establishment with appropriate reports to layer 3 so that it can manage the network routing and QoS (quality of service). These functions are performed autonomously. They are supported by information received from the network management system and the routing and switching functions. Further, these functions are constrained to operate within the predetermined polices established by an overarching network management function.

18.2.3 Neighbor Discovery

The waveform includes provisions for assisted neighbor discovery (AND) and non-assisted neighbor discovery (NND). In AND, the nodes are informed of the locations of other nodes via a

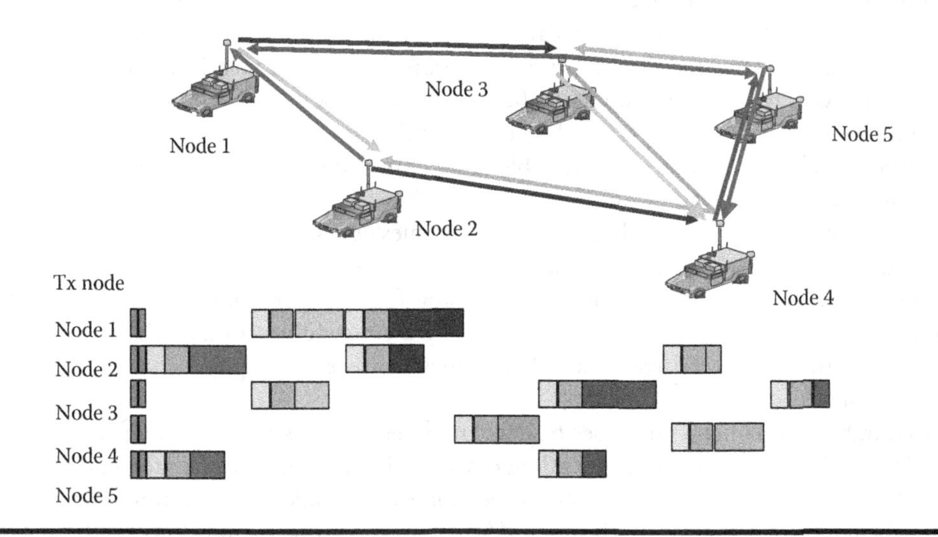

Figure 18.3 MANET operation.

network broadcast. This broadcast is preferentially transmitted by the air tier, but other network broadcast techniques, such as flood routing, could be used. In NND, the neighbors use search techniques and periodic "Hello" transmissions to discover each other. The Hello transmissions minimize the possibility of a hostile receiver intercepting node location by precluding any node transmission that has not first received a signal from a node that is already networked.

18.2.4 Timeslot Classes

The waveform uses a time division duplexing/time division multiple access (TDD/TDMA) channel that provides two classes of timeslots to carry network traffic, MAC-assigned traffic slots, and demand-assigned traffic slots.

When a neighbor is discovered, the nodes negotiate a time and frequency that will not interfere with other nodes and establish two MAC traffic slots to communicate over for the next epoch. These two slots provide a minimum of bidirectional bandwidth between two nodes that is used to request demand-assigned slots for user data, to pass positions for antenna pointing, to update timing advance values needed to overcome distance latencies, and to negotiate modulation or frequency changes to maintain the link quality. In addition, these slots are used to carry some user data.

The demand-assigned traffic slots are used only to send user data. These timeslots are assigned to grant extra bandwidth to the neighbors that need it based on traffic demands. In addition, distributing the traffic capacity in several slots over an epoch instead of concentrating it into a single slot per epoch reduces the average latency of the system.

18.3 Layer 2 Timing

The most recent version of the waveform introduces the concepts of a control epoch and superframe. These are illustrated in Figure 18.4.

The superframe is a relatively long time period which is used to time the exchange of data that does not require rapid updates. This information includes things such as routine status data and node position. The use of the superframe allows shared data fields in the data link layer messages.

The control epoch is the most significant period for the data link layer functions of data link control and MAC. MAC and demand timeslot assignments are defined in terms of the control epoch, as are beam pointing schedules. In addition, the neighbor discovery and contention-based synchronization protocol timing is based on the control epoch.

Control epochs are divided into a variable number of frames. The most recent version of the waveform uses 16 frames within the control epoch. These frames are used to distribute timeslots on a particular link to minimize latency. The first timeslot assigned to a link (the MAC slot as illustrated in Figure 18.4) is randomly assigned to a frame using interference avoidance rules. Subsequent timeslots for that link are assigned using frame preference rules to distribute them in time over the control epoch. These rules basically attempt to use a frame that does not have a timeslot for that link. Interference avoidance rules, however, have precedence over the frame preference rules.

MAC traffic slots contain node-specific data link layer messages followed by any user data that can be included in the allotted timeslot. All nodes are assigned to one MAC traffic slot per control epoch. Thus, the minimum capacity that can be assigned to a link is one minimum length MAC slot per the configured control epoch time. MAC timeslots can be extended in length to accommodate more user data, but it is preferred to provide additional slots to improve the link latency.

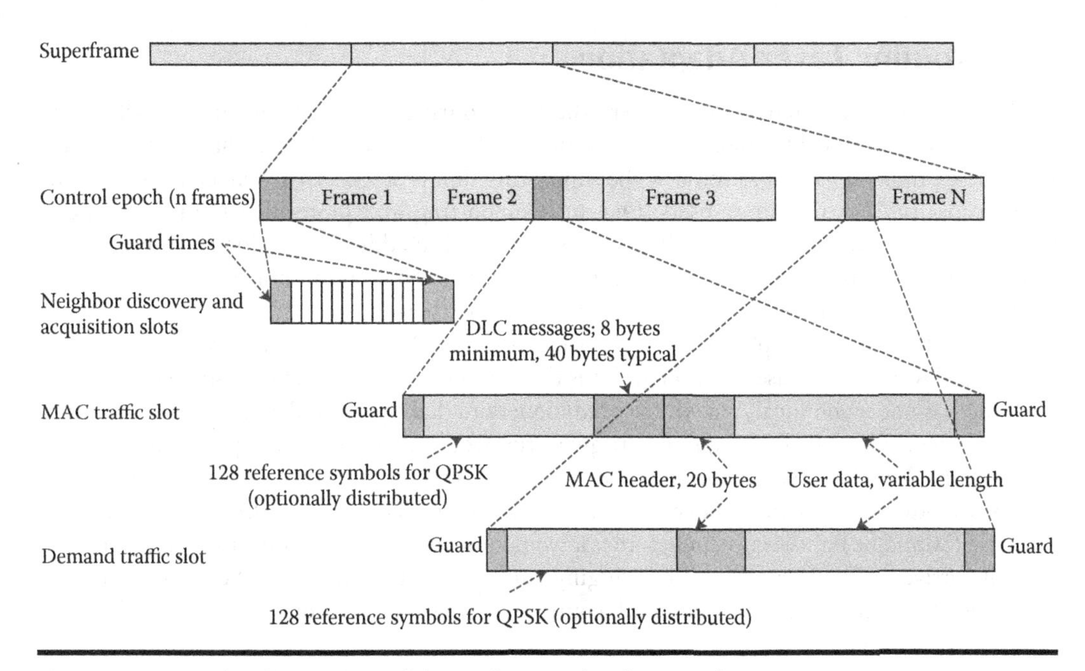

Figure 18.4 Timing hierarchy and frame formats for the waveform.

18.4 Network Throughput

The network throughput that can be achieved in the links that employ the waveform will vary depending on node locations and channel conditions. The node locations will determine the frequency reuse factor that can be obtained in the network. The channel propagation conditions will determine the data rate that can be sustained on each link. The tactical networking waveform data link layer includes dynamic power control and rate adaptation, which monitor the link quality and adjust the link power and modulation to maintain the highest channel data rate achievable under those conditions. The link power control and rate adaptation control are designed to achieve the required link availability and bit error rate.

The waveform uses a hybrid error control scheme that includes channel equalization, forward error correction, and data link layer automatic repeat request. The forward error correction and equalization functions are part of the physical layer.

18.4.1 Frequency Reuse

Frequency reuse improves the efficiency with which the network links utilize spectrum resources. As discussed, client terminals and peer terminals are assigned to noninterfering frequency channels and timeslots during the discovery operation. Co-frequency operation can be used for spatial isolation by means of beam directivity and geographic separation.

There is an additional low data rate mode for operation when operating in a noisy or stressed environment. The waveform is designed to operate in such an environment due to its use of directional antennas and TDD/TDMA in an ad hoc network. However, in some rare cases, this may not provide sufficient protection for particular nodes in an operation. In such cases, the radio can be commanded into a mode which provides substantial additional protection.

18.5 Routing Layer Adaptation

Traditional fixed-site wired networks were the norm during the development of much of the Internet routing protocol technology. The dynamic nature of tactical IP traffic has required analysis of these protocols as well as some of the supporting link protocols that were created to extend higher capacity links to remote users. One such supporting link protocol is the point-to-point protocol over Ethernet (PPPoE) [2]. PPPoE was developed to address a specific need in the commercial Internet world. At the turn of the current century, the digital subscriber line (DSL) service model had a problem. Potential equipment vendors and carriers saw that broadband cable modem or DSL would eventually replace dial-up service, but the hardware cost estimates for devices were well beyond what a home user would pay. Thus the initial focus was on small customers for whom a T1 line was not economical, but who needed more than dial-up could deliver. If sufficient number of customers bought DSL equipment, quantities would drive down the prices to where the dial-up home Internet users might be interested. PPPoE was initially designed to provide a small LAN with individual independent connections to the Internet at large, but also such that the protocol itself would be lightweight enough that it would not encumber the home usage market when it finally arrived. PPPoE succeeded in bringing sufficient volume to make the home user price acceptable. It remained the dominant DSL connectivity mechanism more than a decade later.

18.5.1 PPP over Ethernet (PPPoE) Extensions for Credit Flow and Link Metrics (RFC 5578)

PPPoE is a protocol for establishing and encapsulating sessions between hosts (clients) and traffic-access aggregators (servers) for PPP transport over real or emulated Ethernet [3]. PPPoE works well when both session endpoints have similar bandwidth, forwarding, and buffering capabilities that do not vary over time. As discussed, waveforms with variable bandwidth and limited buffering require extensions to PPPoE. RFC 5578 addresses these improvements with optional extensions to PPPoE that support credit-based session flow control and session-based link metric quality reports. These extensions are designed to support radio systems with point-to-point waveforms.

When the local client (radio) detects the presence of a remote radio neighbor, it initiates a PPPoE session with its local server (router). The radio also establishes a radio link connection with the remote radio over the point-to-point RF link.

The remote radio also establishes a PPPoE session with its local server (router). The radios associate the two PPPoE sessions and the point-to-point radio link protocol (RLP), creating a complete data path. Now a PPP session is established via the PPP IP control protocol (IPCP) as described in RFC 1661. Included in this IPCP exchange is the router IP address. With the exchange of the IPCP IP addresses, each router inserts the remote IP address into its local routing tables.

The PPPoE-based radio-to-router protocol includes messages that define how an external system will provide the device with timely information about the quality of a link's connection. They also include a flow control mechanism to indicate the amount of data the device can forward. The device can then use the information provided in the PPPoE messages to dynamically adjust the interface speed of PPP links.

For example, the tactical networking radio can provide ground-to-ground or ground-to-air communications using similar devices. When the radio picks up a signal from another device, it initiates a PPPoE session with a directly connected device (router). The PPPoE session encapsulates the packets that are relayed over a PPP link between the local and remote devices. The remote radio then forwards traffic to a remote device using an independent PPPoE session. The two devices exchange link control protocol (LCP) and IPCP messages to configure the link and exchange open shortest path first (OSPF) messages to establish the network topology.

Each radio monitors the link for changes in the link bandwidth, quality, and utilization. If any changes are detected, the radios announce the new set of metrics to the respective devices through a PPPoE active discovery quality (PADQ) message, which is an extension to the PPPoE discovery protocol (RFC 2516). The device transforms these metrics into a bandwidth value for the PPP link and compares it to the value currently in use. When the device detects that the difference exceeds a user-specified threshold, it adjusts the speed of the PPP link. OSPF is notified of the change and announces any resulting routing topology changes to its neighbors. This PPPoE interaction between the radio and the router is often termed "radio aware routing" (RAR).

18.5.2 OSPF with MANET Extensions

From an architecture perspective, the waveform is formed at layers 1 and 2 in the Open Systems Interconnection protocol stack. Layer 2 is called the U-MAC (universal media access control) because it has been designed to universally link various routing (layer 3) and physical (layer 1) layers together. Currently, the waveform interfaces with two open architecture layer 3 routing protocols: (a) optimized link state routing (OLSR) and (b) OSPF with MANET extensions. This section discusses the use of OSPF in tactical radio networks.

To optimize the use of OSPF with MANETs, extensions to OSPFv3 were developed resulting in a well-understood routing protocol (OSPF) used in a network topology that is constantly changing and where bandwidth is limited.

OSPF is optimized to tightly couple OSPFv3 with RAR-compliant radios to provide faster convergence and reconvergence through neighbor presence indications and to help determine accurate, real-time link metric costs. Optimizations also include the following: minimizing OSPFv3 packet size by implementing incremental hellos, minimizing the number of OSPFv3 packet transmissions by caching multicast link-state advertisements (LSAs), optimizing overlapping relay functionality to minimize the number of flooded LSAs, and implementing selective peering to reduce the OSPF network overhead by minimizing the number of redundant full adjacencies that an OSPF node maintains.

18.5.3 Radio-Aware Link-Metrics Tuning for OSPF

A RAR-compliant radio reports link-quality metrics to the router and these are used by OSPFv3 as part of the link metrics. Network management adjusts the usage of radio metrics by OSPFv3 in two steps. First, the system engineer with network knowledge configures how the radio-reported bandwidth, latency, resource, and relative link-quality (RLQ) metrics are converted to an OSPFv3 link cost. Second, a hysteresis threshold is configured on this resultant link cost to minimize the propagation of LSAs that report link-metric changes.

OSPFv3 receives raw radio-link data and computes a composite. In computing these metrics, the following factors are considered:

- Maximum data rate—the theoretical maximum data rate of the radio link, in bytes per second
- Current data rate—the current data rate achieved on the link, in bytes per second
- Resources—a percentage (0–100) that can represent the remaining amount of a resource (such as battery power)

- Latency—the transmission delay packets have encountered, in milliseconds [4]

  Latency in tactical networks affects mission achievability. The mission success depends on the ability to access data in "real time" and to predictably make traffic decisions faster than an adversary, or even faster than an adversary can even assess. Thus, using metrics such as hop count or cost as routing metrics is becoming less important. Rather, it would be beneficial to be able to make path selection decisions based on performance data (such as latency) in a mission timely and scalable way.
- RLQ—a numeric value (0–100) representing relative quality, with 100 being the highest quality

Metrics are weighted during the configuration process to emphasize or de-emphasize particular characteristics. For example, if throughput is paramount, weighting the current data rate metric factors more heavily into the composite metric is done. Similarly, metrics that are of no concern from the composite calculation can be weighted with a zero multiplier.

Link metrics can change rapidly, often by very small degrees, which can result in a flood of suboptimal routing updates. In a worst-case scenario, the network churns almost continuously as it struggles to react to minor variations in link quality. A tunable dampening mechanism to configure threshold values allows any metric change below a given threshold to be ignored.

The tunable hysteresis mechanism adjusts the threshold to the routing changes that are propagated when the router receives a signal that a new peer has been discovered, or that an existing peer is unreachable. The tunable metric is weighted and is adjusted dynamically to account for these conditions.

18.5.4 Emerging Topology Management Functionality

As the tactical networking waveform evolves, enhanced capabilities are designed, tested, and either included or not, depending upon market considerations. One interesting capability is advanced topology management.

As discussed earlier, ad hoc networks that employ directional antennas combined with TDMA scheduling significantly increase the scalability of MANETs [5,6]. In DTDMA MANETs, each node executes a non-interfering schedule of transmissions and receptions that exploit the spatial isolation available from the directional antenna to greatly increase the frequency reuse of the network. It has been shown [7] that if the antenna beamwidth is narrow enough and the antennas have adequate sidelobe control, the interference limits are essentially removed and the remaining theoretical constraint on scalability is caused by the per-hop load multiplication that occurs in any multi-hop packet switched network.

Managing the topology of a MANET by judiciously selecting the neighbors of each node allows the network to achieve an optimum balance of throughput, overhead, latency, and robustness [8–11]. The use of a DTDMA protocol increases the complexity and overhead at the data link layer, but provides more capability to manage the topology [12,13] to achieve the best network performance.

A tactical directional MANET (DMANET) has dynamic operation that requires a distributed topology management mechanism that can select in near real-time a nearly optimum set of connections for each node, strictly based on locally available information. Optimization methods that could be employed in such a mechanism include the genetic algorithm (GA) and the set-based particle swarm optimization (SBPSO) algorithm. The relative merit of each algorithm is how quickly (in terms of processing) each converges to a suitable selection of a set of neighbor links.

For topology management, every node in the network defines a local neighborhood, which contains as a minimum all of the other nodes within its "interference range." The topology manager at a given node selects a subset of the nodes in its neighborhood such that the subset maximizes a neighborhood topology metric (NTM) and initiates a link establishment process with the selected nodes in the subset. A resolution protocol is included for cases where this "greedy" approach results in disagreements between nodes on the desirability of link formation.

The topology management optimization problem is a special case of the well-known 0-1 knapsack problem, which has been widely studied and is known to be NP complete. Both GA and SBPSO algorithm are known to be effective for problems of this type. The ability of both algorithms to select a near optimal set of connections in a distributed, real-time topology manager for a directional ad hoc network is a key result for tactical DMANETs.

Both algorithms have been seen to be capable of successfully choosing a high scoring link set. With the NTM calculation limit, the GA consistently outperformed the SBPSO algorithm. However if the limit was relaxed for the SBPSO algorithm search so that the swarm size could be increased by a factor of 2 or more, the SBPSO capability becomes essentially the same as the GA. On the other hand, in tests, the GA performed adequately with less than 2000 NTM calculations per search, whereas SBPSO required at least 10,000 NTM calculations per search before it began to find acceptable solutions. Hence, the GA had a substantial processing load advantage over SBPSO algorithm.

Low internode communications overhead is key for an efficient, scalable cognitive network algorithm. As the number of nodes in the network increases, the amount of coordination communications can grow at a much faster rate than the node count. Bee colony particle swarm optimization (PSO) has the useful feature of keeping the number of communication interchanges low by updating just the local neighbor group. Optimized topology management continues to be researched and developed as the waveform continues to evolve.

18.6 Summary

The increasing demand on spectrum resources has heightened the need for more spectrally efficient radios. The directed-beam antenna is an enabler for significant gains in this area, but such directive-beam antenna radio networks require intelligent adaptations to protocols at the physical, link, and network layers. The tactical networking waveform adapts the radio network to use these extensions to provide a highly capable, dynamic tactical networking capability.

References

1. BelAir Networks, *Capacity of Wireless Mesh Networks*, White Paper, 2006, http://www.belairnetworks.com
2. RFC 2516 A Method for Transmitting PPP Over Ethernet (PPPoE), https://tools.ietf.org/html/rfc2516
3. RFC 5578 PPP over Ethernet (PPPoE) Extensions for Credit Flow and Link Metric https://tools.ietf.org/html/rfc5578
4. https://tools.ietf.org/html/draft-ietf-ospf-te-metric-extensions-08
5. Cain, J. B., T. Bilhartz, and L. Foore, 2003. A link scheduling and ad hoc networking approach using directional antennas. In *MILCOM proceedings. IEEE 2003 Military Networks Conference*, October 2003, Monterey, CA, USA.

6. Olds, K., 2005. The high-band networking waveform: A directional ad hoc networking waveform for battle-command-on-the-move. In *MILCOM proceedings. IEEE 2005 Military Networks Conference*, October 2005, Atlantic City, NJ, USA.
7. Li, P., C. Zhang, and Y. Fang. The capacity of wireless ad hoc networks using directional antennas. *IEEE Trans. on Mobile Computing*, October 2011, 10(10), 1374–1387, ISSN: 1536–1233.
8. Hideaki, T. and L. Kleinrock, 1984. Optimal transmission ranges for randomly distributed packet radio terminals. *IEEE Trans. on Communications*, March 1984, 246–257, ISSN: 0090–6778.
9. Ni, J. and S.A.G. Chandler, 1994. Connectivity properties of a random radio network *IEE Proceedings on Communications*, August 1994, 141(4), ISSN 1359–7019.
10. Timo, R., and L. Hanlen, 2006. MANETs: Routing overhead and reliability. *Vehicular Technology Conference Proceedings, 2006 IEEE Vehicular Technology Conference*, May 2006, Melbourne, Australia.
11. Gowrishankar, S., T. G. Basavaraju,, D. H. Manjaiah, M. Singh,and S. K. Sarkar, 2008. Theoretical analysis and overhead control mechanisms in MANET: A survey. In *Proceedings of the World Congress on Eng.* July 2008, London, UK.
12. Feng, X. and P.R. Kumar, 2004. The number of neighbors needed for connectivity of wireless networks. *Wireless Networks* 10, pp. 169–181.
13. Martin, D., P. Guangyu, and H. K. Jae, 2007. Topology formation in degree-constrained directional antenna networks. In *MILCOM proceedings. IEEE 2007 Military Networks Conference*, October 2007, Orlando, FL, USA.

Chapter 19

Military Applications of Directional Mesh Networking

Todd Mcintyre, Marc J. Russon,
Stephen M. Dudley, and Victor Wells

Contents

19.1 Introduction

Situational awareness, command and control, and the role of timely information have always been essential to the outcome of wars. Though the implementation specifics of network-centric warfare (NCW) have transitioned significantly since its conceptual introduction in the late 1990s [1], it remains perhaps the defining transformation in military affairs of the last decade. NCW "translates information superiority into combat power by effectively linking knowledgeable entities in the battlespace" [2]. In short, military command and control and thus warfighter effectiveness is optimized by getting the right information to the right entity at the right time. The process of creating a network that spans over increasingly large areas of operation, sometimes globally—from rear command echelons to the edge of the fighting area—is often referred to in defense parlance as the global information grid (GIG) [3,4]. Recent global conflicts have demonstrated just how important information flow is in the conduct of modern warfare and the bandwidths necessary to support the information flow. Key to achieving the vision of the GIG is the need to provide robust, high capacity communications networks at all security levels. Air-to-air, high capacity, directional mesh networking is being pursued by defense strategists. One example is the joint aerial layer network (JALN), illustrated in Figure 19.1 [38]. JALN defines a top level architecture critical for solving battlespace connectivity by providing three critical functions: a ubiquitous high capacity backbone (HCB); distribution, access and range extension (DARE), and network transition (gateway) functions. This chapter addresses the unique challenges associated with directional mesh networks and how they differ from commercial networks.[*]

[*] This information consists of L-3 Communications Corporation, Communication Systems-West Division general capabilities information that does not contain controlled technical data as defined within the International Traffic in Arms Regulations (ITAR) Part 120.10 or Export Administration Regulations (EAR) Part 734.7–11.

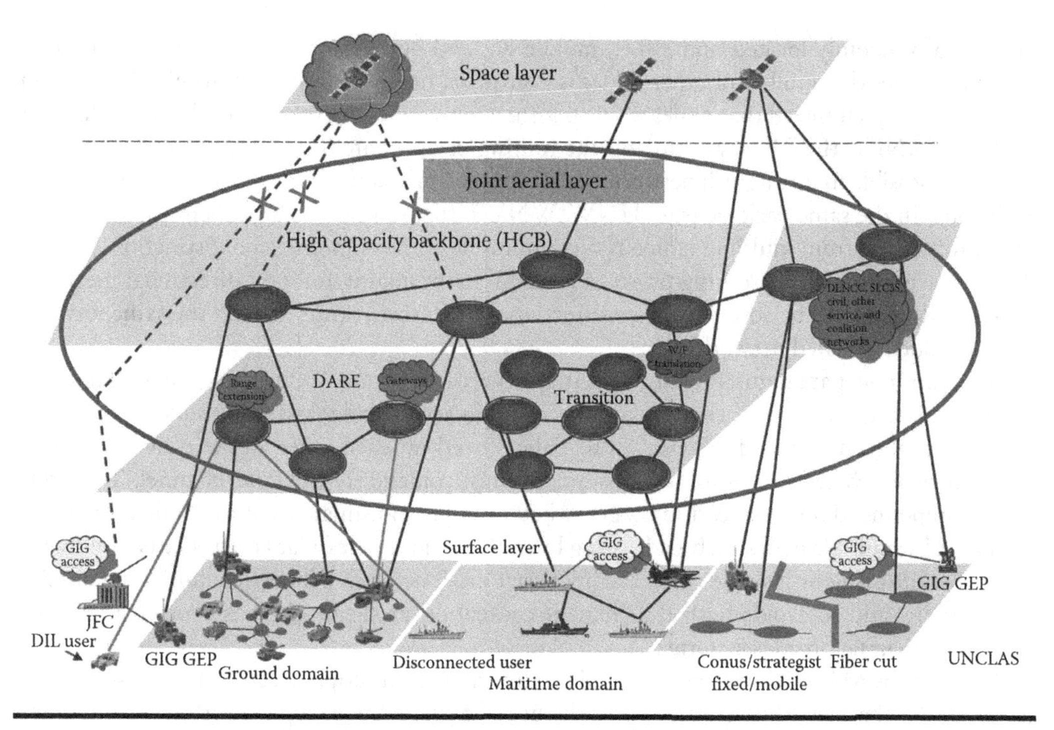

Figure 19.1 Joint aerial layer network (JALN).

19.2 Summary of Benefits of Directionality

Networking, in a military context, has to accommodate the expeditionary nature of the military use cases. Military wireless systems are often called on to provide both local connectivity with neighbors, and long distance connectivity to wired networks. Military systems have the need for higher data rates, at greater distances, than commercial systems. Military systems require wireless data transmission of gigabits per second (Gbps) at ranges of hundreds of kilometers.

The use of high-gain directional antennas significantly reduces the power required to form a data link. It also reduces the spectrum conflicts with neighbors, resulting in the ability to support more users per area. Using directional antennas limits propagation of Radio Frequency (RF) energy and results in more secure communication.

This chapter examines various aspects of a directional mesh network based on common military use cases. The chapter begins by providing applicable use cases. It describes unique aspects and approaches for implementing the physical layer, data link layer, and the network layer. This description is followed by a discussion of ways to reduce complexity. Finally, the chapter discusses problems that need to be resolved to meet future military networking needs.

19.3 Key Differences between Military Directional Mesh Networks and MANET

A great deal of research has been conducted into mobile ad hoc networks (MANET) systems. Here the focus has been on achieving a bottom-up self-organization of networks from a random number of randomly located nodes (i.e., mobile with ad hoc network formation). In a military context, neither the number of other nodes is a surprise, nor is their location (typically) random, nor is it 100% guaranteed that nodes want to form a network with other nodes that they discover.

In a MANET, the RF waveform and the routing protocol are often cooptimized so that it is almost impossible to distinguish between the RF waveform and the routing protocol because they both come in the same package (e.g., HNW, WNW). In a military context, being able to aggregate information from multiple radios is a common need. Maintaining the distinction between the RF waveform and the routing protocol is useful in evaluating military directional mesh networking. Routers can be added to many directional mesh scenarios in exactly the same way that they are added to wired networks.

The common paradigms for mobile wireless networking are WiFi and cellular. Here, the mobile node can always hear the access point or radio access node, and if multiple access points are present, all can be heard. Having all neighbors overhear each other all the time has costs as well as benefits. When they want to transmit, all must contend for the same channel. This problem is compounded for broadcast messages, when multiple transmissions of the same message are overheard by multiple nodes. Rebroadcasting by multiple nodes uses a far larger slice of bandwidth than rebroadcasting by a single node. A directional mesh network deliberately limits the number of neighbors, and so avoids both the channel contention problem, and has a much less severe problem managing broadcast storms.

Many of the MANET systems are based on time division multiple access (TDMA) scheme for accessing the channel. This scheme allows the use of a single frequency for both transmitter and receiver, and allows almost infinite variation in allocating channel access between multiple users or between the transmitter and receiver function. Such a scheme might be considered necessary for network formation because it resolves the rendezvous problem, that is, how a node discovers its neighbors. However, many of these benefits accrue only at short ranges. As the (time) diameter of the network increases, the guard time between time slots must increase proportionally. The prudent response to maintain link efficiency is to increase the time slot duration. However, when allocating traffic between multiple users, that also increases the variable packet delay (jitter). For real-time control traffic, jitter is so important that TDMA cannot be used in many larger network circumstances.

The frequency division multiple access (FDMA) scheme assigns different transmitter and receiver frequencies to different users. One of the typical ways of describing this is as a "gender" since with each pair of nodes, the transmitter and receiver frequencies must be selected so that when one node transmits on one frequency, the other nodes receive on that frequency. The choice of the word gender would suggest that only two are possible. In fact, multiple genders are possible, and directional mesh networks sometimes exploit multiple gender solutions to enhance connectivity and system performance.

19.4 Platform Use Cases

Several topologies can be supported using directional antennas as shown in Figure 19.2.

There are several use cases that can utilize directional mesh networks. We highlight three of them here. These include backhaul/relay networks, tactical mesh networks, and airborne radio access networks. These have been chosen to illustrate different aspects of directional mesh networking. Some general characteristics are highlighted in Table 19.1.

The circumstances that drive toward these use cases are described in the following sections.

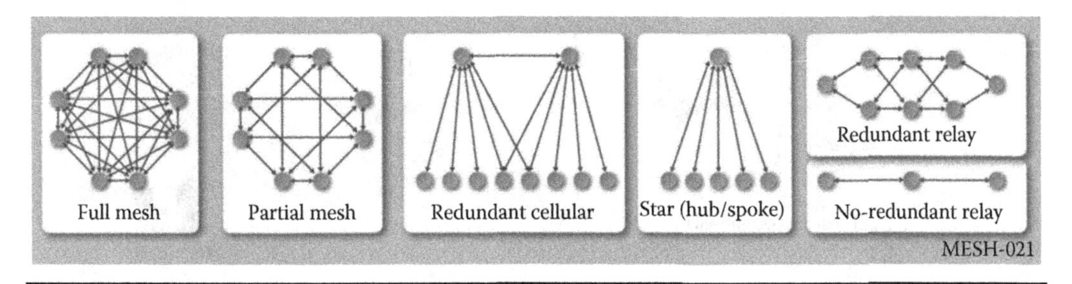

Figure 19.2 Examples of directional mesh topologies.

Table 19.1 Characteristics of Use Cases

Use Case	*Routing Domains*	*Users*	*User Mobility*	*Traffic*	*Range*
Backhaul/ relay	Multiple and isolated	On-platform or tethered ground stations	User tethered to platform/ ground station	Encapsulated messages from constituent networks	Long
Tactical mesh	Single	On-platform	User tethered to platform	Native messages	Short
Airborne radio access network	Single[a]	Off-platform	Mobile from platform to platform	Encapsulated messages	Short

[a] Multiple routing domains could be supported, provided that there are multiple independent authentication and authorization processes, and a backhaul/relay-like network ability to maintain isolation between these networks.

19.4.1 Backhaul/Relay for Airborne Networks

The backhaul/relay use case is based on the need to move information between platforms, and from a platform to a command center where it can be exploited. Both airborne and ground stations can be deployed, but the ground stations behave just like another platform in terms of maintaining the network connectivity. High data rate links and long ranges are often involved.

Figure 19.3 illustrates the encapsulation of traffic from multiple independent networks into the backhaul network. Multiple independent networks can coexist (*dark gray* with dotted lines and *gray* with solid lines). Each is maintained as an isolated routing domain. Traffic from each network is encrypted and encapsulated into messages that are native to the backhaul/relay network. This guarantees the confidentiality of messages within each routing domain. Both airborne and ground platforms may participate in the backhaul/relay network. One or more connections to the wired network may participate in the network. These ground entry points provide a gateway between the airborne network and the wired network on the ground. At some point in the wired network, the traffic from each of the backhaul/relay streams is extracted, decrypted, and delivered to the network to which it belongs.

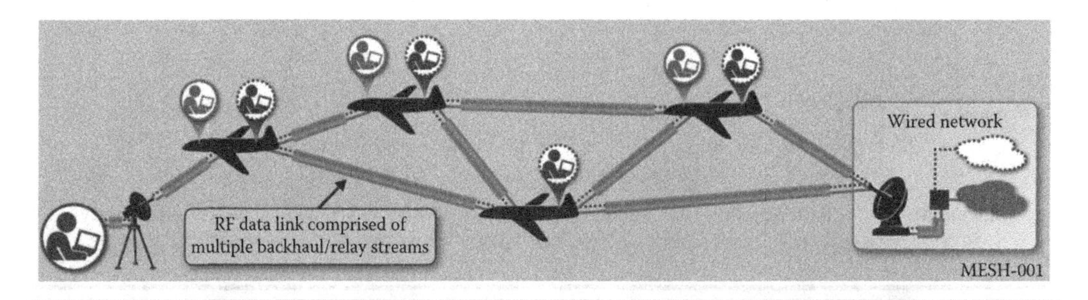

Figure 19.3 The backhaul/relay use case takes traffic from networks on airborne platforms and relays it back to a ground entry point where it can connect to the network to which it belongs.

19.4.2 Tactical Mesh Networks

The tactical mesh use case is based on the need to coordinate the activities of a single group of platforms. Here the network belongs to the network to which the user group belongs. There is no need to encapsulate messages for transport. There is a single routing domain. Although distances could be large, it is more common that distances are relatively small (tens of kilometer as opposed to 100s of kilometer). Tethering the network to a backhaul network or a ground entry point is not a permanent feature of the network. It may be supported from time to time but the expectation is that it operates completely the same whether a connection to the wired network world exists or not.

Figure 19.4 illustrates a typical tactical mesh scenario. Each of the airborne platforms participates, and a single user community on the airplane is supported by the network. Note that a tactical mesh network and other forms of networks may coexist on the same airplane when they have separate antennas and operate at different frequencies.

19.4.3 Airborne Radio Access Networks

An airborne radio access network can be thought of as a collection of cell towers in the sky. It is based on the concept of placing a radio access network on each airplane, much like each cell tower has a radio access network. The idea is the same as for a traditional cellular deployment, to allow mobile users to get access to the network. In this case, the infrastructure is mobile. Relatively high data rates are needed to support the aggregate rate of a large number of users.

Figure 19.5 illustrates the creation of a network from a collection of airplanes and ground vehicles. The airplanes and ground vehicles create and maintain a network for transport of traffic. Each

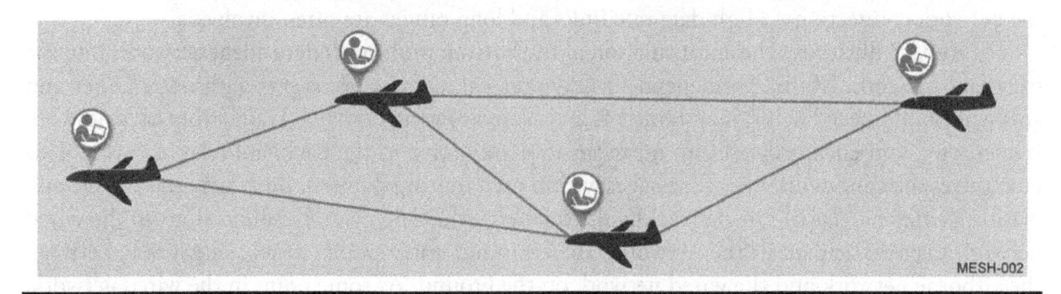

Figure 19.4 The tactical mesh use case allows for the formation of a network between multiple platforms.

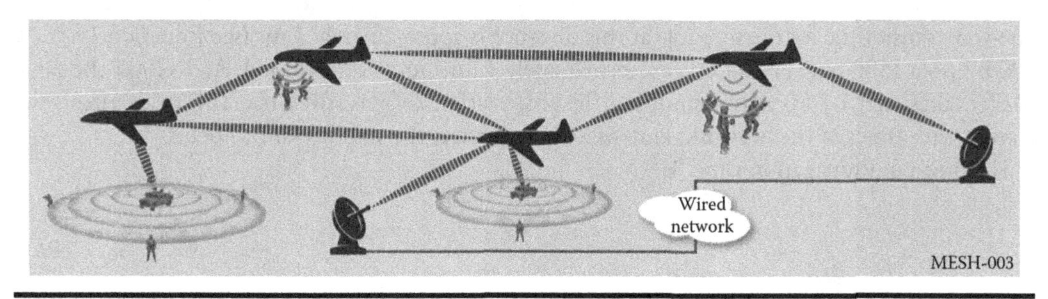

Figure 19.5 An airborne radio access network provides a cellular or cellular-like service to mobile users.

of the mobile users could get network access through any of the platforms. As with the backhaul/relay use case, there is almost always a connection to the wired network. Following the model of a cellular network, this network could require that all users be subscribers to a single service, and so be able to reach all other users on the system by means of a location service. However, since the underlying network is based on the model of encapsulating user traffic (a requirement to manage user mobility), it is also possible to support multiple independent routing domains, each with their own authentication and authorization processes. Under the multiple network model, traffic would be transported over the same RF links but otherwise the two networks would be unaware of each other's presence.

19.4.4 Use Case Summary

Three use cases have been presented. Backhaul/relay systems typically require high-data rates and long-range performance; so, they rely on directional networks. Tactical mesh networks typically support moderate to short internode distances, but can still benefit from directionality by reducing power, and utilizing spatial isolation to increase spectral efficiency. Airborne radio access networks also benefit from the higher data rates and improved spectral efficiency provided by systems with directional antennas.

19.5 Physical Layer Considerations

There are challenges that that must be overcome to maintain a network that simultaneously supports high-data rates and long-range operation. Long-range operation introduces more network latency than typical network protocols can support. Aircraft motion introduces Doppler shifts in the transmitted RF signals, which limits the type of modulation and coding schemes that can be used by nodes in the network. The location of the antenna on the airplane has a significant impact on network topology and performance because the platform will block the RF connection between nodes when certain maneuvers are performed. Finally, getting antennas to track their targets reliably can be a problem. Each of these challenges is described in the sections that follow.

19.5.1 Latency Due to the RF Propagation Time

One of the consequences of long-range RF connectivity is network delay introduced because of the wave propagation time. Airborne network nodes can have separations of up to 300 km. The one-way transit time for the message at this distance is approximately 1 ms (see Equation 19.1). At 100 mbps, a 1500 byte packet takes approximately 12 ms to be transmitted. At 1 Gbps, the same packet takes only 1.2 ms to transmit, but the propagation delay is still 1 ms. This delay affects the convergence times of the network, and, as a consequence, the stability of protocols.

Definition: Propagation time [1]

$$t = \frac{d}{c} \tag{19.1}$$

t = propagation time
d = distance
c = speed of light

19.5.2 Doppler

One of the effects of mobility is of the introduction of frequency shifts in the transmitted signal do to Doppler (see Equation 19.2). As platforms move toward each other, the frequency appears to increase, and as they move away from each other the frequency appears to decrease. This affects the choice of waveform that can be used to provide physical layer connectivity. Waveforms that have good performance in a multipath environment (e.g., Orthogonal Frequency Division Multiplexing (OFDM)) rely on the inherent nulling of the interference from the adjacent carrier in a multicarrier modulation. When there is Doppler, the theoretical nulling does not occur, which introduces self-interference and reduces range and data rate. Compensating for Doppler is a significant design driver for some mobile platforms.

Definition: Doppler shift [2]

$$f' = \frac{f \cdot c}{(v + c)} \tag{19.2}$$

f' = Doppler-shifted frequency
f = emitted frequency
c = speed of light
v = relative speed of two platforms

19.5.3 Antenna Placement

Building a directional mesh network implies having not just one, but perhaps many, antennas. There is a limit to the number of antennas that can be mounted on a single platform. For ground vehicles, there are a limited number of suitable mounting locations. Weight and cosite interference can be an issue. Airborne platforms have all of these issues, plus the need to preserve the platform's aerodynamics. The aerodynamics issue arises because the antenna needs to be mounted where it has a good view in the direction of the intended receiver/transmitter. In fact, safety of flight problems arises when the antenna causes a bump on the aircraft so large that it adversely affects the flight characteristics of the airplane.

Notwithstanding the great strides that have been made with electronically steerable arrays, geometry can often be a problem. When the intended receiver/transmitter is at a low angle to the mounting plane of the antenna, the effective illumination area of the antenna is very low, so antenna gain drops. Putting a radome on the aircraft to house a directional antenna is often the only solution. Where this affects the directional mesh solution is that there are a limited number of suitable mounting locations for a radome.

There are two metrics used to identify the area that the antenna can see: (1) the field of view is defined as the area that the antenna can "see" given the pointing direction. Typically, this is assumed to be at an arbitrary pointing direction and (2) the field of regard is defined as the area that the antenna can see at all possible orientations.

The field of regard is often represented by what is called a shadow mask. An example of a shadow mask is given in Figure 19.6. The shadow mask is specific to the antenna-mounting location on a platform. The antenna-mounting location that produced the shadow mask in Figure 19.6 is given as the *gray* dot shown in Figure 19.7. To interpret the plot, note that the 180° azimuth angle is facing toward the back of the airplane. Given that the mounting location is on the chin of the aircraft, the 0° elevation angle is blocked by the aircraft. Pointing angles of less than −13° elevation are needed to clear the aircraft at the 180° azimuth pointing angle.

Starting from the 180° azimuth angle, and moving toward zero degrees azimuth, the platform blockage is first from the fuselage behind the mounting location (to about 170° azimuth), then by the wings (to about 140° azimuth), and finally by the fuselage in front of the mounting location (to 0°). The shadow mask is symmetrical, since the mounting location is on the center line. At no point can the antenna have an elevation pointing angle higher than 24° without platform obstruction.

In practice, all antenna pointing angles must be referenced to the pitch, roll, and yaw of the aircraft. The pitch is the least intuitive, since "level" flight (i.e., neither climbing nor descending) often means an upward pitch angle of 3–5°. The pitch angle is different for different speeds and engine power levels, so the zero pitch angle must be understood when creating an antenna shadow mask.

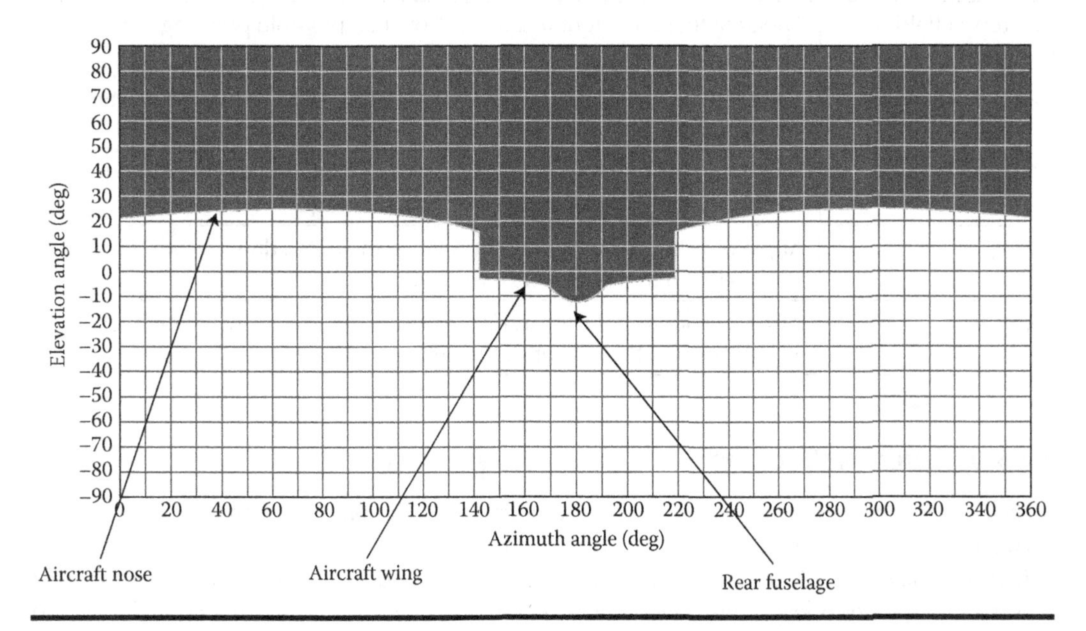

Figure 19.6 This antenna shadow mask represents the field of regard of the antenna placed on the platform in Figure 19.7. (Adapted from DISA, [Online]. Available: http://www.disa.mil/services/enterprise-engineering.)

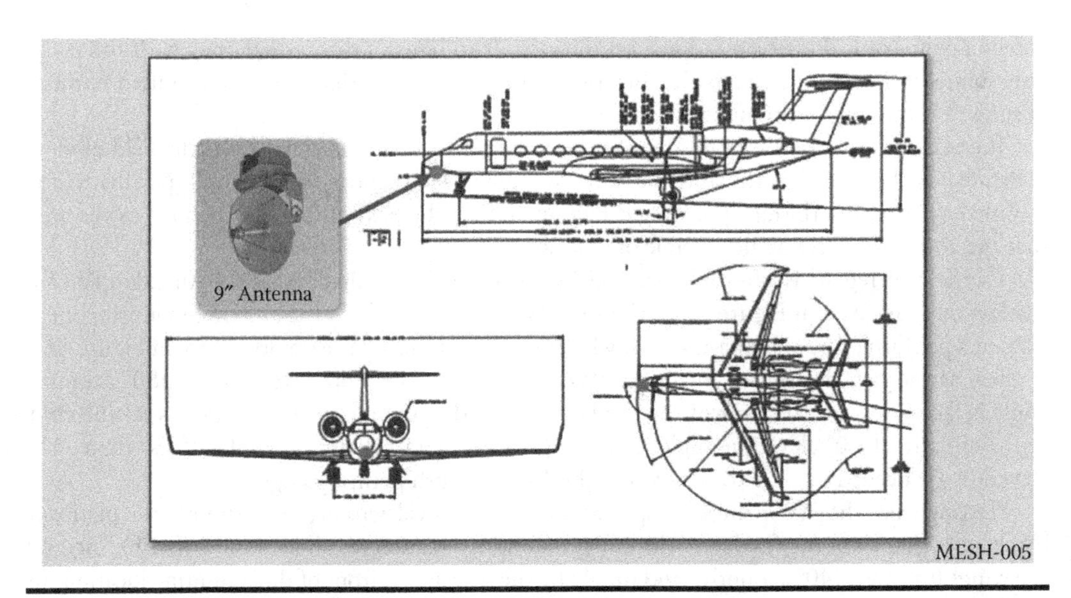

Figure 19.7 The placement of an antenna affects its field of regard. Placing an antenna at the location indicated in *gray* results in the antenna shadow mask shown in Figure 19.6.

Term	*Definition*
Antenna field of view	The area that an antenna can "see" when pointed to a specific position
Antenna field of regard	The area that an antenna can "see" over all possible pointing angles
Antenna shadow mask	A representation of a single antenna's field of regard by azimuth and elevation angle. Typically, this is illustrated on a rectangular grid
Pitch	Angle of rotation of the aircraft relative to an axis drawn through the center of the wing attachments to the fuselage
Yaw	Angle of rotation of the aircraft relative to an axis drawn from bottom to top through the center of the aircraft
Roll	Angle of rotation of the aircraft relative to an axis drawn from back to front through the center of the aircraft

19.5.4 Antenna Tracking

Mobile directional mesh networks have the problem that the platform can obscure the path between nodes. Maintaining antenna tracking to a remote node is not just a question of an individual antenna being able to track the location of the remote node. The antenna tracking function moves to the platform level, and some aspects of antenna pointing, such as which nodes should establish links with which other nodes become network functions.

At the level of the individual antenna, periodically exchanging information about location, speed, and direction of the platform allows the antenna to be deliberately pointed. Pointing significantly improves data link performance over simple tracking, even though the distance that the node travels appears to be quite small.

Example: An airplane traveling at 960 km/h (600 mph) at a distance of 300 km has about 1 ms delay between the time that information is sent, and that information is received. To point the antenna at where the node will be in 1 ms from that time, the error in location would only be 0.5 m.

The issue is not the error in the accuracy of the actual location, but rather, the error in the measurement process. With the advent of global positioning system (GPS) systems, the actual location of the node can be known to within a few meters. However, antenna tracking requires, effectively, the continuous testing of a hypothesis about which shoulder of the antenna pattern is receiving the signal. With both azimuth angle and elevation angle as degrees of freedom, this is not a simple task, and designing algorithms that can reliably distinguish between moving away and moving across the antenna pattern is difficult. So, the convention is to periodically exchange location, speed, and direction information.

Term	*Definition*
Slew rate	The maximum rate at which an antenna can be repointed. This is typically given in degrees of rotation per unit of time

19.6 Data Link Layer Considerations

Perhaps the single most important characteristic of a data link to an RF engineer is the trade between RF power, range, and data rate. Only two of these can be independently selected. Given this trade, the limits on range are affected mostly by how high one can fly, and the curvature of the earth. Supporting multiple links in the same vicinity requires careful spectrum management. Putting multiple links on the same airplane requires careful cosite interference management. The need to manage cosite interference also leads to the use of different frequencies or genders, which deals with how to prevent transmitters and receivers on the same platform from using the same frequency. The overall question of how to share time, frequency, and space resources between users, and between the transmitter and receiver functions of the data link is considered when choosing both the multiple access schemes and the duplexing schemes for the data link. Each of these items is considered in the sections that follow.

19.6.1 RF Power, Range, and Data Rate

Wireless networks are governed by the laws of information theory. For a given bandwidth and signal-to-noise ratio, there is a maximum amount of information that can be decoded. These limits are described by the Shannon–Hartley equation. Although no practical modem implementation achieves these information limits, many come within 1 dB of them. This is close enough that the form of the equation can be used to describe the behavior of the communication process in wireless networks.

Information capacity [4]

$$C = B\log_2\left[1 + \frac{\text{Signal}}{P_{\text{Noise}} + P_{\text{Interfere}}}\right] \tag{19.3}$$

C = Capacity in bits per second
B = Bandwidth in Hertz
Signal = Signal power
P = Power (noise or interference)

The ratio of signal to interference and noise power (SINR), and bandwidth of the signal govern the data rate that can be achieved. At lower data rates, the SINR needed to decode the signal drops. The received power of the signal at the intended receiver (PRX) is affected by the transmit power level at the transmitter (PTX), the nature of the transmit antenna (gain multiplier GTX), the propagation medium (loss at distance d: $L_{\text{path}}(d)$), the distance from the transmitter (d), and the nature of the receive antenna (gain multiplier GRX). If we assume that the losses due to the medium are constant with distance, this relationship can be seen from the Friis transmission equation.

Friis transmission equation modified from Reference 5 is given hereunder:

$$P_{\text{RX}} = P_{\text{TX}} * G_{\text{TX}} * G_{\text{RX}} * L_{\text{path}}(d) \tag{19.4}$$

P_{RX} = Received power
P_{TX} = Transmitted power
G_{TX} = Transmit antenna gain
G_{RX} = Receive antenna gain
$L_{\text{path}}(d)$ = Loss fraction at distance d

The form of these two equations can be used to see the trade space that designers of wireless systems have to work with. Given the medium through which the RF signal propagates, they cannot simultaneously choose the RF power of the transmitter, the range at which they operate, and the data rate. They can, however, choose any two and derive the value of the third.

One of the challenges faced by designers of wireless systems is that the loss function, $L_{\text{path}}(d)$ is highly nonlinear. Even if we assume a propagation channel that has no absorption, it is not possible to keep the cross section of the RF energy constant over distance. The best possible case is that the spread of energy is linear in two dimensions. This best case is the spreading of power over an area that increases as the square of distance. When accommodating other losses in the system, such as absorption, or scattering, or diffraction, the loss function can be much worse than this. A common shorthand to represent the sum of all loss mechanism is by representing them as a path loss coefficient (α) to the ratio of distance. In Equation 19.5, the spreading of power over a reference sphere is suggested by the first component, and the second component, which is raised to the power of α suggests the effect of distance from that reference. If there were no losses, α would be 2, and that is the general assumption for air to air data links, but for ground links the range of values of α is between 3 and 8.

Path loss

$$L = \left(\frac{\lambda}{4\pi d_{\text{ref}}}\right)^2 \left(\frac{d_{\text{ref}}}{d}\right)^\alpha \tag{19.5}$$

L = Loss
λ = Wavelength in meters

d_{ref} = Reference distance for measuring RF power
d = Distance from transmitter

Designers of wireless systems that wish to increase the range or data rate of their system without increasing the transmit power only have choice regarding the type of antenna used on the transmitter, and on the receiver. However, the gains associated with choosing directional antennas can be large under right conditions. In general, the size of the antenna needed to achieve a given amount of gain is highly dependent on the frequency of operation. The relationship between the effective area of an antenna, the operational wavelength, and maximum antenna gain is shown in Equation 19.6.

Antenna effective area [6]

$$A_e = \frac{\lambda}{4\pi} G \tag{19.6}$$

A_e = Effective area of antenna
λ = Wavelength in meters
G = Antenna gain

If we assume equal effective areas for an antenna at 1 and 15 GHz, with wavelengths of 0.3 and 0.02 m, respectively, then the relationship between maximum gain is 0.02/0.3 = 0.0666, or about −11.76 dB. At 1 GHz, a directional antenna would have −11.76 dB less gain than the same size antenna at 15 GHz.

Military applications often have the requirement for longer ranges, even if they do not have a requirement for higher-data rates. The benefits of a higher-gain antenna and smaller size and weight are significant. So, in spite of the poorer propagation conditions, military systems often operate at higher frequencies.

19.6.2 Curvature of the Earth

At all of the frequencies of interest for airborne communications, path loss through the earth itself is effectively infinite. So, the curvature of the earth provides a limit on the range, power, and data rate trade space. The relationship between height of an airborne platform, distance to an earth terminal, and slant angle is shown in Figure 19.8.

19.6.3 Spectrum Management

A significant reason to use directional antennas is their impact on spectrum management. Although the formulas described in Section 19.6.2 would suggest that the designer is free to increase RF power, this action has a cost when there are more than two users of the spectrum in the same area. From Equation 19.4, a decrease in transmitted power has a linear impact on the interference power received by other unintended receivers of the signal. From Equation 19.3, if the sum of noise power and interference power is dominated by interference, then the denominator also drops (about) linearly. The overall impact is that the capacity of the intended link remains about the same. So, the capacities illustrated by the circles in Figure 19.9, are about the same. However, more links can be established so that the system capacity per unit area increases.

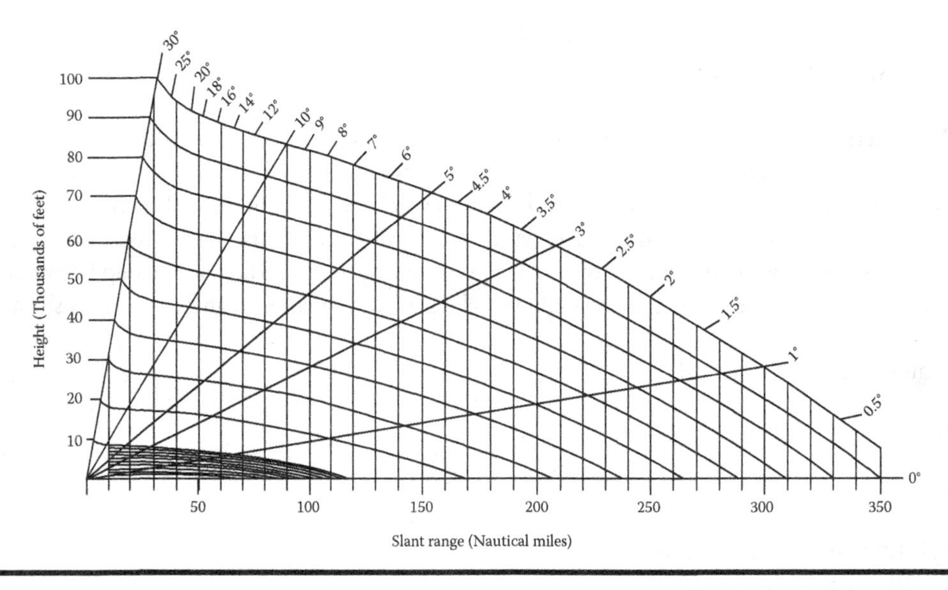

Figure 19.8 The curvature of the earth affects the maximum range of airborne systems. The height above the earth of an airborne platform and the slant range to a position on the earth are shown in relation to the slant angle.

With directional antennas, the gain of the transmitter and receiver antennas varies significantly with offset angle. From Equation 19.4, the reduction in interference power delivered to unintended receivers is linear. An example of the antenna gain patterns for a directional antenna is shown in Figure 19.10. The example shows antenna gain in a polar plot of gain (dB) vs azimuth angle. As can be seen, there is a main lobe, and often there will be multiple side lobes. Given the angular offset from the main lobe to the unintended (or intended) receiver from the transmitter antenna and the pointing angle offset of the receiver antenna, the amount of interference can be estimated.

Defining an exclusion zone for reuse of the same frequency is no longer circular, as it is for an omnidirectional antenna. Nor can a single exclusion zone be defined since the orientation of the unintended receiver's antenna has a significant impact on the level of interference. For a

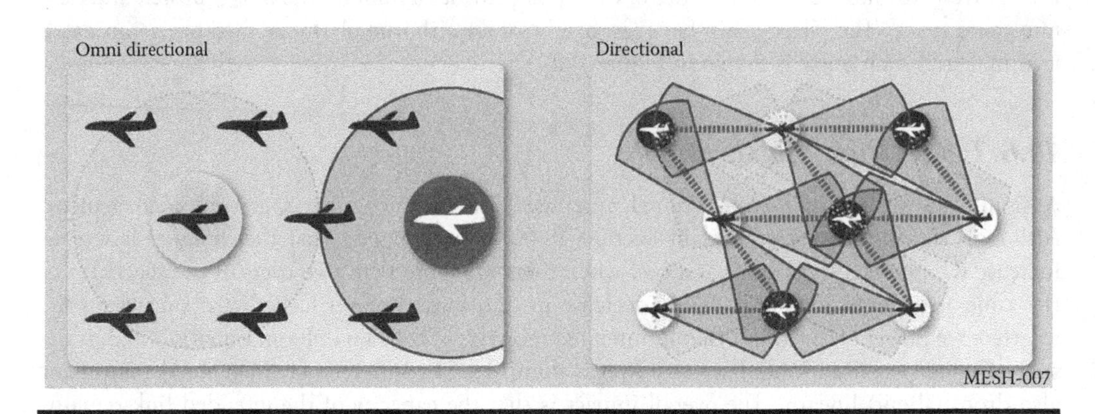

Figure 19.9 Directionality significantly increases the range or data rate and increases spatial utilization.

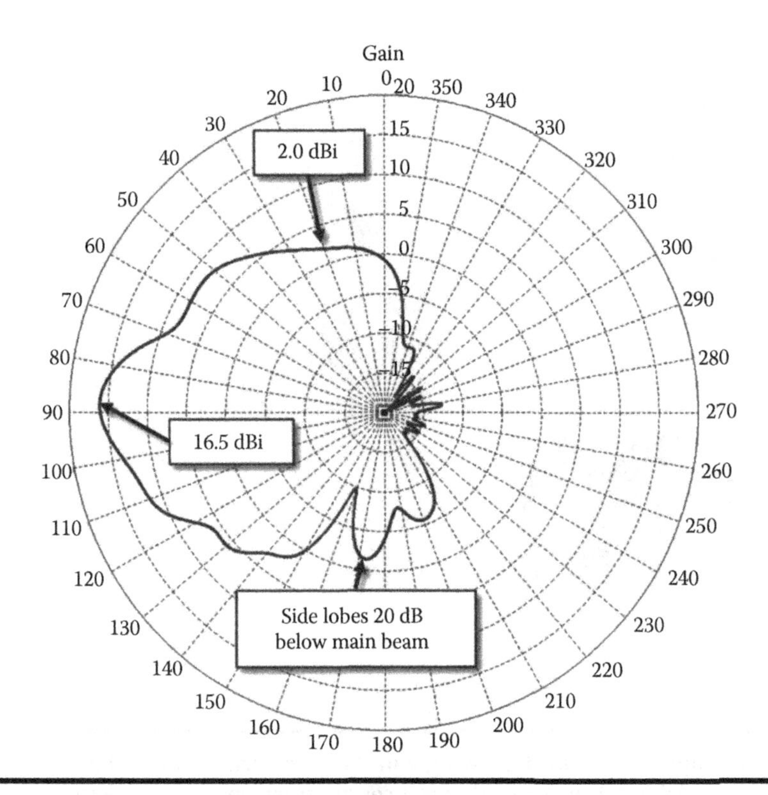

Figure 19.10 Polar plot of circular horn antenna gain by azimuth angle.

given receiver antenna orientation, the exclusion zone will have the same shape as the transmitter antenna pattern if we assume that the path loss is linear. However, there are other considerations, including terrain. Figure 19.11 shows an example of an omnidirectional ground based transmitter where terrain alters the exclusion zone.

There are three general ways that spectrum can be managed. The terms used to describe these ways are underlay, overlay, and interweave. The basic premise of underlay is that interference is avoided, often by controlling power levels or by controlling directionality. The basic premise of overlay is that interference is subtracted out using signal processing techniques with knowledge of the nature of the transmitted signal. The basic premise of interweave is that the spectrum is allocated in a way that users are not using the same allocation of spectrum at the same time. Research in the dynamic spectrum allocation mechanisms has been especially intense for interweave strategies, and has resulted in beacon-based systems for authorizing access to spectrum (e.g., medical body area devices in hospitals), RF sensing before transmitting (Wi-Fi-based systems), and a number of database driven allocation mechanisms. Table 19.2 provides a summary of different mechanisms to improve spectral utilization classified by the nature of the management strategy, and by whether decision making is unilateral or collaborative.

Unilateral decision making implies that the decision is made by each node without consulting with its neighbors. Collaborative decision making implies that nodes in the system consult are both able to communicate with other nodes in the system, and that they actively participate in the strategy. For example, interference cancellation can be enabled at the transmitter by special coding (unilateral approach), or by having the signals of multiple transmitters coordinated so that the receiver does not see the signal.

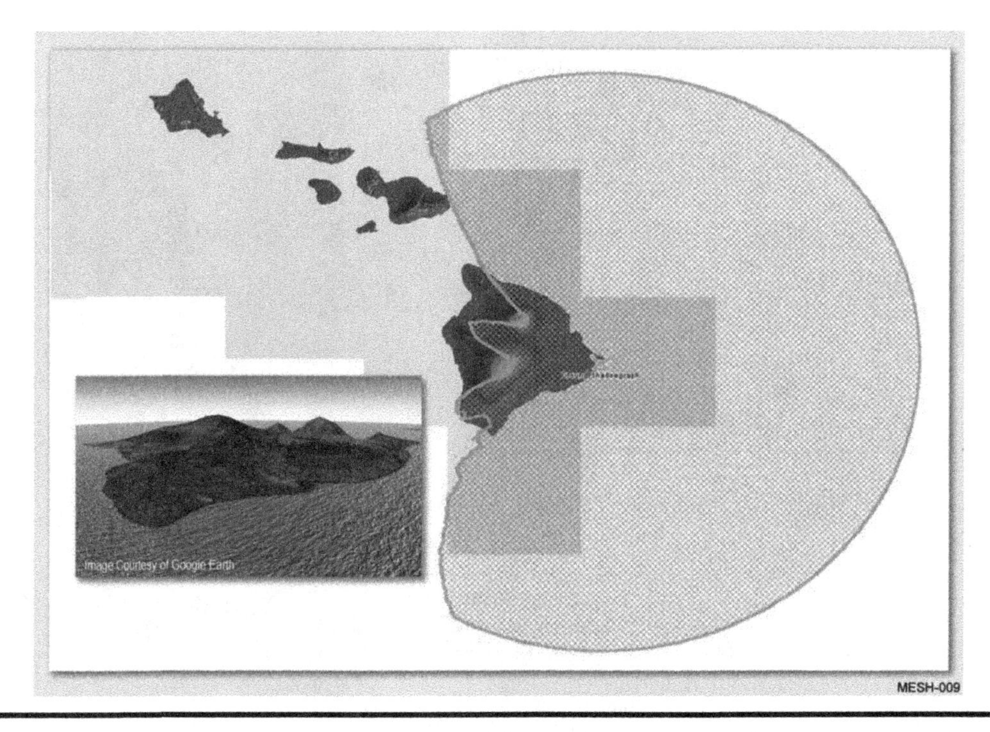

Figure 19.11 Example of terrain shadowing with Google Earth inset of Hawaii. (Adapted from DISA, [Online]. Available: http://www.disa.mil/services/enterprise-engineering).

A significant aspect of spectrum management is knowing which signals are present, which need to be avoided, and the nature of those signals. In general, unilateral decision making implies that the system is doing its own spectrum sensing. Interference cancellation often relies on knowledge of the nature of the interference signal. In a strategy that does not exploit collaboration, each node must be able to detect and classify signals. As can be inferred by the complexity of the spectrum-sensing problem, and the signal classification problem, the decision-making process is too complex to rely on simple strategies that relay on parallel duplication of demodulator streams. This has led to the increasing importance of cognitive radio architecture in building radio systems.

A cognitive radio operates in some form of a cognition cycle. This may be as simple as an observer, orient, decide, act (OODA) loop, or have multiple overlapping cycles that permit both reasoning and learning as illustrated in Figure 19.12.

19.6.4 Cosite Interference

Having multiple antennas on the same airplane leads to a condition where it is possible for them to interfere with each other. If they are using the same receive frequency as any of the other transmitters, the probability of interference is very high. Part of the issue that drives the need for frequency separation is that the RF front ends of the receivers have imperfect filters. If the transmit and receive frequencies are too close, the RF filter of the receiver will not be able to discriminate sufficiently well between the desired and undesired signals. What works well in practice is to ensure that none of the receivers are using the same frequency as any of the transmitters.

Table 19.2 Spectrum Management Can Exploit Overlay, Underlay, or Interweave Strategies, and Involve Either Unilateral or Collaborative Decision Making

Spectrum management strategies		Cancel interference @ TX	Cancel interference @ RX	Successively decode	Reduce spectral footprint	Reduce power	Spread spectrum	Repoint antenna	Narrow antenna beam	Manage bream nulls	Relocate spatially	Relocate temporally	Relocate spectrally	Put radio to sleep	Share sensing functions	Combine traffic
Unilateral decisions	Overlay	X	X	X												
	Underlay				X	X	X	X	X	X						
	Interweave										X	X	X			
Collaborative decisions		X	X	X	X	X	X	X	X	X	X	X	X	X	X	X

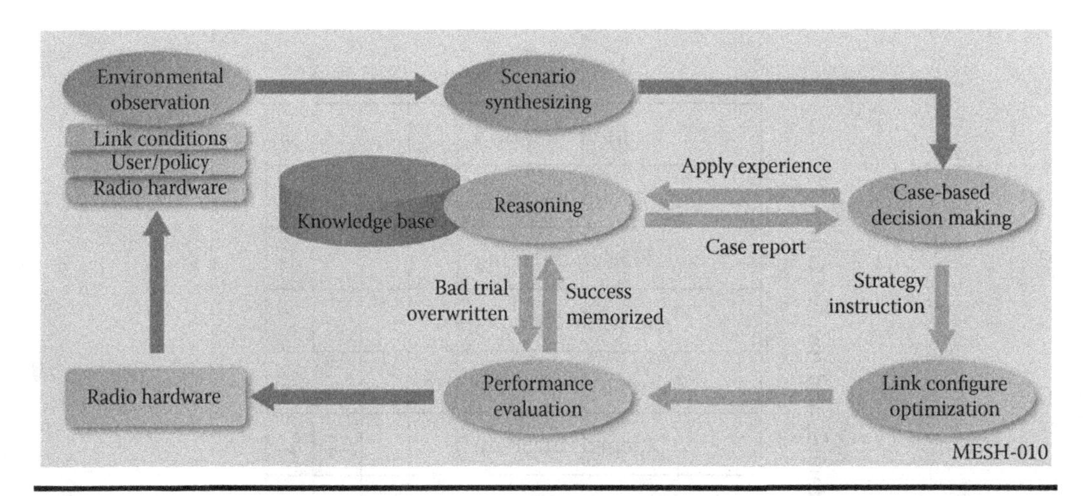

Figure 19.12 A cognition cycle can exhibit both reasoning and learning. (Adapted from L-3 Communications, 2014. In *CDL Frequency Management Training,* Fort Sam Houston, Texas.)

There is ongoing research in this area to reduce the filtering requirement by using signal processing to subtract out the undesired signal. Since the undesired signal is colocated on the same platform, knowledge of the signal sent by the other transmitter can be used to enhance this process. (In spectrum management parlance, this is an overlay technique to mitigate interference.)

19.6.5 Gender

With all of the antennas on a platform under the constraint to use a different receive frequency than any of the transmit frequencies, the simplest mechanism is to have all of the transmitters transmit on one frequency, and all of the receivers receive on a different frequency. This introduces the concept of a gender. Prior to the introduction of networking, only a single pair of genders was ever considered. Each platform in the chain of data links would be either transmit on "A," and receive on "B," or they would transmit on "B," and receive on "A."

With the advent of networking, however, the concept of multiple genders has been introduced. The impetus for this development is the observation that gender partitions the set of nodes. No node can communicate with a member of its own gender. For a two-gender case, this restriction is severe because it means that for any three neighbors, at least one pair will not be able to form a data link.

Gender is a shorthand for the set of rules that prevent the same frequency from being used on the same platform at the same time by both transmitters and receivers. From this perspective, the concept of gender applies equally well to time duplexing as it does to frequency duplexing. Extending the number of genders from two to three or four increases the number of possible connections, and means that we could use the terminology: triplexer or quadplexer instead of duplexer to describe the RF components that provide this capability.

If we define gender from the perspective of the transmit frequency, then all directional antennas will use the same transmit frequency on one node, but each link may use a different receive frequency. From this perspective, gender assignment is a map-coloring problem. Gender is shown in Figure 19.13 as tile coloring. In this system, any neighbor can communicate directly with any neighbor that has a different tile color (i.e., there are no restrictions in this

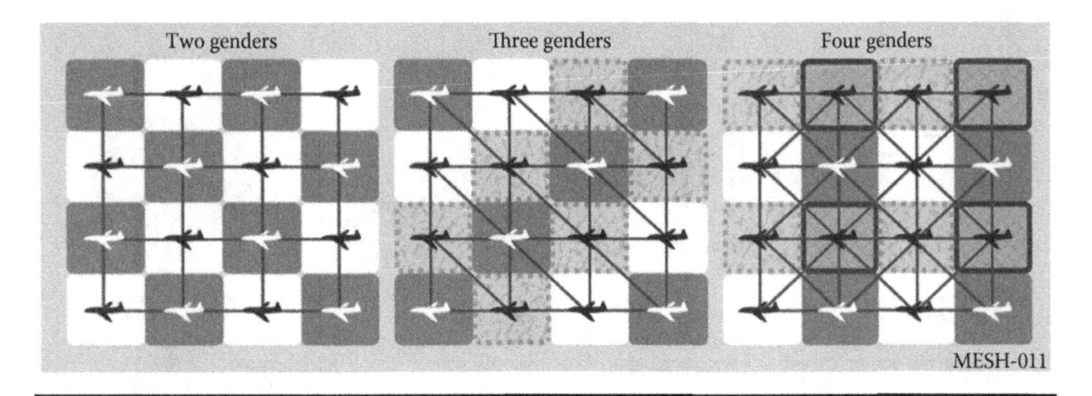

Figure 19.13 Gender assignment is a map-coloring problem. Increasing the number of genders increases the number of interconnection options.

assignment). Each node's transmit frequency is determined by the mapping of the tile coloring to frequency, and its receive frequency is determined by the transmit frequency of the neighbor. (Note that the same thought process also applies to using time instead of frequency to designate the gender.)

The number of frequency-based genders that can be achieved in practice is determined by the filter characteristics of the RF system. For a timeslot-based system, the number is determined by time synchronization constraints. This has two components. The first is the worst case time offset between nodes within the network. The second is the worst case time offset from when messages are sent and when they arrive at the neighbor node. This time guard band usually gets larger as the end to end diameter of the network increases. However, there is one exception. That is a hub/spoke situation, such as might exist for a radio access network. Here the timing offset on which a spoke node can transmit to be guaranteed to reach the hub node within the timeslot window can be calculated in advance.

Term	*Definition*
Duplexer	A device that permits the selection of transmit and receive frequency for use on the same antenna from a set of two frequencies
Triplexer	A device that permits the selection of transmit and receive frequency for use on the same antenna from a set of three frequencies
Quadplexer	A device that permits the selection of transmit and receive frequency for use on the same antenna from a set of four frequencies

19.6.6 Multiple Access Schemes

As a way of introducing the concept of multiple access schemes, it is best to consider the different ways that access to the channel can be granted to a transmitter and one or more receivers at a time. The communications channel is first partitioned in space, and then in time and frequency. Coding can be used to provide a mechanism to separate signals using the same frequency

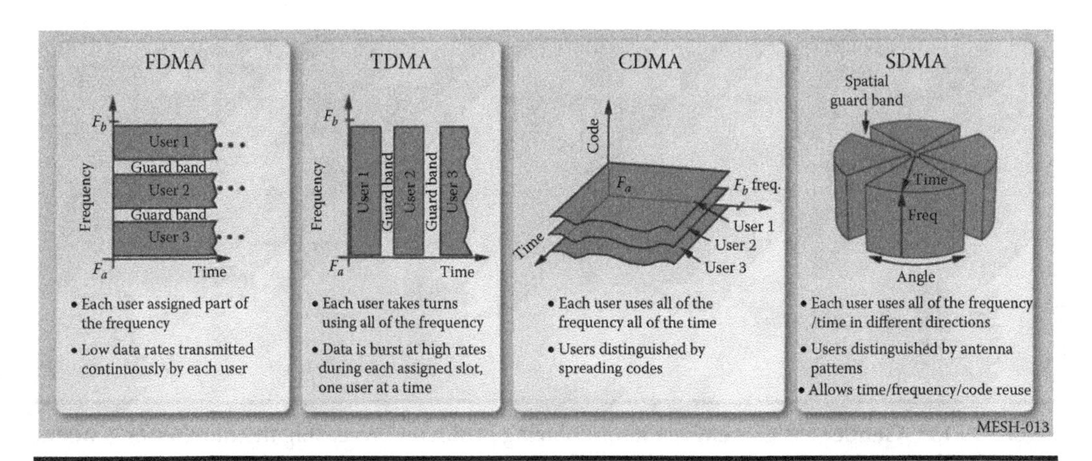

Figure 19.14 Multiple access schemes are based on how channel resources are used.

and time resources. Partitions can also be created for other characteristics such as polarization. Visualizations of FDMA, TDMA, code division multiple access (CDMA), and the impact of using directional antennas are illustrated in Figure 19.14.

In FDMA, each user is assigned a frequency on which one user can transmit and other users can listen. Transmission is normally continuous over time. In TDMA, the same frequency is used by all users, who separate it into timeslots in which each user transmits. CDMA uses coding to separate the channels; all users use the same frequency and transmit all the time. In carrier sense multiple access—collision detect (CSMA–CD) time is used to separate users, but there is no separation into timeslots. Users contend for access to the channel and use protocols to both announce their intention to begin using the channel and detect collisions on those announcements.

Two separate, but related concepts are the multiple access scheme and the duplexing scheme. Multiple access is about how the channel resources are allocated between users, whereas duplexing is about how the channel resources are allocated between the transmit and receive function. Specifying how the system works requires understanding both concepts. For example, time division duplexing (TDD) and frequency division duplexing (FDD) schemes can be used with any of the multiple access schemes.

From the perspective of the system, increasing directionality leads to either increased range or increased data rates. It also leads to decreasing connectivity, the number of other nodes that can use the channel to communicate, and increasing spatial diversity. Put in another way, it leads to a larger number of possible partitions to the area in which the channel is used since the interference to neighbors in a separate partition is reduced.

There is no specific choice for a "correct" multiple access scheme for all directional networks. However, for specific cases, some will be better than others. Longer distances tend to favor FDMA and CDMA over TDMA and CSMA, mostly because they use highly directional antennas. TDMA and CSMA schemes are highly sensitive to a number of neighbors. All neighbors must hear each other for their protocols to work. Only if the antenna gain is sufficiently low that all neighbors can be reached within the same beam can their benefits be brought to bear. So, high gain RF antennas (<5°), and lasers will almost never be able to benefit from TDMA or CSMA protocols. Lower-gain antennas, such as those used in the radio access network portion of an airborne radio access network often use TDMA or CSMA approaches.

Term	*Definition*
Duplexing scheme	The policy by which channel resources are allocated between the transmit and receive function of a communications link. Time division and frequency division are the most common
Multiple access scheme	The policy by which channel resources are allocated between multiple users of the same channel. Time division, frequency division, and code division are the most common

19.6.7 Summary of RF Data Link Layer Considerations

We are able to operate close to the Shannon limit of information capacity. The maximum distances and data rates of a system can be determined by well-known functions. If we consider only maximum distances, these are limited only by the curvature of the earth.

Good spectrum management allows networks to have higher capacities. Without directional antennas, long-range systems would interfere with each other so much that contention for the channel would limit the maximum data rates.

The cosite interference problem arises because the maximum isolation achievable between transmit and receive portions of the RF system is smaller than the difference in signal between a nearby receiver, and one that is far away. Resolving this problem often leads to the creation of "genders." The pairing of a transmit frequency and a different receive frequency makes it simpler to understand and manage the process of assigning transmit and receive frequencies. A two-gender solution does not permit full connectivity in a mesh network. The gender assignment to achieve full connectivity is a map-coloring problem. No two adjacent nodes can be assigned the same color.

The channel that is used to connect nodes in a network has time, frequency, and spatial dimensions. Nodes in the network cannot transmit at the same time and frequency, and in the same space without interference. Multiple access scheme and duplexing schemes provide classification mechanisms for strategies for assigning these resources between users and between the transmit and receive functions.

19.7 Network Layer Considerations

Important aspects of building a directional mesh network are whether the network will need to interact with the Internet, and the degree of Internet protocol (IP) mobility. The following is a brief review of IP basics, including a discussion of how the Internet is formed as a means of interconnecting a set of autonomous systems (ASs).

19.7.1 IP Basics

There are several basic assumptions of IP networking that are pertinent to a discussion of full mobility. One basic assumption of IP networking is that there is a distinction between a host and a router. The host originates and terminates IP traffic. The router forwards that traffic. Another is that the IP address is both an identity, and a location. Each host has an IP address assigned to it that is used by corresponding nodes (i.e., nodes who correspond with it) to indicate that it is the intended recipient of the message. Routers interpret this IP address as a network location, in a manner similar to the way we interpret a street address as a location. They maintain forwarding tables that allow them to look up the current location of that IP address, and use that information to select the forwarding path over which the message is sent. A routing protocol is used to populate those forwarding tables.

It is possible that each router could maintain a forwarding table that contained a list of individual IP addresses and the data link that should be used to forward traffic to them. For small networks, this is a viable strategy. However, in building the Internet, it was necessary to minimize the number of entries in these forwarding tables. So, the practice of creating subnetworks (subnets) and of aggregation of subnets into larger subnets was adopted so that the entire subnetwork could be summarized as a single entity.

The Internet is partitioned into a collection of ASs. IP addresses are assigned by the Internet Assigned Numbers Authority (IANA) [8] to exactly one of those ASs. The protocol that they use to talk between them is called border gateway protocol (BGP) [9]. Each Internet router is required to maintain a list of all of those assignments. That list is very large (500,000+), and is growing all the time (see http://www.cidr-report.org/as2.0/ and http://www.cidr-report.org/v6/as2.0/ [10,11]). Updating that list at every Internet router is time consuming. It can take days for a new route to be propagated throughout the world.

19.7.2 IP Mobility Protocols

The aforementioned aspects of IP networking affect mobility in several ways. First, IP addresses are assigned to specific organizations. Second, IP addresses end up being assigned to a specific geography by virtue of their need to be physically tied to routers. When a node or a network becomes mobile, it must overcome the fact that traffic from corresponding nodes will be sent by default, not to their current location, but to the network location of the AS to which the IP address was assigned. That AS becomes responsible for delivering the message. If the node is not currently within the AS, that means a protocol is needed to forward that traffic (and that something in the AS at the assigned location for that IP address had better know-how to forward traffic to that node).

A pictorial representation of the relationship between the Internet, ASs, and any mobility infrastructure is shown in Figure 19.15. Making the system work requires paying attention to data forwarding in three separate domains. The first is between a corresponding node in another AS and the gateway to the mobility infrastructure. In the example, this is represented by the collection of AS subtended from the Internet, and by the presence of a gateway within the AS marked as GRN in the figures. The second is within the mobility infrastructure. In the example, this is represented by the lines interconnecting a set of airplanes. The third is from the mobility infrastructure to mobile users (i.e., hosts) that are infrastructure subscribers. In the example, this is represented by the separation of the person icons from the interconnected set of airplanes.

To handle the first domain, the Internet community has created a suite of IP mobility protocols. This includes mobile IPv4 [12], mobile IPv6 [13], and network mobility (NEMO) [14,15]. The underlying assumption in all of these is that a stable, wired, Internet exists. Hosts can move from location to location and get a forwarding address from a local router (care-of address) in any of the AS. They have a permanent home network that has an agent (a router) whose function is to forward traffic destined for them. The protocol has the mobile node learn the forwarding address

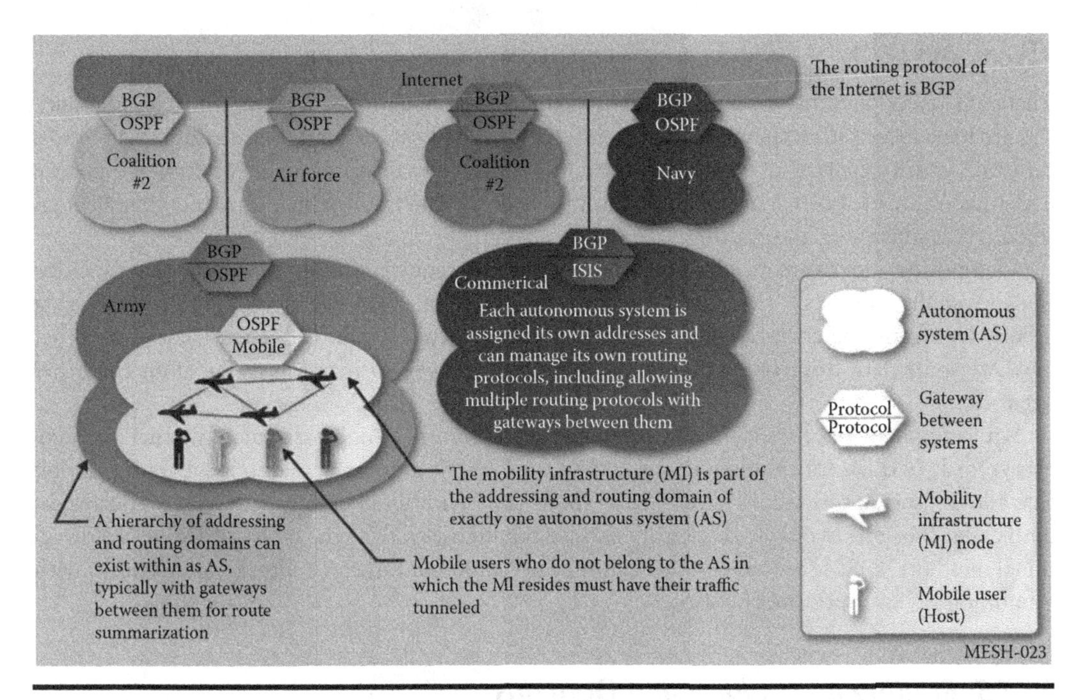

Figure 19.15 A mobility infrastructure belongs to the routing and addressing domain of a single autonomous system.

form the local router, and then pass this back to the home agent. The home agent then encapsulates all of the messages and forwards them using a tunnel to the forwarding address.

A related effort in the Internet community is the host identity protocol [16]. This allows the host to maintain multiple IP addresses, one that is its identity, and another that is its location. Forwarding is based on the location IP address. Like IP mobility protocols, it requires that the location IP is one that belongs to the wired network hierarchy so that corresponding nodes can use that address to forward traffic.

19.7.3 Mobility Infrastructure

For mobile nodes in a directional mesh networks, only some have a wireless link to a node in the wired IP network. The directional mesh wireless network extends from one or more gateway locations. All of the directional mesh nodes can be considered as part of the mobility infrastructure.

Forwarding within the mobility infrastructure has two aspects. The first relates to how messages between nodes that belong to the mobility infrastructure network are forwarded. The second relates to how messages to the infrastructure subscribers are forwarded. When a message arrives at the mobility infrastructure from an infrastructure subscriber, the radio access node that receives the message must know to which node within the mobility infrastructure the message should be forwarded. That means that the mobility infrastructure must be able to track the association between infrastructure subscribers and mobility infrastructure nodes. The mobility infrastructure must also know how to forward messages to another mobility infrastructure node; thus, it must be able to track other nodes that belong to the mobility infrastructure as well.

19.7.4 Security Enclaves

There are many circumstances in a military context where communications with specific entities are forbidden. Often they have different security classifications and so are said to belong to a different security enclave or domain. Partitioning of the networks into ones that can and cannot exchange traffic breaks one of the key assumptions of the Internet, namely that everyone belongs to the same addressing and routing domain.

Composing a network to allow transport of traffic requires the insertion of what is called "cross-domain solutions." There are two forms that these take. One is an encryption device that allows data from one enclave to be tunneled over the network of a different enclave. The other is where some data from one network is processed in a way to make it available on the other network.

An example of the first kind of cross-domain solution is a high assurance Internet protocol encryptor (HAIPE). An example of the second kind of cross-domain solution is a voice solution that allows users in one security enclave to talk directly with users in another security enclave. In both cases, the need to deal with security enclave boundaries affects the network aspect of a directional mesh network. It becomes not just a single network, but rather a collection of networks with a common RF transport mechanism.

19.7.5 Routing and Addressing Domains

The network aspect of a directional mesh network is affected by the number and nature of routing and addressing domains. Duplicating addresses is forbidden within a single network, so any time that an address space is duplicated, it must belong to a different routing domain. The Internet designates a private address space in addition to the public address space where duplication is allowed. It handles the potential routing problem by requiring that Internet routers MUST drop all packets from a private address domain. In contrast to a public IP address that must be purchased, users can freely assign IP addresses from the private address space to their computers. Because nearly every manufacturer of home routers has their equipment assigned addresses from the private address range 192.168.X.X (where the Xs indicate any number from 0 to 255) [17,18], the number of home computers in the world with the same IP address is large.

Private address domains are nearly as prevalent in military networks as they are in the rest of the world, and for many of the same reasons. One of the simplest reasons for using private addresses is that there is no reason to pay for a public IP address for a piece of equipment that will never be connected to the Internet. Even beyond this, the simplicity of being able to physically remove equipment from one platform and reuse it in another without the need to reconfigure it is fairly compelling.

Having multiple routing and addressing domains affects the networking solution for the Backhaul use case. A great deal of attention is paid to designating the placement of: HAIPE devices, network address translation devices, or other cross-domain solutions to allow messages from one routing domain to be transported in another routing domain.

19.7.6 Routing Protocol Efficiency

Most military applications of directional mesh networks are for mobile use cases. At the same time, the prevalence of the IP, has led to the standardization of IP networking within these networks. This causes several problems because physical mobility can break some of the fundamental assumptions of IP networking. One of those assumptions is that the IP address of a host is its network location. A second assumption is that the IP address of a host is its identity.

These two can be at odds with each other where mobility is concerned. One strategy for getting a mobile host to align its address with its network location is to give it another IP address as a "care of" forwarding address. This is the strategy employed by IP mobility protocols. Another strategy is to separate the identity and location information, such as is pursued with the host identity protocol (RFC 4423). However, all these protocols rely on the assumption that the network to which the host is attached has a stable network (location) addressing scheme.

Where this causes a problem is that, with a mobile network, the amount of forwarding information that needs to be exchanged increases as the square of the number of nodes that need to be tracked. That is, each added node needs to learn the forwarding information for every other node in the network and vice versa. What decades of research into MANET has proven is that this does not scale [19–21].

In a wired IP network, scalability is managed because of strategies that are built into routing protocols. The most basic of these strategies is the role separation between host and router, that relieve the burden of information tracking from the majority of nodes in the network (i.e., routers only—no hosts). Role separation is also used to distinguish between routers. Routers that are designated as some form of "border" router provide summaries of the networks which they can reach as opposed to the complete list of individual addresses.

In a mobile network, forming up ad hoc nodes into a hierarchy that facilitates summarization is generally not possible. Where much of the research into mobile directional mesh networks has focused is in the area of broadcasting or flooding information. A great deal of progress has been made into the question of how duplication can be reduced. The nature of flooding is to duplicate a message that arrives on one interface and broadcasts it out all of its other interfaces. The amount of duplication depends on the number of interfaces at each node. If the network can form loops, then there is a possibility that the duplicated message will be duplicated again, leading to a cascade of duplicated messages and eventual network failure.

The most basic way that this can be handled is by limiting the number of hops that any single message can be propagated. Under this assumption, the total number of message would be the duplication ratio raised to the power of number of hops. That is to say if we have three interfaces on every node, each incoming message would be duplicated two times. If the maximum hop count is 10, then there could be as many as 210 or 1024 message hops. That number is more or less independent of the size of the network.

The next way that this can be handled is by testing each message as it arrives, and only broadcasting new messages. This method of managing flooding is illustrated in the left-hand side of Figure 19.16. As shown, there are a number of redundant messages.

Two strategies to eliminate redundancy have gained interest. In the first, the mesh is deliberately turned into a tree by identifying a set of nodes called relay nodes (multipoint relay [22,23] or overlapping relay [24]). These are the only nodes that have the broadcasting role. In the second, a connected dominating (CD) set is formed. In a CD set (CDS) approach, a connected set of nodes is identified so that all nodes are no more than one node away from that set.

One of the reasons for interest in these strategies is that they can be implemented without central control. They are suited to a MANET style of deployment, and do not require role specialization. The same algorithm can work at each node.

As can be seen from the message count in Figure 19.16, both the relay and CDS approaches create a tree. By definition, the number of links in a tree is always one less than the number of nodes; so, both these approaches provide an optimal solution from the perspective of

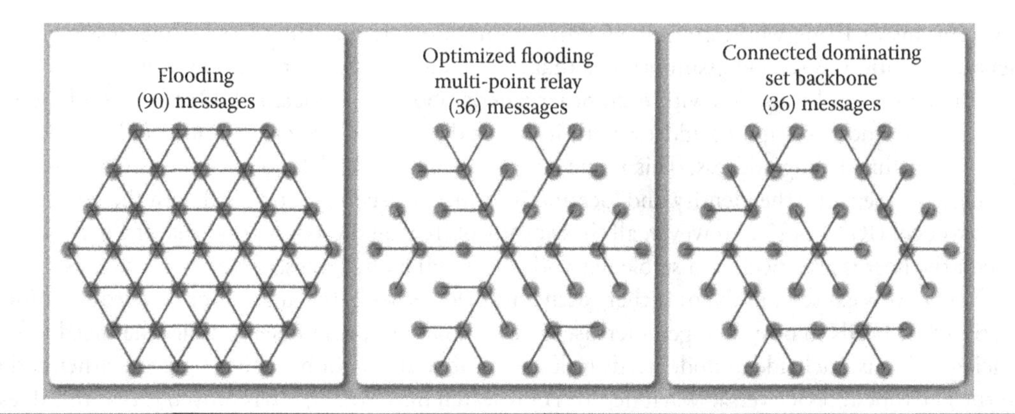

Figure 19.16 Optimization of flooding has led to both multipoint relay and connected dominating set strategies to significantly reduce the number of messages that need to be sent.

message count. A number of routing protocols are based on one of these philosophies (e.g., RFC 5614 Mobile Ad Hoc Network (MANET) Extension of Open Shortest Path First (OSPF) Using Connected Dominating Set (CDS) Flooding [24] and RFC 5820 Extensions to OSPF to Support Mobile Ad Hoc Networking [23]).

Summarizing the message counts for these approaches, we have 36 messages for a tree topology, 90 messages for flooding with message duplication testing, or 1024 messages without testing. The behavior of the tree-based approaches is linear. The behavior of the duplicate message elimination process is highly dependent on topology, but is almost always much higher than a tree. Limiting hop count by itself leads to exponential growth in the number of hops allowed.

19.7.6.1 Network State Dissemination

Counting messages is one way of looking at the problem of routing protocol efficiency. The implicit assumption of message counting is that every node needs to have the information in a single message. There are other strategies that do not make this assumption. A brief examination of those strategies will be given as an illustration of the type of research that is being conducted into mesh networking that can have impacts on the directional mesh networks used by military users.

One example is hazy sighted link state routing [25]. Here, the assumption is that not all network state information is needed all of the time. Instead of broadcasting all of the information all of the time, the frequency of updating information about an individual node is based on the distance from that node. The underlying assumption is that mobile nodes do not randomly and spontaneously appear at new locations in zero time. Information about the general direction in which a node can be found is valid much longer than information about the complete path.

Ant colony routing takes a slightly different tack [26]. Network state information can become redundant over time. Information about the probability of a direction being on the route to a destination is either increased or reduced in strength by the transactions on the path. Packets are sent into the network to explore (look for nodes), and results are returned along the exploration path. After a node is discovered, rather than broadcasting to all directions to confirm the continued presence of that node, those packets are broadcast in a constrained manner, beginning with knowledge of where the node used to be. This reduces the number of broadcast message by not sending them into places where the network is pretty sure that the node is not present. (Although if it discovers that the node is not present there, it can increase the search area.) (http://en.wikipedia.org/wiki/Routing_in_delay-tolerant_networking#cite_note-vahdat2000-6).

Epidemic routing takes a third tack [27]. No knowledge of topology is retained, so broadcasting network state (e.g., topology updates) is not attempted. The implicit assumption is that messages do not need to have an optimal routing, nor is latency or latency variation critical. Messages are exchanged with neighbors who pass the message on to their neighbors, just like medical epidemics result from contagion. Epidemic routing is not very efficient, but if the probability of passing the message on is varied, it does not always result in bad behavior. For example, nodes that are highly mobile, and have a high probability of being physically close to other nodes in the network could be highly susceptible to becoming a message carrier. Or, nodes that are on a backbone path could have a much higher probability of carrying the message than those that are not on the path. Messages that need to be given to everyone as quickly as possible could have the contagion probability increased to 1, and become a flooding event. At the other extreme, the network need not even be continuously connected to permit messages to be delivered as epidemic routing is disruption tolerant.

All of these strategies have advantages for some use cases. Strategies to improve the performance of data forwarding in a mobile environment have to consider the degree of mobility, the degree of connectivity when considering how much network state information is disseminated, and how often that information is disseminated. There is no silver bullet with regard to resolving the problem of data forwarding for mobile networks. It is an area of active research that we hope will continue to provide us with better solutions.

19.8 Network Management

Network management can be split into two functions: managing network elements and managing system dynamics. Although not absolute in the division of functionality, two factors weigh heavily in the division of functionality between the two. One is the time horizon for events to be acted on, and the other is the complexity of interaction between all the system elements and across all the layers.

19.8.1 Managing Network Elements

Managing network elements is based on well-established functions from commercial enterprise network management systems, often referred to as FCAPS. FCAPS means fault management, configuration management, account management, performance management, and security management. It is a two-way relationship. The network elements onboard the airplanes must report their condition to a network manager for situation awareness. The network manager must make decisions based on that awareness and push policies to those network elements for enforcement of behaviors.

A hierarchy of network management functions is created in a system to be able to manage the network elements, the network to network interactions, and the services that must be supported across those network elements [28]. A service-oriented architecture [29] is exploited to reduce the complexity of managing these elements. An example illustrating the relationship of airborne networking components is shown in Figure 19.17.

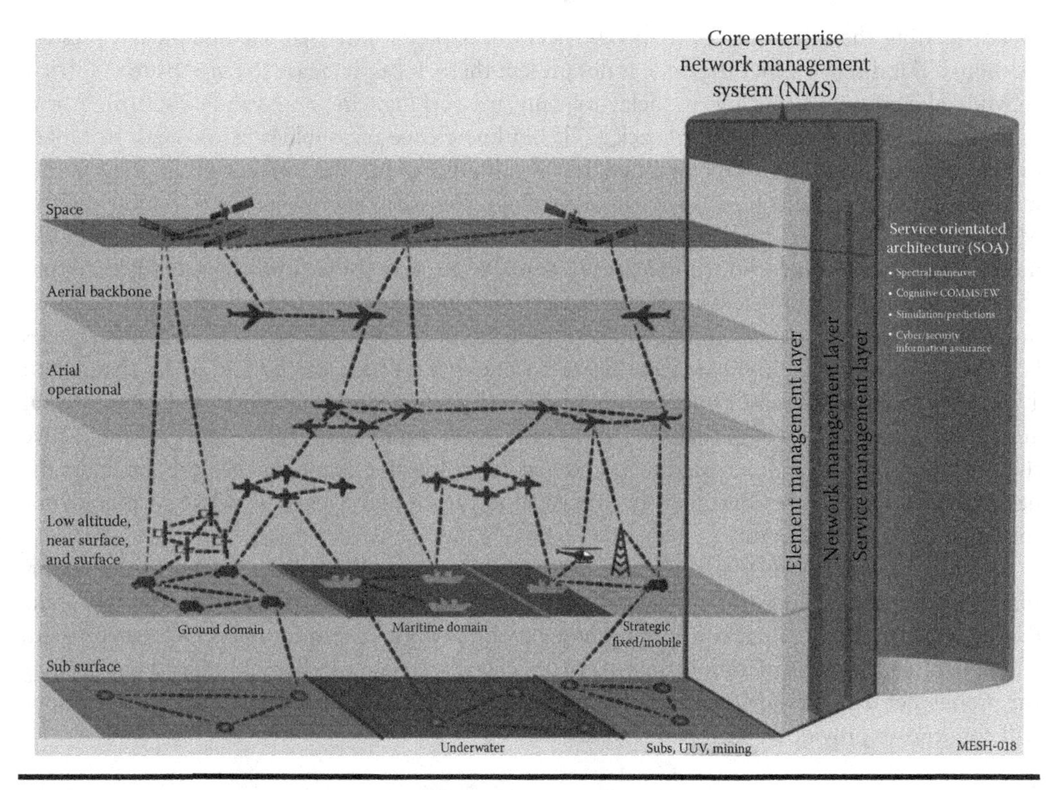

Figure 19.17 Management functions span different systems.

19.8.2 Managing System Dynamics

Directional mesh networking in an airborne military net-centric [30] environment creates unique challenges not found in the stable wired infrastructure. Standard networking protocols and architectures use a layered architecture approach. The layered open systems interconnection (OSI) model [31] was designed to allow for new protocols and services to be implemented at the layers while preserving well-defined interfaces to the layers above and below. This architecture method works well in relatively static wired network infrastructures. However, this methodology is less efficient and generally less optimal in a dynamic wireless ad hoc environment because of the lack of feedback and information flow across protocol layers.

A cross-layer architecture allows better and more dynamic optimization of the overall network communication system by utilizing global information across all network layers. Unlike wired networks where management of the network topology is independent of managing the individual data links, an airborne network must explicitly decide which nodes will be connected by which radio data links.

A directional mesh network must manage and coordinate spectrum allocation with its own radios as well as with the peer radios to which it connects in the network mesh. This is a dynamic process as nodes join, leave, and change positions within the mesh. Depending on the waveform type the radio employs, this management process also involves frequency and gender selection, deconfliction, and switching.

In the dynamic environment of wireless airborne directional mesh networks line of sight (LOS) data links can be blocked both their own platform, and by terrain for low-flying platforms.

These blockages are typically managed by higher level cross layer management functions that predictively and proactively change links and move antennas to optimize the network topology to achieve specific network optimizations goals, that is, typically connectivity, and availability. In addition to blockage control, the network manager monitors link margins between all sets of connected radios. This allows optimization of range and data rates between all connected data links and allows the system to proactively change link connections to balance and achieve overall network optimization.

The selection of an appropriate forwarding algorithm is dependent on the number of nodes, links, and the dynamics of a given network. Different purposes for deploying the airborne network result in different optimal approaches. As a result, the use of cognitive agents embedded at both the system and platform levels is becoming more prevalent.

19.8.3 Discovery

A directional mesh network requires antenna pointing. One of the consequences of using directional antennas is that finding a neighbor requires that both endpoints are pointed along, or almost along, the same vector at the same time. For a directional mesh, a discovery protocol is almost always required, since it may take tens of seconds, even with a good search process, for a ground station to find an airplane when it knows the region of the sky in which it should be, and the airplane knows the region where the ground station is located. For airborne-to-airborne communications, the problem is more difficult, since the simple assumptions about region of sky to search cannot always be made.

Discovery protocols for a directional mesh network need to be able to operate at ranges that equal or exceed the range at which a link can be formed. Since the discovery process needs to work even if the antenna is not pointed directly at the potential neighbor, the use of a directional antennal for the discovery protocol is almost never considered. Discovery data links are almost always low data rate link because they must operate at long ranges, and without the benefit of directional antennas.

The use of a low data rate link leads to relatively simple discovery protocols that do not require a large number of message exchanges between potential neighbors before the decision is made to point antennas. The information exchanged is generally the identity, location, speed, and direction of the mobile node. Exchanging speed and direction information simplifies the process of antenna pointing, particularly where the antenna must be mechanically steered, and the slew rate of the antenna takes one or more seconds to reverse its pointing direction.

19.8.4 Security

Rules for managing security of networks are encoded into standards like X.805 [32]. This standard defines three different security planes that exist within each network. These are user (data) plane, control plane, and management plane. One of the most basic rules of security is that each of these planes is isolated from the others as much as is physically possible. Where it is not possible to maintain isolation, data guards are used to provide guarantees that the system will operate as intended.

With regard to guarantees provided by security-related processes, three are commonly cited. These are

Confidentiality
Integrity
Availability

Confidentiality implies that the information in a message is not allowed to be read by an unauthorized agent, including agents that might need to process the message along the forwarding path. Integrity implies that the message is received unaltered from its original condition. Availability implies that the message will be received at the intended destination, equipment failure conditions excepted.

These rules affect the ways that hardware is designed for directional mesh networks. A particular problem is ensuring that the data-forwarding engines (e.g., router of switches) can be controlled by the management plane without introducing security vulnerabilities to attack from the user plane. This requires accommodations in all the three planes at the same time.

19.9 Dealing with Complexity

Designing a system that is robust requires complexity reduction. The traditional mechanism to reduce complexity is standardization of the interfaces between components. Then, instead of a possibly different interface between every pair of components, there is a single interface to every component individually. The pictorial representation of this process is illustrated in Figure 19.18. From this perspective, a standardization effort has the potential of converting a process whose complexity scales as N^2–N to one whose complexity scales as N.

There are a number of complexity reduction mechanisms in play for directional mesh networks. These include:

- Layered protocols

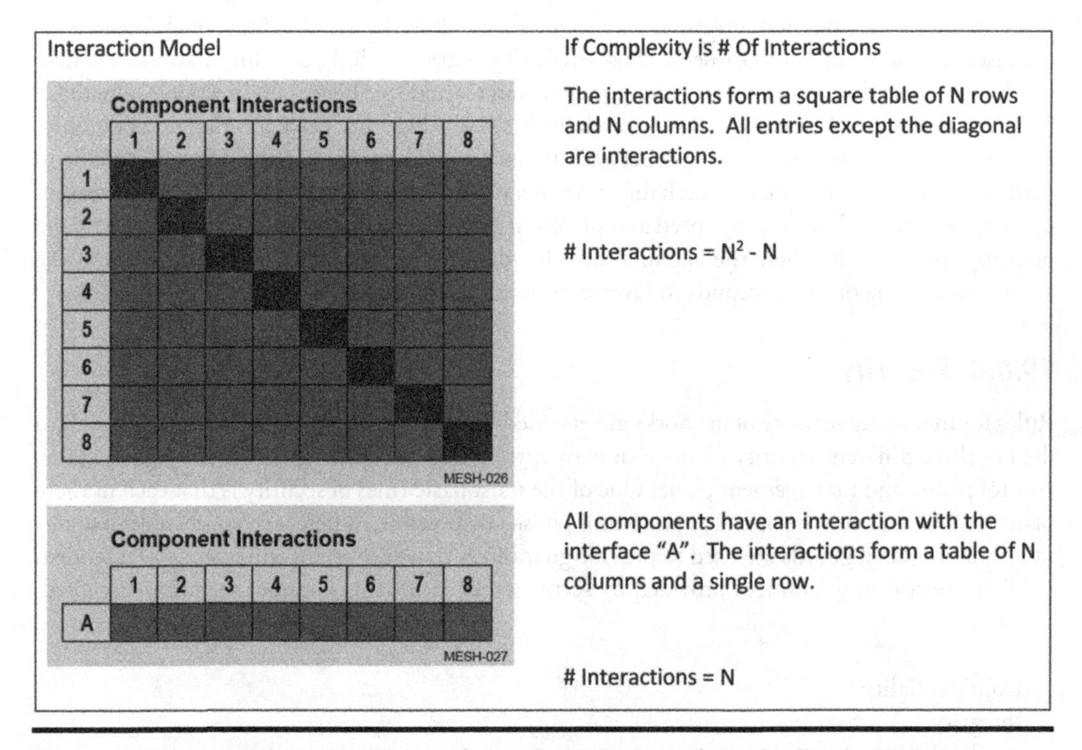

Figure 19.18 Pictorial representation of complexity impact of standardization.

- Partitioning of network interactions into user plane, control plane and management plane segments
- The definition of network to network gateways

Where mobile directional mesh networks have faced difficulties is both that (1) some of the underlying assumptions used in creating the standard protocols are not valid in a mobile scenario and (2) many of the off-line decisions made when building a network have become online decisions.

As an example, in a wired network, the distinction between a router and a host is an off-line decision. However, for mobile networks, particularly MANET, that distinction has disappeared, meaning that every node has to manage routing information instead of only a small fraction of nodes.

The problem of defining a routing protocol that operates efficiently is related to the problem of how much information needs to be dynamically managed. If there are "N" nodes in the system, the amount of information that needs to be tracked is $N^2 - N$. That scales as the square of the number of nodes. However, if there was a way to limit the amount of information that each node needs to track to its neighbors to a fixed number (n), then the scaling becomes linear. The effect and calculations are illustrated in Figure 19.19.

In general, it is not possible to have a dynamic system scale in this way. However, it has already been noted that at a high level, there is very little about flight plans that is ad hoc. There are two scenarios where the N^2 scaling problem can be converted to a linear scaling problem. Both are related to the interaction of flight plans. The first case is where airplanes are placed on-station using a racetrack-holding pattern. The relationship of airplanes to each other from a network-location perspective remains more or less constant, except for the relationship with nearest neighbors. The second case is where flight patterns have small groups of airplanes arriving in sequence, and can be thought of as a carousel. Both cases are illustrated in Figure 19.20. In both cases, there is a requirement to actively track the dynamics of neighbors. If the system also tracks which neighbors are downstream from it in the topology then, by default, all other nodes are upstream.

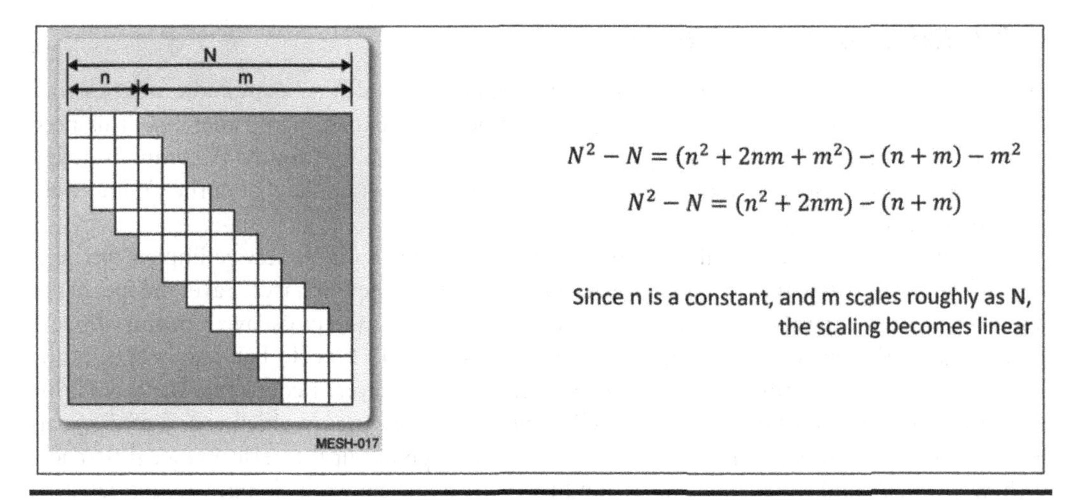

Figure 19.19 Linear scaling for routing can be achieved if the system can limit the number of neighbor nodes that must be tracked.

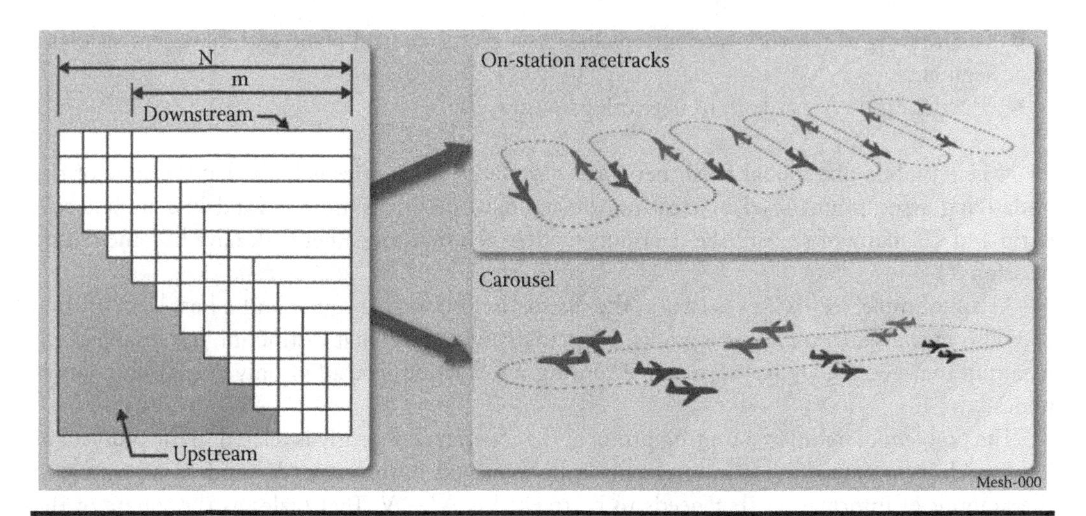

Figure 19.20 Racetrack holding patterns and carousel flight plans provide an opportunity to change the scaling of airborne network routing form N^2 to linear.

The impact on the complexity of interactions is shown in the table on the right-hand side of Figure 19.20. Squares indicate the individually tracked associations. Summarization of all the downstream nodes into a single tracking element is illustrated by the rectangles to the right of the diagonal. The blank area to the right is treated as a single default assumption.

In response to the needs for mobile wireless networks, several other standardization efforts have been undertaken. These include

Radio-to-router interface
Netcentricity and gateway definition

A short treatment of each of these follows.

19.9.1 Radio-to-Router Interface

Cross layer network management and control necessitate the ability to share radio metrics with network control information as well as higher layer network management functions. This need necessitates a standard interface and protocol between the radio and router. A common design question arises with respect to radio-to-router interfaces that is not unlike that of the OSI network layer architecture.

Layer independence with well-defined interfaces allows for multiple service implementations at each level while still maintaining interoperability based on those interfaces. Layer independence does impose costs in terms of information fidelity, performance, and cross-layer optimizations. A recent study at the Massachusetts Institute of Technology (MIT) Lincoln Laboratory specifically analyzed and compared three radio-to-router interfaces [33]. Layered control versus monolithic control provides the benefits of interoperability with disparate systems which is one of the goals of the researchers in this paper. However, the authors also point out the performance differences of these protocols and their ability to respond to physical and network changes appropriately in highly dynamic environments. Another cost of interoperability and service independence is the need to reduce the set of manageable parameters to a smaller subset of actual radio and network parameters. This is especially true with military radios where low-level functionality from each manufacturer differs widely and not all level parameters are made available externally to the radio. In addition to interoperability, performance is a key challenge with the emerging radio-to-router standards. The dynamics of the system determine the latency requirements for receiving status on radio and network metrics and the ability to command changes based on current network optimization goals. The performance problem becomes more acute and grows exponentially as the number of links and nodes increase in the overall network.

19.9.2 Net-Centricity

Joint forces network management for directional mesh networks face many interoperability challenges. Most military radios are developed for program-specific requirements and each service develops their own unique networks. These challenges have led to current JALN Department of Defense (DoD) initiatives to architect and define architectures and management tools to provide commonality between the joint services and the multiple disparate radio systems in use today [34,35]. The government is also promoting the open mission systems (OMS) architecture model as a potential means to increase interoperability and functionality between systems.

A primary objective of the DoD net-centric mandate and vision was to provide a convergence of protocols and interoperability at the IP network layer. The net-centric objective alone is not sufficient to meet interoperability and convergence of the multiple disparate radio systems found in the military. The most significant challenge with the net-centric objectives is the management of multiple disparate radios utilizing multiple different waveforms and link layer protocols. Perhaps, the second biggest challenge is providing for and managing multiple security domains.

A promising method for achieving interoperability between network systems involving multiple waveforms is the use of gateway systems. Gateways systems ideally employ multiple software-defined radios (SDR) so that radio resources can be allocated and instantiated as the network management scenario dictates.

Network management abstractions and translators are another mechanism for achieving interoperability that is transparent to the individual radios. The data distribution service (DDS) is emerging as an intraprocess and internode communication standard. Translation services have also shown their utility in providing protocol transparency for such services such as network management protocols and network messaging.

19.10 Looking Forward

The problems of scalability in mobile directional mesh networks are severe. In spite of decades of research into MANET protocols, the largest mobile directional mesh networks only accommodate ~100 nodes at vehicular-to-pedestrian speeds (DARPA WNAN program as reported on page 5 of Reference 36), and most can manage only a few tens of nodes. There is a need to look forward. We consider four cases in this regard.

Software defined networking (SDN)
Platform reconfiguration
Service negotiation plane
Achieving full mobility

Platform reconfiguration looks at the issue of how radios are connected to routers and other network elements such as firewalls, HAIPE, and translators. Nowadays, SDR-are becoming more capable. Permanently wired connections restrict the flexibility of using those radios for different purposes when those purposes require removing network elements or adding network elements.

The service negotiation plane looks at the issue of network independence. A reconfigurable platform opens up the possibility of disconnecting the radio from a particular router, and creating multiple independent network topologies. However, there is no standard way for networks to communicate their intent to radio resources.

Achieving full mobility looks at the problem of IP mobility protocols, which are defined only for a direct attachment to a wired network that connects to the Internet. Having fully mobile networks whose connection to the Internet is dynamically negotiated through other mobile networks is totally undefined.

19.10.1 Software-Defined Networking

One of the recent changes to the IP networking world has been the introduction of SDN [37,39]. Advances in large enterprise network systems, server farms, and cloud infrastructures have generated the need to create more flexible and easily upgradable network hardware. These features solve many routing and security problems in a military network.

In simple terms, SDN refers to the concept of the separation of the control plane from the from the data plane in networking equipment. Existing network equipment tightly integrates the control software with the data plane hardware, making it difficult to apply new technology. The nature of the difference between a traditional routing paradigm and an SDN paradigm is illustrated in Figure 19.21.

With a traditional routing paradigm, there is one owner of the tables that drives the switching fabric. Those tables are data that is part of an application. At most, one application can have control of those tables at one time. The application is based on running a fixed algorithm (that has strict stability constraints). That algorithm depends on a routing protocol that runs between neighbors in the system to provide it with information. The system is brittle because changing the algorithm at one node can easily upset the stability requirements of the rest of the network.

With an SDN approach, the interface to the tables that drive the switch fabric is via a protocol, not internal to an application. That introduces new possibilities. Forwarding tables can be partitioned and control assigned to multiple entities, and those entities are no longer required to be colocated on the same node. That allows some traffic to be treated with different algorithms than other traffic. It also allows for the simple dissemination of policies into the network for either forwarding or security purposes.

The backhaul use case, the tactical mesh use case, and the airborne radio access network use case all have situations where specific types of interactions require specific treatment. An exciting aspect of SDNs is that a wide array of abstractions and application programming interfaces (APIs) are being developed commercially and in academia. Many of these control mechanisms provide the ability to rapidly design, test, and validate new network middleware in wireless ad hoc directional networks including but not limited to: unique MANET routing protocols, traffic engineering mechanisms, firewalls, and other security mechanisms.

SDN also makes it easier to deploy artificial intelligence based approaches to optimize the network. In contrast to the general Internet scenario, user behaviors for military networks are well known. And, as opposed to an almost infinite flavor variation, military use cases can be more easily classified than the general case. The SDN paradigm provides an infrastructure to dynamically

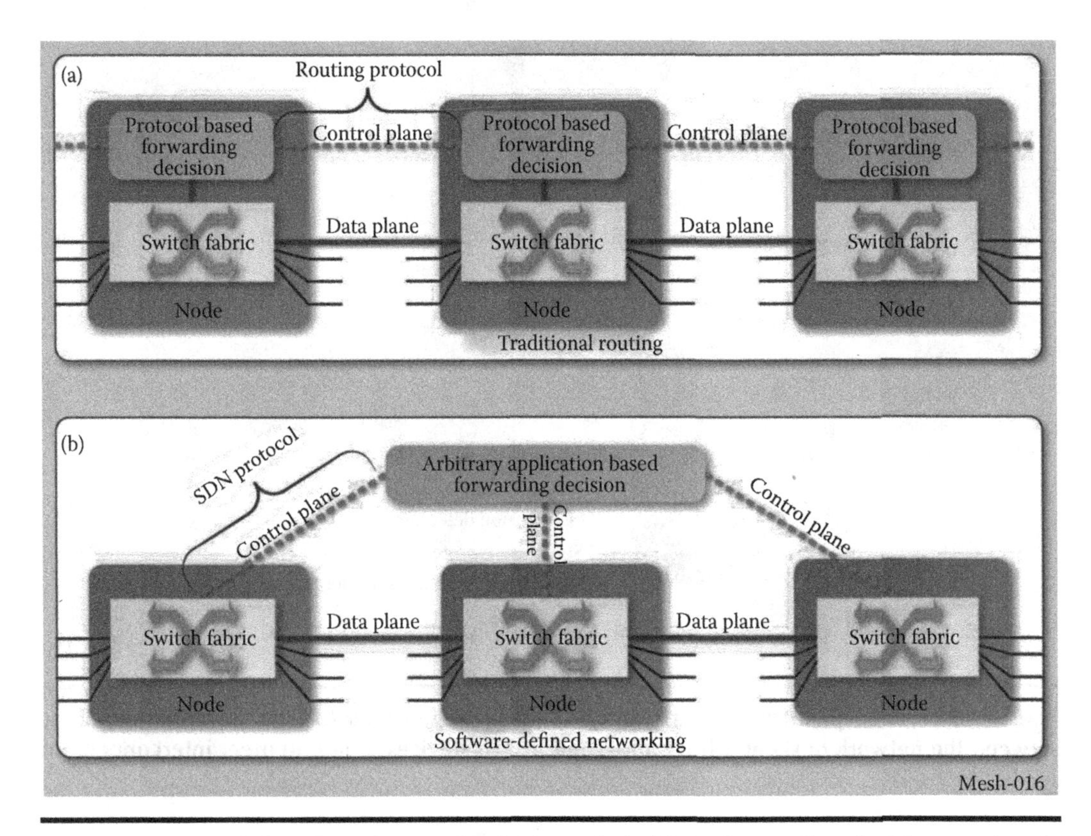

Figure 19.21 Traditional routing (a), binds the switch fabric forwarding decisions to data in a locally stored decision-making application. Software-defined networking; (b) separates the decision-making application from switch fabric control.

implement forwarding rules sets on a case-by-case basis, and the mechanisms to flexibly accommodate a variety of decision-making algorithms.

19.10.2 Platform Reconfiguration

The system configuration problem arises because of the way that equipment is wired on an airplane. Most airborne platforms also contain full-fledged networks with multiple hosts supported by a router or switch. Their radios either contain an internal router, or are permanently wired to a router. Once a router is wired to a radio, it remains wired for decades (rewiring an airplane is an expensive multiple month operation). Often the airplane is also wired with devices that are placed on the airplane for a specific mission. Using the platform for a slightly different mission requires that the physical connectivity of the platform be amenable for use in that new mission. The problems for new missions are the devices that must sit in line between networks like HAIPE, firewall, translators, etc. The typical wiring for these kinds of devices is illustrated in Figure 19.22. The problem is that there is no mechanism to achieve the rapid reconfiguration to meet the needs of slightly different missions.

A built-in reconfiguration capability would allow the system resources to be reused in different ways. For example, the advent of SDR has made it possible for the radio hardware to behave in different ways. An SDR could provide a radio access network service, even when its originally

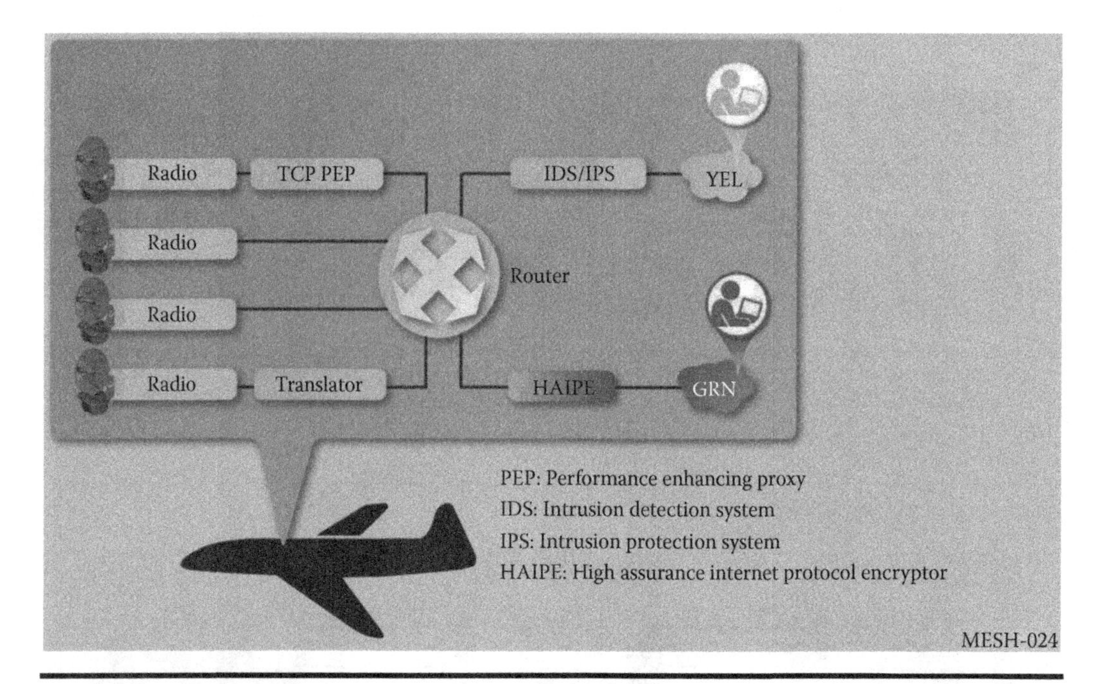

Figure 19.22 The typical wiring of a directional mesh uses a router to provide interconnections between the network hosts and the radios, and inserts devices in line to meet interconnection needs.

intended purpose was to support a tactical mesh. Being able to repurpose that radio so that it is no longer connected to its normal user community on the airplane but gets used as a radio access network node leads to the concept of using a reconfiguration switch between the radios and on-board networks. An example of a reconfiguration switch is given in Figure 19.23, configured for the same services as in Figure 19.22.

19.10.3 Service Negotiation Plane

One of the issues mentioned earlier with military networks is the number of security enclaves that must be accommodated. Traditionally, the authorization to form a data link has been enforced by the use of encryption keys. Platforms to which compatible keys are assigned are implicitly assumed to be authorized to form data links with, and to become part of the network to which the radio is attached. As we move toward more interesting networking configurations, the issue of security enclave boundaries becomes more important.

If we have a mobile platform that has some reconfiguration capability, it would be able to allocate the radio resources to a different user community on the airplane, or even to a user community that was not on the airplane, such as when the platform is used to provide a bent-pipe relay service on behalf of some other network. Under those circumstances, the simple fact of having an encryption key cannot be construed to mean authorization to make a connection or that a connection should be made. Likewise, forcing all users to have an encryption key, even for the case when the desire is to provide a relay service on behalf of another network significantly restricts the use cases that the platform can support.

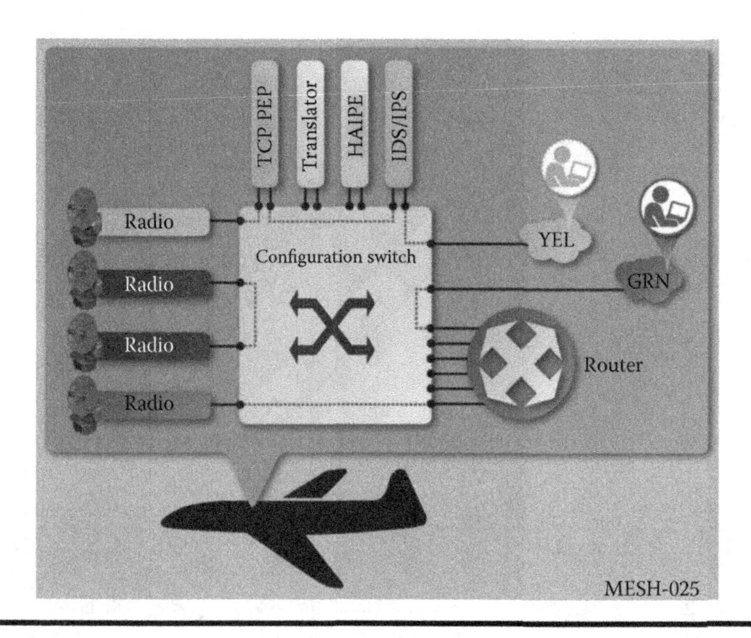

Figure 19.23 The use of a reconfiguration switch makes it possible to use platform resources in different ways.

The problem with conducting an authentication and authorization process after the radio link has been formed is that the radio interface is typically hard wired to a router. The problem is illustrated in Figure 19.24. Here, it is desired to exploit the unused radio resources of the platform that hosts the yellow network to provide a relay service for the green network. However, since the yellow network and the radios are hard wired to the router, the commanders are forbidden from attempting to form those data links.

Security rules for networks can be found in documents like X.805 [32]. X.805 defines a user data plane, a control plane, and a management plane as a way to partition the elements of a network. From a security perspective, a different security enclave is a different network entirely. It cannot share a data plane, a management plane, or a control plane. It can be inferred from this that neither the authentication and authorization mechanisms cannot reside entirely in the management plane of either of the two networks, nor can the addressing scheme or routing protocols of the data plane resources of the two networks be involved. The problem of message exchange is severe.

Where this line of reasoning leads is to the need for a special kind of a cross-domain solution to allow this kind of traffic to be exchanged. Because the purpose of this communication is to authorize access to some kind of service, it could be called a service negotiation plane. Like all cross-domain solutions, it would allow a very restricted message set to be exchanged between networks. In this case, the exchange would be between the management plane of each network, and the service negotiation plane. The service negotiation plane would enforce the message set restrictions, but would allow the two networks to conduct a negotiation about what kind of service each would offer or request from the other.

The service negotiation plane interface would necessarily reside on the radio, and not be a separate network component. The reason for this is that if it resided as a network element of one of the networks, it could be either by-passed, or attacked on those networks. The interface would need to be up full time facing both directions since either end may wish to change the service offered

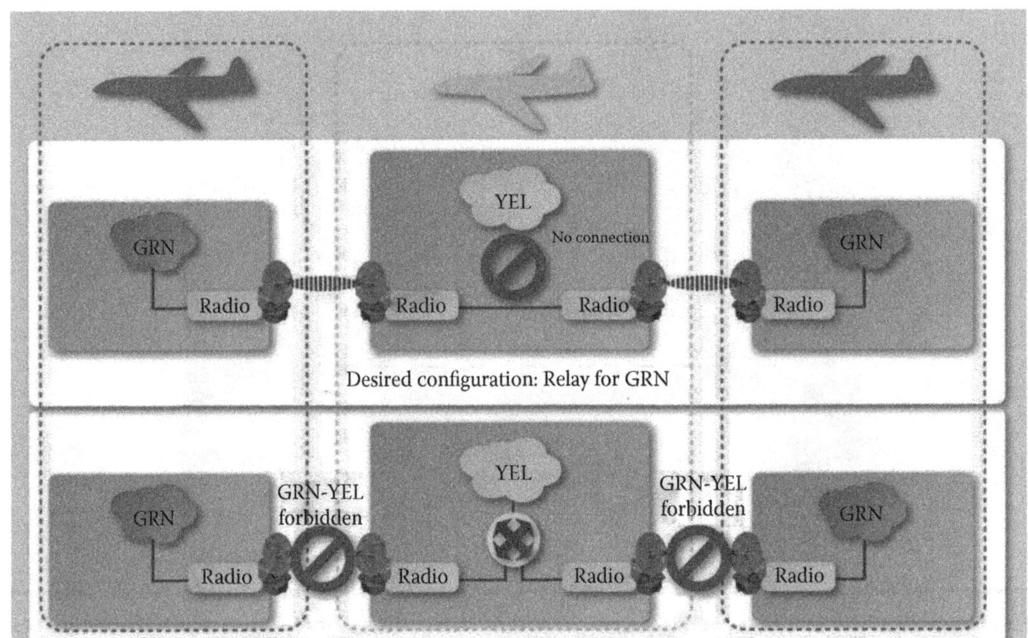

Figure 19.24 The problem of reusing radios for multiple purposes arises because of hard wiring radios to routers. The GRN network wants to have a relay service from the platform housing the YEL network. If the actual configuration is B, then the GRN network cannot get a relay service from the platform that houses the YEL network because that would force the merger of the GRN and YEL networks, which is forbidden.

or provided, including terminating the session while notifying the other party. A conceptual view of this arrangement is illustrated in Figure 19.25. Communications can be established between the platform hosting the green network, and the platform hosting the yellow network to request that the yellow network be configured for a particular kind of service (such as the bent-pipe relay service shown as a dotted line on the right-hand side). The communications path between authorization agents of the two platforms is illustrated as a *black* line. Note that a discovery service would also be connected to the service negotiation plane; so, initial requests could come from that direction as well.

The concept of a fully mobile network is slightly different than the concept of IP mobility. In the IP mobility paradigm, there is a permanent wired network (the Internet) from which all other mobile entities are subtended. The network is not mobile, per se. Rather, the individual hosts are mobile (mobile IP), or a mobile platform with an entire network of hosts is mobile (NEMO).

Achieving full mobility of the network would imply that the network could operate equally well untethered from the Internet as it could when tethered. Fully mobile networks could offer transport services to each other, could advertise whether they had access to other networks, and could provide forwarding services on behalf of hosts that subscribed to the service they offered.

Several features already discussed in this section would enable realization of the same. One feature is the ability to reconfigure the interconnection between radios and network elements such as routers or switches (such as was discussed in the section on disaster recovery scenarios). Another

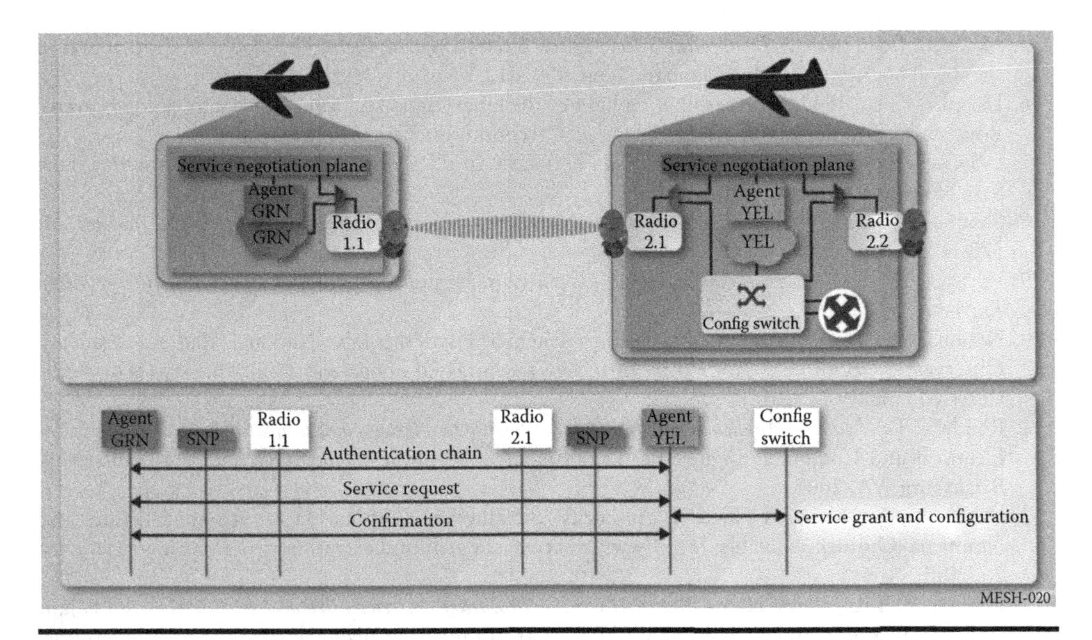

Figure 19.25 A service negotiation plane would permit messaging between different enclaves to allow local configuration of the system (say, to provide a bent-pipe relay service).

feature is the ability for one mobile network to negotiate services with another mobile network, or with a node that provided an entry point to the Internet or some other wired network (service negotiation plane). Since data forwarding would not be based on sharing a common routing protocol between the two networks, an SDN approach is quite likely.

Allowing the existence of individual hosts within another mobile network to be known outside that network would be an explicit decision made during the service negotiation process (as opposed to being implicitly assumed as part of the routing protocol). For one network to blindly tunnel all traffic from another network through it to some sort of gateway would be possible, just as it would be possible for one network to allow direct delivery of messages to hosts within the network. Maintaining the scalability of this type of a system would not be based on the necessity of knowing a path to every other node, or of knowing the full network state. Both the amount of network state information that needs to be exposed between mobile networks and the frequency of dissemination are significantly different than for a MANET approach. So, new paradigms would be needed to support this mode of operation.

References

1. Cebrowski, A. and J. Gartska, 1998. Network-centric warfare: Its origin and future. In *Proceedings,* vol. Dodccrp.org, Jan.
2. ALberts, D., J. Gartska, and F. Stein, 2000. *Network Centric Warfare: Developing and Leveraging Informaitn Superiority.* 2nd ed. p. 2, dodccrp.org, Feb.
3. DISA, [Online]. Available: http://www.disa.mil/services/enterprise-engineering.
4. Raduege, C. D., 2009. "GIG 2.0 Global Information Grid," National Defense Industrial Association, available on Web http//www.ndia.org/DoDEntArchitecture/Pages/presentationsDoD09.aspx. In *DoD Enterprise Architecture Conference,* Jun 1–4, 2009, St. Louis, MO.

5. Maxwell, J. C. 1865. *A Dynamical Theory of the Electromagnetic Field*, Philosophical Transaction of the Royal Society of London, January 1, pp. 459–212.
6. Doppler, C. A., 1842. On the coloured light of the binary stars and some other stars of the heavens. Royal Bohemian Society of Sciences, Prague, Czechoslovakia.
7. L-3 Communications, 2014. CDL FlyPlan Software. In *CDL Frequency Management Training*, Fort Sam Houston, Texas.
8. Shannon, C. E., 1948. A mathematical theory of communications, *Bell System Technical Journal*, 27, 379–423.
9. Friis, H. T., 1946. A Note on a Simple Transmission Formula, *Proceedings of the Institute of Radio Engineers*, 34(5), 254.
10. National Radio Astronomy Observatory. Antenna Fundamentals. National Radio Astronomy Observatory, [Online]. Available: http://www.cv.nrao.edu/course/astr534/AntennaTheory.html (accessed January 14, 2015).
11. Rondeau, T. W., 2007. *Application of artifical intelligence to wireless communications.* PhD dissertation Electrical and Computer Engineering Department of Virginia Polytechnica and State University, Blacksburg, VA, 2007.
12. IANA. Intenet Assigned Numbers Authority, the Internet Corporation for Assigned Names and Numbers, [Online]. Available: http://www.internetassignednumbersauthority.org/ (accessed January 26, 2015).
13. Rekhter, Y., T. Li, and S. Hares, 2006. *RFC 4271 "A Border Gatreway Protocol 4 (BGP-4)"*. IETF, Jan.
14. Bates, T., P. Smith, and G. Huston. CIDR Report. [Online]. Available: http://www.cidr-report.org/as2.0/ (accessed January 26, 2015).
15. Huston, G. "IPv6 CIDR Report," [Online]. Available: http://www.cidr-report.org/v6/as2.0/ (accessed January 26, 2015).
16. Perkins, C., 2010. *RFC 5944 "IP Mobility Support for IPv4, Revised"*. IETF, Nov.
17. Perkins,C., D. Johnson, and J. Arkko, 2011. *RFC 6275 "Mobility Support in IPv6"*. IETF, Jul.
18. Leung, K., G. Dommety, V. Narayanan, and A. Petrescu, 2008. *RFC 5177 "Network Mobility (NEMO) Extensions for Mobile IPv4"*. IETF, April.
19. Tsirtsis,G., V. Park, and V. Narayanan, 2012. *RFC 6626 "Dynamic Prefix Allocation for Network Mobility for Mobile IPv4 (NEMOv4)"*. IETF, May.
20. Moskowitz, R., P. Nikander, P. Jokela, and T. Henderson, 2008. *RFC 5201 "Host Identity Protocol"*. IETF, April.
21. Rekhter, Y., B. Moskowitz, D. Karrenberg, G. J. de Groot, and E. Lear, 1996. *RFC 1918 "Address Allocation for Private Internets"*, IETF.
22. Hinden, R. and B. Haberman. 2005. *RFC 4193 "Unique Local IPv6 Uncast Addresses"*, IETF.
23. Cesar, A. S., B. McDonald, I. Stavrakakis, and R. Ramathan, 2002. On the scalability of ad hoc routing protocols. In *IEEE INFOCOM*, New York.
24. Gupta, P. and P. R. Kumar, 2000. The capacity of wireless networks, *IEEE Transactions on Information Theory*, 46(2), 388–404.
25. Conti, M. and S. Giordano, 2007. Multihop ad hoc networking: The theory, *IEEE Communications Magazine*, 45(4), 78–86.
26. Clausen, T. and P. Jacquet, 2003. *RFC 3626 "Optimized Link State Routing Protocol (OLSR)"*. IETF, Oct.
27. Roy, A. and M. Chandra, 2010. *RFC 5820 "Extensions to OSPF to Support Mobile Ad Hoc networking"*. IETF, Mar.
28. Ogier, R. and P. Spagnolo, 2009. *RFC 5614 "Mobile Ad Hoc Network (MANET) Extensions of OSPF Using Connected Dominating Set (CDS) Flooding"*. IETF, Aug.
29. BBN, "BBN Technical Memorandum No. 1301 "Hazy Sighted Link State (HSLS) Routing: A Scalable Link State Algorithm", 31 Aug 2001. [Online]. Available: http://www.ir.bbn.com/documents/techmemos/TM1301.pdf (accessed January 26, 2015).
30. Wikipedia, "Ant colony optimization algorithms," [Online]. Available: http://en.wikipedia.org/wiki/Ant_colony_optimization_algorithms (accessed January 26, 2015).
31. Becker, D. and A. Vahdat, 2000. Epidemic routing for partially connected ad hoc networks. *Technical Report CS-2000-06*, Dept of Computer Science, Duke University, April.

32. Elmasry, G. F., M. Jain, R. Welsh, K. Jakobowski, and K. Whittaker, 2011. Network management challenges for joint forces interoperability, *IEEE Communications*, 49(10), pp. 81–89.
33. The Open Group. Service Oreinted Architecture, [Online]. Available: http://www.opengroup.org/subjectareas/soa (accessed January 26, 2015).
34. Wikipedia. Net-centric, [Online]. Available: http://en.wikipedia.org/wiki/Net-centric (accessed January 26, 2015).
35. Wikipedia. OSI Model, [Online]. Available: http://en.wikipedia.org/wiki/OSI_model (accessed January 26, 2015).
36. ITU. ITU-T Recommendation Series X: Data Networks and Open System Communications, 10/2003. [Online]. Available: http://www.itu.int/ITU-T/recommendations/rec.aspx?rec=7024 (accessed January 26, 2015).
37. Cheng, B.-N., J. Wheeler, and L. Wyster, 2012. Radio-to-router interface technology and its applicability on the tactical edge, *IEEE Communications Magazine*, 50(10), 70–77.
38. Schug, T., C. Dee, N. Harshman, and R. Merrell, 2011. Air force aerial layer networking tranformation initiatives. In *MILCOM*, November 2011, Baltimore, MD, pp. 1974–1978.
39. Wikipedia, Software-Defined Networking, [Online]. Available: http://en.wikipedia.org/wiki/Software-defined_networking. (accessed January 26, 2015).

Chapter 20

Collaborative and Opportunistic Content Dissemination via Directional Antennas

Yong Li, Wei Feng, Li Su, Xiang Chen, and Depeng Jin

Contents

Abstract

With the popularity of mobile data and applications, we introduce the system of collaborative and opportunistic mobile content dissemination with directional antennas, where the content is distributed within the network by device-to-device opportunistic communications and the network nodes are equipped with directional antennas. In such a system, opportunistic contact between mobile devices carried by humans or vehicles, is capable of providing high-bandwidth communication capacity for content dissemination via short-range communication technologies like millimeter wave communications, Ultra-Wide Bandwidth (UWB), and so on, which is known as opportunistic collaborative content dissemination.

In this chapter, we first introduce the system overview of collaborative and opportunistic mobile content dissemination with directional antennas. Then, we introduce the physical-layer technologies to support this kind of short-range opportunistic transmissions via directional transmissions, with the technologies of beamforming and efficient codebook design. Then, we introduce the collaborative and opportunistic mobile content dissemination system and to analyze the different mobility models used in these two different systems and analyze its impact on the directional transmission design. Finally, using a fluid approximation, we analyze the system performance of collaborative and opportunistic mobile content dissemination with directional antennas. Specifically, we derive a theoretical model to depict the system performance of content dissemination time. The accuracy of the proposed analysis is confirmed by simulation results, which also show that the directional antenna performs better than the omnidirectional antenna in our considered scenario, especially when the antenna beam is well scheduled with small beamwidth and high beam steering rate.

20.1 Introduction

Recently, interests on large-scale vehicular ad hoc networks have grown significantly [1], as more and more human and vehicles are equipped with smart devices to provide communication capacities. Many applications of smart phones and vehicular networks are also emerging, including mobile video steaming, automatic collision warning, remote vehicle diagnostics, emergency management and assistance for safely driving, vehicle tracking, automobile high speed Internet access, and multimedia content sharing. Specially, in terms of vehicular communications, in the United States, Federal Communications Commission has allocated 75 MHz of spectrum for dedicated short-range communications in vehicular networks [2], and IEEE is also working on the related standard specifications. Efficient content dissemination is a key issue for many vehicular network applications, such as content publishing for safety information and entertainment data [3]. These content dissemination applications might be supported within the existing wireless infrastructure, such as WiFi and third generation, but the coverage issue and economic consideration from both service providers and end users do not make such a solution efficient and viable. Another reason is that the wireless infrastructure may be congested, or even damaged in disaster. Consequently, content dissemination through vehicular ad hoc networks is highly desired. Since human networks and vehicular networks are highly mobile and sometimes sparse by nature, it is difficult to maintain a connected network to distribute the content [4]. Opportunistic contact between devices, by contrast, is capable of providing high-bandwidth communication capacity for content dissemination, which is known as opportunistic collaborative vehicular content dissemination [4].

In this chapter, we analyze the system performance of opportunistic content dissemination, where the content can be transmitted only when two devices come into the communication range of each other. Specifically, we use the vehicular network as an example to study this problem. Therefore, the *vehicular mobility* model and the *communication range* are two of the most important factors that influence the system's performance. We have collected a large vehicle trace in Beijing, and validate a suitable analytical mobility model based on this *Beijing* trace in this study. With regarding the issue of transmission range, we consider the applicability of directional antennas that have more than one steerable directional beam to improve the system's achievable performance. Although there exist some recent works on content dissemination [5,6,3,7], to the best of our knowledge, this is the only study on the performance of the collaborative content dissemination system under realistic vehicular scenarios using an accurate mobility model and directional antennas. Our contributions are as follows.

- We introduce the Lévy-walk model [8] for the vehicular mobility, and validate this model on the *Beijing* trace, which is the largest available vehicular mobility trace. This model is used to setup a realistic vehicular mobility environment for the performance evaluation of collaborative vehicular content dissemination.
- In order to enable the proximity directional communication, we propose a codebook design and corresponding training procedure based on circular antenna array with a two-layer structure. It allows every pattern type to own its corresponding codebook and highly directional antenna radiation pattern. Moreover, to enable multicast transmission, we propose a transmit and receive multicast beamforming scheme to promote the physical layer communication efficiency.
- We provide a theoretical model to analyze the content dissemination speed for collaborative vehicular content dissemination by a fluid approximation, and use extensive simulation results obtained under realistic vehicular settings to validate the accuracy of our model. Furthermore, we evaluate the influence of the directional antenna on the performance of the vehicular content dissemination system by both analytical and simulation results. Our results confirm that the directional antenna performs better than the omnidirectional antenna.

The rest of this chapter is organized as follows. We introduce the system overview in Section 20.2. In Sections 20.3 and 20.4, we design the physical layer directional communication technologies in terms of codebook design, training procedure, and multicast beamforming. Then, in Section 20.5, we use a fluid approximation, and analyze the system performance of collaborative and opportunistic mobile content dissemination with directional antennas. Finally, we conclude in Section 20.6.

20.2 System Overview

20.2.1 System Description and Model

Figure 20.1 illustrates the concept of collaborative vehicular content dissemination, where the goal is to disseminate the content to a group of subscribers through opportunistic communication. As not all the nodes are willing to participate in the content dissemination, there exist two types of vehicle nodes in the system, known as *helpers* and *subscribers*. Helpers are willing to buffer the content in their storages, and to further transmit the content to other helpers or subscribers.

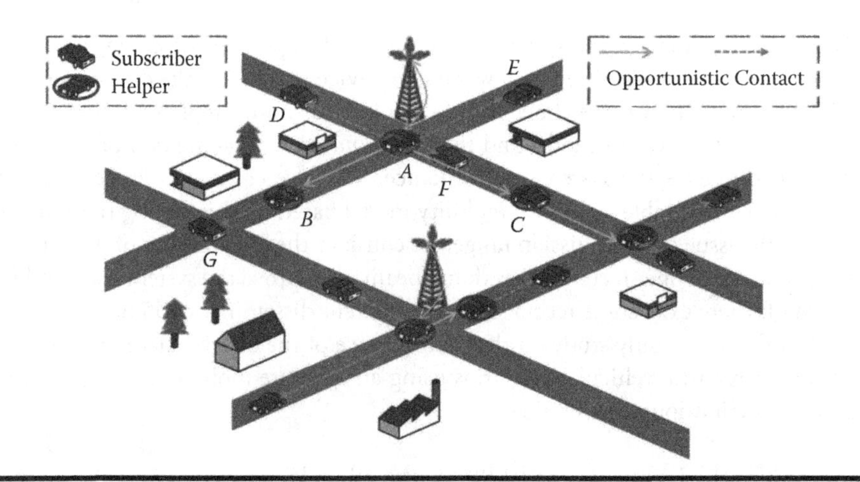

Figure 20.1 System overview of the collaborative vehicular content dissemination. The infrastructure network first transmits the content to some helpers, which then disseminate the content to other encountered helpers (solid line) or encountered subscribers (dashed line) through opportunistic communication.

Subscribers are only interested in receiving the content and will not transmit the content to other nodes. Nodes are equipped with omnidirectional or directional antennas to transmit or receive packets of the content. At the beginning, a small number of helpers will obtain the content from the content source. Content dissemination may be split into two phases, namely, the disseminating phase, during which the content is transmitted among the helpers, and the receiving phase, when a subscriber finally receives the required content from a helper. Considering Figure 20.1 again, in the disseminating phase, helper *A* transmits the content to helpers *B* and *C* by opportunistic contact. Helpers *B* and *C*, after received the data from *A*, will carry and forward the data to other helpers or subscribers later. In the receiving phase, subscribers *D*, *E*, and *F* receive the packet from helper *A*, while subscriber *G* receives the packet from helper *B*.

20.2.2 Vehicular Mobility

Existing mobility models, such as random walk and random waypoint, cannot realistically represent the collaborative vehicular content dissemination system, where wireless devices are attached to vehicles and vehicular mobility patterns influence the system performance significantly. Let us consider the two-dimensional (2D) vehicular mobility defined by a sequence of steps that a vehicle travels. A step is denoted by a tetrad (l, ϕ, v, τ) during which a vehicle travels a flight followed by a pause, where $l > 0$ is the flight length, ϕ is the direction of the flight, v is the mobility velocity, and τ is the time duration of pause called the pause time. Thus, step n is defined by $(l_n, \phi_n, v_n, \tau_n)$. Assume that the vehicle starts its first step at time $t = 0$. It chooses a direction ϕ_1 randomly from the uniform distribution in the range [0, 360°] and has a uniformly distributed velocity v_1, as well as chooses a flight length l_1 and a pause time τ_1 according to certain probability distributions. Consequently, in step 1, the vehicle moves the flight of the length l_1 at the direction ϕ_1 with the velocity v_1. It then stops for the pause time τ_1 during which the vehicle stays at the location where the current flight ends. After the pause, it chooses another step, and the process repeats. Clearly, the accuracy of a vehicular mobility model is determined by the accuracy of the distributions for flight length l and pause time τ.

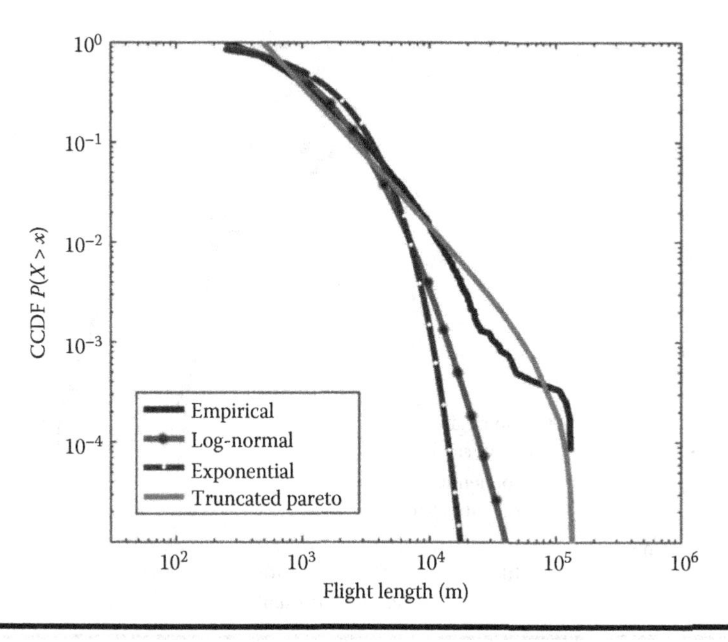

Figure 20.2 Comparison of the empirical flight-length distribution extracted from *Beijing* trace with the fitted log-normal, exponential, and upper-truncated Pareto distributions.

We use the largest available real vehicular trace, the *Beijing* trace which we collected ourselves, to study the distribution of flight length and pause time by curve fitting techniques, which is widely used in the mobility modeling [9–11]. This trace contains the mobility track logs obtained from 2700 participating taxis carrying global positioning system (GPS) receivers during the whole month of May in 2010. To obtain this trace, we utilized the GPS devices to collect the taxi locations and timestamps and general packet radio service modules to report the records every one minute for moving taxis. Based on the *Beijing* trace, we utilize the angular model of Reference 11 to extract the data of flight length and pause time, and subsequently to study the distributions of flight length and pause time. Figure 20.2 shows the empirical complementary cumulative distribution function (CCDF) of flight length, where we also applied the maximum likelihood estimation to fit three known distributions—the exponential, log-normal, and truncated Pareto [12]—to the data. We observe that the truncated Pareto has the best fit to the empirical CCDF among the three distributions. Similarly, Figure 20.3 compares the empirical CCDF of the pause time extracted from the trace with the three fitted known distributions, where it can again be seen that the truncated Pareto distribution provides the best fit to the empirical pause-time distribution. More specifically, the mean squared error (MSE) between the empirical distribution and the truncated Pareto distributed flight length is 17.2%, while the MSE for truncated Pareto distributed parse time is 15.5%. In comparison, for example, for the exponential distribution, the MSE values are 34.9% and 318.3% for the flight length and parse time, respectively. This experiment suggests that both the pause time and flight length follow the truncated Pareto distribution. Therefore, we use the Lévy walk [8] to model the *vehicle* mobility. Note that the existing work [11] has also validated that the Lévy walk can accurately model the *human* mobility by demonstrating that the truncated Pareto distribution can also fit the flight length and parse time of human mobility very well.

In summary, in our vehicular mobility model, the direction and velocity follow the uniform distributions, while the Lévy-walk model selects the flight length and pause time randomly with

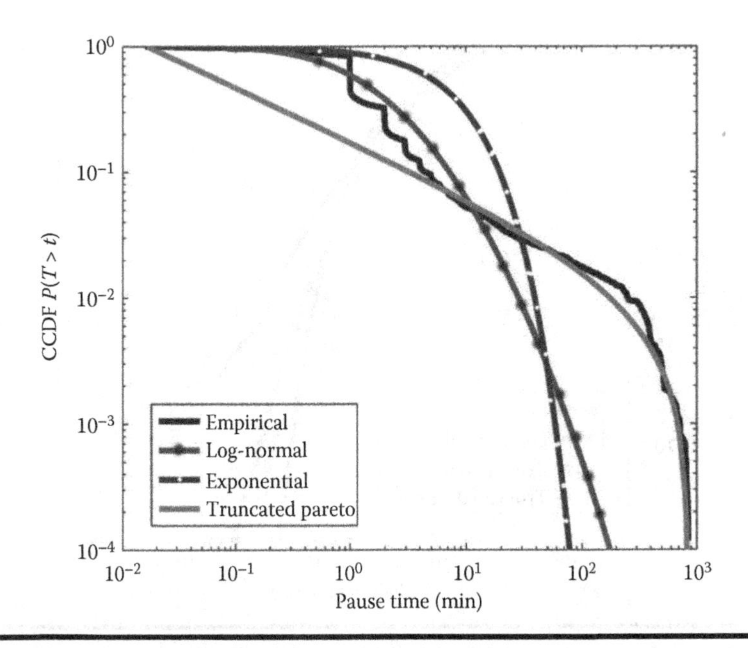

Figure 20.3 Comparison of the empirical pause-time distribution extracted from *Beijing* trace with the fitted log-normal, exponential, and upper-truncated Pareto distributions.

the truncation factors $\xi^f_{\min}$ and $\xi^f_{\max}$ for l and $\xi^p_{\min}$ and $\xi^p_{\max}$ for τ, respectively, according to the Lévy distribution with the exponent parameter ε, whose characteristic function is defined by

$$f(x) = \frac{1}{2\pi}\int_{-\infty}^{+\infty} e^{-jxz-|cz|^{\varepsilon}}\,dz \tag{20.1}$$

where $j = \sqrt{-1}$ and c is a scaling coefficient. Specifically, the initial location of each vehicle is randomly chosen from a uniform distribution in the defined area. At step n, the tetrad $(l_n, \phi_n, v_n, \tau_n)$ is generated randomly according to the corresponding distributions. If the drawn duplet (l_n, τ_n) does not pass the truncation checking, that is, $l_n < \xi^f_{\min}$, or $\tau_n < \xi^p_{\min}$, or $l_n > \xi^f_{\max}$, or $\tau_n > \xi^p_{\max}$, then it is discarded and another duplet is regenerated. This procedure is repeated until the mobility pattern of the whole network is obtained.

20.2.3 Antenna Orientation

The antenna dynamics of node i are denoted by a duplet $A_i = (\vartheta_i(t), \theta_i(t))$, where the antenna orientation $\vartheta_i(t)$ is related to the beam steering direction, while the antenna beamwidth $\theta_i(t)$ is related to the antenna patterns. Both the beam steering and antenna patterns can change according to the system requirements. However, there are practical limitations in directional antenna implementation, and hence we model the antenna under realistic settings [13]. The beamwidth θ_i is chosen from the set {15°, 30°, 45°, 60°, 90°, 180°}. Given a θ_i, there are $720/\theta_i + 1$ beam patterns, one for omnidirectional beam and $720/\theta_i$ for directional beams, each with an approximately $(\theta_i)°$ half-power beamwidth. Each directional beam is overlapping with the next beam and rotated by $(\theta_i/2)°$

to the next, and all the $720/\theta_i$ beams cover the 360° circle. We use an extensively used approximate model [14] for the antenna gain. Given the beamwidth θ_i, the main lobe gain, denoted by g_m, is defined as

$$g_m(\theta_i) = \frac{4}{\tan^2(\theta_i/2)} \tag{20.2}$$

where g_m is obtained as the maximum beamwidth with no energy leakage. As the antenna direction changing is often caused by the upper layer traffic or packet sending requests, whose rate is usually modeled by Poisson process, we assume the antenna orientation ϑ_i changes its direction following the Poisson process with rate r_ϑ. If $r_\vartheta = 0$, the antenna orientation never changes. We define three policies for the beam steering: random steering (RS), circle steering (CS), and polling steering (PS). In RS, nodes randomly choose new beams from the feasible set of antenna patterns when Poisson changing events occur. CS chooses the next beam which is not overlapped with the current beams clockwise, while PS selects the next beam which is overlapped with the current beams clockwise. Since helpers need to transmit the content to as many as possible subscribers, it is better for a helper to transmit the content in all directions by using an omnidirectional antenna. By contrast, a subscriber needs to receive the content from one of the helpers. Therefore, it can use a directional antenna to point to one of the helpers, or it may still use an omnidirectional antenna. In our study, we investigate whether a directional or an omnidirectional antenna is more beneficial in this context.

20.2.4 Content Transmission

When the content transmission occurs is decided by the physical propagation model. For any node, we can use the triplet $S(t) = (\mathbf{x}(t), \vartheta(t), \theta(t))$ to represent its state at time t, where $\mathbf{x}(t)$ is the node's position which is determined by the mobility model, $\vartheta(t)$ and $\theta(t)$ are the antenna orientation and beamwidth, respectively. Suppose that, at time t, vehicle i has the state $S_i = (\mathbf{x}_i, \vartheta_i, \theta_i)$, that is, the vehicle is at the position $\mathbf{x}_i$ and its antenna is pointing in the direction ϑ_i with the beanwidth θ_i, while another vehicle j has the state $S_j = (\mathbf{x}_j, \vartheta_j, \theta_j)$. Imagine that i is a helper which has already obtained the content, while j is either a helper or subscriber without the content. The content transmission from i to j will happen only when node j can capture the signal sent by node i with a power above a certain threshold denoted by ψ. In vehicular opportunistic networks, nodes are usually spare and, furthermore, no end-to-end path exits between nodes. From this point of view, the free space path model is adequate for the physical communication channel. Therefore, our propagation model uses the following equation [14] to compute the received power

$$P_r(S_i, S_j) = \frac{P_t \cdot \lambda^2 \cdot d_{\text{ref}}^2 \cdot G_t(A_i, \mathbf{x}_i, \mathbf{x}_j) \cdot G_r(A_j, \mathbf{x}_i, \mathbf{x}_j)}{4\pi^2 \cdot |\mathbf{x}_i - \mathbf{x}_j|^4} \tag{20.3}$$

where P_t and P_r are the transmit and receive powers, respectively, λ is the wavelength, G_t and G_r are the gains of the transmit and receive antennas, respectively, while d_{ref} is a reference distance given by $d_{\text{ref}} = 2D/\lambda^2$ with D being the maximum antenna dimension. The gains $G_t(A_i, \mathbf{x}_i, \mathbf{x}_j)$ and $G_r(A_j, \mathbf{x}_i, \mathbf{x}_j)$ depend on the antenna patterns and the relative positions of two nodes. Recalling the definition of g_m in Equation 20.2, the expression for G_t is given by

$$G_t(A_i, \mathbf{x}_i, \mathbf{x}_j) = \begin{cases} g_m, & \vartheta_i \cdot \Delta\mathbf{x}_{ij} \le \cos(\theta_i/2), \\ 0, & \vartheta_j \cdot \Delta\mathbf{x}_{ij} > \cos(\theta_j/2), \end{cases} \tag{20.4}$$

where $\Delta\mathbf{x}_{ij} = \mathbf{x}_i - \mathbf{x}_j$. The expression for $G_r(A_j, \mathbf{x}_i, \mathbf{x}_j)$ is similar. If $P_r(S_i, S_j) \ge \psi$, node j can receive the content from node i.

20.3 Codebook Design for Beamforming

A beamforming protocol including codebook design and training method has been given in IEEE 802.15.3c standard. Codebook refers to several sets of antenna weight coefficients, and training stands for the process finding the best beam pair. The 3c codebook only uses phase modulation without amplitude adjustment, which reduces the power consumption of radio frequency (RF) devices. The 3c training procedure consists of three stages (DEV-to-DEV linking, sector-level searching, and beam-level searching), which saves time compared with complex enumeration search accomplished in a single stage. The protocol defines three beamforming pattern types: quasi-omni pattern, sector, and beam. However, the 3c codebook is only designed for the beam. The other two pattern types, that is, quasi-omni pattern and sector, do not have corresponding codebooks [15]. Besides, the gain loss due to codebook limitations is relatively high, and training setup time is still long. Some improvements on beamforming algorithm (especially on codebook design) have been proposed, such as References 15 and 16. Further, Reference 15 provides codebooks for all three pattern types, and Reference 16 helps us regarding how to reduce gain loss. Nevertheless, their beams still point to an unwanted direction besides the expected one due to limitations of linear antenna array.

In this section, a beamforming algorithm including codebook design and corresponding training method is proposed, which is based on a two-layer circular antenna array structure. It allows all three beamforming pattern types to have their own codebooks. Beams are highly directional and are evenly distributed in the range of 360° because of using circular antenna array. Besides, the two-layer structure helps to follow the training procedure faster. According to performance evaluations, the designed algorithm shows great advantages in the aspect of gain loss, robustness, and antenna radiation patterns compared with the 3c standard.

20.3.1 Codebook Design

The beamforming system model is shown in Figure 20.4. At the transmitter signals are up converted to RF signals and transmitted by antenna array, while an opposite procedure takes place at the receiver. We can observe that each antenna element has a complex weight coefficient. All weight coefficients of the antenna array form an antenna weight vector (AWV) [17]. The main goal of beamforming is to find the best transmit and receive AWV pair to achieve better channel performance according to certain criterion, for example, maximizing received power.

Codebook is a matrix. Each column of the matrix is an AWV and it refers to a certain beam pattern. Each row of the matrix stands for weight coefficients of an antenna element [18]. Codebook defined in the 3c standard can be seen in Reference 19.

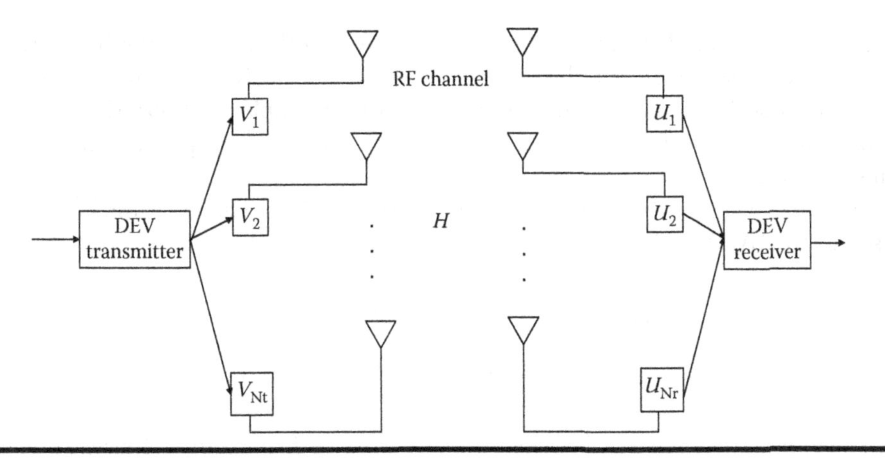

Figure 20.4 Beamforming system model.

Array factor of the circular antenna array is defined as:

$$A_k(\theta) = \sum_{n=0}^{\widetilde{N}-1} w(n,k) e^{j(2\pi/\lambda) r \cos(\theta - (2\pi n/\widetilde{N}))}, \quad k = 0,1,2,\ldots,K-1, \tag{20.5}$$

where w, n, and k are codebook matrix, row, and column index of the matrix, respectively; $\widetilde{N}$ and K are numbers of antenna elements and beam patterns, respectively; λ is the wave length, and r is the radius of circular antenna array [20]. Circular antenna array has a feature that when a weight vector is circularly shifted by an element, the beam pattern keeps the original shape but rotates an angle of $2\pi/\widetilde{N}$. Therefore, the beams of circular codebook are evenly distributed and they share the same properties such as antenna gain.

A codebook design for all three beamforming pattern types is given. We propose a two-layer circular antenna array structure, as shown in Figure 20.5. The circles depict the outside layer while

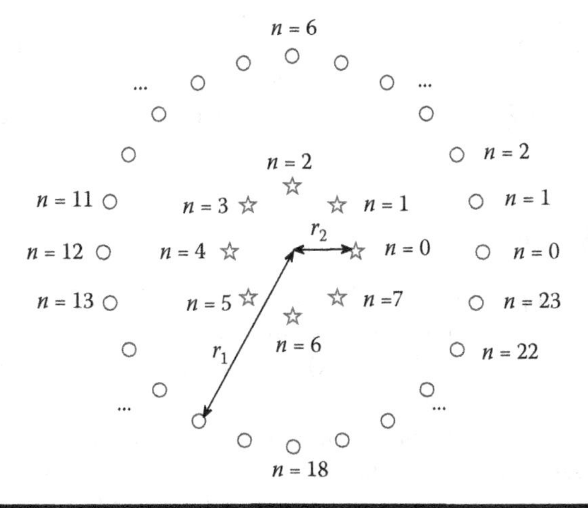

Figure 20.5 Two-layer antenna array structure (with 24 outside layer elements).

the stars stand for the inside layer, and antenna elements of both layers are numbered from $n = 0$. The outside layer is designed to produce beams. Its array radius r_1 is $N\lambda/16$, where N denotes the number of antenna elements of the outside layer, and N is a multiple of 8 (more than 8). The inside layer is used to generate sectors and quasi-omni patterns. It consists of eight antenna elements and its radius r_2 is $\lambda/2$.

To maximize array factor value on the strongest emission direction, also known as main response axis (MRA) [16], the proposed beam codebook is defined as:

$$w(n,k) = e^{-j(2\pi/\lambda)r_1 \cos((2\pi \times \mathrm{mod}((n-k),N))/N)},$$
$$n = 0,1,2,\ldots,N-1; \; k = 0,1,2,\ldots,K-1; \; N = K. \tag{20.6}$$

Each column of the matrix is a circular shift result of the first column. Thus we only need to calculate the first column and then perform cyclic shift operations instead of calculating every AWV for different beams.

Sectors and quasi-omni patterns are produced by the eight inside layer antenna elements. The sector codebook is as follows:

$$w(n,k) = e^{-j(2\pi/\lambda)r_2 \cos((2\pi \times \mathrm{mod}((n-k),8))/8)},$$
$$n = 0,1,2,\ldots,7; \quad k = 0,1,2,\ldots,7. \tag{20.7}$$

The codebook for quasi-omni pattern is defined as:

$$w(n,k) = \begin{cases} 0, & (n-k) \text{ is odd} \\ e^{-j(2\pi/\lambda)r_2 \cos(2\pi \times \mathrm{mod}((n-k),8)/8)}, & (n-k) \text{ is even} \end{cases}$$
$$n = 0,1,2,\ldots,7; \; k = 0,1. \tag{20.8}$$

Formulas 20.6 through 20.8 are codebooks for beam, sector, and quasi-omni pattern, respectively. There are two quasi-omni patterns and eight sectors in all cases. The shape of each quasi-omni pattern is almost the same as the combination of four sectors, which is depicted in a more detailed manner in Section 20.4.2. This relation between shapes helps training procedure faster and more accurate.

20.3.2 Training Stage

Training stage is also an important part of beamforming besides codebook design. It refers to the methods of searching a beam to achieve the best link quality by using signal-to-noise ratio (SNR), Signal to Interference and Noise Ratio (SINR), or channel capacity as judging criterion. In this study, we choose SNR as the criterion.

In IEEE 802.15.3c beamforming protocol, training procedure contains three stages: DEV-to-DEV linking, sector-level searching, and beam-level searching. DEV-to-DEV linking tries to find the best quasi-omni pattern pair, while sector-level searching and beam-level searching are aimed at finding the best sector pair and beam pair, respectively. The proposed circular antenna array training can also be divided into these three stages.

Stage 1: DEV-to-DEV linking: In DEV-to-DEV linking stage, outside layer elements are turned off while those of inside layer are turned on. From Figure 20.6a, we can find that there are altogether two quasi-omni patterns, which can nearly cover the whole range of 360°. The full line figure depicts quasi-omni pattern 0 while the dotted line depicts quasi-omni pattern 1.

The training procedure of this stage is as same as that of 3c training. In this study, we assume the system is a symmetric antenna system (SAS). There are two quasi-omni patterns at the transmitter (Tx-quasi) and two quasi-omni patterns at the receiver (Rx-quasi). The transmitter sends training sequences (TSs) using 2 Tx-quasi, and the receiver uses 2 Rx-quasi to receive TSs from every Tx-quasi. The receiver records SNR values of every quasi-omni pattern pair and selects the best one, as well as sends the feedback information about the best pair to the transmitter. After $2 \times 2 = 4$ training processes, the best quasi-omni pattern pair can be found.

Stage 2: Sector-level searching: Using 20.7, eight sectors can be generated with intervals of 45° and they are numbered from S_0 to S_7. Four of the sectors have the similar shape with quasi-omni pattern 0 (Figure 20.6b) and the other four are shaped like quasi-omni pattern 1 (Figure 20.6c). We assume that Tx-quasi 0 and Rx-quasi 1 are the best match in previous stage. In sector-level searching, the receiver keeps the best Rx-quasi, and the transmitter uses four sectors that correspond to the shape of the best Tx-quasi, as shown in Figure 20.6b. After that, we can find the

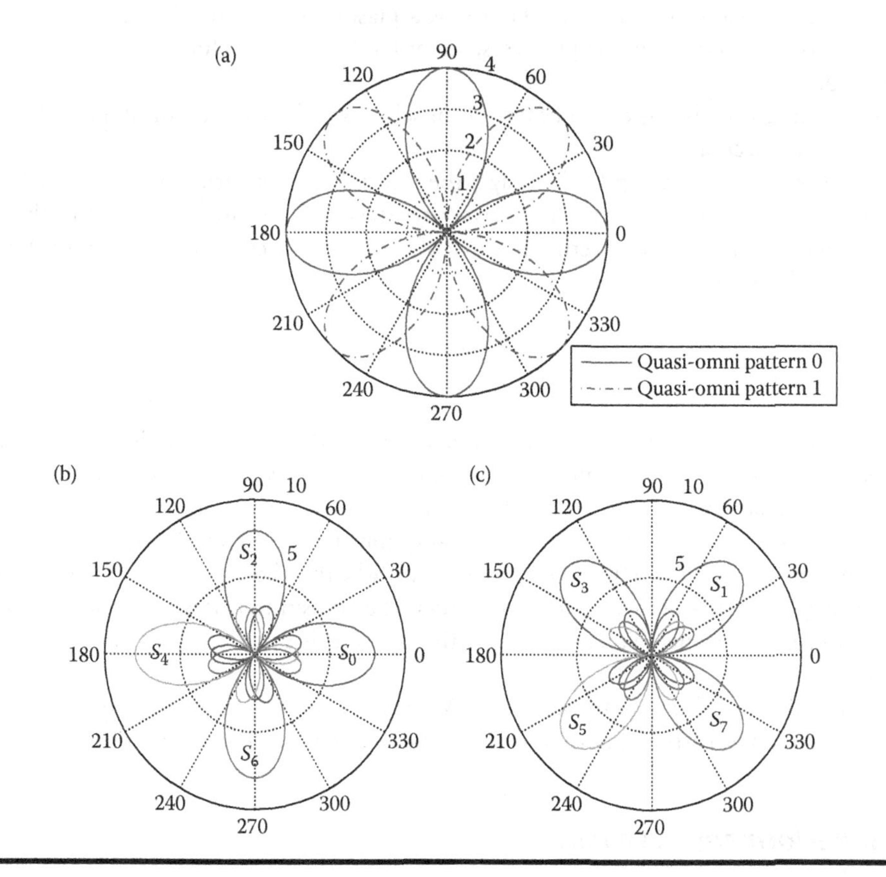

Figure 20.6 Coarse beam selection (using eight inside layer elements): (a) Quasi-omni patterns of circular codebook; (b) Sectors corresponding to the best Tx-quasi; and (c) Sectors corresponding to the best Rx-quasi.

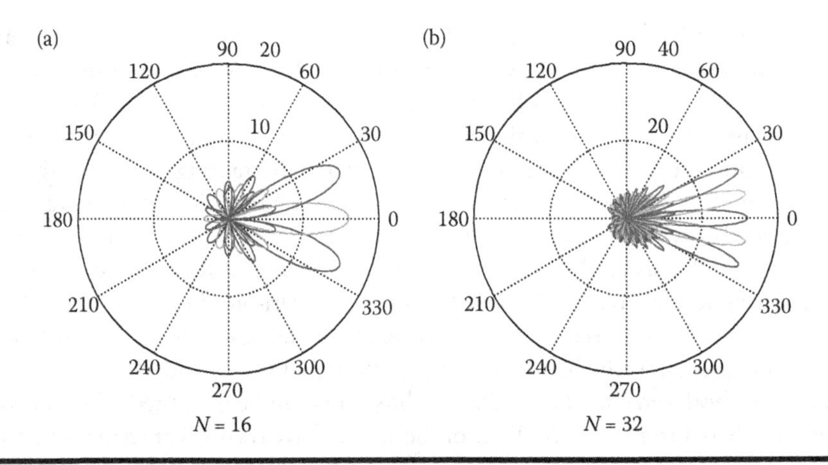

Figure 20.7 Beam-level searching of circular codebook. (a) Beam pattern when $N = 16$, and (b) beam pattern when $N = 32$.

best Tx-sector. Then the transmitter keeps sending the best Tx-sector while the receiver uses four sectors that correspond to the shape of the best Rx-quasi (Figure 20.6c). The best Rx-sector can be selected. After $2 \times 4 = 8$ training processes, sector-level searching is finished and the best sector pair is found.

Stages 1 and 2 can also be called coarse beam selection, because we can find a sector (coarse beam) within 45° accuracy.

Stage 3: Beam-level searching: In this stage, the transmitter first turns on the outside layer elements and turns off the inside layer. The receiver uses the best Rx-sector and the transmitter sends a certain number of fine beams covering the area of the best Tx-sector. The criterion to choose the number of fine beams is as follows:

$$N_{\text{beam}} = 2^{\text{round}(\log_2 N)-3} + 1 \tag{20.9}$$

where N_{beam} refers to the number of selected fine beams, and function round() returns the nearest integer to its argument. The center one of the fine beams has the same MRA direction with the best sector. Examples can be seen in Figure 20.7, which shows the criterion under the circumstances of $N = 16$ and $N = 32$ with the assumption that the azimuth angle of MRA direction of the best sector is 0°. After finding the best Tx-beam, the receiver turns on the outside layer elements and turns off the inside layer. Then the transmitter uses the best Tx-beam and the receiver uses fine beams to receive TSs. At the end of this stage, the best fine beam pair can be selected.

Stage 3 is also called fine beam selection. We can get beams with increased directivity and accuracy, which can be further improved with the increase in the number of outside layer elements.

20.3.3 Performance Evaluations

Evaluations are performed in terms of antenna radiation patterns, antenna directivity, training setup time, and so on. The analysis of simulation results and comparison between the proposed beamforming method and 3c protocol are as follows.

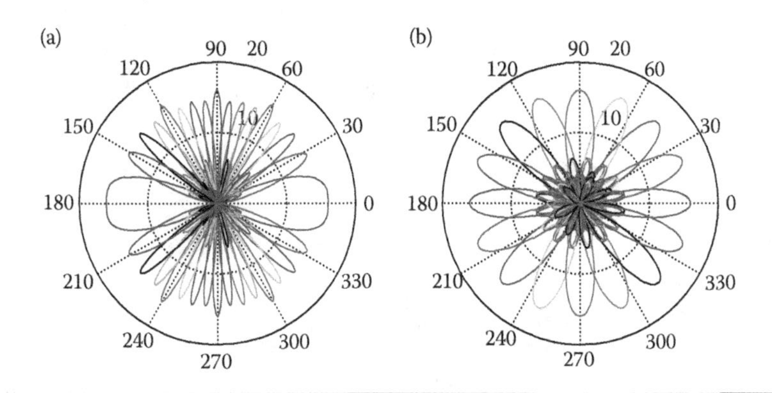

Figure 20.8 2D antenna patterns of 3c and circular codebook ($N = K = 16$): (a) 3c codebook; (b) Circular codebook.

In reality the antenna radiation pattern is a 3D picture. However, in order to facilitate the study of antenna pattern, the main cross section of 3D picture (also called main 2D antenna pattern) is often used to analyze beam properties [21].

Figure 20.8 shows the 2D antenna beam patterns of 3c and circular codebook when $N = K = 16$. It is clear that beams of circular codebook are evenly distributed in the range of 360° and all beam widths are equal to each other. The antenna pattern of linear antenna array is symmetrical with respect to x-axis; so, almost every beam points to an additional unwanted direction besides the expected direction. However, circular antenna array is able to generate beams that only focus on one direction.

The 3D antenna pattern of circular codebook is shown in Figure 20.9a. All beams of circular codebook are similar to it, and they are just like stretching corresponding main 2D antenna patterns along z-axis. The shape of 3c codebook beams varies and Figure 20.9b is an example, which shows that some 3D antenna patterns of 3c beams are shaped like a funnel. The pattern spreads to many directions along z-axis, and the direction of the beam becomes unclear. We can also see that the vertical view of Figure 20.9b is totally different from its main 2D antenna pattern on x–y plane, as shown in Figure 20.9c and 20.9d.

In a word, circular codebook generates more directional beams than 3c codebook no matter in the case of 2D or 3D antenna radiation patterns.

Antenna directivity measures the power density the antenna radiates in a certain direction, versus the power density radiated by an ideal isotropic antenna with the same total power. The relation between antenna gain G and directivity D is:

$$G = eD, \quad e \in [0,1] \tag{20.10}$$

where e is antenna electrical efficiency, and it is always assumed to be 1 when doing theoretical study. Generally the maximum antenna directivity is used to analyze antenna properties and it refers to the directivity on MRA direction, which can be written as:

$$D_0 = \frac{\max_\theta |A(\theta)|^2}{\mathbf{w}^{\mathrm{H}}\Omega\mathbf{w}} \tag{20.11}$$

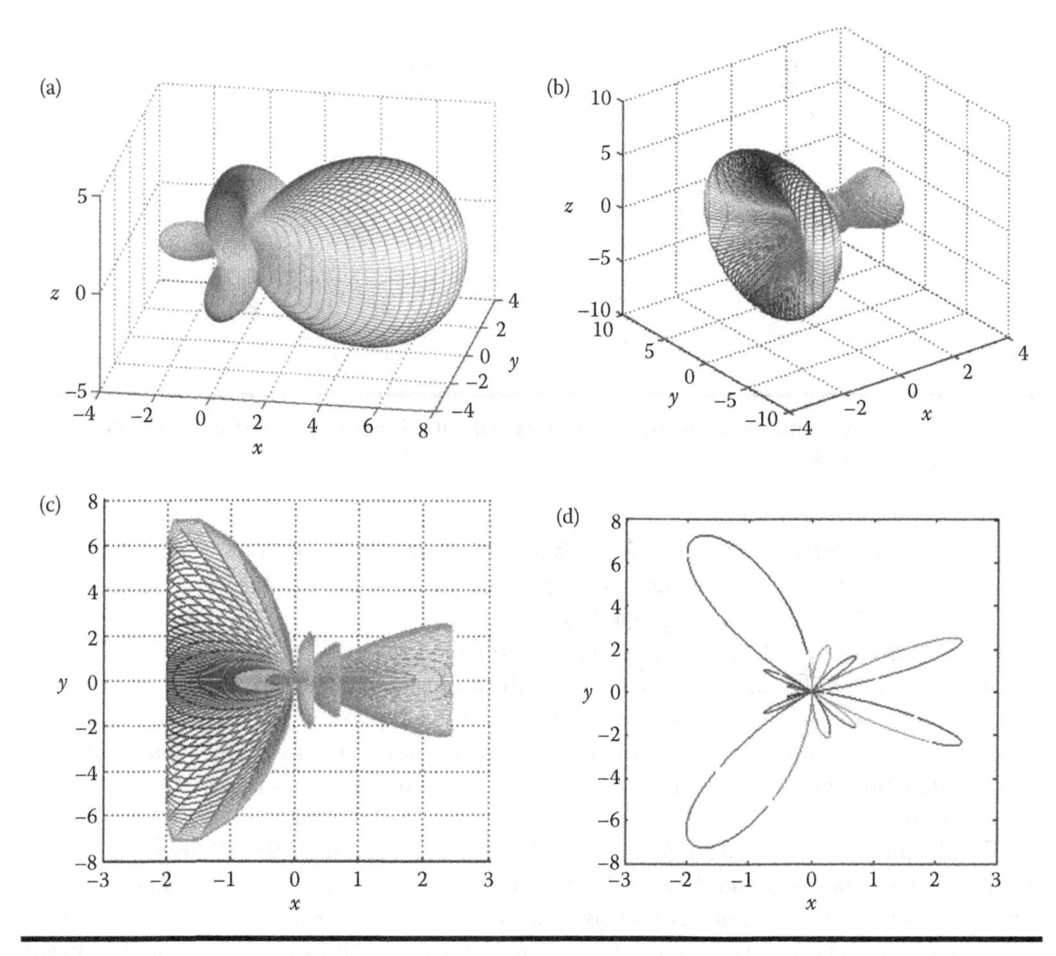

Figure 20.9 3D antenna patterns: (a) 3D antenna pattern of circular codebook; (b) 3D antenna pattern of 3c codebook; (c) Vertical view of 3D antenna pattern in (b); and (d) Main 2D antenna pattern of (b).

where $\max_\theta|A(\theta)|^2$ means the square of the array factor value on MRA, $\mathbf{w}^H$ stands for Hermitian transpose of weight vector $\mathbf{W}$, and Ω is a matrix whose definition is related to the structure of antenna array. D_0 is decided by the property of antenna array and codebook design.

Let $N = 16, 24, \ldots, 56$, the comparison of directivity values between 3c codebook and circular codebook is shown in Figure 20.10. Since the shape of beams varies in 3c codebook, D_0 changes according to the beam. D_0 of beams generated by circular codebook is a constant when antenna element number is fixed, and it is higher than the minimum value of D_0 in 3c codebook, but lower than its maximum value. In other words, the antenna gain of circular codebook is relatively lower than 3c codebook, but the gap between them is quite narrow considering both maximum and minimum D_0 of 3c beams.

Since the number of beams in antenna codebook is limited, the full antenna gain can only be achieved in certain directions, not every direction. For example, the antenna gain at the azimuth angle of 0° is larger than that of 10° when $N = 16$. Gain loss at the worst case of direction (mostly along the intersection of two beams) is analyzed here. Antenna gain is proportional to the square of array factor, thus it is easy to calculate gain loss using minimum and maximum values of array factor.

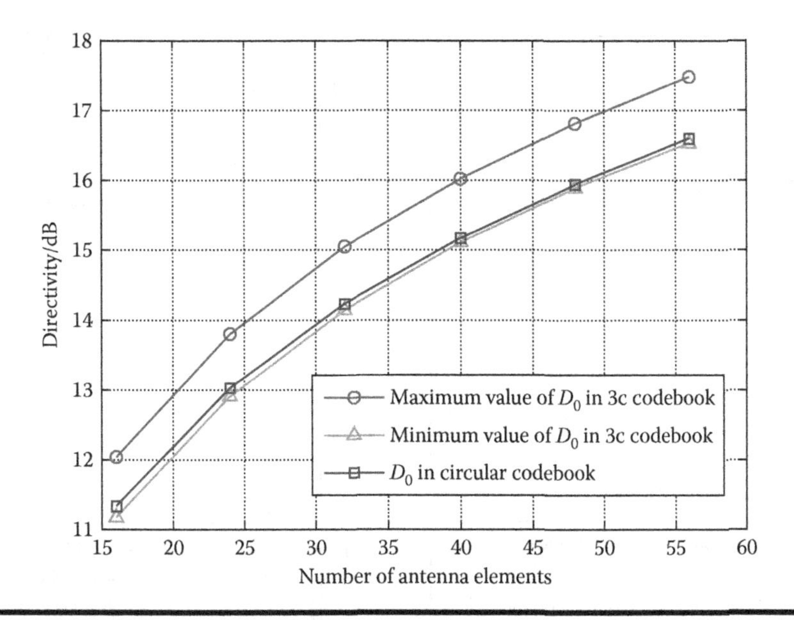

Figure 20.10 Directivity comparison between 3c and circular codebook.

The simulation result of gain loss due to codebook limitations of two codebooks is shown in Figure 20.11. From the figure we can conclude that because of codebook limitations, the gain losses of circular codebook at the worst direction are about 3.5 dB compared with full antenna gain, while those of 3c codebook are around 5 dB. Circular codebook has less gain loss than 3c codebook.

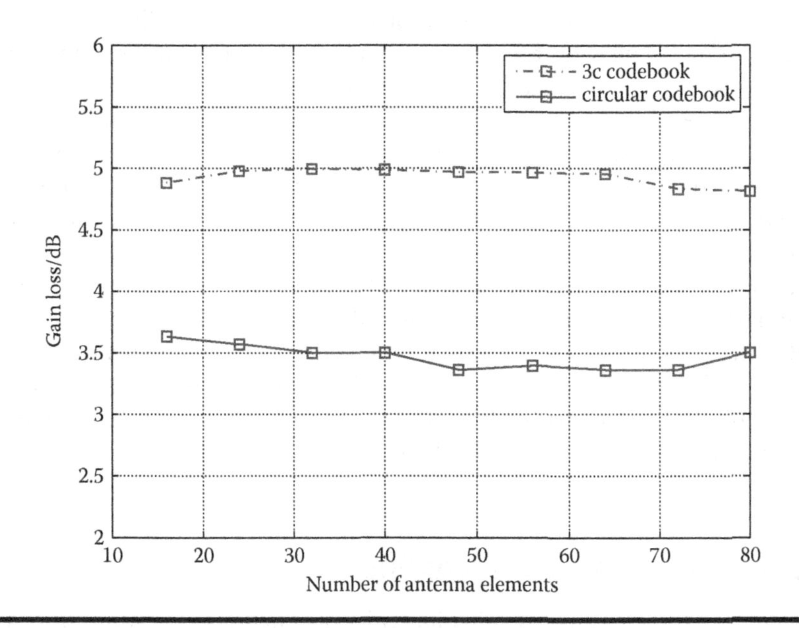

Figure 20.11 Gain loss due to codebook limitations.

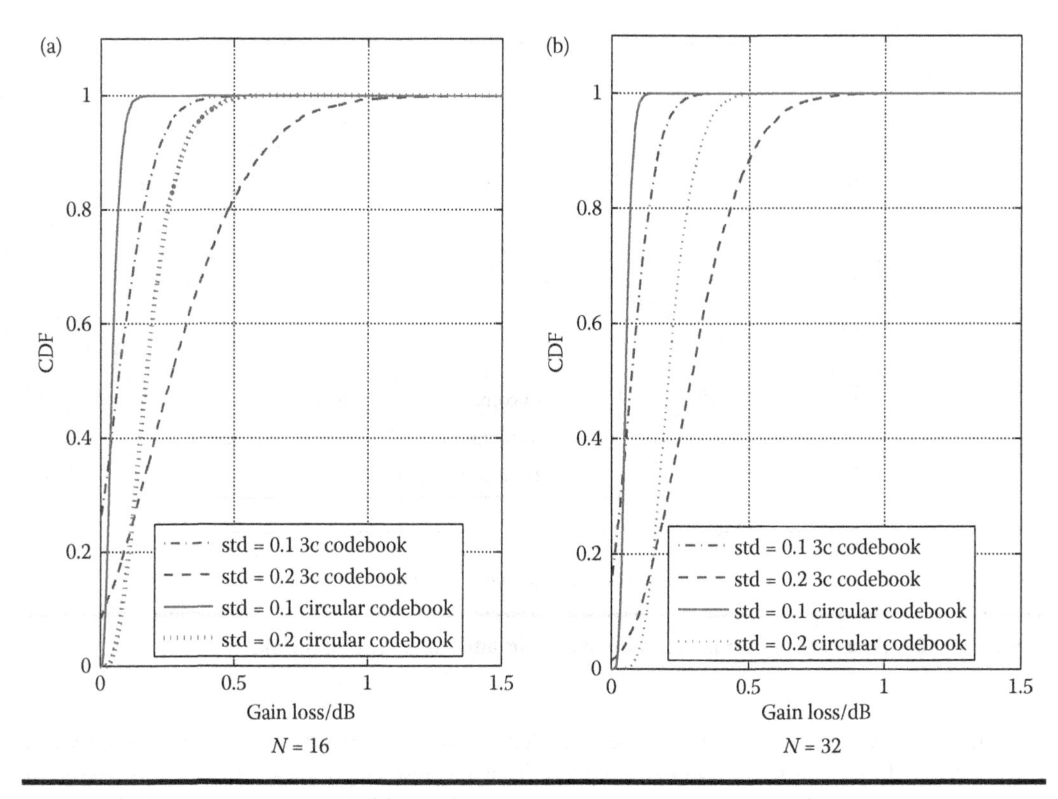

Figure 20.12 Gain loss due to phase shift errors. (a) CDF of gain loss when $N = 16$ and (b) CDF of gain loss when $N = 32$.

In the codebook design proposed in this study, weight coefficients are generated only by phase modulation. Since errors are likely to exist in phase modulation of millimeter wave frequency band, the gain loss due to phase shift errors is an important part of codebook property analysis.

From Reference 22 we know that shift errors are in proportion to absolute phase angle. Here phase shift errors are assumed to follow normal distribution, whose mean is 0 and standard deviation (std) has two cases, that is, 0.1 and 0.2. Because of phase shift errors, the antenna radiation pattern is different from that of correct phase modulation. The cumulative distribution function (CDF) diagram of antenna gain loss on MRA direction due to phase shift errors are shown in Figure 20.12.

The figure shows that the CDF curve of circular codebook has a faster convergence to 1 than 3c codebook. For example, when $N = 16$ and std = 0.1, the gain loss of circular codebook is less than 0.2 dB with 100% probability, while that of 3c codebook is around 0.4 dB (Figure 20.12a). The performance of both codebooks is improved with the increase in the number of antenna elements (Figure 20.12b). The circular codebook has less gain loss due to phase shift errors and its performance is more robust than 3c codebook under the same conditions.

20.4 Beamforming for Multicasting

In order to improve the content dissemination efficiency, multicast beamforming is needed to mitigate path loss and maintain directional data transmission as well as high link quality. So far, no specific beamforming scheme for multicasting is considered in international millimeter wave standards such as IEEE 802.15.3c and IEEE 802.11ad. To our knowledge, conventional multicast beamforming model lies in long-term evolution (LTE) standard which supports enhanced multimedia broadcast multicast service (EMBMS) [23,24]. In the above network, the transmitter has a multiantenna array, while each receiver has a single isotropic antenna. Two coordinated transmit multicast beamforming design problems have been proposed and investigated, including quality of service (QoS) beamforming and max-min fair beamforming [25,26]. The former refers to a problem that minimizes the total transmitted power while maintaining a target received SNR for each user, and the latter is aimed to maximize the minimum received power among all users. Both optimization problems are Non-deterministic Polynomial-time (NP) hard and need complicated techniques such as semidefinite relaxation (SDR) plus randomization to obtain optimal or quasi-optimal solutions, which are difficult to be applied in practical hardware designs. Besides, both solutions require perfect channel estimation, which is a tough work in high-frequency systems. Thus, the existed multicast beamforming scheme is not suitable for 60-GHz-mm-wave system.

In this section, we propose a multicast beamforming scheme employed in a system incorporating devices each with a multiantenna array. Due to antenna beamformers at both transmitter and receiver sides, the beams become highly directional so that interference from other users is largely reduced. The proposed algorithm is of low complexity, and a constraint is made that the beamforming weights can only be produced by phase shift without amplitude adjustment for practical use. Improvements in the aspect of data throughput are shown in simulation performance.

20.4.1 Communication Model

According to IEEE 802.11ad standard, the basic network element in 60-GHz system is personal basic service set (PBSS), which is consisted of a PBSS control point (PCP) and several stations (STAs) [27].

In conventional multicast scenarios in LTE, the transmitter is equipped with a multiantenna array while each user has a single isotropic antenna. Thus the interference between different STAs needs to be considered, and it is impossible to process multicasting directly between STAs. To solve these problems, we adopt a multicast model comprising one transmitter with N_t antenna elements and K receivers each with N_r antenna elements. Every multiantenna array can generate AWV, which is comprised of a plurality of complex weight coefficients corresponding to antenna elements [21]. In practical millimeter wave communication systems, it is difficult and complicated to operate current amplitude adjustment of each antenna element. Therefore, we only use phased antenna arrays. It should be noticed that multicast beamforming can be done directly between STAs in our model, thus both PCP and STA can serve as multicast transmitter. For the ease of illustration, Figure 20.13 shows a typical example of the proposed multicast beamforming model, where PCP serves as the transmitter and all STAs are in one multicast group. The goal of multicast beamforming is to find the best transmit and receive AWVs to achieve high link quality.

In unicast beamforming scenarios, the PCP transmits the message to all STAs one by one after they send requests for the same content. We use STA_k to represent the kth STA, and use link k to denote the channel between PCP and STA_k ($k = 1, 2, \ldots, K$). $\mathbf{H}_k$, $\mathbf{v}_k$ and $\mathbf{u}_k$ denote $N_r \times N_t$ channel matrix, $N_t \times 1$ transmit AWV and $N_r \times 1$ receive AWV of link k. The relation between received signal y_k and transmitted signal s is:

$$y_k = \mathbf{u}_k^H \mathbf{H}_k \mathbf{v}_k s + \mathbf{n}_k \tag{20.12}$$

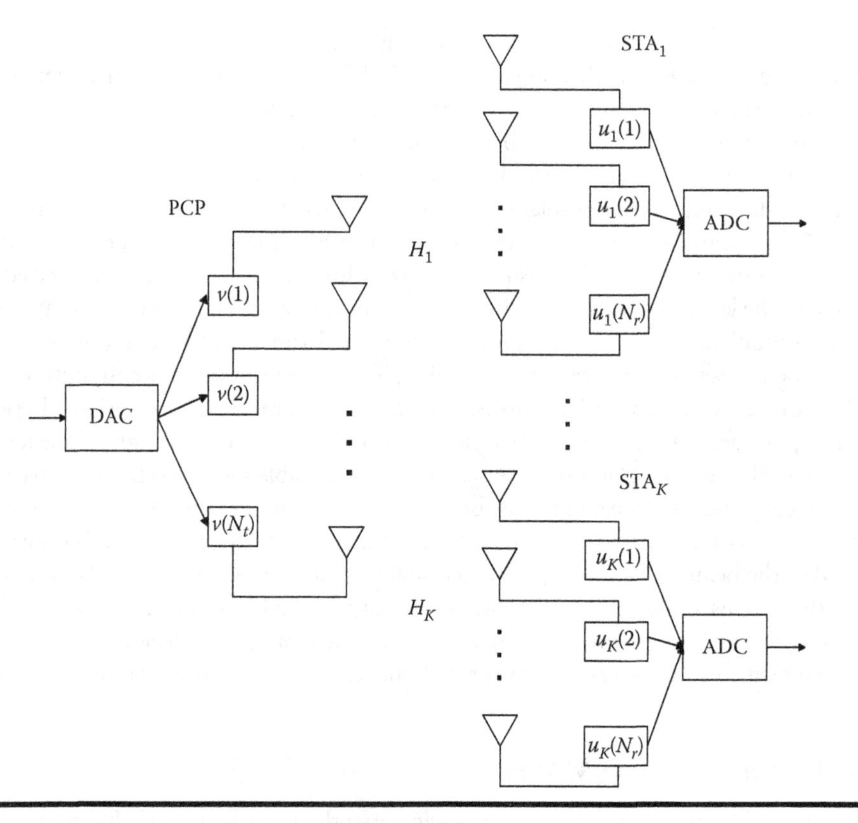

Figure 20.13 Multicast beamforming system model.

where $\mathbf{u}_k^H$ means Hermitian transpose of $\mathbf{u}_k$, and $\mathbf{n}_k$ is zero mean additive white Gaussian noise (AWGN) vector [28]. The total throughput of the PBSS is:

$$\mathrm{V}_{\text{unicast}} = \frac{KS}{\sum_{k=1}^{K} t_k} = \frac{KS}{\sum_{k=1}^{K} S/C_k} = \frac{K}{\sum_{k=1}^{K} 1/C_k} \tag{20.13}$$

where S is the data size of signal s, while t_k denotes data transmission time of link k. C_k is the channel capacity of link k adopting unicast beamforming:

$$C_k = W\log_2(1+|\mathbf{u}_k^H\mathbf{H}_k\mathbf{v}_k|^2|s|^2/\sigma^2) \tag{20.14}$$

where W is system bandwidth, and σ^2 is noise power.

In multicast beamforming scenarios, the PCP delivers the same message to all STAs simultaneously using $N_t \times 1$ multicast transmit AWV $\mathbf{v}$. Total data throughput is defined as:

$$\mathrm{V}_{\text{multicast}} = \frac{KS}{\max(t_1', t_2', \ldots, t_K')} = K\min(C_1', C_2', \ldots, C_K') \tag{20.15}$$

where t'_k denotes data transmission time of link k, and C'_k stands for the channel capacity of link k adopting multicast beamforming:

$$C'_k = W\log_2(1+|\mathbf{u}_k^H\mathbf{H}_k\mathbf{v}|^2|s|^2/\sigma^2) \tag{20.16}$$

It is clear that maximizing the minimum SNR among all STAs helps increase the total multicast throughput.

20.4.2 Multicast Beamforming Scheme

The proposed multicast beamforming scheme consists of two stages. In the first stage, the AWV pair for unicast beamforming of each link is found. In the second stage, transmit and receive AWVs for multicasting are obtained based on the results from previous stage.

Stage 1: Finding unicast transmit and receive AWVs: In order to maximize received signal power of link k in Equation 20.12 subject to power constraints, that is, $\|\mathbf{v}_k\|^2 = N_t$ and $\|\mathbf{u}_k\|^2 = N_r$, the optimal solution is:

$$\begin{aligned}\mathbf{v}_{k,opt} &= \sqrt{N_t}\,\mathbf{e}_{\max}(\mathbf{H}_k^H\mathbf{H}_k),\\ \mathbf{u}_{k,opt} &= \sqrt{N_r}\,\mathbf{e}_{\max}(\mathbf{H}_k\mathbf{H}_k^H),\end{aligned} \tag{20.17}$$

where $\mathbf{e}_{\max}(\mathbf{H}_k^H\mathbf{H}_k)$ and $\mathbf{e}_{\max}(\mathbf{H}_k\mathbf{H}_k^H)$ stand for normalized eigenvectors corresponding to the maximum eigenvalue of $\mathbf{H}_k^H\mathbf{H}_k$ and $\mathbf{H}_k\mathbf{H}_k^H$, respectively [28].

According to mathematical analysis, the following equations are satisfied [29]:

$$\begin{aligned}&\lim_{m\to\infty}\underbrace{(\mathbf{H}_k^H\mathbf{H}_k)\times\cdots\times(\mathbf{H}_k^H\mathbf{H}_k)}_{m\text{ pairs}}\mathbf{t} = \alpha\mathbf{v}_{k,opt},\\ &\lim_{m\to\infty}\mathbf{H}_k\times\underbrace{(\mathbf{H}_k^H\mathbf{H}_k)\times\cdots\times(\mathbf{H}_k^H\mathbf{H}_k)}_{m\text{ pairs}}\mathbf{t} = \beta\mathbf{u}_{k,opt},\end{aligned} \tag{20.18}$$

where α and β are constant values, and $\mathbf{t}$ is an $N_t \times 1$ nonzero arbitrary vector. It is also good to know that the above convergence can be achieved when m is not big [18]. We presume that the channel matrices of uplink and downlink are mutual Hermitian conjugate.

The process of finding the best unicast beamforming AWV pair between PCP and STA_k is shown in Algorithm 1, which is based on the principle in Equation 20.18, and can avoid channel estimation and high complexity eigenvalue decomposition [29].

Since only phase shift is adopted in our beamforming model, every AWV vector needs to be element-normalized. For example, this process of $N_t \times 1$ vector $\mathbf{t}$ and $N_r \times 1$ vector $\mathbf{r}$ can be depicted as:

$$\begin{aligned}\mathbf{t}(j) &\leftarrow \mathbf{t}(j)/|\mathbf{t}(j)|, \quad j = 1,2,\ldots,N_t,\\ \mathbf{r}(i) &\leftarrow \mathbf{r}(i)/|\mathbf{r}(i)|, \quad i = 1,2,\ldots,N_r.\end{aligned} \tag{20.19}$$

Algorithm 1 Finding the best unicast AWV pair of link k

Require: Maximum iteration number m_{max}.
Ensure: Transmit AWV $\mathbf{v}_k$ and receive AWV $\mathbf{u}_k$.

1: Initialize m with 0; Initialize $\mathbf{t}$ with an $N_t \times 1$ non-zero arbitrary vector.
2: **repeat**
3: The PCP uses element-normalized vector of $\mathbf{t}$ as transmit AWV, sending N_r TRN-R sequences [27] to STA$_k$. Meanwhile, STA$_k$ uses $N_r \times N_r$ identity matrix $\mathbf{I}_{N_r}$ as receive beamforming matrix. Specifically, the receive AWV for the ith ($i = 1, 2, \ldots, N_r$) TRN-R sequence is the ith column of $\mathbf{I}_{N_r}$. STA$_k$ selects the channel response of main path, which is then multiplied by the corresponding transmit signal in the ith TRN-R sequence. Update the ith element of received vector with the above product.
4: Update $\mathbf{r}$ by the received vector ($\mathbf{r} \leftarrow \mathbf{H}_k\mathbf{t} + \mathbf{n}_k$).
5: STA$_k$ uses element-normalized vector of $\mathbf{r}$ as transmit AWV, sending N_t TRN-T sequences [27] to the PCP. Meanwhile, the PCP uses $N_t \times N_t$ identity matrix $\mathbf{I}_{Nt}$ as receive beamforming matrix. Specifically, the receive AWV for the jth ($j = 1, 2, \ldots, N_t$) TRN-T sequence is the jth column of $\mathbf{I}_{N_t}$, and the PCP chooses the product of channel response of main path and its corresponding transmit signal in the jth TRN-T sequence as the jth element of received vector.
6: Update $\mathbf{t}$ by the received vector ($\mathbf{t} \leftarrow \mathbf{H}_k^H \mathbf{r} + \mathbf{n}_k$).
7: After normalizing the element of vector $\mathbf{t}$ and $\mathbf{r}$, update $\mathbf{v}_k$ with $\mathbf{t}$ and $\mathbf{u}_k$ with $\mathbf{r}$.
8: Update m: $m \leftarrow m + 1$.
9: **until** $m = m_{max}$

Algorithm 1 is performed between the PCP and every STA, after which the best AWV pair can be found for every single link.

Stage 2: Finding multicast transmit and receive AWVs: There is little difference between unicast and multicast beamforming at the receiver side. Hence, update the multicast receive AWV of each STA with the ones obtained in Algorithm 1.

With the purpose of balancing the transmit power on every direction, initial transmit AWV for multicasting is written as:

$$\mathbf{w} = \frac{1}{\sqrt{P_1}}\mathbf{v}_1 + \frac{1}{\sqrt{P_2}}\mathbf{v}_2 + \cdots + \frac{1}{\sqrt{P_k}}\mathbf{v}_k \tag{20.20}$$

where P_k is the received power when using $\mathbf{v}_k$ and $\mathbf{u}_k$ as AWV pair ($P_k = |\mathbf{u}_k^H \mathbf{H}_k \mathbf{v}_k|^2 |s|^2$ theoretically).

In order to raise the minimum received SNR among all users, Algorithm 2 is designed for further transmit multicast beamforming.

Algorithm 2 Finding the best multicast transmit AWV

Input: Maximum iteration number l_{max}; Coefficient γ in the interval (0,1).
Output: Multicast transmit AWV $\mathbf{v}$.

1: Initialize l with 0; Initialize $\mathbf{v}$ with element-normalized vector of $\mathbf{w}$.
2: **repeat**
3: The PCP uses $\mathbf{v}$ as transmit AWV, and each STA keeps using its corresponding receive AWV $\mathbf{u}_k$. Select the minimum received power of all users and return its corresponding STA index k^*.
4: Update $\mathbf{v}$ with:

$$\mathbf{v} \leftarrow \gamma\mathbf{v} + (1-\gamma)\mathbf{v}k^*,$$
$$\mathbf{v}(i) \leftarrow \mathbf{v}(i)/|\mathbf{v}(i)|, \quad i = 1, 2, \ldots, N_t.$$

5: Update l: $l \leftarrow l + 1$.
6: **until** $l = l_{\max}$

The operation in Algorithm 2 is aimed at leading the multicast transmit beam to approach the beam direction with minimum received power. In other words, it is a practical solution to make the minimum SNR of all links as high as possible.

The PCP can generate multicast transmit beam that points to multiple directions at the same time, which can highly increase the efficiency and throughput of the wireless network. Since multiantenna arrays are adopted by all users, the above beamforming scheme is feasible for multicasting between STAs directly.

20.4.3 Performance Evaluation

We perform several simulation tests of throughput properties in different beamforming scenarios. Performance comparison and analysis are depicted as follows.

In this part, we assume that each path between PCP and STA_k is directional line-of-sight (LOS) path. With reference to Reference 30, the channel matrix can be written as:

$$\mathbf{H}_k = a_k \mathbf{e}_r(\theta_{rk})\mathbf{e}_t(\theta_{tk})^H \tag{20.21}$$

where a_k is channel attenuation coefficient, and can be depicted as $a_k = \sqrt{N_t N_r}\lambda/4\pi d_k$ according to Friis propagation law (λ is wavelength and d_k is the distance between PCP and STA_k). $\mathbf{e}_t(\theta_{tk})$ and $\mathbf{e}_r(\theta_{rk})$ are Vandermond vectors with normalized power related with angles between physical path and antenna arrays at the transmitter and the receiver, respectively.

First we consider the simplest scenario in multicast beamforming, that is, the PBSS comprises one PCP and two STAs. All devices adopt phased uniform linear antenna arrays and the antenna element spacing is half wavelength [20]. Let $N_t = 16$, $N_r = 8$, $\gamma = 0.8$, $m_{\max} = 15$ and $l_{\max} = 20$. The distance between PCP and STA_1 is $10\sqrt{2}$ m while that between PCP and STA_2 is $4\sqrt{3}$ m. The channel bandwidth $W = 1.76$ GHz, signal power $|s|^2 = 1$ and noise power $\sigma^2 = 10^{-6}$, as well as $\mathbf{t}$ in Algorithm 1 is initialized with a vector of all ones.

Channel matrix $\mathbf{H}_1$ is generated, leading the transmit beam to point to 60° and receive beam to focus on 67.5° between PCP and STA_1 in unicast scene. Similarly, $\mathbf{H}_2$ is designed to make the transmit beam and receive beam point to 120° and 90°, respectively, between PCP and STA_2 in unicast scenario. The antenna beam patterns of the above situations are show in Figure 20.14 and Figure 20.15, respectively.

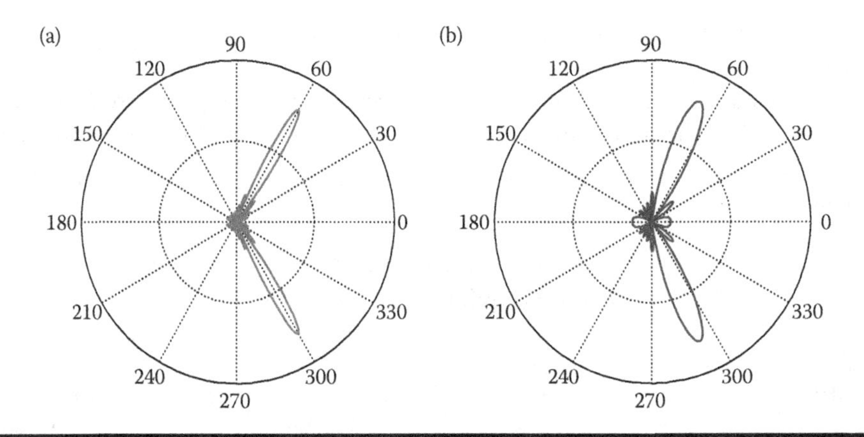

Figure 20.14 Beam patterns between PCP and STA_1 in unicast scenario: (a) Transmission beam; (b) reception beam.

In unicast scenario, the channel capacity of link 2 is around 1.5 times larger than that of link 1. It is obvious that the attenuation of link 1 is severe than that of link 2 due to longer distance between PCP and STA_1. Therefore, in order to compensate for higher path loss of link 1, more beam energy should be allocated to link 1 in multicast beamforming. Figure 20.16 illustrates that the multicast transmit beam points to two directions, that is, 60° and 120°, and the beam power in the direction of link 1 (60°) is higher than that of link 2, which validates the above analysis.

Figure 20.17 depicts the total throughput when changing the number of STAs in 5000 random cases. In each case, d_k obeys independent uniform distribution between 5 m and 15 m. When the number of users increases, the throughput changes little under the circumstances of unicast beamforming. However, the throughput grows rapidly when choosing proposed multicast beamforming scheme. We define the process of generating AWV $\mathbf{w}$ as sum-up algorithm, while the process in Algorithm 2 as iteration algorithm whose initial $\mathbf{v}$ is generated randomly. It can be seen that sum-up algorithm achieves better performance than iteration algorithm, none of which is better than the proposed algorithm.

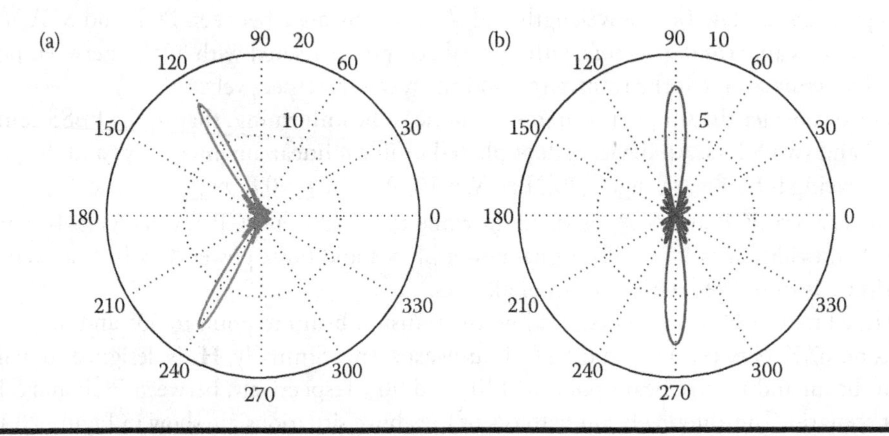

Figure 20.15 Beam patterns between PCP and STA_2 in unicast scenario: (a) Transmission beam; (b) reception beam.

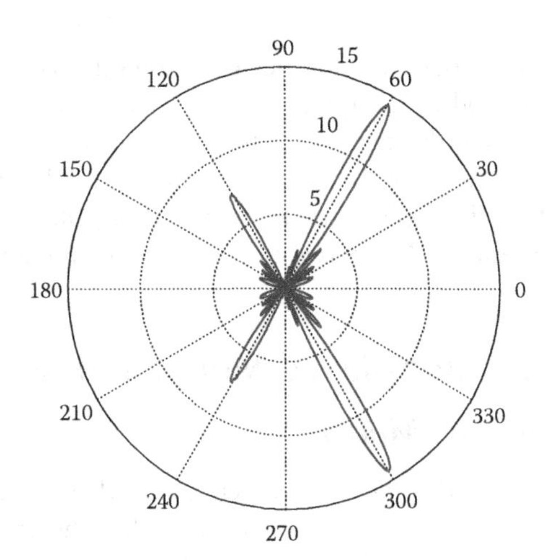

Figure 20.16 Transmit beam in multicast beamforming.

According to Reference, the problem of maximizing the minimum SNR of all users with transmitted power constraint is modeled as quadratically constrained quadratic programming (QCQP) problem. In our proposed system, it can be depicted as

$$\begin{aligned} \max_{\mathbf{V}\in\mathbb{C}^{N_t\times N_t}} \min_{k} \; & \frac{|s|^2}{\sigma^2}\operatorname{trace}(\mathbf{VP}_k)_{k=1}^{K} \\ \text{subject to: } & \operatorname{trace}(\mathbf{V}) = N_t, \\ & \operatorname{rank}(\mathbf{V}) = 1, \\ & \mathbf{V} \succeq 0, \end{aligned} \tag{20.22}$$

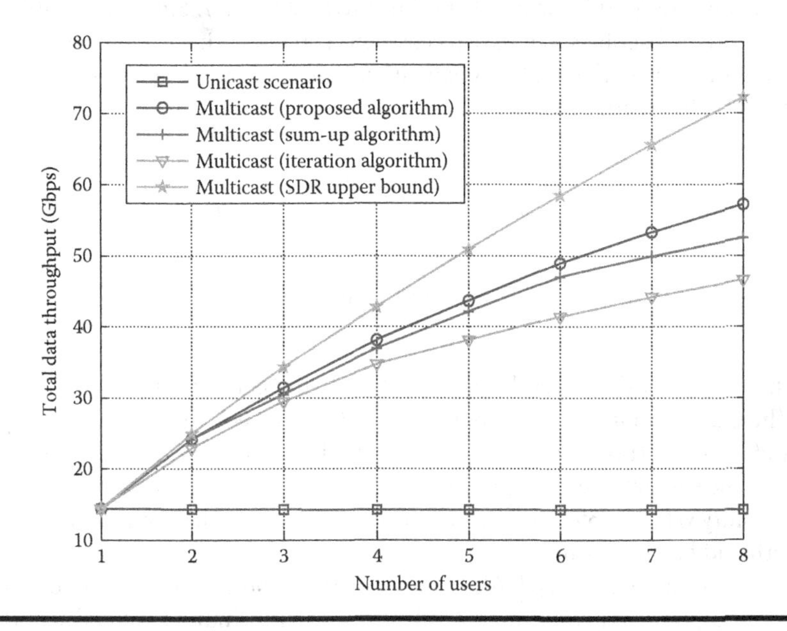

Figure 20.17 Throughput comparison in multiuser network.

where $\mathbf{V} = \mathbf{v}\mathbf{v}^H$ and $\mathbf{P}_k = (\mathbf{u}_k^H \mathbf{H}_k)^H (\mathbf{u}_k^H \mathbf{H}_k)$, in which $\mathbf{u}_k$ denotes the receive AWV of link k obtained in Algorithm 1. An upper bound of the minimum received SNR can be obtained by applying SDR technique, that is, dropping the constraint that the matrix $\mathbf{V}$ should be rank one [31]. The above model is then converted to a semidefinite programming (SDP) problem and can be solved via MATLAB toolbox SeDuMi [32,33]. The upper bound of total throughput can be further obtained and its simulation result is shown as the line with stars in Figure 20.17.

20.5 Fluid Approximation-Based Model and Results

20.5.1 Dissemination Time Model

We note that when a node with the content contacts with other nodes without the content, it will disseminate the content. The status of whether a node has the content in the opportunistic vehicular network can be viewed as a stochastic process that is controlled by opportunistic contact events. Consequently, the content dissemination is a complicated stochastic process consisting of a large number of component processes. We employ a fluid approximation model [34] to analyze this highly complex content dissemination system. It is well known that the fluid approximation is incapable of describing the dynamics of this stochastic system [34]. However, it allows us to replace this stochastic process by "joining" all the nodes to form a deterministic process, and the result obtained by this approximation is known to be close to that of the underlying stochastic process in the expectation sense.

The content dissemination process starts at time $t = 0$ when some helpers obtained the content from the source. A node with the content is called an *infected* node. Assume that the total number of vehicles in the system is N and, furthermore, there are H helpers and S subscribers. Then, the fraction of the helpers, denoted by φ_h, is $\varphi_h = H/N$, while the fraction of the subscribers, denoted by φ_s, is $\varphi_s = S/N$. Define $h(t) = N_h(t)/N$ as the proportion of the helpers that have received the content at time t, where $N_h(t)$ is the number of infected helpers at t. Similarly, let $s(t) = N_s(t)/N$ be the proportion of the subscribers that have received the content at time t, where $N_s(t)$ is the number of infected subscribers at t. The fluid approximation describes the dynamics of the system by the following ordinary differential equations:

$$\frac{d}{dt} h(t) = \zeta(\varphi_h - h(t))h(t) \tag{20.23}$$

$$\frac{d}{dt} s(t) = \eta(\varphi_s - s(t))h(t) \tag{20.24}$$

where ζ is the contact rate between helpers, while η is the contact rate between a helper and a subscriber. These contact rates depend upon node mobility speeds and antenna propagation characteristics, and are proportional to the new area covered per unit time [35]. Under the omnidirectional antenna, the contact rate is proportional to $\pi R \upsilon$, where R is the communication range and υ the node mobility velocity [35]. In the directional antenna case, the contact rate also depends on the beamwidth and beam steering rate.

Based on the mean field theory [36], Equations 20.23 and 20.24 correspond to the random node mixing assumption and are asymptotically valid when the number of nodes in the system is large. Combining Equations 20.23 and 20.24 yields

$$\frac{dh(t)}{ds(t)} = \frac{\zeta}{\eta} \frac{\varphi_h - h(t)}{\varphi_s - s(t)}$$

which leads to

$$s(t) = \varphi_s - \frac{\varphi_s - s_0}{(\varphi_h - h_0)^{\eta/\zeta}} (\varphi_h - h(t))^{\eta/\zeta} \tag{20.25}$$

where $s_0 = s(0)$ and $h_0 = h(0)$. We now solve Equation 20.23 to obtain $h(t)$ explicitly. Note that

$$\frac{1}{(\varphi_h - h(t))h(t)} = \frac{1}{\varphi_h} \left(\frac{1}{\varphi_h - h(t)} + \frac{1}{h(t)} \right)$$

Therefore, from Equation 20.23 we have

$$\frac{dh(t)}{\varphi_h - h(t)} + \frac{dh(t)}{h(t)} = \varphi_h \zeta dt$$

which explicitly yields

$$h(t) = \frac{\varphi_h h_0}{h_0 + (\varphi_h - h_0)e^{-\varphi_h \zeta t}} \tag{20.26}$$

By substituting Equation 20.26 into Equation 20.25, we explicitly obtain

$$s(t) = \varphi_s - \varrho \left(\varphi_h - \frac{\varphi_h h_0}{h_0 + (\varphi_h - h_0)e^{-\varphi_h \zeta t}} \right)^{\eta/\zeta} \tag{20.27}$$

where $\varrho = (\varphi_s - s_0)(\varphi_h - h_0)^{-\eta/\zeta}$. Define T_ω as the time at which the proportion $(1 - \omega)$ of the subscribers have received the content. In other words, only the proportion ε of the subscribers have not yet received the content at time T_ω. We refer to T_ω as the dissemination time, which depicts the content dissemination speed, and we use T_ε as a metric of the system performance. From the definition of the dissemination time, we have $\varphi_s(1 - \omega) = s(T_\omega)$. According to Equation 20.27, we obtain

$$e^{-\varphi_h \zeta T_\omega} = \frac{\beta h_0}{(\varphi_h - \beta)(\varphi_h - h_0)}$$

where $\beta = (\varphi_s \omega / \varphi_s - s_0)^{\zeta/\eta}$. Therefore,

$$T_\omega = \frac{1}{\varphi_h \zeta} \ln \left(\frac{(\varphi_h - \beta)(\varphi_h - h_0)}{\beta h_0} \right) \tag{20.28}$$

20.5.2 Numerical Results

The simulated system covered an area of 2000×2000 m^2 with time steps $n = 60 \times 60 \times 2$. We randomly used 70% of the network nodes as subscribers, and the remaining 30% as helpers. The content source randomly selected 10% of the helpers and disseminated the data to them at $t = 0$. The vehicular mobility discussed in Section 20.2.2 was adopted, in which ϕ_n obeyed the uniform distribution in [0, 360°), and v_n followed the uniform distribution in [8,34] m/s, while l_n was generated according to Equation 20.1 with $\varepsilon = 1.5$, $c = 2.5$, $\xi^f_{\min} = 5$ m and $\xi^f_{\max} = 1000$ m, and t_n also obeyed the Lévy walk (20.1) with $\varepsilon = 1.5$, $c = 2.5$, $\xi^p_{\min} = 30$ s and $\xi^p_{\max} = 600$ s. All the antennas, whether omnidirectional or directional, had the same transmit power. We set the antenna gain to let the communication range of two omnidirectional antennas being 60 m, and calculated the gain and communication range of directional antennas according to Equations 20.3 and 20.4. In order to obtain credible and reliable results, we simulated the system with the specific settings of node mobility and antenna 100 times to obtain the simulated content dissemination time T^*_ω where $\omega = 0.1$. Then, we plotted the simulation results by averaging over the 100 different runs, and also plotted their confidence interval in all figures. By extracting the contact rates ζ and η, we also calculated the theoretical content dissemination time T_ω using Equation 20.28. This enabled us to investigate the accuracy of our proposed model for content dissemination time by comparing T^*_ω and T_ω. Furthermore, we analyzed how the directional antennas with different beam steering policies, beam steering rates, and beamwidths influence the system performance.

The content dissemination times of three beam steering policies as function of the number of nodes are shown in Figure 20.18, where the simulation results are averaged over 100 runs. As the number of nodes N increases, the content dissemination time T^*_ω decreases. The reason is obvious. A larger N means a higher node density, since the system area is constant, which in turn leads to more opportunistic contacts to disseminate the content. The RS policy needs the longest time to disseminate the content, while the CS policy achieves the shortest data dissemination time, under the same settings of nodes and antenna beams. Specifically, the CS reduces the content dissemination time by 51% and 32%, compared with the RS, for $N = 100$ and 500, respectively. This result shows that if we schedule the beam steering, rather than RS, significant performance enhancement can be achieved. The performance enhancement of the CS policy over the PS policy is about 23% to 2%. Thus, in the beam scheduling, better system performance can be achieved by changing the beam to let the antenna swap more area and avoid overlapping the already covered area. It can be seen that, in the directional antenna based opportunistic content dissemination system, designing an appropriate beam scheduling algorithm is important. From Figure 20.18, we can see that the theoretical results of T_ω are very close to the simulation results of T^*_ω, which validates the accuracy of our content dissemination model (20.28).

Next, we studied the influence of beam steering rate r_ϑ under the RS policy with a fixed beamwidth $\theta = 45°$, and the results obtained are shown in Figure 20.19. Again, the simulation results T^*_ω agree with the theoretical ones T_ω, and similar observations to those for Figure 20.18 can be drawn regarding the number of nodes. As the beam steering rate increases, the content dissemination time reduces significantly. For example, the dissemination time for $r_\vartheta = 40$ is only 30% and 50% of those with $r_\vartheta = 10$ and $r_\vartheta = 25$, respectively. However, it should be pointed out that the beam steering rate is limited by real system antenna implementation, and the performance enhancement is obtained at the cost of energy leakage.

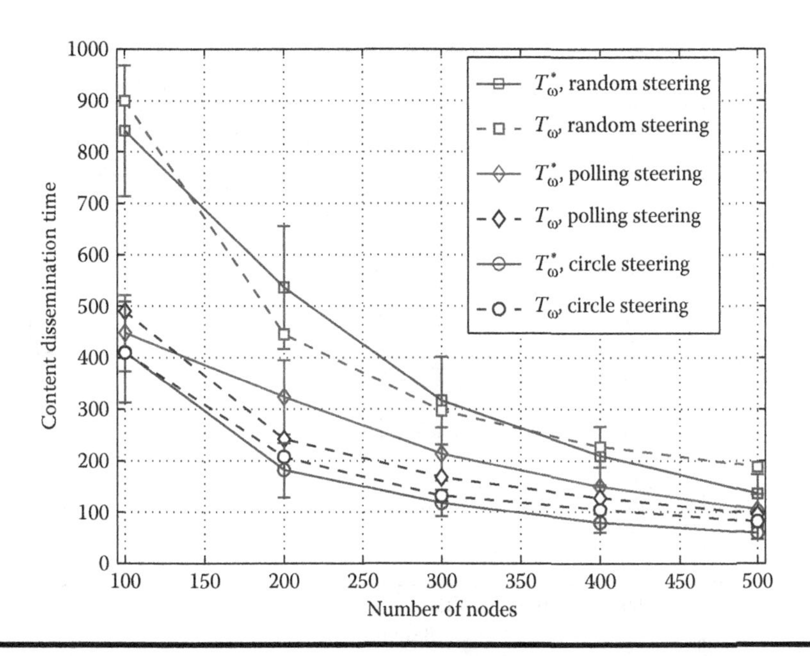

Figure 20.18 Content dissemination time as function of the number of nodes for three beam steering policies with beamwidth $\theta = 45°$ and beam steering rate $r_\vartheta = 20$, where dashed curves are the fluid model based results given by Equation 20.28, while solid curves are the simulation results with the vertical bars indicating the standard deviation.

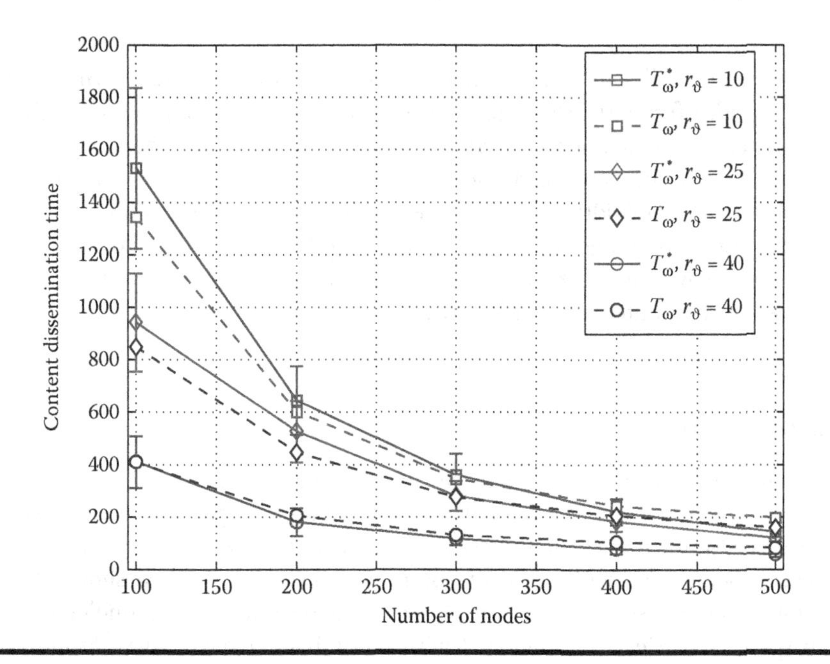

Figure 20.19 Content dissemination time as function of the number of nodes for different beam steering rates r_ϑ with beamwidth $\theta = 45°$ and RS, where dashed curves are the fluid model based results given by Equation 20.28, while solid curves are the simulation results with the vertical bars indicating the standard deviation.

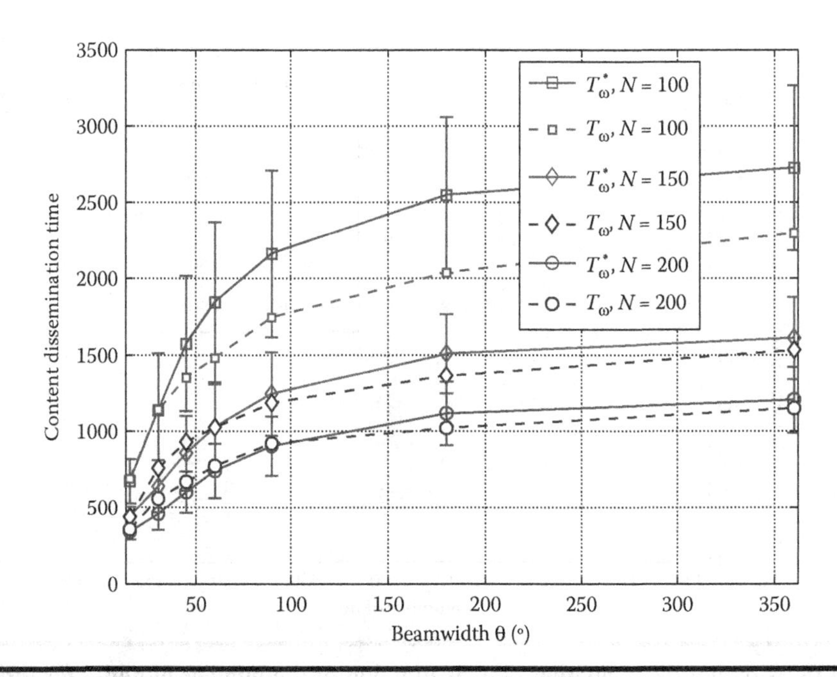

Figure 20.20 Content dissemination time as function of the beamwidth varying from 15 to 360° for different numbers of system nodes with $r_\vartheta = 10$ and RS, where dashed curves are the fluid model based results given by Equation 20.28, while solid curves are the simulation results with the vertical bars indicating the standard deviation.

The results of content dissemination time as function of beamwidth are shown in Figure 20.20. When the beamwidth $\theta = 360°$, the antenna is omnidirectional. From Figure 20.20, it can be seen that the direction antenna offers superior system performance over the omnidirectional one. Moreover, as the beamwidth is reduced, the content dissemination time is also reduced. In other words, the smaller the beamwidth, the larger the achievable performance enhancement. The reason is that the transmission range is enlarged by reducing the beamwidth, and the beam is steered to enable the nodes cover more new area which in turn creates more communication contacts. Thus, the content is distributed more efficiently.

20.6 Conclusion

This chapter analyzed the performance of collaborative content dissemination with the aid of directional antennas by utilizing the vehicular network as an example. Our contributions include validating the Lévy-walk model for vehicular mobility and deriving a fluid approximation for studying the collaborative vehicular content dissemination system. We have shown that, with the aid of directional antennas supported by the physical layer beamforming technologies, the content propagation speed is accelerated, compared with omnidirectional antennas. Simulation results have confirmed the accuracy of our proposed model. Our ongoing work is further investigating beam scheduling algorithms as well as the beamwidth and steering rate control problem in more complicated and realistic scenarios to fundamentally reveal the benefit or loss of directional antennas for general mobile wireless networking.

Acknowledgments

This work is supported by the National Basic Research Program of China (973 Program) (No. 2013CB329105), National Nature Science Foundation of China (No. 61301080, No. 61171065, and No. 61273214), National High Technology Research and Development Program (No. 2013AA013501 and No. 2013AA013505), Chinese National Major Scientific and Technological Specialized Project (No. 2013ZX03002001), and Chinas Next Generation Internet (No. CNGI-12-03-007).

References

1. Khabazian, M., S. Aissa, and M. Mehmet-Ali, 2011. Performance modeling of message dissemination in vehicular ad hoc networks with priority, *IEEE J. Selected Areas in Communications*, 29(1), 61–71.
2. Zhao J. and G. Cao, 2006. VADD: Vehicle-assisted data delivery in vehicular ad hoc networks. In *Proc. 25th IEEE INFOCOM*, Barcelona, Spain, April 23–29, pp. 1–12.
3. Johnson, M., L. De Nardis, and K. Ramchandran, 2006. Collaborative content distribution for vehicular ad hoc networks. In *Proc. 44th Allerton Conf. Communication, Control, and Computing*, Monticello, IL, September. 27–29, pp. 751–760.
4. Câmara, D., N. Frangiadakis, F. Filali, and C. Bonnet, 2011. Vehicular delay tolerant networks. In M. M. Cruz-Cunha and F. Moreira, Eds., *Handbook of Research on Mobility and Computing: Evolving Technologies and Ubiquitous Impacts*. IGI Global, 2011, pp. 356–367.
5. Gao, W. and G. Cao, 2011. User-centric data dissemination in disruption tolerant networks. In *Proc. 30th IEEE INFOCOM*, Shanghai, China, April 10–15, pp. 3119–3127.
6. Costa, P., C. Mascolo, M. Musolesi, and G. Picco, 2008. Socially-aware routing for publish-subscribe in delay-tolerant mobile ad hoc networks, *IEEE J. Selected Areas in Communications*, 26(5), 748–760.
7. Reich, J. and A. Chaintreau, 2009. The age of impatience: Optimal replication schemes for opportunistic networks. In *Proc. 5th ACM Int. Conf. Emerging Networking Experiments and Technologies*, Rome, Italy, December. 1–4, pp. 85–96.
8. Mantegna, N. 1991. Lévy walks and enhanced diffusion in Milan stock exchange, *Physica A: Statistical Mechanics and its Applications*, 179(2), 232–242.
9. Karagiannis, T. J.-Y. Le Boudec, and M. Vojnovi, 2010. Power law and exponential decay of intercontact times between mobile devices, *IEEE Trans. Mobile Computing*, 9(10), 1377–1390.
10. Zhu, H., L. Fu, G. Xue, Y. Zhu, M. Li, and L. M. Ni, 2010. Recognizing exponential inter-contact time in VANETs. In *Proc. 2010 IEEE INFOCOM*, San Diego, CA, March 14–19, pp.1–5.
11. Lee, K., S. Hong, S. J. Kim, I. Rhee, and S. Chong, 2009. SLAW: A new mobility model for human walks. In *Proc. 28th IEEE INFOCOM*, Rio de Janeiro, Brazil, April 19–25, pp. 855–863.
12. Aban, I. B., M. M. Meerschaert, and A. K. Panorska, 2006. Parameter estimation for the truncated pareto distribution, *J. American Statistical Association*, 101(473), 270–277.
13. Navda, V., A. P. Subramanian, K. Dhanasekaran, A. Timm-Giel, and S. R. Das, 2007. MobiSteer: Using steerable beam directional antenna for vehicular network access. In *Proc. 5th Int. Conf. Mobile Systems, Applications and Services*, San Juan, Puerto Rico, June 11–14, pp. 192–205.
14. Ramanathan, R. 2001. On the performance of ad hoc networks with beamforming antennas. In *Proc. 2nd ACM Int. Symp. Mobile Ad Hoc Networking & Computing*, Long Beach, CA, October 4–5, pp. 95–105.
15. Chen, L., Y. Yang, X. Chen, and W. Wang, 2011. Multi-stage beamforming codebook for 60GHz WPAN. In *Proc. ICST 2011*, August, Palmerston, New Zealand, pp. 361–365.
16. Zou, W., Z. Cui, B. Li, Z. Zhou, and Y. Hu, 2011. Beamforming codebook design and performance evaluation for 60GHz wireless communication. In *Proc. ISCIT 2011*, October, Hangzhou, China, pp. 30–35.

17. Ramachandran, K., N. Prasad, K. Hosoya, K. Maruhashi, and S. Rangarajan, 2010. Adaptive beamforming for 60 GHz radios: Challenges and preliminary solutions. In *Proc. mmCom'10*, September, Chicago, Illinois, USA, pp. 33–37.
18. Yong, S. K., P. Xia, and A. Valdes-Garcia, 2011. *60GHz Technology for Gbps WLAN and WPAN: From Theory to Practice*, Published Online: September 29, 2010. DOI: 10.1002/9780470972946.ch1 by John Wiley & Sons, Ltd.
19. *IEEE Standard for Information Technology Part 15.3: Wireless Medium Access Control (MAC) and Physical Layer (PHY) Specifications for High Rate Wireless Personal Area Networks (WPANs)*, IEEE Standard 802.15.3c, 2009.
20. Xue, Z., W. Li, and W. Ren, *2011. Antenna Array Analysis and Synthesis*, Beijing, China: Beihang University Press.
21. Gross, F. 2005. *Smart Antennas for Wireless Communications (With MATLAB)*, New York, NY: McGraw-Hill Professional.
22. Park, J., H. Kim, W. Choi, Y. Kwon, and Y. Kim, 2002. V-Band reflection-type phase shifters using micromachined CPW coupler and RF switches, *IEEE Journal of Microelectromechanical Systems*, 11(6), 808–814.
23. Mechanna, O. and N. D Sidiropoulos, 2013. Joint multicast beamforming and antenna selection, *IEEE Transactions on Signal Processing*, 61(10), 2660–2773.
24. Mechanna, O. N.-D Sidiropoulos, and G.-B. Giannakis, 2012. Multicast beamforming with antenna selection. In *Proc. SPAWC 2012*, June, Cesme, Turkey, pp. 70–74.
25. Sidiropoulos, N. D. 2006. Transmit beamforming for physical-layer multicasting, *IEEE Transactions on Signal Processing*, 54(6), pp. 2239–2251.
26. Xiang Z. and M. Tao, 2013. Coordinated multicast beamforming in multicell networks, *IEEE Transactions on Wireless Communications*, 12(1), 12–21.
27. *Standard for Information Technology-Wireless MAN Medium Access Control (MAC) and Physical Layer (PHY) Specifications-Enhancements for Very High Throughput in 60 GHz Band*, IEEE standard 802.11ad, 2012.
28. Wennstrom, M., M. Helin, A. Rydberg, and T. Oberg, 2001. On the optimality and performance of transmit and receive space diversity in MIMO channels, *IEEE Seminar on MIMO: Communications Systems from Concept to Implementations*, December, London, UK. pp. 1–6.
29. Xiang, P., S. K. Yong, J. Oh, and C. Ngo, 2008. Multi-stage iterative antenna training for millimeter wave communications. In *Proc. GLOBECOM 2008*, November, Crete Island, Greece. pp. 1–6.
30. Tse, D. and P. Viswanath, 2005. *Fundamentals of Wireless Communication*, Cambridge, England: Cambridge University Press.
31. Wang, Z. 2011. Research on transmit multicast beamforming algorithms, M.S. thesis, Beijing Jiaotong University, Beijing, China.
32. Sturm, J. F. 1999. Using SeDuMi 1.02, a MTALAB toolbox for optimization over symmetric cones, *Optim, Meth. Softw.*, 11–12, pp. 625–653.
33. Fang, S. and W. Xing, 2013. *Conic Linear Optimization*, Beijing, China: Science Press.
34. Altman, E., F. De Pellegrini, and L. Sassatelli, 2010. Dynamic control of coding in delay tolerant networks. In *Proc. 29th IEEE INFOCOM*, San Diego, USA, March 14–19, pp. 1–5.
35. Altman, E., V. Kavitha, F. De Pellegrini, V. Kamble, and V. Borkar, 2011. Risk sensitive optimal control framework applied to delay tolerant networks. In *Proc. 30th IEEE INFOCOM*, Shanghai, China, April 10–15, pp. 1–9.
36. Chaintreau, A., J.-Y. Le Boudec, and N. Ristanovic, 2009. The age of gossip: Spatial mean field regime. In *Proc. 11th ACM Int. Joint Conf. Measurement and Modeling of Computer Systems*, Seattle, WA, June 15–19, pp. 109–120.

Chapter 21

The Evolution of Directional Networking Systems Architecture

Matthew Sherman

Contents

Abstract

Directional networking is a critical technology for achieving spectrum efficiency. It also has other key advantages for military applications such as reduced probability of detection and improved antijam capabilities. While directional networking is just catching on commercially in technologies such as IEEE 802.11ad, the U.S. Department of Defense (DoD) has been actively investing in this technology for many years. The technology has been deployed for some time in systems such as highband networking radio. The U.S. DoD and numerous defense companies continue to invest in this technology as it evolves to meet the emerging needs of concepts such as the joint aerial layer network. This chapter reviews past activities and anticipated directions for military directional networking in the future.

21.1 Introduction

A critical problem for the U.S. Department of Defense (DoD) and military systems in general is radio frequency (RF) spectrum efficiency. As the data needs of military systems continually increase (see Figure 21.1) and pressure mounts to share or outright release spectrum to the commercial world, DoD must shoehorn more and more data into less and less spectrum. In addition, military systems differ from commercial systems in their needs.

One key difference is the need to rapidly relocate infrastructure as combat missions move from one area or region to another. Sufficient time is not available to lay fiber optical cables or develop the backbone infrastructure that commercial systems depend on for their services. So military systems have come to rely on technologies like mobile ad hoc networking (MANET) that allow systems to rapidly reconfigure wireless networks to provide backbone capabilities to systems that are continually being relocated or are on-the-move (OTM). Commercial systems rarely require this.

Another way that military systems differ from commercial systems is their level of concern for intentional interference—jamming. While for various reasons, commercial system has started to recognize the possibility and potential need to counter jamming (e.g., truckers with global positioning system [GPS] jammers, sensitive power grid control systems, public safely in the event of terrorism, etc.) military systems have long accounted for the fact that their enemies might intentionally jam radio frequencies to prevent command and control and other critical functions from being performed. In addition, military forces often do not want their current location to be known, whereas commercial systems generally do not care if their location is known. Military systems may rely on "low probability of detection" (LPD) techniques to enable them to communicate while limiting the probability that their presence will be detected.

Unfortunately, spectrum is a finite natural resource that cannot be manufactured. However, a number of techniques exist that can be applied to improve efficient use of spectrum. Advanced modulation and coding techniques are one approach to increase the number of bits of information per Hertz of spectrum (bits/Hz) which is a measure of efficient spectrum use. Recently the idea of "dynamic spectrum access" has been added to the arsenal of techniques for improving spectrum efficiency. With this technique, frequency use is actively monitored rather than simply assigned. If a frequency is assigned to one user or service, but is not currently in use, then another

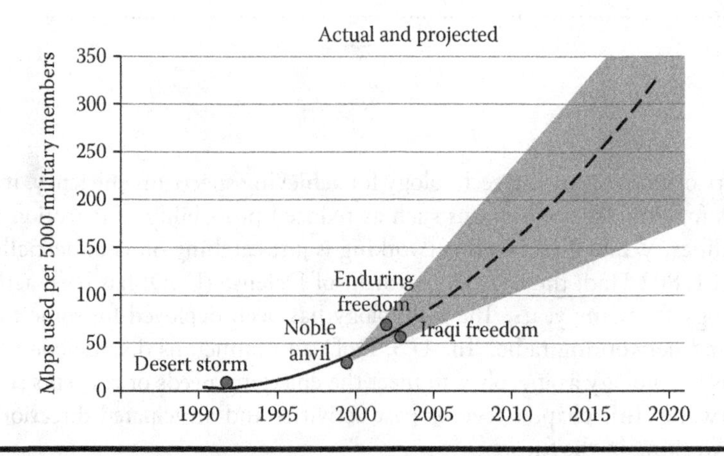

Figure 21.1 Explosion in DoD spectrum requirements. (Adapted from Electromagnetic Spectrum Strategy, 2013. A Call to Action, US Department of Defense, September 11, 2013 (Accessed 17/12/2014) http://www.defense.gov/news/dodspectrumstrategy.pdf.)

user or service can be assigned on a temporary basis allowing greater overall use of the assigned frequency.

Yet another way we can increase efficient use of spectrum is through spatial reuse. A classic example is in cellular systems where a frequency is used in one cell, and then reused in another cell that is far enough away that the two cells cannot interfere. A number of newer techniques improve spectrum efficiency through spatial reuse.

Fundamentally, three orthogonal "aspects" of spectrum exist that we use to organize how we access it—frequency, time, and space. If two destinations exist at different points in space, it is possible to use the same spectrum to deliver information to both points in space simultaneously. One approach to doing this that has become very popular is called multiple input, multiple output (MIMO) antenna technology. By taking advantage of the fact that two antennas in different locations have a different fading environment, it is possible to increase the capacity of the system (or its resiliency) by understanding the properties of the channel and coding multiple streams of data on the same frequency. MIMO technology has been adopted for many commercial systems, but has been less popular in military systems to date.

While adding antennas can achieve an increase in efficiency, another approach is to use high gain directional antennas. These antennas limit their radiation in all directions except for the desired one. What this means is that two directional antennas can be collocated (pointing in different directions), and each simultaneously deliver data to users at two different locations. Many cellular systems apply this approach to increase their spectrum efficiency. At base stations they will often use "sector" antennas, each pointing in a different direction around the base station to simultaneously deliver data to different subscribers on the same frequency. Directional antennas are also used for "point-to-point" (PTP) links used in fixed (nonmobile) radio systems. This has the advantage that many PTP links can be setup in the same area using the same frequency without interfering with each other.

A key point is that the reuse of frequency may have an upper limit within a given area with sectored or PTP antenna technology. When we become concerned with the density of spectrum use in an area, the measure of spectrum efficiency is different than if we are only concerned with the efficiency at a single point in space. If you think of the earth as a sphere, where we only need to deliver data on the surface of the sphere, then an appropriate measure of spectrum efficiency would be bits/Hz/meter squared (m^2) or bits /Hz/m^2.

While military systems certainly do take advantage of concepts such as sectored antennas and PTP links, these techniques require fixed infrastructure. Military environments especially when units are OTM may not be fixed for long enough periods of time to apply these techniques. Recall the concept of a MANET. A MANET would typically be applied with omni (nondirectional) antennas. While this enables many nodes to hear each other and dynamically form networks, it limits the spectral efficiency of those networks. But if you were to add directional antennas to each node that are dynamically steered to the intended destination, many of the nodes could communicate simultaneously using the same frequency. This is the concept of directional networking.

Directional networks offer several advantages over the standard MANET. For one, several nodes in close proximity that would otherwise have interfered with each other if they had multiple simultaneous transmitters using omni antennas now are unlikely to interfere. This means spectrum efficiency can be greatly increased in the bits/Hz/m^2 sense. Theoretically it can increase by 2 pi/beamwidth [2]. Because the receiver listens in only one direction at a time, the probability of being interfered (or jammed) from other directions is greatly reduced. And finally because the system only transmits in one direction at a time, the probability of detection is reduced, improving LPD.

While these are all important advantages, they do come at a cost. Probably, the most obvious cost is the costs of the antennas themselves. Directional antennas are almost always more expensive than omni antennas. Also, since you do not know which direction you will need to communicate in advance, you either need a large number of antennas to cover all directions, or you need to be able to rapidly steer a directional antenna further increasing system costs. Typical MANET scheduling protocols are tailored to the fact that everyone in a neighborhood can hear each other. They need to be modified to account for the fact that only a subset of neighbors can hear transmissions in a given direction. MANET protocols also leverage the omnidirectional antenna assumption by using broadcast and multicast techniques for their control protocols. Modifications must be made if all transmissions are to be via directional antennas.

While these challenges are substantial, they are not insurmountable. Often the additional investment is justified by the benefits offered, and some military directional networking systems have been developed and deployed. The rest of this chapter is organized as follows:

- Section 21.2 discusses the "early days" in directional networking when concepts were initially developed and demonstrated.
- Section 21.3 discusses the evolution of highband networking waveform (HNW) which is the primary directional networking system deployed by the military today.
- Section 21.4 discusses some other directional networking efforts such as DirecNet®.
- Section 21.5 discusses possible future directions in military directional networking.
- Section 21.6 provides a summary and some conclusions.

21.2 The Early Days (DARPA FCS-C)

Research on the use of directional antennas in MANET goes back to at least 1984 [3,4]. But the first substantial demonstration of a military directional MANET grew out of the Defense Advanced Research Projects Agency's (DARPA) efforts for Future Combat System—Communications (FCS-C). The FCS program was developing a networked "system of systems" that would serve as the next generation of army communications targeting the year 2020 [5]. The FCS program concept of operations (CONOPS) at that time required that a unit with about a dozen networked platforms should control a large area of responsibility (on the order of 10 × 25 km). Figure 21.2 shows this general concept. While there were recommendations to evolve this CONOPS [5] networking all the elements was seen as critical to achieving this goal in any case. DARPA worked to develop several enabling technologies focused on advanced battlefield networking.

The networking desired (generally expressed as "information exchange requirements" or IER) for FCS implied increased spectrum requirements. LPD and AJ need further increased the spectrum required since often spectral efficiency is traded to achieve LPD and AJ. A core capability FCS-C would leverage to address the increased data networking needs was to be directional networking. But a general desire also existed to shrink the size, weight, and power (SWaP) of the communication systems to enable their deployment on smaller platforms. So in addition to directional antennas, another key FCS-C enabler was the use of higher radio frequencies into the millimeter wave (mmw) band. Specifically, 38 GHz was considered.

Use of higher radio frequencies often goes hand in hand with directional networking. Directional antennas are typically larger in SWaP than omnidirectional antennas at the same frequency. However, all antenna structures grow smaller as the operating frequency goes higher.

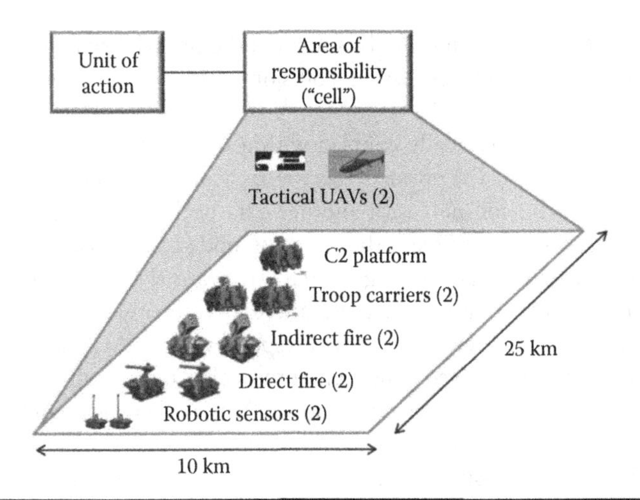

Figure 21.2 Notional FCS Program Area of Responsibility ("Cell"). (From Paul Sass and James A. Freebersyser, 2002. In *Proc. SPIE 4741, Battlespace Digitization and Network-Centric Warfare II*, 9 (August 6). Copyright 2002 SPIE. Redrawn and reprinted, with permission.)

Thus using higher frequencies is a way of managing the SWaP of directional systems as well as addressing aggregate wireless networking needs by making more spectrum available.

While a lot more millimeter wave spectrum is available than in lower bands the spectrum comes with considerable constraints. At lower bands, propagation effects are such that a link can be formed in a non-line-of-sight (NLOS) environment where antennas cannot directly "see" each other. Typically, millimeter wave systems only operate with direct line-of-sight (LOS) and antennas must be able to see each other to form a link. Another concern is that less hardware is deployed in millimeter wave bands so components can be more expensive and less integrated. Finally, various atmospheric effects (such as rain) have greater impact.

Ultimately, a multiband system (that leverages the various benefits of different bands and works around the drawback of any single band) is desirable. DARPA chose to work toward a dual band system which used highly directional antennas at 38 GHz, and moderately directional antennas at lower bands (20 MHz to 3 GHz). They also worked toward other critical technologies such as RF information assurance techniques and beam-steered agile antennas for both high and low bands.

While many technologies are required to make directional networking work, much of the research conducted focuses on the wireless protocols required. A good summary of some of the key protocols applied in FCS-C can be found in Reference 6. The protocol set described is called "utilizing directional antennas for ad hoc networking" (UDAAN) and primarily focuses on the media access control (MAC) and networking (NET) layers. However a "complete" directional networking system is presented—the first time such a presentation was made along with field test results.

Most of the directional networking research at that time had been focused on the MAC layer. Directional carrier sense multiple access (CSMA) was being researched heavily, and the UDAAN MAC was based on CSMA with collision avoidance (CA). A CSMA/CA MAC first senses the media to see if it is vacant. If so, it assumes it is allowed to transmit. The "CA" comes from the use of backoff techniques to help avoid collisions. After a transmission (or if the media is sensed as busy/occupied) the MAC will wait a period of time (the backoff) to allow other radios the opportunity to use the media. Each radio will backoff a random amount. Whichever radio has the shortest backoff gets to use the media. This reduces the probability that multiple nodes try to transmit at the same time after a transmission (collision avoidance) and also helps with fair access to the media.

Several other protocols are typically coupled with a CSMA MAC. They include request-to-send (RTS) and clear-to-send (CTS) protocols, as well as acknowledgment (ACK) protocols. Data packets might be large and a collision with another data packet could cost a lot of time on the media. A short RTS packet can first see if the receiving node can correctly hear the transmitting node and reserve the media for transmitting the data. If heard the receiving node can transmit a short CTS indicating that the channel is reserved and data transmission can proceed. After transmission of the data packet, the receiving node will respond with an ACK so the transmitting node knows the data was correctly received. The ACK protocol is often coupled with retransmission protocols so that if the ACK is not received the transmitting node will try and send the data again at a later point. This helps to achieve reliable transmission.

So a typical CSMA transmission sequence would be RTS-CTS-DATA-ACK. The UDAAN Directional MAC (D-MAC) leveraged these protocols but made certain unique enhancements. It used different backoff methods depending on what triggered the need for backoff (e.g., media busy, no CTS received, no ACK received, etc.). It also tightly coupled power control with the directional transmissions. To achieve high network capacity, systems are generally required to transmit the minimum amount of power required to achieve reliable transmission of a packet.

Many CSMA systems (such as 802.11) incorporate a concept called a network allocation vector (NAV). The NAV is used to track RTS, CTS, and other messages that reserve the media and track how long the network (media) will be occupied. Each node maintains its own NAV (which will typically be different at each node as each node will hear different traffic than other nodes based on propagation). For a directional system, it is best to maintain a "directional NAV" which tracks not just the fact that the media is busy, but what direction the media is busy in relative to the node. UDAAN couples power control to the directional NAV as well.

Figure 21.3 summarizes the UDAAN control protocols. A backoff in UDAAN is called a "forced idle" (FI). As shown in Figure 21.3 several types of FI exist: media busy (FI-busy), no CTS received (FI-noCTS), no ACK received (FI-noAck), and ACK received (FI-Ack). The behavior for FI-NoAck would be most typical of a CSMA system such as 802.11. Most CSMA systems

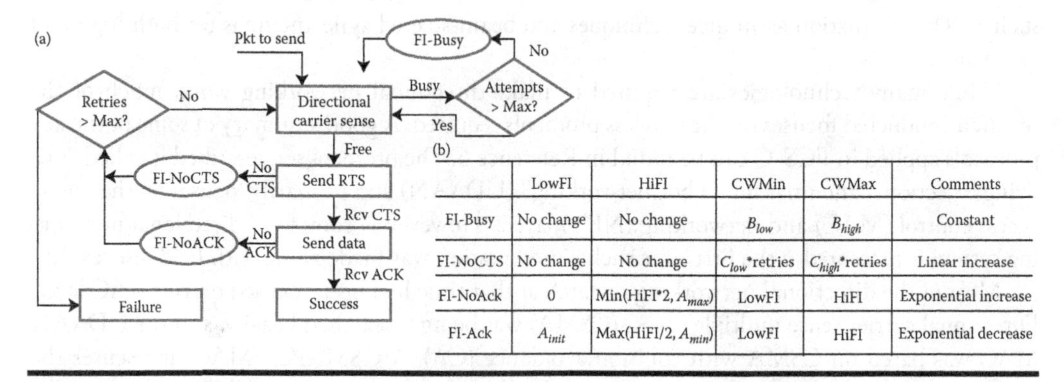

	LowFI	HiFI	CWMin	CWMax	Comments
FI-Busy	No change	No change	B_{low}	B_{high}	Constant
FI-NoCTS	No change	No change	C_{low}*retries	C_{high}*retries	Linear increase
FI-NoAck	0	Min(HiFI*2, A_{max})	LowFI	HiFI	Exponential increase
FI-Ack	A_{init}	Max(HiFI/2, A_{min})	LowFI	HiFI	Exponential decrease

Figure 21.3 UDAAN D-MAC control protocols. (a) High-level flow chart of D-MAC and (b) forced idle control. (Redrawn and reprinted, with permission, from Ramanathan, R. et al. Ad hoc networking with directional antennas: A complete system solution, *Selected Areas in Communications, IEEE Journal,* 23(3), 496–506. Copyright 2005 IEEE.)

maintain a contention window (CW). When they need to backoff, they draw a random number from 0 to the value of CW (the maximum required backoff). Depending on events on the media, the value of CW can change. Typically, a minimum value (CW_{min}) and a maximum value (CW_{max}) are predetermined.

For 802.11, if an error occurs on the media, the value of the CW will double. Since multiple errors result in multiple doublings (a factor of 2^N for N errors) this is called exponential backoff. So if a data packet is lost (no ACK), UDAAN doubles the size of the window the next backoff is chosen from. If a packet is successfully received (ACK received) UDAAN cuts the size of the window into half (normal 802.11 protocols would reduce the window to the smallest value allowed, CW_{min}, once a packet is correctly received). If an RTS or a CTS is lost UDAAN performs a linear increase of the backoff window. And for the case where the medium is detected busy, no change in the values of the backoff windows occurs.

In contrast, 802.11 would perform exponential backoff for all these cases. An important point is that the various parameters used to tune the UDAAN backoff protocols were determined through simulation of the FCS-C directional networking system. So they were optimized for the purpose of directional networking—specifically the system implemented under FCS-C.

While the wireless protocols employed are important, critical to this work were the adaptive antenna systems themselves. Engineers frequently partition protocol sets into layers—transport layer, networking layer, and so on. The antenna and its control system can be thought of as an "RF" layer. Adaptive directional antennas broadly fall into two categories—switched and steered. Switched antennas have a static set of beam patterns which are dynamically selected on a packet by packet basis. Steered antennas do not have a fixed pattern but can be pointed in any direction. Of course, many variations and combinations of both exist. Null steering antenna (usually coupled with beam steering) can place a null on a specific interferer. Switched beams can be implemented with each beam coming from a separate physical antenna, or with a set of antennas combining their signals with a fixed set of complex weightings on each antenna. Often a separate omni antenna is included, or there is a mode that forms an omni beam with reduced gain. The UDAAN protocol was designed to operate with a switched beam antenna together with an omni antenna. But extensions to use UDAAN with a beam steering antenna are straight forward.

Given a receiving and transmitting antenna, multiple combinations are possible assuming each has beamforming and omni antennas. The four combinations are (transmit omni, receive omni) termed by UDAAN "no beamforming" or N-BF, (transmit beamforming, receive omni) termed T-BF in UDAAN, (transmit omni, receive beamforming) or R-BF, and (transmit beamforming, receive beamforming) TR-BF. UDAAN used all combinations EXCEPT R-BF. For the UDAAN protocols, RTS and CTS are received in omni mode (they could be transmitted in BF mode). DATA and ACK would typically be in TR-BF mode.

UDAAN used a directional NAV (D-NAV) that incorporated power control information. Typically in the 802.11 protocol [7] if the NAV says the channel is busy no transmissions are allowed. But the RTS and CTS packets in UDAAN contained enough information that a specific power could be computed where it was safe to transmit even though the D-NAV indicated the channel was busy. Table 21.1 shows the system parameters used.

If no NAV constraints exist, when an RTS is transmitted it is transmitted at a nominal power determined by a configuration parameter. If the RTS fails, the transmit power is gradually increased till it succeeds. After that, a response packet's (CTS, DATA, or ACK) transmit power is determined by this equation:

$$(P - R) + T + Mt$$

Table 21.1 UDAAN Parameters for D-NAV Operation

Parameter	*Description*
P	Power transmitted (in header of received packet)
T	Receive threshold (in header of received packet)
A	Antenna in use (in header of received packet)
R	Received power (computed on packet reception)
Mt	Allowed transmit margin to overcome variations in propagation (configuration parameter)
Mvcs	Allowed margin for propagation variations when computing NAV allowed Tx power (configuration parameter)

This, of course, assumes a roughly symmetric channel. P–R gives the loss in the channel. T is the threshold power that must be received to correctly decode the packet. So the transmitting node must transmit enough power to overcome the channel with enough remaining power to be correctly received. With ideal power control, P – R + T would suffice. But because of time lags, measurement errors, and so on, some amount of margin (Mt) must be added to the transmission power for reliable reception.

When a node is idle, and it hears an RTS or CTS it must set its NAV based on what is heard. The RTS/CTS will give a time duration for the NAV to be set. In 802.11, a node would not be permitted to transmit during the time duration the NAV is set regardless of whether the media appears to be unoccupied (clear). This is done to address the "hidden node" problem where the responding node (e.g., sending a CTS in response to the RTS) may be too far away for potentially interfering nodes to hear. If they rely purely on what they sense on the media to decide whether they can transmit, they may transit on top of the responding node causing a failed reception. Because the NAV protects the responding node as if you could hear it on the media, this NAV-based protocol is sometimes called "virtual carrier sense" (VCS).

In the case of directional systems, you may not need as much protection as in omni systems. The antenna systems provide a lot of rejection in most directions. So, the idea of a "directional NAV" was created [8]. In addition to a duration field, the packet would include a direction field. Now the receiver of a packet would only suppress transmissions if they would potentially interfere with nodes in that direction. UDAAN took this one step further by computing an "allowed power" and associating it with the NAV. The allowed power is dependent upon the exact propagation loss and threshold power required at the receiver to receive the packet. Specifically, the allowed power was computed as

$$(\mathrm{P}-\mathrm{R})+\mathrm{T}-\mathrm{Mvcs}$$

Note that in this case, margin is subtracted from the allowed power, rather than added as when computing the required transmitted power to transmit a packet. The key is that you want to protect the node that sent the RTS (or other packet requiring a response) so you need to allow for the fact that propagation conditions could be better when you transmit than when you received the RTS. Note that it is possible to receive multiple RTS. With each RTS, you would compute the allowed power associated with the NAV and only keep the smaller of the current power associated with the

NAV and the newly computed power. When transmitting, you would compare the required transmit power to the allowed power (accounting for directional antenna gain in that direction) and decide if the packet can be transmitted. If not, the packet would be deferred until the NAV expired.

In addition to these protocols many more were specifically developed to support UDAAN directional networking. A lot of work was done on link characterization. A link characterization module in the protocols collected statistics on links and used them to predict the power required particularly for the first packet in a sequence of packets. Note that while the focus in this chapter has been on the RTS-CTS-DATA-ACK sequence, UDAAN also supported RTS-DATA, DATA-ACK, and DATA only sequences. The required power was characterized using adaptive filtering on estimates of the power required to send prior packets. These were derived based on a series of calculations using position knowledge and antenna pointing information some of which was encoded in the packet headers. The link characterization also provided quantized versions of the amount of energy required to transmit data on each link up to the network (routing) layer to assist in deciding what link to use.

For routing, UDAAN used Hazy-sighted link state routing (HSLS) [9]. This protocol was used to update the routing tables in a forwarding module which handled both data and control traffic. Different modules generating control traffic would also provide radio profiles for use in determining how their control traffic should be treated. Packets also used a type of service, ToS, field to determine their queuing treatment (similar to use of the ToS field for Internet Protocol [IP]). For data traffic the routing module would provide multiple ToS setting dependent routing tables. These tables also included radio profiles for data transmission. Simulations of the UDAAN protocol can be found in Reference 6. A few key results are shown here (see Figure 21.4).

First UDAAN made a comparison of the impact of switched versus steered antennas in the system. Simulations for both switched and steered antennas were run with varying antenna gain (and corresponding antenna beamwidth). As gain goes up, of course, the beamwidth gets narrower. For a steered system performance generally improves with narrower beamwidth. However for the switched beam system, the number of beams was held constant in the simulation. As beamwidth gets narrower, this can effectively lead to gaps in coverage since the size of the sector that must be covered by each antenna beam is fixed, and eventually the beamwidth will be narrower that the coverage required.

This effect shows up in Figure 21.4a. If you look at the lower curve for the switched antenna as the antenna gain increases (beamwidth gets narrower) a point is reached where increasing the gain hurts performance more than it helps. However, steered antennas always outperform sectored antennas. The reason why is that the steered antenna always points the maximum gain at the intended destination, while the switched beam antenna must sometimes operate off the peak of the beam. The performance loss is bounded by the gain at the edge of a sector (often 3–6 dB less than the peak gain). As seen in Figure 21.4a, the performance difference can be substantial. However, if the number of beams were allowed to grow as the beamwidth narrowed to maintain the gain at the edge of a sector the performance should track the performance of a steered antenna closely.

Another interesting result is seen in Figure 21.4b. This compares the performance of a 26 dBi steered antenna to a 0 dBi omni antenna. The steered antenna improves throughput in the network by a factor of roughly 8 to 10. For a 26 dBi antenna you would expect about an 8–10° (full as opposed to half) beamwidth. Based on Reference 2 one might expect a capacity improvement on the order of 360/10 or about 36. Clearly this was not the case, and other factors must be at play that limit the gains possible compared to the theoretical maximum.

The third plot (Figure 21.4c) shows the impact of virtual carrier sense (VCS) which is to say the UDAAN power aware D-NAV. For these simulations the number of nodes simulated is

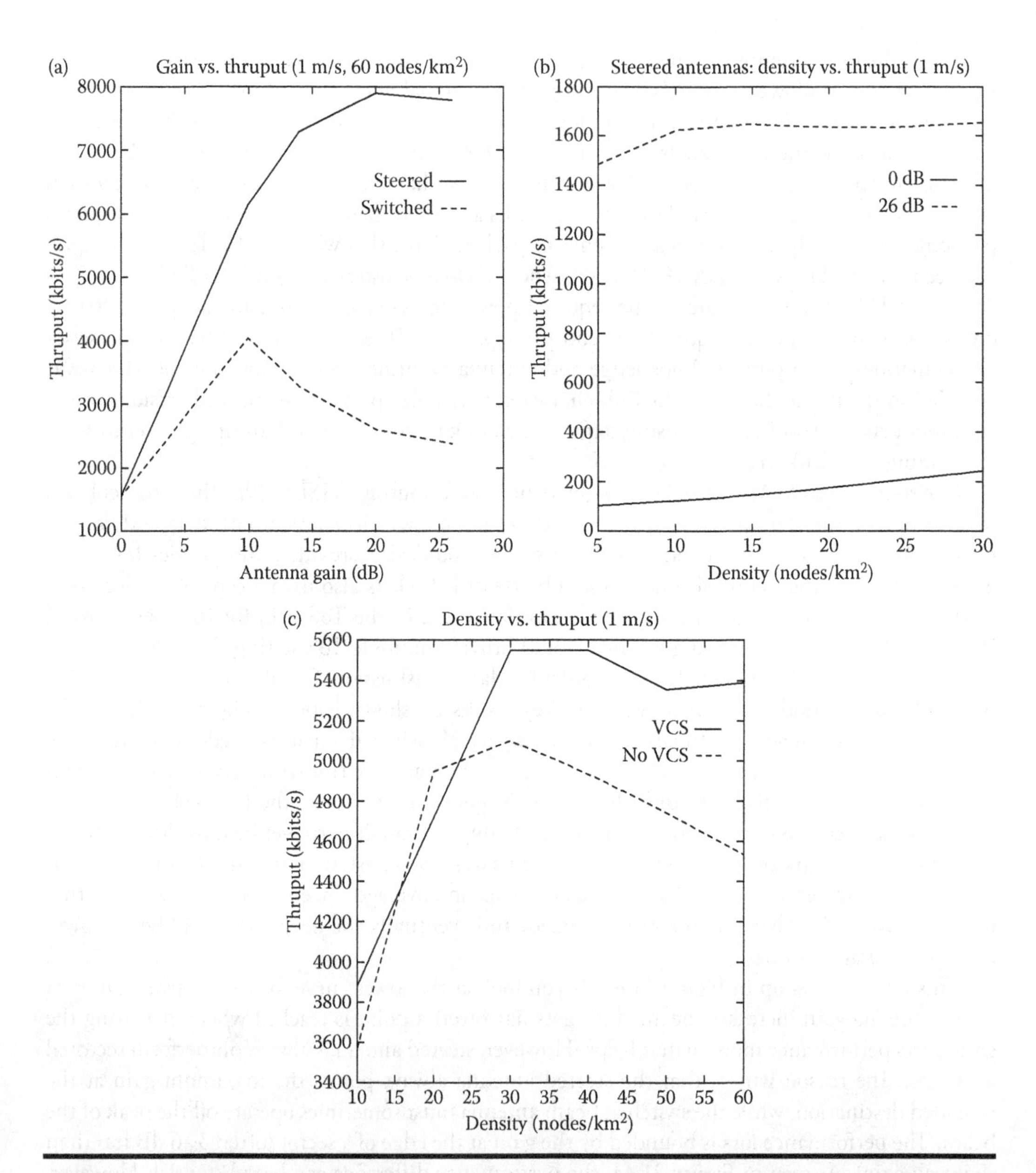

Figure 21.4 UDAAN simulation results. (a) Effect of antenna gain, (b) throughput dependence on density, and (c) VCS vs. no-VCS. (Redrawn and reprinted, with permission, from Ramanathan, R. et al. Ad hoc networking with directional antennas: A complete system solution, *Selected Areas in Communications, IEEE Journal*, 23(3), 496–506. Copyright 2005 IEEE.)

constant, but the density is increased by restricting the area over which the terminals can move. When the density is low, the performance difference between VCS and no VCS is small. This is because in a given node's vicinity, there are not a lot of nodes to talk to, and the chance of colliding with a packet from another node is low. Hence the D-NAV offers little advantage. But as the networks get denser, the probability of a collision goes up substantially. Then the VCS offers a clear advantage over the no VCS case. At 60 nodes/sq. km the advantage is about 17.3%.

Figure 21.5 Equipment for FCS-C directional networking experiments. (a) Row of demo 1 vehicles and (b) Demo 2 vehicle. (Redrawn and reprinted, with permission, from Ramanathan, R. et al. Ad hoc networking with directional antennas: A complete system solution, *Selected Areas in Communications, IEEE Journal*, 23(3), 496–506. Copyright 2005 IEEE.)

In addition to simulations a number of outdoor experiments were performed with the FCS-C equipment developed and UDAAN. Figure 21.5 shows the equipment used.

In the first set of experiments/demo (Figure 21.5a) there were four 90° antennas each with 10 dBi peak gain and 6 dBi gain at the edge of their sectors and a 6 dB omni antenna. Operation was at 2.4 GHz only. The performance of this system with UDAAN was compared to a simpler set of MANET protocols running over the omni antennas in the vehicles and UDAAN "soundly beat" the omni system. The second experiment/demo (Figure 21.5b) added 38 GHz antennas to five of the vehicles and a helicopter with 38 GHz and 2.4 GHz capability to the scenario. Also a 2.4 GHz antenna added that pointed up to talk to the helicopter. The 38 GHz system had 22 antenna beams each with a peak gain of 16 dBi. They could also be combined for an omni response of about 0 dBi.

As noted, this was really the first substantial experiment the military had done with directional networking and many lessons were learned. These included

- Sidelobe and backlobe antenna performance can affect performance of the overall system.
- Gain at the edge of a switch beam sector should be better than for any omni antenna also available to the system.
- Use of three-dimensional antennas models is essential for good performance and is improved by tracking/modeling of vehicle pitch and yaw.

21.3 The Evolution of HNW

The next evolution of military directional networking was called HNW. The roots of HNW could be traced back to experimental activities conducted by the Naval Research Laboratory (NRL). The objective of the NRL activities was to support "Network-Centric Operations via Littoral extension of the network from navy ships to the forces ashore" [10]. The experimental system developed by NRL used time division multiple access (TDMA) rather than the CSMA-based protocols used in the DARPA FCS-C experiments. One of the most difficult problems to solve in directional networking is called the "discovery" problem. Before you can talk to another node you have to know where they are relative to your location. This can be very difficult if both nodes use directional antennas for all transmissions and receptions. To simplify the discovery problem for the NRL experiments, omni antennas were used for discovery.

It is illustrative to look at the protocols used for the NRL experiment when considering HNW. The scheduling of media access for a TDMA system is much different than for a CSMA system. This scheduling can be expressed as a "graph coloring" problem where the nodes of a graph are the nodes in the network, and the lines between them are the required links. For the classic omni (broadcast) MANET each node is given a color. The color represents a slot or set of slots. A classic question would be "what is the minimum number of colors that can be used in the graph such that none of the links between nodes interfere with each other?"

A more detailed mathematical explanation of the graph coloring approach can be found in Reference 11. However, time slots could be organized in a recurring pattern where each recurrence is termed an epoch. A key point is that the scheduling problem is a "2-hop" problem. If a node is connected to a "red" node (the red node uses a "red" slot to broadcast information) it cannot be connected to second red node. If it were connected to another red node then both red nodes would broadcast at the same time and the transmissions would interfere with each other. If a node is connected to a red node, it cannot itself be a red node. Because TDMA systems are typically half duplex (can only transmit or receive at a given moment) if a node were red then if its neighbor were also red it would be transmitting when it needs to be receiving from its neighbor.

The constraints in this problem for a directional MANET are quite different. In the omni MANET, a single node transmitting impacts all its two-hop neighbors. It is presumed that all nodes "broadcast" which is to say their transmission must be heard by all their immediate neighbors. If one neighbor fails to hear the transmissions, then the transmission fails. In the case of an ideal directional system, by nature only one of all the nodes immediate neighbors will hear the transmission. In such a case, the constraints on the slots are best expressed by coloring the links rather than the nodes. And the constraint becomes that no one node may have more than one line connecting to it colored the same color. It this case, the color represents a "pair" of slots where in one slot node A transmits to node B, and in the other slot node B transmits to node A. Thus it represents a bidirectional unicast connection, versus in the broadcast omni MANET where the color represent a broadcast unidirectional connection. These paradigms are illustrated in Figure 21.6.

In Figure 21.6, a "centralized' scheduling algorithm is assumed to get to the minimum number of colors. This is to say that the algorithm required full knowledge of the links at each node in the network (although it could be run independently at each node and reach the same answer). So in Figure 21.6b the nodes with the greatest number of neighbors are the most constraining. Nodes 2 and 4 require at least three colors each with one color in common. Once those are assigned it is

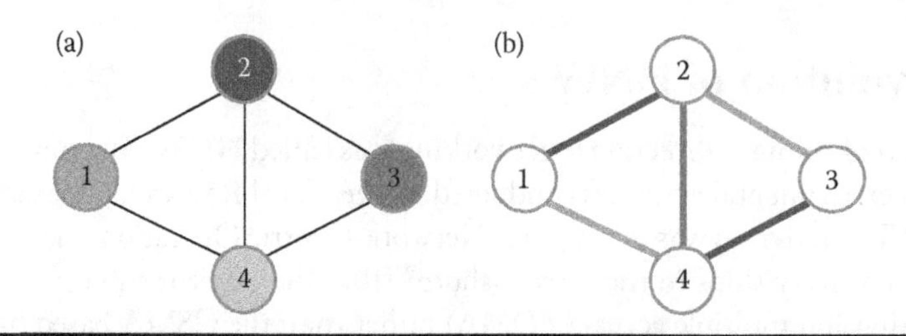

Figure 21.6 Example graph colorings representing different slot assignment constraints. (a) Four-node broadcast MANET requiring four colors (slots) and (b) four-node directional MANET requiring three colors (slots).

necessary to check if the existing color assignments are consistent at the neighboring nodes. So, if both 2 and 4 had selected green to talk to 1, then there would be a conflict. If 2 selects red, then 4 must select green. However, if 2 selected green then 4 could have selected red. So more than one allowed solution exists, but to arrive at a solution requires consideration of the state at all the neighbor nodes. We also see that fewer colors are required in the directional case than in the omni broadcast case. This means that greater reuse of the slots is achieved and this reflects in higher spectral efficiency.

Since propagating the knowledge required to optimize and rearranging slot assignments as MANET topologies evolve can be difficult the NRL work targeted a slot assignment protocol that could be run in a distributed fashion with a limited number of nodes involved in the interactions. The directional networking assumptions can simplify the scheduling approach, but other constraints need to be considered. The problems considered before were concerned with finding the minimum number of slot required. By not requiring the minimum number of slots the scheduling constraints are considerably relaxed.

So in Figure 21.6b if nodes 2 and 4 were already connected to nodes 1 and 3 and now wanted a connection between themselves, there is some chance that the color "blue" shown in the diagram may already have been assigned by one of them to say node 3. That connection would first need to be torn down and reassigned impacting three nodes rather than just the two nodes trying to form a link. However, if five colors were available, it would be guaranteed that nodes 2 and 4 could negotiate a new link without having to impact any existing link.

The key here is that if you are purely concerned with connectivity (rather than throughput) then a simple approach to assigning slots is to restrict the maximum number of neighbors a node can have to N, and then use at least $2N$–1 slot sets. While a directional node may have line of sight to many nodes (more than N), in a dense network it would be impractical to connect to all those nodes at once for routing purposes. So it is sensible to restrict the number of connections (active neighbors). If two nodes can have at most N neighbors and want to add a new neighbor, they can each have at most N–1 slots assigned (otherwise, they could not add another neighbor). So between them (if there were no duplicates) they would be using at most $2N$–2 slots. As long as at least one more slot is available, they can always find a slot to link with and not have to impact any of their neighbors.

The analysis above assumes an ideal directional antenna where only the intended receiving destination hears the transmission. But real antenna systems cannot guarantee this, so a larger number of extra slots are required, and the slot assignment protocols for this project were more complicated. Also, this approach only guarantees connectivity with a fixed amount of capacity and does not address dynamic traffic loads.

The general framing (epoch) structure used for the NRL experiments is shown in Figure 21.7. Further, N slots were used organized in repeating epochs. The number slots and sizes are

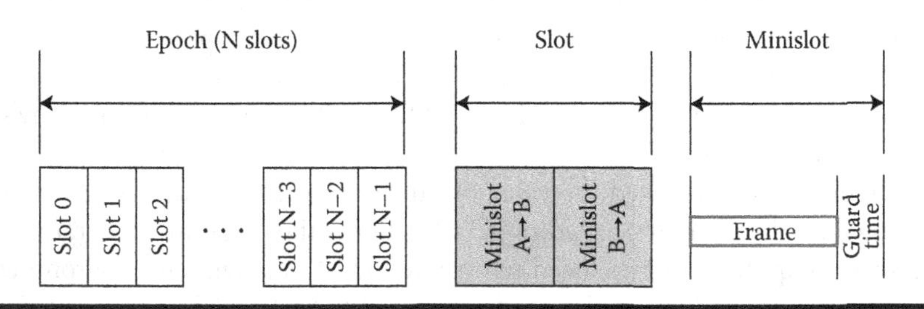

Figure 21.7 Epoch and lower-framing structures for NRL experiments.

configurable on startup but a nominal configuration would use a 100 ms epoch consisting of 25 slots each 4 ms long. Each slot was further subdivided into minislots. In the first minislot, one node in the link would transmit. In the second minislot, the other node would take its turn. Because of propagation delay, timing uncertainties, and so on, each minislot had a guard time allocated to ensure proper functioning of the system. In each minislot would be a "frame" of traffic that could include multiple higher layer packets.

To establish links, a separate omni control channel was used. Each timeslot of the epoch was allocated on the omni channel for a single node to transmit control information. All other nodes would listen during that slot. Note that for this slot structure to be useful, it is necessary that all nodes in the network (or wishing to be in the network) know where slots begin and end. The easiest solution for this synchronization problem is to assume that all nodes have GPS timing available. However many systems are sensitive about that dependency. For these discussions we presume GPS quality timing is available without discussing how it is achieved.

Two types of messages were exchanged on the omni links—hello messages and directional channel allocation control messages. Hello messages would broadcast basic information about a node such as its ID, location, neighbors, and current slot allocations. The channel allocation messages were used to negotiate directional links between nodes. On the directional links, each link would be assigned a semipermanent (SP) slot that would be held fixed as long as possible until it became unreliable for some reason. Then, as needed to satisfy traffic requirements additional demand assignment (DA) slots would be assigned to support the link. The DA slots would be released if traffic needs no longer warranted them.

For networking, the well-known optimized link state routing (OLSR) was used for routing traffic in the system. Very little change was required to accommodate this, except that OLSR broadcast messages needed to be replicated on each directional link. Priority queuing was also used on each link to address QoS.

While some simulations are available for the system described, and testing was planned [10] no published test results are known to this author. However these experiments clearly laid the ground work for what would become HNW. The Harris Corporation was the contractor conducting the NRL experiments, and continued development of this technology on their own. The first adopter of this technology was the US Army Warfighter Information Networking—Tactical (WIN-T) program.

The WIN-T program was created in 2002 to address a key gap in the communications capabilities at the time [12]. The army was relying on the mobile subscriber equipment (MSE) system for its tactical communications. The MSE primarily provided telephone and switching services. But during some key conflicts (Desert Storm in 1992, as well as Operation Enduring Freedom in Afghanistan in 2001) it was found that MSE could not keep up with the pace of battle. Units kept moving, and it took too much time to establish and disestablish the MSE equipment. A more mobile system was required, ideally one that would work while units were OTM. In addition, the army was moving toward network-centric warfare. This was the concept that giving the warfighter an "information advantage" would translate into tactical and strategic advantages on the battlefield. Recalling the FCS vision presented earlier, 'WIN-T is the "network" in the army's "network-centric" future combat system' [13].

The WIN-T program recognized the value of directional networking. They were committed to implementing high throughput networking OTM. To get the range and data rates desired high gain antennas were required. Add the need to operate OTM in a dynamic environment and it is clear that directional networking is the answer. Accordingly, WIN-T incorporated directional networking into the three-tiered architecture shown in Figure 21.8 [14].

Space layer
Utilizes available transponded MILSATCOM (e.g., Ka-band) and commercial (e.g., Ku)

Airborne layer
WCP – WIN-T communications payload

Ground layer
TCN – Tactical communications node
NOSC – Net ops and security center
MCN – Modular communications node
JGN – Joint gateway node*
PoP – Point of presence
TR – Tactical relay
STT+ – satellite transportable terminal
Wireless
*Provides access to non-WIN-T networks

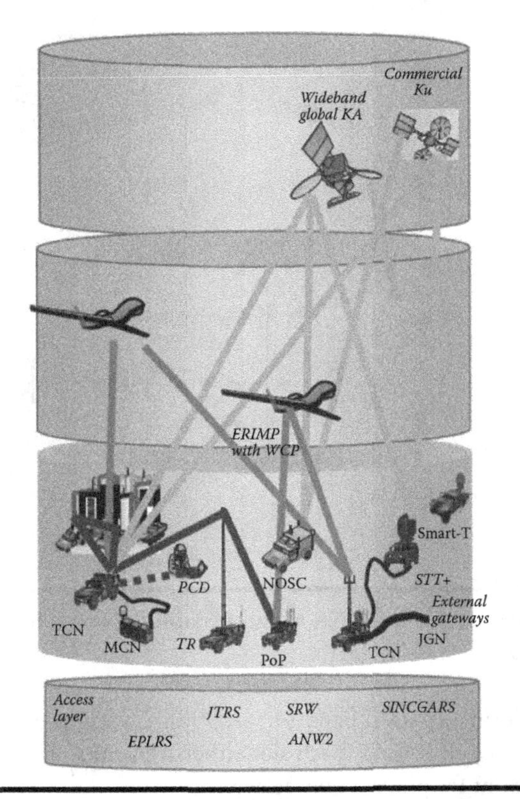

Figure 21.8 WIN-T multitier architecture. (Redrawn and reprinted, with permission, from Ali, S. R. and R. S. Wexler. Army Warfighter Network-Tactical (WIN-T) Theory of Operation, In *Military Communications Conference, MILCOM 2013–2013*, November 18–20, pp. 1453, 1461. Copyright 2013 IEEE.)

In Figure 21.8, the three gray cylinders represent the three tiers—ground, air, and space. The three tiers compose the backbone network and underneath them are access networks provided by other systems and waveforms. While each of the nodes in the diagram plays an important role in the network, the focus for this discussion is on the yellow and green links. These are being realized with the HNW waveform. The yellow links are at Ku Band, and the solid green links at C Band. Hence (like FCS-C) a dual band system is used and is realized with the highband networking radio (HNR).

Pictures of the HNR can be seen in Figure 21.9 [15]. HNR is a deployed directional networking system that uses the HNW waveform. On the left of Figure 21.9 is a vehicle-mounted system. However, the antenna can also be mounted on a mast for fixed or at-the-halt operation as shown on the right of Figure 21.9. Airborne configurations are also available. Additional variants in smaller form factors are now available as well [16].

A high-level description of the HNW waveform is found in Reference 17. Numerous similarities between HNR and the early NRL system exist. These include use of a TDMA waveform structure, OSPF-based networking, and the use of switch beam antennas. The HNW system still establishes semipermanent slots and then dynamically adds and tears down additional slots as needed to satisfy traffic needs. However, enhancements and changes exist as well.

As presented in Reference 17 HNW operates at data rates from 6 to 54 Mbps where the NRL system operated at a peak of 11 Mbps. Figure 21.10a shows typical data rate versus range

Figure 21.9 The HNR radio. (Adapted from LeBlanc, D., 2008. *Army Communicator Summer*, 33(3).)

performance at C band (4.5–4.99 GHz) using data rate adaptation based on Es/No. Minislots are no longer used. Each slot either transmits or receives. Slots do still incorporate guard time to account for propagation delays.

Perhaps, the largest single change from the NRL experiments is the use of in-band discovery. Discovery is one of the most difficult problems for directional networking systems. For two nodes to find one another, they have to blindly align their transmit and receive beams and successfully send a message. To support a discovery function HNW adds some "special" slots to their epoch

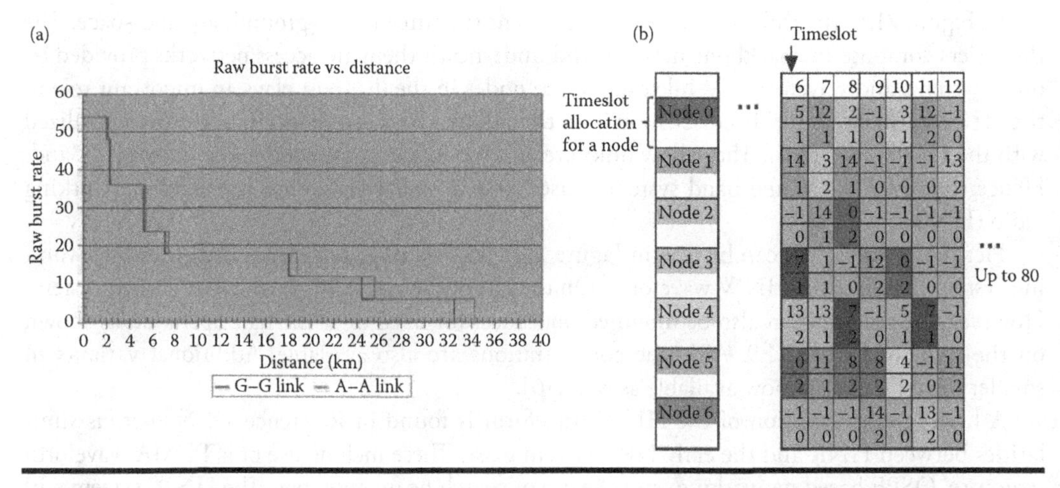

Figure 21.10 Simulation results. (a) Data rate vs. range at C band and (b) sample time slot allocation. (Redrawn and reprinted, with permission, from Griessler, et al. Modeling architecture for DTDMA channel access protocol for mobile network nodes using directional antennas. In *Military Communications Conference, 2007. MILCOM 2007.* October 29–31, pp. 1, 6. Copyright 2007 IEEE.)

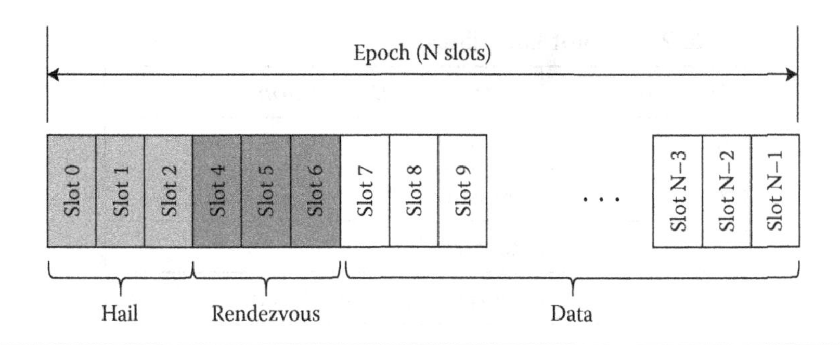

Figure 21.11 Typical epoch structure.

structure. The two types of slots are called "hail" slots and "rendezvous" slots. The number of hail and rendezvous slots is programmable, but a typical number might be three slots of each. Figure 21.11 shows a typical epoch structure for HNW.

The discovery process presumes all nodes are synchronized to the epoch and slot prior to attempting discovery. As with the NRL system, hello packets are transmitted for discovery. Nodes randomly decide if they will transmit in a hail slot or listen for hails and pick a slot to transmit in. A similar process is followed for responding in a rendezvous slot. On each attempt the nodes use a different antenna beam until all beams have been tried. Then, they repeat. Because hello messages (and their responses in rendezvous slots) are so short, multiple transmissions may be possible in hail and rendezvous slots. Using this technique with a reasonable number of beams all new neighbors can be found fairly quickly.

If a node hears a hello from a new node, they check the signal quality on the hello, and if sufficient they may send an invite message to the transmitting node in a rendezvous slot. The invite would include a list of possible transmit and receive data slots the hailing node could use to setup a link. On receiving an invite, a node will check its local status (existing allocations, potential interference from a link with the new node, etc.) and decide if a link should be formed. If yes, it will respond to identify receive and transmit slots that are acceptable and a link will be established.

Each used slot will contain a data frame, and to make optimal use of the available bandwidth multiple IP packets may be packed into a single frame. HNW also allows for segmentation and reassembly of packets so that all the available transmission bandwidth can be used. Since HNW slots no longer include paired bandwidth as with the NRL system, nodes must keep track of what function to perform in each slot. Figure 21.10b shows an example time slot allocation. The allocation shows 13 slots starting with slot 6 of an epoch. It could be presumed that slots 0 to 5 were the hail and rendezvous slots so they would not be allocated. For each node ID (such as node_0) there are two rows of data. If a link is using that slot, the top row indicates which node the link is with. So slot 6 is being used by node_0 to communicate with node_5. If the node address in the top row is −1, then the slot is unassigned. The second row indicates what function the node must perform in the slot. So under the 5 in slot 6 we see a 1 for node_0. This means the node should transmit. Table 21.2 shows the encodings used for the slot use field:

Some quick observations can be made. First if Node_0 is transmitting to Node_5 in slot 6, then Node_5 should indicate that it is receiving from Node_0 which it does. It will be seen that four nodes are allocated to transmit during slot 6. While the network configuration is unknown, this suggests a fair amount of reuse of spectrum in the system. The army states spectral reuse for HNW of up to 15× and spectrum efficiency up to 2.2 bits/Hz [18].

Table 21.2 Slot Encodings

Encoding	*Description*
0	No action
1	Transmit slot
2	Receive slot
3	Pending assignment as transmit slot
4	Pending assignment as a receive slot

Source: Griessler et al., 2007. Modeling architecture for DTDMA channel access protocol for mobile network nodes using directional antennas. In *Military Communications Conference. MILCOM 2007.* October 29–31, pp. 1, 6. Copyright 2007 IEEE.

21.4 Other Directional Networking Systems

HNW is currently the best known directional networking system deployed today. However, other developments exist. The Boeing Company has invested in a waveform called "directional network waveform" (DNW). This waveform was first tested in 2004 [19] and had additional testing as recently as 2010 [20]. But little seems to be publicly known about this waveform. Other military waveforms have some directional networking capabilities but little is known about them publicly as well.

The most public and active activity for military directional networking has been the DirecNet® Task Force. DirecNet was formed in 2006 as an industry consortium of defense contractors. Its goal was to define an open standard for a directional networking waveform. In all 12 member organizations including the U.S. Army and U.S. Navy participated. DirectNet members published some papers [21,22] and made numerous public presentations on their standards efforts. A brief summary of their activities is provided here.

DirecNet defined the concept of a theater area network (TAN) and a CONOPS (operational view) as shown in Figure 21.12 [21]. A TAN as defined by DirecNetl covers a larger geographic area than other networking constructs such a metropolitan area network (MAN) and so required a different moniker. DirecNet will support air–air, air–ground, and ground–ground links. A defining factor for a TAN is that all terminals in the TAN would be controlled by a common waveform and security management system but the terminals might not all interoperate with each other directly.

DirectNet set out to define a suit of related standards (see Figure 21.13). Several have been completed, but a number still remain. DirecNet includes two main technical working groups—one working on the waveform and one working on the network management. At this time it is not clear if DirecNet will achieve industry acceptance.

21.5 Future Directions in Military Directional Networking

Directional networking continues to evolve, and a number of innovations are ongoing. One innovation may be redefining what directional networking means. In the past, it was assumed that a fast beam steering or switching antenna was required for directional networking. Those antennas

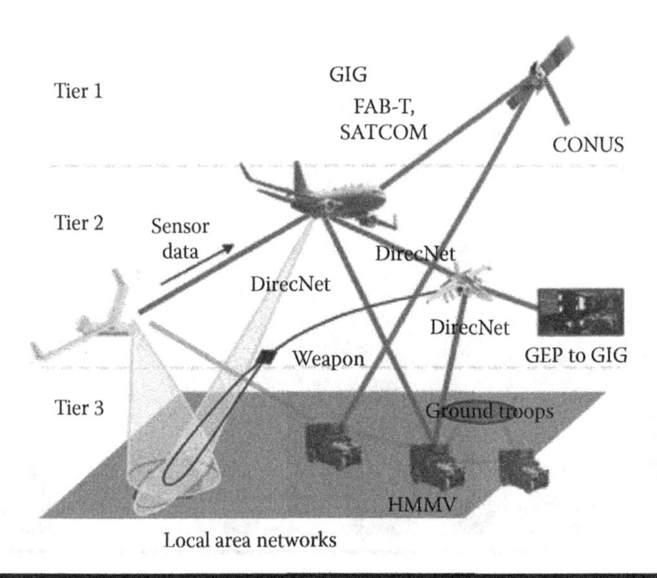

Figure 21.12 The DirecNet Operational View (OV-1). (Redrawn and reprinted, with permission, from Sonnenberg, J., Davidson, S. A., and Sherman, M. The DirecNet network management architecture. In *Military Communications Conference, MILCOM 2013–2013*, November 18–20, pp. 1756, 1761. Copyright 2013 IEEE.)

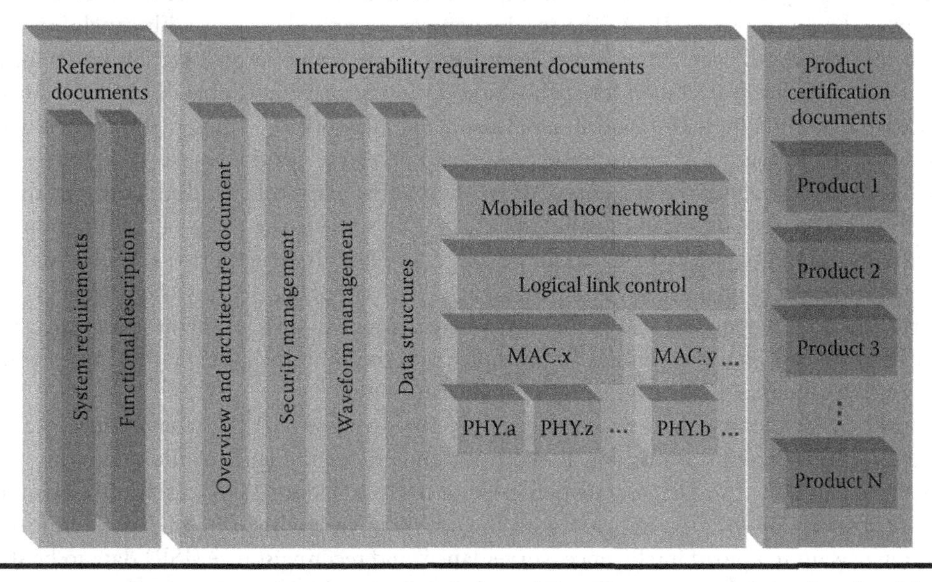

Figure 21.13 DirecNet standards and related documents. (Redrawn and reprinted, with permission, from Olds, K. et al. The DirecNet™ standard reference architecture: A roadmap for interoperability. In *Military Communications Conference, 2011—Milcom 2011*, November 7–10, pp. 2105, 2110. Copyright 2011 IEEE.)

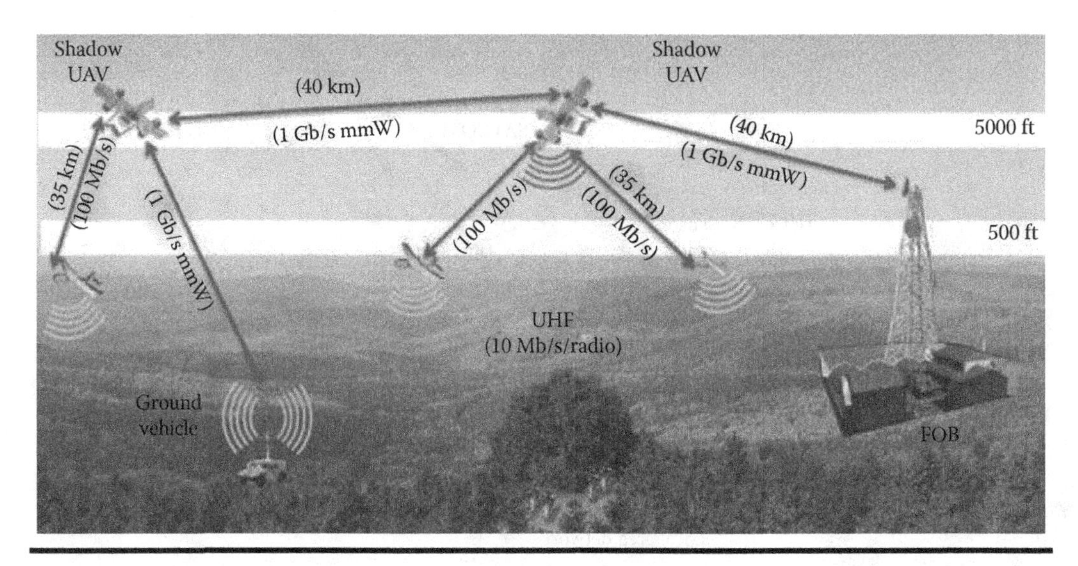

Figure 21.14 Original DARPA Mobile Hotspots CONOPS. (Adapted from Mobile Hotspots, Broad Agency Announcement, Defense Advanced Research Projects Agency, DARPA BAA 12–23, February 10, 2012. https://www.fbo.gov/utils/view?id=93c07c7951854002eab64cc50 58d9828 [Accessed 2/7/2015]).

are very expensive, and limit the use of directional networking to higher echelons in the theater. Ideally one might hope that every vehicle and soldier would have the benefit of directional networking techniques to enable high throughput and spectral efficiency.

One approach to reducing the cost is to use simpler dish antennas and protocols. The DARPA Mobile Hotspots program is actively working in this area [23]. The original CONOPS for that program is shown in Figure 21.14. The motivation was to provide troops with capabilities comparable to what fourth generation provides today. While a mobile base station could be provided, finding sufficient backhaul for such a base station is near impossible. So Mobile Hotspots proposed providing a high bandwidth aerial backbone network (>1 Gbps) with connectivity to various broadband sources of information as well as a "hotspot" service for troops below.

What is different about the hotspots approach is that the platform would support four simultaneous links and that a fast-switching antenna was not presumed. The needs of the program could be met with dish antennas. In order to find enough bandwidth to support the 1 Gbps data rate, the system planned to operate at E-band frequencies (71–86 GHz). Because the frequency was so high, this shrank the required size of the dish antenna which allows for further cost reduction in the antenna system. However, unlike with simple point to point links, the system must handover links between antennas as the aircraft moves and maneuvers.

Consider that the notional hotspots system planned that each UAS would carry two small pods each supporting two links, one at the front and one at the back of the pod along with a hotspot (see Figure 21.15). The aircraft may need to make a "Figure 21.8" to station keep and service a set of troops on the ground below with the hot spot. Meanwhile a more distant node on the ground may want to upload intelligence, surveillance, and reconnaissance (ISR) data to be shared with the troops under the hotspot. The ISR node might initially be serviced by the front antenna on the aircraft. But eventually the antenna servicing the ISR node may be blocked by the pod itself as the plane does its Figure 21.8 and the back of the plane faces the ISR uplink node. This would break the ISR link interrupting the traffic.

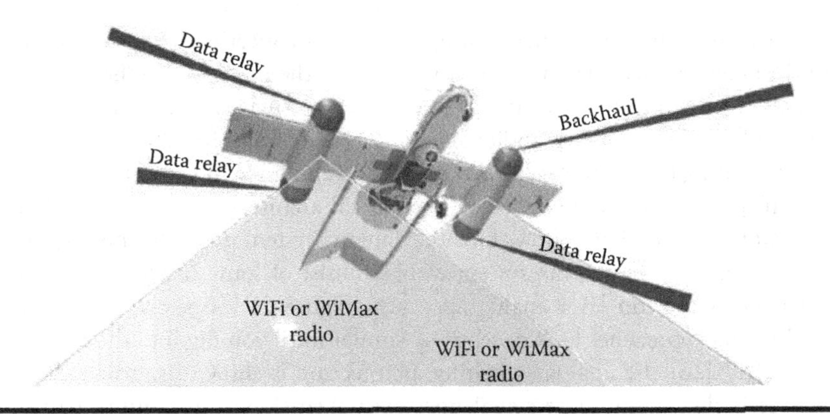

Figure 21.15 Notional drawing of DARPA Hotspots Airborne Platform. (Adapted from Mobile Hotspots, Broad Agency Announcement, Defense Advanced Research Projects Agency, DARPA BAA 12–23, February 10, 2012. https://www.fbo.gov/utils/view?id=93c07c7951854002eab64cc5058d9828 [Accessed 2/7/2015]).

To service the ISR node without interruption it is necessary to hand off from one antenna to another in the same sense your cell phone would be handed off from one cell tower to another without dropping a call. To do this seamlessly it is necessary to have a "make before break" capability. The antenna systems are engineered so that they have overlapping coverage. The link may start on the front dish of the left pod, and before the signal would be blocked, the back left antenna would acquire the signal and start decoding the same ISR stream. At the network layer, duplicate packets would be discarded. Once the new link is established, the link on the front dish would be torn down leaving it free to establish other links as needed.

This functionality is not trivial in that it requires algorithms capable of predicting when a link will be lost so that a new one can be established first. At the network layer a "topology manager" is normally used to control and track how the network is connected. It would now have to track the aircraft motion and run predictive algorithms to decide when to bring up a new link and disestablish existing links. Discovery and acquisition using narrow beam antennas on both sides of the link must also be addressed. All these factors need to be accounted to enable this style of a directional network.

While this may not seem as glamorous as the fast beam steering in systems such as HNW, it is directional networking and it can meet the needs of many important military networking applications. The key limitation is that that any given node can only support a few links. But some important networks are "sparse" networks with only a few nodes to be connected. One possible application of this style of directional networking would be the joint aerial layer network (JALN). This is a system of systems concept meant to provide supplemental bandwidth to existing SATCOM systems as well as replacement capacity should those SATCOM systems become unavailable [24,25]. The backbone network would only have a limited number of nodes at any one time.

Of course, this type of directional network is not limited to the E-band. It works at lower bands (such as Ku) and even at optical bands. Also limiting each node to three or four links does not necessarily limit the scale of the network. If you have three links already, and a new node wants to join the network, one of the links can be broken to establish a link with the new node, but then the new node can establish a link with the node whose link was broken. In that way a fully connected network is maintained, though more hops are required to get between nodes.

Another area where directional networking is evolving concerns the push of digital technology toward the antenna. Fast-switching antennas are still the preferred method to service a large number of nodes where links must be time shared. Antennas such as used for HNW today include a large number of antenna elements and analog switches to obtain full coverage. However, much of the analog switching can be replaced with digital switching technology.

Figure 21.16 [26] shows example analog and digital architectures for switched beam antennas with 32 beams. Figure 21.16c shows how the antenna system might be mounted on a vehicle. For ease of installation, the antenna system would consist of four subarrays each covering 90°. Figure 21.16a shows the top level analog approach. Figure 21.16b shows the top level digital approaches. The two approaches look deceptively similar until you dig into the details.

In Figure 21.17 [26] the analog switching architecture is shown in greater detail. The first thing we notice is that many more signal lines exist then shown in Figure 21.16b. While the number of beams may be identical to the number of elements, that does not mean that only one element is excited at a time. Sidelobe and other requirements may require that several elements must be excited or combined to properly form a beam. So "switched" beam antennas often are a combination of switched beam with some characteristics of steered arrays such as complex weighting of each element. However for these types of switched beam antennas the weights are often fixed and identical for each switched beam position. While this simplifies the design relative to a true beam steering antenna, it leads to a more complex design than for a pure switched beam system as a set of weight must be permuted for each of the switch settings.

In Figure 21.17 we see a single centralized PA that feeds a splitter which creates four unequally weighted paths. All four paths must be switched to one of the four subarrays resulting in 16 analog runs that must be distributed across the four subarrays and the vehicle. Inside each subarray we see eight throw switches are used to distribute each of the four signal paths to any of the eight elements. Finally a four-throw switch is used to select one of the four signals (or no signal at all) at each element.

It will be noted that this design requires a large number of fairly lossy components. The large number of components drives manufacturing cost and the additional loss limits the performance of the antenna system both on receive and transmit modes. While this general approach has drawbacks, it does get the job done and till recently has been commonly applied to realize switched beam antennas for directional networking.

In contrast, the digital switching subarray approach is shown in Figure 21.18 [26]. This concept leverages recently developed RF integrated circuits (RFIC) that are now commercially available. These RFIC perform up/down conversion on multiple channels for a wide range of frequencies. Rather than a corporate PA/LNA as in the analog design, a separate PA and LNA (the LNA is integrated in the RFIC) is used at each antenna element. Because the RF switching losses in front of the PA are now very small, the PAs used at each element can be significantly lower power for an equivalent performance as compared to the all analog design. A low-cost system on a chip (SoC) that includes an FPGA and processor is used locally for beam combining/control and to implement a standard Ethernet interface.

The digital approach has many advantages over the analog design. The fact that RF losses are substantially lower in the system means that power dissipation can be significantly reduced. Another consequence of reduced loss is much improved gain/noise temperature (G/T). This is a parameter used to characterize the antenna's receive performance. If the same antenna designs are used at both ends of the link, both the reduced losses on transmit and the improved G/T on receive mean that the same link can be closed with much less power for the digital design than used in the analog switched design.

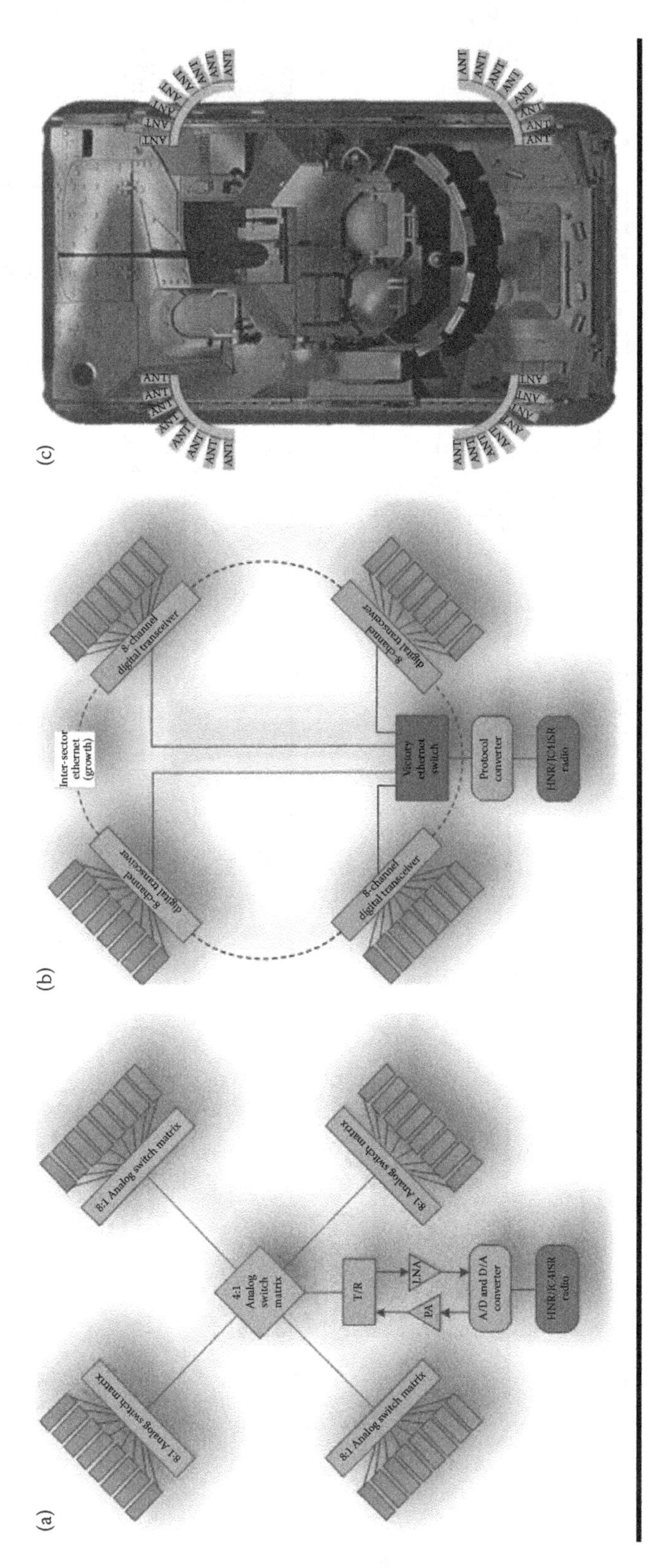

Figure 21.16 Example architectures for analog and digital switched antennas. (a) Analog switched antenna; (b) digital switched antenna; and (c) example install on military vehicle.

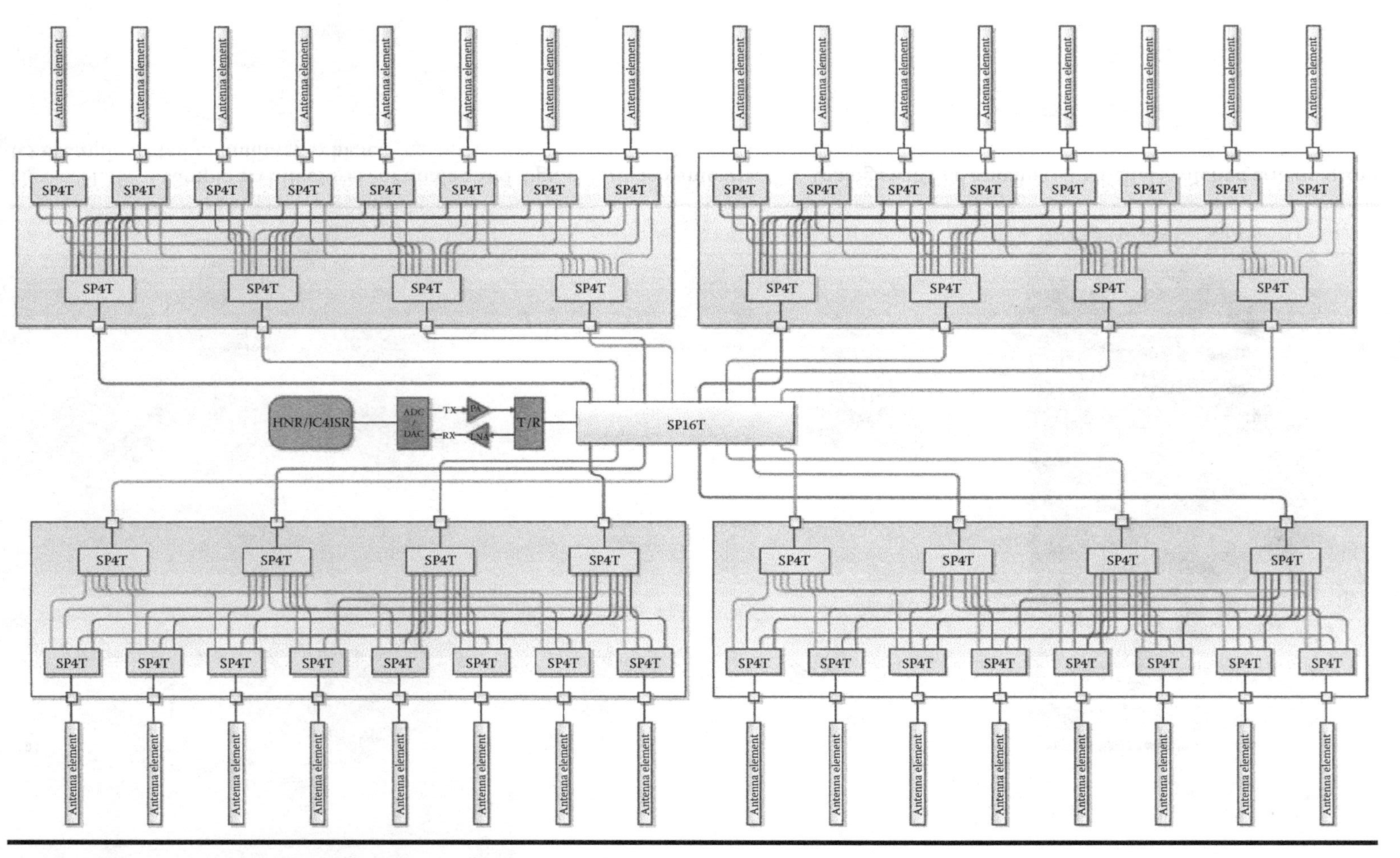

Figure 21.17 Detailed architectures for analog beam switch.

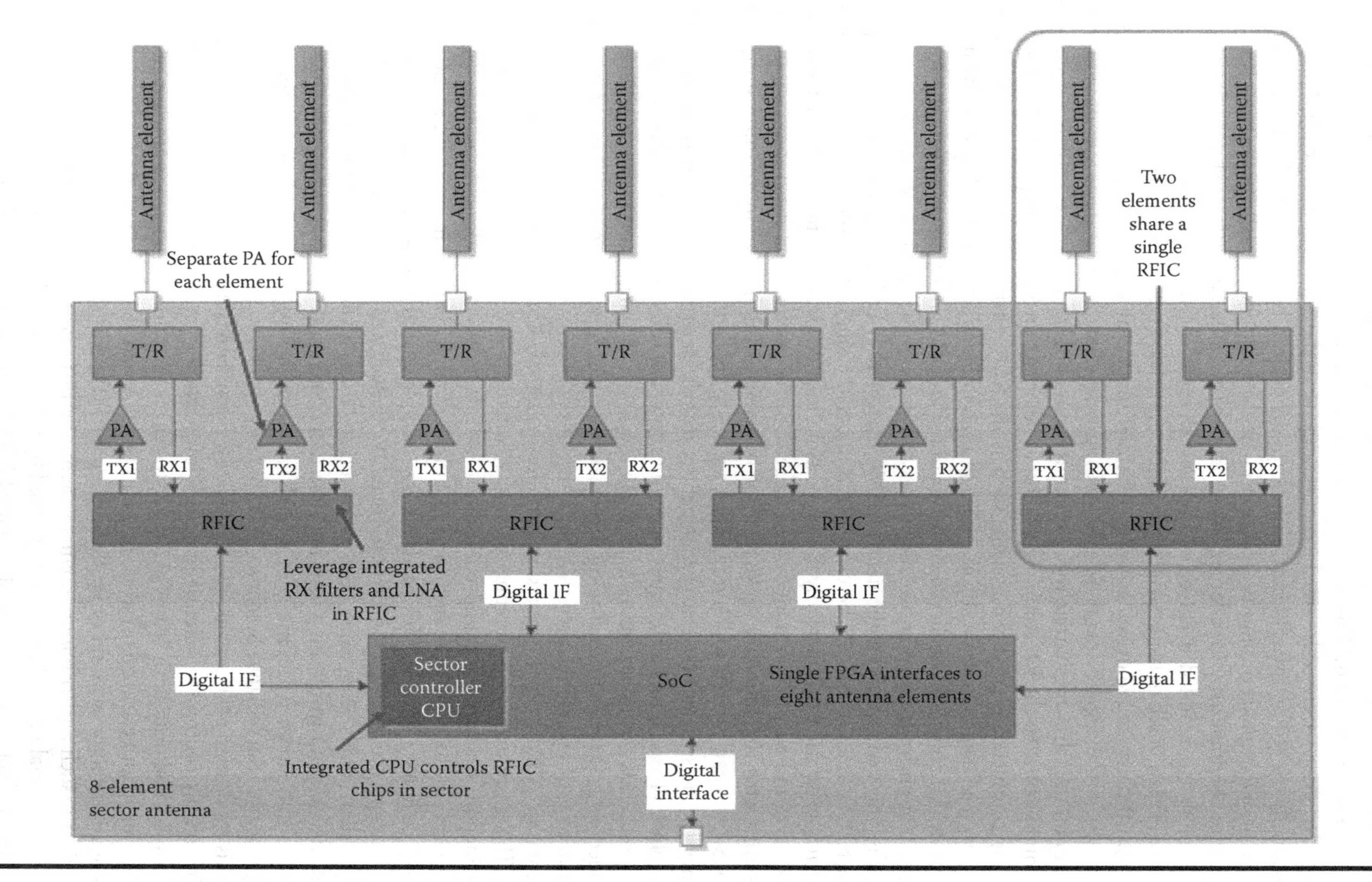

Figure 21.18 Detailed architectures for digital switched antenna subarray.

The RFIC typically has a very broad operating range. If coupled with sufficiently broadband antenna elements this means the digitally switched antenna system can operate over a wide frequency range or set of bands. This would be more difficult to achieve with the analog switching approach. The digital array is also much more versatile than the analog. Because each element can be individually excited and weighted, a multibeam system can be realized with the same equipment. The primary impact of multibeam operation is the additional signal processing in the SoC. This can often be handled with a larger FPGA in the SoC. The SoC tracks Moore's law so that more capability can be added at a given cost point with each year. RF components do not typically follow Moore's law. And, of course, the most important advantage would be the cost.

The cost differences are driven by several factors. As already noted, many less discrete components exist in the digital design. Manufacturing costs directly track parts counts. However, how common the parts are matters too. Analog RF components do not see as high a demand as digital components. So the RF components tend to be pricier on a per a component basis as well. Leveraging commercially used components such as the RFIC further accentuates that difference since commercial demand is typically much larger than military demand which drives down component costs. All and all it is believed that directional antenna architectures will gradually trend to designs that push the digital processing closer and closer to the antenna elements.

Another interesting trend is the use of MIMO/sparse array techniques. MIMO and adaptive antenna technology may loosely be categorized under the single moniker of "smart antennas." MIMO has been used commercially for some time to increase robustness and capacity. Both adaptive array and MIMO techniques are focused on the antenna, and it is sometimes difficult to determine what the difference is. An array of 10 antennas can be used for beamforming or for MIMO. So how can you tell which techniques are being applied?

Generally the spacing between the elements can help determine which set of techniques is being applied. Phased arrays techniques normally attempt to fully control the radiation pattern in all directions. This generally requires an element spacing of less than the wavelength of operation (λ) divided by 2. If the element spacing is greater than that, "grating lobes" will occur making it impossible to fully control the pattern in all directions at once.

In contrast, MIMO signal processing looks for the signal at each antenna element to be decorrelated. This would require an antenna spacing (d) where $d > \lambda/2$. But recently developments blur this distinction. An example would be multiuser MIMO (MU-MIMO) or distributed MIMO. But this term can have different meanings to different people. For instance IEEE Std. 802.11ac-2014 [27] includes a MU-MIMO mode where multiple spatial streams are encoded such that different streams are intended for different users. While such a capability would also be useful for military system (and is being considered in projects such as DARPA's 100G activities [28]) another form of MU-MIMO would be where multiple users collaborate to send the same information to a single destination. An example of a DARPA program investigating this is Computational Leverage Against Surveillance Systems (CLASS) [29].

In this application of smart antenna technology, the antennas from the individual users are too far apart to fully control the antenna pattern. However, they can control the pattern at multiple points in space and place both beams and nulls in specific directions. As such the antennas act very much like an adaptive antenna system. It is also possible that the individual transmissions could be encoded differently as would typically be done for MIMO antennas.

So when is such a distributed antenna system a MIMO versus an adaptive antenna system? And is it directional networking? It is suggested by this author that if the signals at each antenna are encoded differently, then it is a "distributed MIMO" application. If only RF parameters such as phase, amplitude, or time delay are varied, the system should be termed a "distributed adaptive

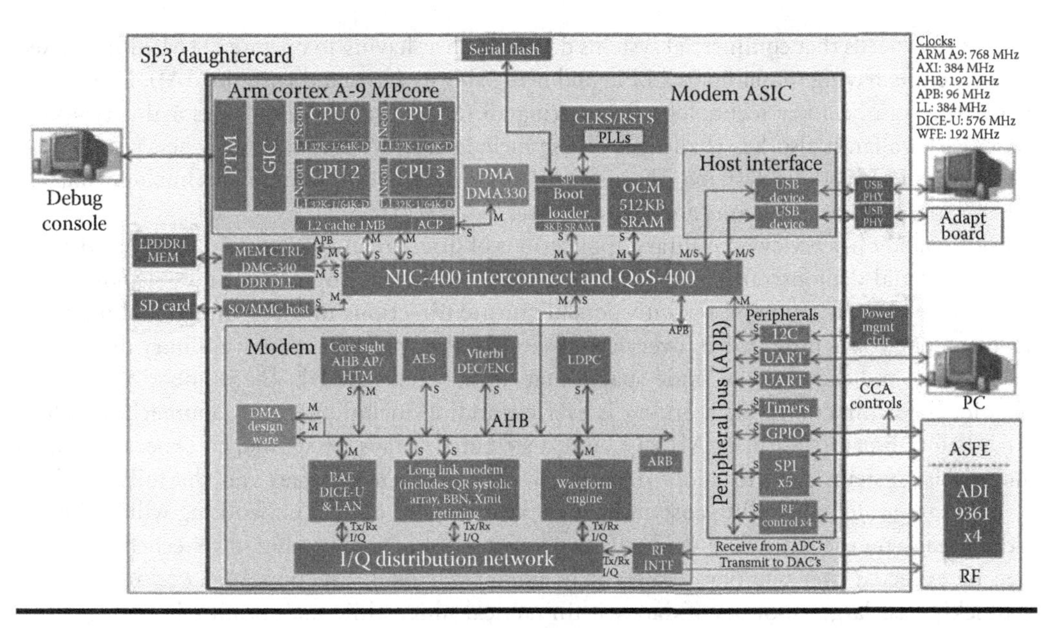

Figure 21.19 CLASS Modem ASIC Block Diagram. (Adapted from Computational Leverage Against Surveillance Systems (CLASS) Phase 2, Broad Agency Announcement, Defense Advanced Research Projects Agency, DARPA-BAA-13-47, January 03, 2013. https://www.fbo.gov/utils/view?id=c8741611700312610978bafbcbe4c69a [Accessed 2/8/2015]).

array." Perhaps these days it is simplest to just refer to all these technologies as "smart" antennas and not worry about the nuances. As for the term "directional networking" it is suggested that if the received signal strength at multiple points in space are being controlled that it be considered a form of directional networking.

Bringing directional networking techniques down to the individual soldier can be a challenge. As noted, directional antennas are innately expensive. Making them low cost enough that you can deploy one or more per soldier would be difficult. One could argue that DARPA CLASS is attempting to do just this in that they are applying distributed antenna technology that controls the antenna pattern in specific directions leveraging individual antenna elements each carried by a different soldier. To make it affordable, DARPA is developing a special ASIC to support the signal processing required (see Figure 21.19 [29]). It includes a great deal of functionality including a quad core ARM processor, a systolic array, LDPC decoder, and so on.

Perhaps another way of bringing directional networking directly to the soldier would be through the use of 60 GHz technology being developed commercially. Two standards (IEEE Std. 802.15.3c [30] and IEEE Std. 802.11ad [31]) include support for directional networking at 60 GHz. Use of this technology commercially can potentially drive down the cost of 60 GHz components. It may be possible to leverage that technology to develop military versions that would be useful for high data rate soldier systems that are both high AJ and LPD.

21.6 Summary and Conclusion

Directional networking is a critical technology for military systems today. Even more than commercial systems, military systems need to be efficient in their use of spectrum. In addition military systems have needs that commercial systems do not, such as having to operate OTM without fixed infrastructure, having to maintain a LPD and being able to withstand jamming (AJ). Directional networking can be a key technology for meeting all these needs. While directional networking has much to offer, it also has challenges. These include cost, SWAP, and the general complexity of directional networking systems. These serve to limit the application of this technology today to upper echelons of military units and strategic asset.

This chapter has reviewed military applications of directional networking. Starting with the first substantial demonstrations in DARPA FCS-C to the development of now deployed systems such as HNW, and ending with possible future directions military directional networking may take. It has provided an overview of some of the protocols used in military directional networking, and some of the trade spaces that must be considered. The number of enabling techniques for directional networking is ever expanding including recent approaches such as make-before-break, Multiuser MIMO, and distributed arrays. It should be expected that as commercial systems start to adopt directional networking techniques and the military finds new approaches to controlling cost and complexity that directional networking will find more and more use on the battlefield gradually penetrating down to lower and lower echelons. It can even be expected that one day soldiers may employ personal directional networking systems to quickly pass large amounts of data within tactical units while maintaining high degrees of LPD and AJ.

References

1. Electromagnetic Spectrum Strategy, 2013. A Call to Action, US Department of Defense, September 11, 2013 (Accessed 17/12/2014) http://www.defense.gov/news/dodspectrumstrategy.pdf
2. Yi, S., Y. Pei, and S. Kalyanaraman. 2003. On the capacity improvement of ad hoc wireless networks using directional antennas. In *4th ACM MobiHoc*, Annapolis MD, pp. 108–116.
3. Chang, J. F. and C. J. Chang, 1984. Optimal design parameters in a multihop packet radio network using random access techniques. In *Proc. of GLOBECOM '84*, November, Atlanta, GA, pp. 493–497.
4. Chang, Chung-Ju, and Jin-Fu Chang, 1986. Optimal design parameters in a multihop packet radio network using random access techniques, *Computer Networks and ISDN Systems*, 11(5), 337–351.
5. Paul Sass and James A. Freebersyser, 2002. FCS communications technology for the objective force. In *Proc. SPIE 4741, Battlespace Digitization and Network-Centric Warfare II*, 9 August 6, Orlando FL, 9.
6. Ramanathan, R., J. Redi, C. Santivanez, D. Wiggins, and S. Polit, 2005. Ad hoc networking with directional antennas: A complete system solution, *Selected Areas in Communications, IEEE Journal*, 23(3), 496–506.
7. IEEE Standard for Information technology—Telecommunications and information exchange between systems Local and metropolitan area networks—Specific requirements Part 11: Wireless LAN Medium Access Control (MAC) and Physical Layer (PHY) Specifications. In *IEEE Std 802.11-2012 (Revision of IEEE Std 802.11-2007)*, pp. 1, 2793, March 29 2012.
8. Takai, M., J. Martin, A. Ren, and R. Bagrodia, 2002. Directional virtual carrier sensing for directional antennas in mobile ad hoc networks. Presented at the *ACM MOBIHOC*, Lausanne, Switzerland, June 2002.
9. Santivanez, C. and R. Ramanathan, 2001. *Hazy Sighted Link State (HSLS) Routing: A Scalable Link State Algorithm*. Cambridge, MA: BBN Technologies, August BBN Tech. Memo BBN-TM-1301.
10. Cain, J. B., T. Billhartz, L. Foore, E. Althouse, and J. Schlorff, 2003. A link scheduling and ad hoc networking approach using directional antennas. In *Military Communications Conference. MILCOM '03*. 2003 IEEE, October 13–16, 2003, Vol. 1, pp. 643, 648. www.dtic.mil/get-tr-doc/pdf?AD=ADA467467

11. Lloyd, E. L. and S. Ramanathan, 1992. On the complexity of distance-2 coloring, Computing and Information, 1992. *Proceedings. ICCI '92., Fourth International Conference*, May 28–30, Toronto, Ontario, Canada, pp. 71, 74. 12. http://en.wikipedia.org/wiki/PM_WIN-T (Accessed 31/1/2015).
13. Via, MG D. L. 2008. WIN-T—Enabling battle command for the Joint Warfighter through total system Life Cycle Management. *Army Communicator Summer* 33(3) (Accessed 31/1/2015) http://www.signal.army.mil/index.php/army-communicator-archives/277-2008/152-vol33-no3-army-communicator-fall-2008, pp 3.
14. Ali, S. R. and R. S. Wexler, 2013. Army Warfighter Network-Tactical (WIN-T) Theory of Operation, In *Military Communications Conference, MILCOM 2013–2013 IEEE*, November 18–20, pp. 1453, 1461.
15. LeBlanc, D., 2008. HNW—WIN-T Increment 2 LOS solution, *Army Communicator Summer*, 33(3) 29.
16. High-Capacity Wireless Networking fly sheet, Harris Corporation, #519849 VPB d0301, January 2012 (Accessed 31/1/2015) http://govcomm.harris.com/solutions/products/defense/high-cap-wireless-net.asp
17. Griessler, Peter, J. B. Cain, Hanks, and Ryan, 2007. Modeling architecture for DTDMA channel access protocol for mobile network nodes using directional antennas. In *Military Communications Conference, 2007. MILCOM 2007.* IEEE, October 29–31, pp. 1, 6.
18. Mann, MAJ T., 2008. JTRS/WIN-T networking waveform quick reference sheets, *Army Communicator Summer*, 33(3), 74.
19. Boeing Supports Test of Network-Centric Targeting Capability at JEFX '06, news release, The Boeing Company, September 20, 2004. http://boeing.mediaroom.com/2006-05-11-Boeing-Supports-Test-of-Network-Centric-Targeting-Capability-at-JEFX-06
20. Boeing Successfully Demonstrates Directional NetWork System for US Fleet Forces Command, news release, The Boeing Company, June 23, 2010 (Accessed 2/1/2015). http://boeing.mediaroom.com/2010-06-23-Boeing-Successfully-Demonstrates-Directional-NetWork-System-for-US-Fleet-Forces-Command
21. Sonnenberg, J., Davidson, S. A., and Sherman, M., 2013. The DirecNet network management architecture. In *Military Communications Conference, MILCOM 2013–2013 IEEE*, November 18–20, pp. 1756, 1761.
22. Olds, K., Cole, R., Lord, B., Duke, M., Sherman, M., Spaulding, J., and Boyd, J., 2011. The DirecNet™ standard reference architecture: A roadmap for interoperability. In *Military Communications Conference, 2011 - Milcom 2011*, November 7–10, pp. 2105, 2110.
23. Mobile Hotspots, Broad Agency Announcement, Defense Advanced Research Projects Agency, DARPA BAA 12–23, February 10, 2012. (Accessed 2/7/2015) https://www.fbo.gov/utils/view?id=93c07c7951854002eab64cc5058d9828
24. Salanitri, D Aerial networks enable joint forces communication through air, ground, sea, US Air Force, December 16, 2014. (Accessed 2/7/2015) http://www.af.mil/News/ArticleDisplay/tabid/223/Article/467749/aerial-networks-enable-joint-forces-communication-through-air-ground-sea.aspx
25. Magnuson, S U.S. Forces Prepare for a 'Day without Space', National Defense Industrial Association February 2014. (Accessed 2/7/2015) http://www.nationaldefensemagazine.org/archive/2014/February/Pages/USForcesPreparefora%E2%80%98DayWithoutSpace%E2%80%99.aspx
26. Sherman, M., Horihan, G., and Cooper, D. M., 2014. Switched Beam Directional Antenna Approaches & Enabling Technology. In *IEEE Military Communications Conference 2014, Panel—Advances in Directional Antenna Technology and Directional Networking.* October 06.
27. IEEE Standard for Information technology—Telecommunications and information exchange between systems—Local and metropolitan area networks—Specific requirements—Part 11: Wireless LAN Medium Access Control (MAC) and Physical Layer (PHY) Specifications Amendment 5: Television White Spaces (TVWS) Operation, IEEE Std 802.11af-2013 (Amendment to IEEE Std 802.11-2012, as amended by IEEE Std 802.11ae-2012, IEEE Std 802.11aa-2012, IEEE Std 802.11ad-2012, and IEEE Std 802.11ac-2013), pp. 1, 198, February 21, 2014
28. 100 Gb/s RF Backbone (100G), Broad Agency Announcement, Defense Advanced Research Projects Agency, DARPA-BAA-13-15, 03 January 2013. (Accessed 2/8/2015) https://www.fbo.gov/utils/view?id=93245636db4470c17d853e5a3c80270b

29. Computational Leverage against Surveillance Systems (CLASS) Phase 2, Broad Agency Announcement, Defense Advanced Research Projects Agency, DARPA-BAA-13-47, January 03, 2013. (Accessed 2/8/2015) https://www.fbo.gov/utils/view?id=c87416117003126l0978bafbcbe4c69a
30. IEEE Standard for Information technology—Telecommunications and information exchange between systems—Local and metropolitan area networks—Specific requirements. 2009. Part 15.3: Wireless Medium Access Control (MAC) and Physical Layer (PHY) Specifications for High Rate Wireless Personal Area Networks (WPANs) Amendment 2: Millimeter-wave-based Alternative Physical Layer Extension, IEEE Std 802.15.3c-2009 (Amendment to IEEE Std 802.15.3-2003), pp. c1, 187, October 12.
31. ISO/IEC/IEEE International Standard for Information technology—Telecommunications and information exchange between systems—Local and metropolitan area networks—Specific requirements-Part 11: Wireless LAN Medium Access Control (MAC) and Physical Layer (PHY) Specifications Amendment 3: Enhancements for Very High Throughput in the 60 GHz Band (adoption of IEEE Std 802.11ad-2012), ISO/IEC/IEEE 8802-11:2012/Amd.3:2014(E), pp. 1, 634, March 14, 2014.

Index

A

B

D

E

I

M

N

O

S

U

V

W

X